Membrane Receptors, Dynamics, and Energetics

NATO ASI Series

Advanced Science Institutes Series

A series presenting the results of activities sponsored by the NATO Science Committee, which aims at the dissemination of advanced scientific and technological knowledge, with a view to strengthening links between scientific communities.

The series is published by an international board of publishers in conjunction with the NATO Scientific Affairs Division

A	**Life Sciences**	Plenum Publishing Corporation
B	**Physics**	New York and London
C	**Mathematical and Physical Sciences**	D. Reidel Publishing Company Dordrecht, Boston, and Lancaster
D	**Behavioral and Social Sciences**	Martinus Nijhoff Publishers
E	**Engineering and Materials Sciences**	The Hague, Boston, Dordrecht, and Lancaster
F	**Computer and Systems Sciences**	Springer-Verlag
G	**Ecological Sciences**	Berlin, Heidelberg, New York. London,
H	**Cell Biology**	Paris, and Tokyo

Recent Volumes in this Series

Series A: Life Sciences

Membrane Receptors, Dynamics, and Energetics

Edited by

K. W. A. Wirtz

Laboratory of Biochemistry
State University of Utrecht
Utrecht, The Netherlands

Plenum Press
New York and London
Published in cooperation with NATO Scientific Affairs Division

Proceedings of a NATO Advanced Study Institute on
Membrane Receptors, Dynamics, and Energetics,
held August 17–30, 1986,
on the Island of Spetsai, Greece

Library of Congress Cataloging in Publication Data

NATO Advanced Study Institute on Membrane Receptors, Dynamics, and
    Energetics (1986: Nisos Spetsai, Greece)
    Membrane receptors, dynamics, and energetics.

    (NATO ASI series. Series A, Life sciences; v. 133)
    "Proceedings of a NATO Advances Study Institute on Membrane Recep-
tors, Dynamics, and Energetics, held August 17–30, 1986, on the Island of
Spetsai, Greece"—T.p. verso.
    Published in cooperation with NATO Scientific Affairs Division."
    Includes bibliographies and index.
    1. Cell receptors—Congresses. 2. Biological transport—Congresses. I.
Wirtz, Karel W. A. II. North Atlantic Treaty Organization. Scientific Affairs
Division. III. Title. IV. Series. [DNLM: 1. Cell Membrane—physiology—con-
gresses. 2. Receptors, Endogenous Substances—physiology—congresses.
QH 603.C43 N279m 1986]
QH603.C43N385   1986   574.87'5   87-10150
ISBN-13: 978-1-4684-5337-9   e-ISBN-13: 978-1-4684-5335-5
DOI: 10.1007/978-1-4684-5335-5

A Division of Plenum Publishing Corporation
233 Spring Street, New York, N.Y. 10013

PREFACE

    A NATO Advanced Study Institute on "Membrane Receptors,
Dynamics and Energetics" was held in order to consider recent
developments in membrane receptor research. Presentations and
discussions focussed on signal transduction mechanisms (e.g.
G-proteins, transducin, Ca^{2+}) and on the structure and function of some selected receptors (i.e. low density lipoprotein
receptor, acetylcholine receptor, growth factor receptors,
toxin receptors). These topics were put in the larger context of current knowledge on the structure and function of
membranes; connections between different fields of research
were established by in-depth discussions of energy transduction and transport mechanisms. Specialized membrane systems
were discussed including chloroplasts and chromaffin granules.
Functional relationships with specific lipids (e.g. phosphoinositides, lipid A) were major topics of discussion.
    This book presents the content of the major lectures and
a selection of most relevant posters presented during the
course of the Institute. This book is intended to make the
proceedings of the Institute accessible to a larger audience
and to offer a comprehensive account of those topics of receptor, membrane and bioenergetics research that were at the
center of interest during our stay on the Island of Spetsai
from August 17 to 30, 1986.

January 1987   K.W.A. Wirtz
                                     Utrecht, The Netherlands

CONTENTS

THE AMPLIFYING TRANSDUCTION CASCADE TRIGGERED BY RHODOPSIN IN VISUAL RECEPTOR CELLS BIOCHEMICAL AND BIOPHYSICAL APPROACHES

Marc Charbre and T. Minh Vuong[*]

Laboratoire de Biophysique Moléculaire and Cellulaire

(UA CNRS 520), DRF-CENG, B.P. 85, 38041, Grenoble, France

After a long rivalry with Hagins'"calcium hypothesis" cyclic GMP is now solidly established as the cytoplasmic transmitter of the visual excitation process in the vertebrate photoreceptor[1,2] : in these cells cGMP directly controls the conductance of the Na^+ channels in the plasma membrane ; in this respect, the visual transduction process differs from the usual pathway of cyclic nucleoside dependent kinase activation, found in the transduction process of many hormonal or neuronal signals. In many other respects the light sensitive and the hormone sensitive systems present striking similarities. The Rhodopsin-Transducin-cGMP phosphodiesterase cascade parallels exactly that of hormone receptor-G protein-AMPcyclase. It became clear a few years ago that transducin, the GTP binding protein of the visual system, is a member of the growing family of G proteins responsible for the coupling of hormonal or neuronal membrane receptors to their various intracellular effectors : AMP cyclase, phosphodiesterase or phospholipase specific for phosphatidylinositol hydrolysis. This analogy between the transducin cascade and the other G protein mediated processes had led a few years ago to the suggestion that rhodopsin might be considered as a special type of hormone receptor[3]. The retinal would be the equivalent of a hormone, that remains permanently bound to its receptor, but in the inactive "11 cis" conformation. Light, by photoisomerizing this latent ligand to its active "all trans" conformation, activates the protein just like the binding of the hormone does to its receptor. Viewing rhodopsin as an hormone receptor is no longer met with amused skepticism from endocrinologists when recently the ß adrenergic receptors and the acetylcholine muscarinic receptor sequences are published[4,5] : close structural analogies and significant sequence homologies are observed between these receptors and rhodopsin : the secondary structures of all these receptors present the same general arrangement with respect to the membrane and appear to contain the same number of transmembrane α-helical segments as well as similar serine and threonine rich C-terminal end, which are recognised by kinases specific for the liganded or (photoactivated) conformation of the receptor[6]. Sequence analogies have even been detected in the hydrophobic segments which, in rhodopsin are known to contribute to the retinal binding site. Furthermore, those equivalent charged residues that are thought to be involved in ligand binding by the muscarinic receptor show a high degree of evolutionary conservation among the various forms of rhodopsin. The kinship is therefore not restricted to the part of the receptor which couples to the G protein, but appears to extend to the receptor part which interacts with the ligand. The three types of receptors- rhodopsin, β adrenergic and muscarinic-clearly originate from a common ancestor ; one might reasonably expect such a pattern to be exhibited by many other receptors.

The visual transduction cascade may therefore be taken as a model of general interest to the problem of intracellular coupling of membrane receptors to signal molecules. For such studies, the rod cell of the vertebrate retina offers substantial advantages over most hormone or neurotransmitter sensitive cell lines : the extreme segregation of the components of the phototransduction chain in the retinal rod outer segment facilitates biochemical work, as the 5 major proteins involved in the control of the cGMP cascade represent more than 90 % of the total protein content of this organelle. Their purification in large quantities is comparatively easy and is further facilitated by the fact that transducin and the PDE in the rod outer segment are only weakly attached to the membrane and can be eluted without the help of detergents. The other major advantage of this system is the triggering of the reaction by a light flash, which allows precise control of the exact number of receptors photoexcited and of the duration of excitation. Furthermore, the quasi-crystalline ordering of the disk membrane bearing the receptor molecule in the rod, opens up the possibility of fast physical measurements on the native structure, such as neutron diffraction and near infrared light scattering. This gives access to fast kinetic studies of the various steps of the transduction cascade in-situ. The concepts of free floating membrane receptors, collision coupling on the membrane surface with intermediate GTP-dependent proteins and multistep amplification can be tested quantitatively on the retinal rod. The kinetics and high amplification gain are already well determined, the fast turn-off mechanisms and regulation processes are now being analysed. In this respect the retinal system is also as a valuable model : the phosphorylation of rhodopsin by a kinase which is permanently active but specific only for the photoexcited form of the receptor has been documented more than 10 years ago. This now appears as a general regulation process for inactivating liganded membrane receptors.

In this report, we first recapitulate briefly the structural and biochemical data concerning the main components of the transduction chain and present the basic triggering and amplication schemes. Then, two biophysical approaches used to measure the kinetics and stoichiometry of amplification will be detailed, and finally we shall discuss the regulation and turn-off mechanisms.

THE COMPONENTS OF THE TRANSDUCTION CASCADE

Up to now, 5 proteins have been well characterized and purified, which are involved in the process. Their main biochemical parameters are listed on table 1. The last important component of the transduction chain, the cGMP-dependent Na channel of the plasma membrane has not yet been isolated and purified, it has been characterized only by electrophysiology. Fig. 1 sketches the respective locations of the various proteins in the native structure.

TABLE 1. THE MAJOR PROTEINS IDENTIFIED IN THE RETINAL ROD OUTER SEGMENT

Proteins (Symbol)	Relation to Membrane	Molecular Weight (kDaltons)	Stoichiometry versus Rhodopsin	Equivalent Molar Concentration in Cytoplasm
Rhodopsin (R)	Intrinsic	39 +2 (glycocong.)	1	–
Transducin (T = Tα+Tβ+Tγ)	Peripheral or soluble	=80 (39+37+6)	$=10^{-1}$	=500 μM
"48 k" or Arrestin or S Antigen (A)	soluble	=50	10^{-1}	500 μM
cGMP Phospho-diesterase (PDE) + inhibitor (I)	Peripheral	=180 (88+84+13)	$=10^{-2}$	=50 μM
ATP dependent R* Kinase (K)	Soluble	68	10^{-3}	5 μM
"Rim Protein" (P)	Intrinsic	240	$=3.10^{-3}$	–

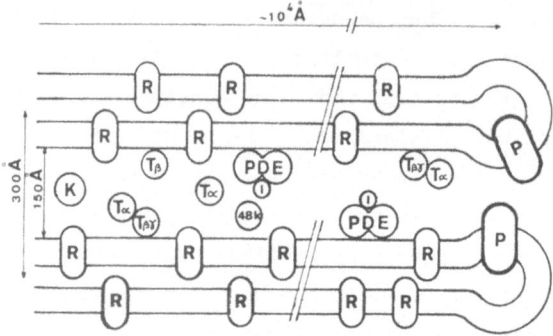

FIG. 1. The locations of the proteins in the interdisc cleft.

Rhodopsin (R)

Structural modeling from the known amino acid sequence predicts that this integral membrane protein is folded into seven α-helices spanning the disc membrane [7] . The model is in agreement with all the previous biochemical and biophysical data. The chromophore site is buried in a central hydrophobic core. All the known sites of interaction with other proteins are on the hydrophilic cytoplasmic face which constitutes no more than 1/4 of the protein. The same proportion of the protein protudes inside the disc. This face has a larger negative charge but has no known function besides its possible contribution to the binding of intradiscal calcium (see Fig. 2).

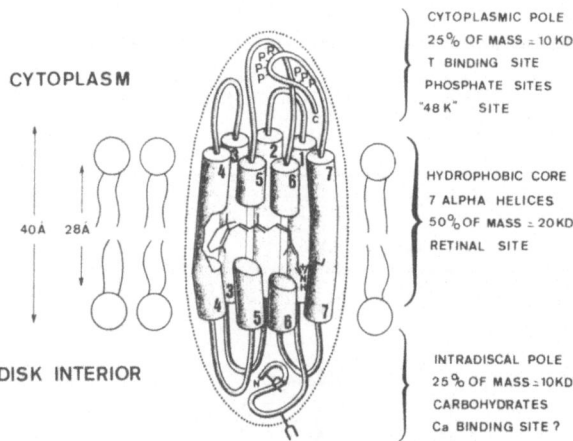

FIG. 2. Gross structure of rhodopsin (adapted from a figure kindly given by P. Hargrave).

Transducin (T)

Transducin is made up of two functionally differentiated subunits, $T\alpha$ (39kD), which bears the guanine nucleoside site, and $T_{\beta\gamma}$, made up of two polypeptide chains (37kD and 6kD). There is about one copy of $T\alpha.T\beta\gamma$ for every ten rhodopsins. In the dark, at the ionic strength of the cytoplasm, T is peripherally bound to the disc membrane[8]. The binding to the membrane is sensitive to the ionic strength and is probably not specifically to rhodopsin, except in that rhodopsin contributes to the electric charge of the membrane. At low ionic strength, T is released as a complex T_αGDP-$T_{\beta\gamma}$. The association between the subunits is not very strong ; T_αGDP and $T_{\beta\gamma}$ can be dissociated by various mild treatments (high pH, high Mg^{2+}) and independently isolated, but the isolated subunits no longer bind to the membrane. The exchange of the GDP for GTP or its non-hydrolyzable analogs markedly changes the properties of the T_α subunit ; it dissociates from $T_{\beta\gamma}$ at normal pH and ionic strength and becomes highly soluble.

ATP-dependent kinase (K)

This protein specifically catalyzes the phosphorylation of photoexcited rhodopsin. It is a soluble protein, apparently present in low concentrations of $\sim$ 1 per 1000 rhodopsins (though it might be partially lost in the ROS isolation procedure). Upon illumination, K binds to R*, and this property can be used for its isolation. It is composed of a single polypeptide of 68 kD.

cGMP Phosphodiesterase (PDE)

PDE is peripherally bound to the disc membrane and present at $\sim$ 1 copy per 100 rhodopsins. Like T, it can be extracted at low ionic strength, although there is no light-dark dependence. It is composed of two catalytic subunits (88 kD and 84 kD) whose peptide maps are remarkably similar and one or possibly two heat stable, trypsin-sensitive, inhibitory subunits of 13 kD. The soluble enzyme is a stable, holoenzyme of these three or four subunits.

48K or Arrestin, or S Antigen

This is a very soluble protein, present in the cytoplasm over the whole volume of cell, including the inner segment of the rod. Indeed, in the dark adapted retina, most of the Arrestin appears stored in the large volume inner segment. This protein was also known as "S Antigen", a name given by eye-immunopathologists in relation with its involvement in the generation of an autoimmune disease of the retina[9]. But its physiological role, as discussed later, is that of a semi permanent blocker of photoexcited rhodopsin, which justifies its abundance. After illumination arrestin is found to bind quantitatively onto the photoexcited rhodopsin, which induces its depletion from the inner segment.

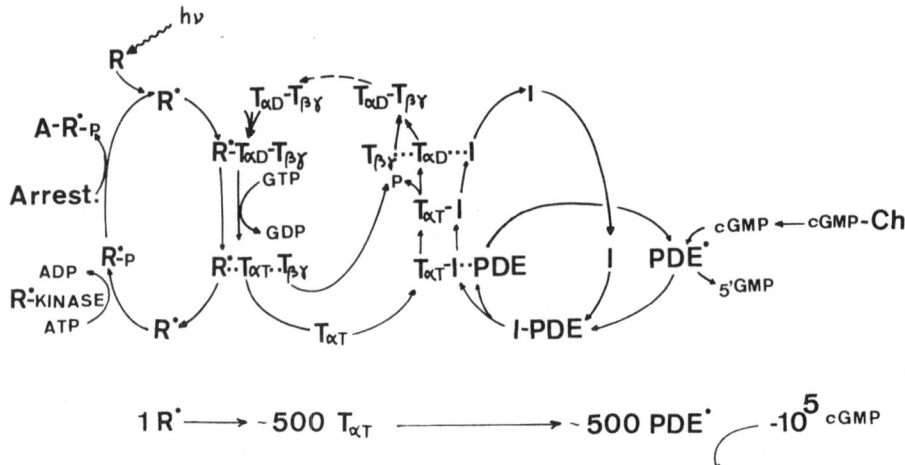

FIG. 3. The cGMP cascade of visual transduction R = Rhodopsin. R* = photoexcited Rhodopsin (Meta II Rhodopsin). R*-Kinase = Kinase which phosphorylates specifically R*. A = Arrestin, which binds specifically to phosphorylated R* and blocks the R*-Transducin coupling. TαD = α subunit of Transducin with GDP bound. TαT = α subunit of Transducin with GTP bound. I = 13 KD inhibitory subunit of the cGMP phosphodiesterase (PDE). cGMP-ch = cGMP dependent Na^+ channel of the plasma membrane.

THE BASIC SCHEME OF THE CASCADE : TRIGGERING AND AMPLIFICATION

A rather complete and complex scheme of the amplifying cascade is shown on fig. 3. One distinguishes a central amplification loop, mediated by transducin, coupled on one side to the receptor cycle and on the other side to the effector enzyme, the cGMP phosphodiesterase. The key amplification step is the catalytic interaction of the photoexcited receptor molecule R* with multiple copies of transducin, allowing the fast sequential activation by GTP of hundreds of transducin molecules by one R*. It must by emphasized that one R* couples sequentially to hundreds different transducin molecules but that for example in the response to a short flash, a given transducin molecule is activated only once : the amplification results from a high stoichiometry of interaction of many T with one R*, and not from a fast shuttling of T between R* and the PDE. Indeed, the activation of one PDE complex requires the binding of one subunit of activated transducin : the PDE will remain active only as long as the transducin is bound. By contrast with the R*-T interaction which is very short lived ($\sim 10^{-3}$ sec.) the T-PDE interaction last as long as one needs to keep the PDE active. It would not help to have the transducin shuttling back very fast toward R*. The amplification at this stage results from the very high turn over rate of hydrolysis of cGMP by the transducin-activated PDE.

The Triggering

The photoisomerization of retinal is the primary event, occuring in picoseconds. It leads to a very short lived, highly excited state of rhodopsin called Bathorhodopsin.

At this stage, the all-<u>trans</u> chromophore is higly constrained by the protein which has not yet changed its conformation. In the point charge model of Honig[10], the chromophore is in close proximity of charged groups in the protein, and its isomerization implies the separation of an ion pair. This is correlated with a 35 kcal/mol energy uptake in the transition[11], to be compared with 0.5 kcal/mol for the <u>cis-trans</u> isomerization of free retinal. This is more than 60 % of the energy of the absorbed photon, a very high efficiency for photoenergetic conversion. However, the function of the visual pigment (in contrast with that of bacterio-rhodopsin for example) is to memorize and transmit the information of the capture of a photon, not to store its energy. The high energy barrier provides here a very efficient protection against spurious triggering of the system by "thermal noise". In later dark reactions, this stored energy will force conformational changes in the protein to relax the strain on the chromophore.

The conformational changes of the protein lead to the active R* state through a cascade of successive steps, characterized by their absorption spectra, which extends over about one millisecond. This seems long enough to allow for major rearrangements in the protein structure, but physical techniques have failed to demonstrate large structural changes of the protein backbone[12]. The active state, R*, has been definitively identified with the "meta II rhodopsin" spectral state[13,14]. It is characterized by its enzymatic properties : two specific sites have developed on the cytoplasmic surface, a site of binding and catalytic activation for transducin[15], and a site of multiple phosphorylation by an ATP-dependent kinase near the C terminal end of rhodopsin[16]. This R* state remains stable for minutes, that is a time much longer than required for the physiological response : this implies the existence of a fast "blocking" process to turn off the stimulation.

First amplification step : R*-catalysed activation of transducin

Transducin in its "resting state", with GDP bound, is a holoenzyme TαGDP-Tβγ, weakly bound to the disk membrane on which it probably diffuses freely, without any affinity of binding to dark- adapted rhodopsin : but it has a high affinity of binding to R*, forming a one to one R*-Tα.GDP-Tβγ complex. The subunit Tα, which bears the guanoside binding site is the one that recognizes R*, but the presence of the Tβγ subunit is necessary for the proper binding to take place.

Binding to R* modifies the conformation of Tα and essentially opens the nucleoside site : the GDP that was occluded in the site and thus essentially non exchangeable on the time scale of hours, becomes very rapidly exchangeable (exchange time $< 10^{-3}$ sec.) with other guanine nucleosides present in the medium, e.g. GDP or GTP[17]. If guanine nucleosides have been suppressed, the GDP site will be emptied since the affinity of GDP for its binding site is of the order of 10^{-5}M. Then, the binding of Tα to R* will become very tight ; the complex will be virtually undissociable. In the presence of guanoside nucleosides, GDP and GTP have comparable affinities for the opened site. A GDP/GDP exchange is ineffective, but the binding of a GTP into the site modifies the structure of Tα which now looses its affinity for R* as well as for Tβγ : Tα.GTP becomes soluble, leaving Tβγ on the membrane and liberating R* which is made available for another R*-T.GDP interaction. Non hydrolyzable analogs of GTP, like GTPγS, induce the same process : this indicates that the energy from hydrolysis of the γ phosphate bond is not required at this stage. Neither R* has provided energy for the exchange : as a catalyst it must have been released unchanged, other-whise it would be unable to interact in sequence with a large number of T.GDP molecules. One must assume that T.GDP is driving the reaction with energy acquired from GTP hydrolysis in a preceding turn of the cycling process. Indeed proteolytic studies demonstrate that Tα.GDP is more unfolded than Tα.GTP.

Once Tα.GTP has dissociated from R* and from Tβγ, the nucleoside site occludes again and the nucleoside becomes non exchangeable. This is more easily seen with a non hydrolysable GTP analog.

Second amplification step : PDE activation

Tα.GTP is the active subunit which interacts with the phosphodiesterase : it binds to the inhibitory subunit of PDE and takes it away from the catalytic subunit, which then becomes active :

$$Tα.GTP + PDEαβγ \longrightarrow Tα.GTP-PDEγ + PDEαβ.$$

The complex Tα.GTP-PDEγ can be physically isolated. Although Tα.GTP is soluble, the complex is membrane bound, under physiological conditions and extractable only at low ionic strength.

In reconstitution experiments with purified transducin and PDE, the activation of PDE lasts as long as the complex Tα.GTP-PDEγ is stable, that is until the GTP in Tα is hydrolyzed to GDP. Then the affinity of Tα for PDEγ decreases and Tα.GDP recovers its high affinity for binding to Tβγ :

$$Tα.GDP PDEγ + Tβγ + PDEαβ \longrightarrow Tα.GDP-Tβγ + PDEαβγ$$

But the intrinsic GTPase activity is very slow, the active Tα.GTP state has a lifetime on the order of 10 sec. This is clearly too long to account for the necessary fast turn off of the PDE under physiological conditions. This will be discussed later.

The biochemical approaches, mainly binding studies of the various interacting components initiated by Kühn[15] and nucleoside exchange studies by Fung et al.[18], were capital for developping this concept of the amplifying cascade. They did not provide however much information on the kinetics of the various steps and on the translocation of the various component occuring in situ. This could be obtained from physical measurements to be discussed now.

Near infrared turbidimetry on ROS fragment suspensions

The simple physical technique of light scattering, especially in its most rudimentary form of turbidimetry, does not seem a priori a very attractive method to study a light-sensitive reaction cascade in a complex membrane system. The sample's sensitivity to visible light is not a problem. It is very easy, especially with solid-state detectors, to work in the near infrared, around 900 nm, where rhodopsin photosensitivity is nonexistent. The main problem lies with the physical analysis, even in the simplest case of dilute suspensions, when the interference between neighboring ROS fragments in the suspension can be neglected. The wavelength of the light is of the same order of magnitude as the dimensions of the ROS fragments. No theory exists to account quantitatively for the very small light scattering changes observed. The approach must then be semi-empirical, and based initially on reconstitution experiments to determine which component is critical to the generation of the various signals.

Indeed, various groups had observed light-scattering transients of the order of 10^{-3} to 10^{-2} (in $\Delta I/I$) upon flash illumination of ROS in the absence of GTP, and had tried to analyse them before it was realized that they might be correlated with the activation of the transducin cascade. However, it had been noticed that their amplitude does not remain linear, but that it saturates when more than 10 % of the total rhodopsin in the ROS is excited per flash. The critical observation was that these transient light-scattering changes are highly sensitive to the presence of GTP. These transients were then viewed as empirical signals monitoring unknown steps of the GTP-dependent, light-triggered reactions, and attempts were made to determine their dependence on the composition of reconstituted systems, on the amount of R* formed, and on the nucleosides or analogs added[24]. Three types of signals were originally identified in suspensions of cattle ROS fragments or in reconstituted membranes and were studied in transmission.

1. A so-called "rhodopsin signal" or R* signal, of very small amplitude ($\Delta I/I \sim 10^{-4}$ for a few percent R* per flash), corresponds to a very fast transmission increase observed during strong flashes when the disc membranes have been stripped of all peripheral proteins or when

all the other signals dependent on the presence of transducin have been saturated. It remains proportional to the amount of R* produced, per flash, up to total bleaching.

2. A "binding signal" that corresponds to a decrease in transmission ($\Delta I/I \sim 10^{-3}$ for a few percent R* per flash), is observed with native ROS or reconstituted membranes, that include rhodopsin and transducin, but not the other peripheral proteins. Its kinetics are complex, with fast (~ 20 msec) and slow ($\sim$ seconds) components. It saturates when the total number of R* formed upon a flash or a succession of flashes equals the number of transducin molecules present in the ROS or reconstituted system : hence a saturation around 10 % photoexcitation in ROS, which corresponds to the native stoichiometry of transducin versus rhodopsin. Therefore, this signal monitors the step R* + T $\longrightarrow$ R*T, which corresponds to the tight binding of transducin to illuminated membranes in the absence of GTP[15]. The rise time of the signal gives an upper limit to the time course of the reaction, and its saturation characteristics indicate that there is a one-to-one stoichiometry of binding of T with R*.

3. A "dissociation signal", replaces the binding signal when one is in the presence of GTP or analogs. It is of opposite polarity (that is, it corresponds to an increase in transmission), of slightly larger amplitude, even for very small flashes, and of very different saturation characteristics. Already detectable after a flash that photoexcites 10^{-5} of the rhodopsin, it saturates at about 10^{-3} photoactivation. From its dependence on the presence of transducin and on the concentration of GTP or analogs, it was concluded that this signal monitors the complete reaction :

$$R* + T.GDP \longrightarrow R*-T.GDP \xrightarrow[\text{GDP}]{\text{GTP}} R*-T.GTP \longrightarrow R*+T.GTP$$

The signal is insensitive to the presence of the PDE and the other peripheral proteins. The amplification, that is, the number of exchange reactions catalyzed by one R* on successive T.GDP, is obtained directly from the comparison of the saturation characteristics of the binding signal versus those of the dissociation signal. Typically, 5 % R* is required for half-saturation of the binding signal, and only 0.05 % R* for half-saturation of the dissociation signal. In the first reaction, one R* binds one T ; in the second reaction, one R* processes 100 T.

Even without an understanding of the physical origin of the signals, it is clear that that rise time sets an upper limit to the reaction time course ; hence the estimate that after a flash photoexciting a few percent of the rhodopsin, the total activation of transducin is completed within less than 100 msec.

Light-scattering studies on oriented ROS : reaction rate in the intact structure

To determine the origins of the signals and to gain information on the structural perturbations involves, new experiments were conducted in which three improvements were made[25] : (a) the use of structurally intact frog ROS ; (b) their magnetic orientation ; and (c) the selection of a particular combination of orientation and scattering angles ($\theta, 2\theta$ geometry), which takes advantage of the quasi-crystalline structure of the ROS by discriminating between the possible origins of the signal based on purely geometrical arguments.

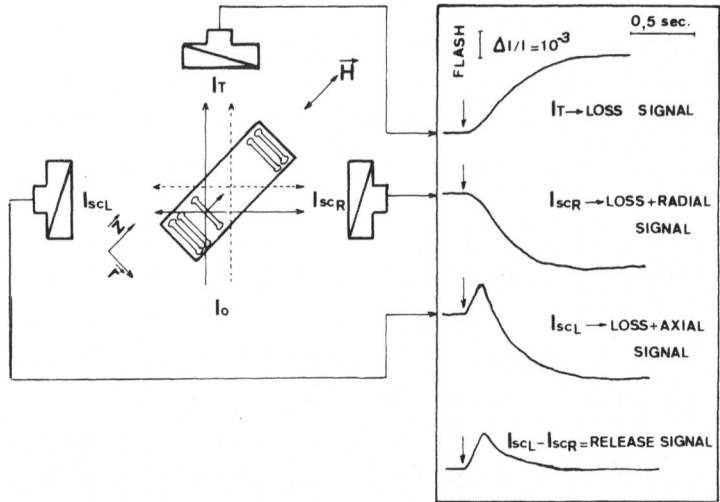

FIG. 4. Schematics of the light-scattering measurements on oriented frog ROS in the $\theta, 2\theta$ geometry. The signals are sketched as they are observed upon a flash photoexciting 5×10^{-4} of the rhodopsin in the presence of GTP.

The frog ROS are oriented at 45° with respect to the incident beam, the scattered light is detected at 90° on both sides, and the transmitted light is simultaneously measured by a third detector (see Figure 4). The outer cell membrane of the ROS is made permeable to GTP and other solutes by shearing through a needle. This treatment fragments the ROS, but it was verified by neutron diffraction that in the case of frog ROS it does not perturb the native ordering of the discs within the fragments. This ordered stacking defines within the ROS a cylindrical coordinate system (r, z).

The amplitude ΔA of the light wave scattered by a structural element ΔV of scattering density p(V) is the product of an angular factor $f(\theta)$ with the scattering power p(V) ΔV and with a phase factor exp (iϕ) : $\Delta A = f(\theta).p(V).exp(i\phi)\Delta V$. The total scattered intensity is defined by the amplitudes and relative phases of the scattered waves. In the Raleigh-Gans approximation, the phases are only dependent on the light path length from the incident wave front to the scattered wave front. Upon a change of shape or internal structure of the ROS without gain or loss of scattering material, the scattering elements are displaced within the particle, and only the relative phases of the scattered waves will change. In the ROS there are two particularly interesting types of displacements corresponding to the cylindrical coordinates : the radial displacement along r, that is to say, parallel to the disc surface, and the axial displacement in the z direction, which corresponds, for example, to a release of scattering material (protein) from the disc into the interdiscal cytoplasm or with a change in disc spacing. It is easily seen in Figure 4 that, in the special geometry chosen, for a light wave scattered toward the right detector the path length, and thus the phase, is independent of the radial coordinate r of the scattering element. It is entirely defined by z. Axial displacements of scattering elements will not change the intensity of the light scattered on the right detector but will only change it on the left detector. Conversely, axial structural changes will induce changes of scattered light intensity exclusively on the right detector.

Complementary information is obtained from the simultaneous measurement of transmitted light on the forward detector, which subtends a very small solid angle. It therefore measures only the transmitted light I_T, if one neglects the scattered light collected. This is equivalent to a measurement of the integral of the light scattered over all angles : $S = I^o - I_T$, where I^o is the incident light intensity. The main contribution to S comes from the very intense forward angle scattering, which is rather insensitive to the shape and internal structure of the ROS, as at vanishing scattering angle it depends only on its total scattering power. A change of shape or structure without gain or loss of scattering material or without change of index of refraction affects the small angle scattering, which is predominant in S, comparatively less than the large angle scattering measured at 90°. Therefore in such a case, one expects the relative change on S measured by the forward detector to be much smaller than that observed on any of the 90° detectors. On the other hand, a uniform loss of scattering material or a change of index of refraction of the ROS without shape change has the same relative effect on the scattering amplitudes at all angles, as it does not affect the phases or the angular factor. In that case, the relative change on S should be equal to that observed on the large angle detectors.

In this $\theta, 2\theta$ geometry, the light-scattering transient observed in the presence of GTP or analogs is complex and displays more components than the "dissociation" signal measured by transmission on unoriented ROS fragments. A new rapid and transient component, which has been termed the "release signal", is seen (on the left detector only) superimposed on a slower component termed the "loss signal" (seen on all three detectors), which corresponds to the previously observed "dissociation" signal. The "loss" signal is strictly isotropic ; it is observed as a decrease of scattered intensity of exactly the same amplitude on both lateral detectors and as an increase of transmitted light, which corresponds to a comparable change of S. These features are indicative of a light-scattering change that results from a loss of scattering material from the ROS into the solution. The "loss" signal is slower with frog ROS than with the smaller cattle ROS fragments, and its kinetics depend on the ROS fragmentation. Small frog ROS fragments separated from larger ones by centrifugation display a more rapid loss signal, as would be expected for a faster loss of protein from the ROS interior in the more fragmented ROS. Indeed, Kühn[8] had demonstrated that, at the ionic strength of the isotonic Ringer used in these experiments, upon illumination in the presence of GTP, the Tα subunit of transducin is selectively solubilized. Furthermore, the "loss" signal has, at saturation, the order of magniture expected for the loss from the ROS fragment of a protein mass corresponding to the pool of Tα.

This, and other evidences, prove that the "dissociation" signal of Kühn et al.[24] is due to the loss of Tα into the solution. Therefore, this signal does not monitor accurately the kinetics of the R*-catalyzed GTP/GDP exchange on transducin, since its rise time is limited by the rate of diffusion of Tα.GTP along the narrow interdiscal space and across the leaky plasma membrane. In the large frog ROS fragments, a lag of more than 100 msec is observed before the rise of the loss signal, at 5.10^{-4} bleaching, that is near saturation.

By contrast, the "release" component is rapid and highly anisotropic. It appears as a positive "bump" superimposed on the early phase of the loss signal, exclusively on the left-side detector. It is best observed by taking the difference between the left and right detectors (see Figure 4). It requires GTP, is observed upon very low flashes, and reaches saturation for a few 10^{-4}R*/R, like the loss component. It presents no time lag (< 5 msec) on the flash, and has a rise time on the order of 100 msec for 5.10^{-4} R*/R. It is not seen at all on the right-side detector (left/right ratio > 20) and is hardly detectable on the forward detector ; these are the characteristics of a light scattering change resulting from an axial change of structure of the ROS without loss of scattering material. In contrast to the loss signal, the rise time of the "release" component is independent of the fragmentation of the ROS ; but its decay correlates with the rise of the loss signal. Further evidence demonstrates that this signal is correlated to the release of Tα.GTP from the disc membrane into the cytoplasmic space. It therefore reveals, without a diffusion lag, the rate of Tα.GTP release upon the completion of the GTP/GDP exchange step. From the slope of the release signal, it was be estimated that in the intact structure of the ROS, at low illumination (10^{-4} R*/R), each R* activates transducin molecules at the rate of about one per millisecond[25].

Time resolved neutron diffraction study

The quasi crystalline organisation of the disc membrane within the ROS allows X rays and neutron diffraction studies. However, it is only with neutrons that measurements on dilute ROS suspensions in D_2O are feasible. Early studies[25] using magnetically oriented suspensions and H_2O-D_2O contrast variation had already provided, 10 years ago, quantitative data about the location of rhodopsin with respect to the lipid bilayer, which have since been proven to be correct. Light induced shift of protein mass at the cytoplasmic surface of the disc had also been observed, and tentatively attributed to the light induced conformation change on rhodopsin - since, at the time rhodopsin was considered as the only protein present in significant concentration in the rod. Dependences of the effect on the illumination rate and the presence of GTP, had not been studied.

In the new context of the transducin cascade, this needed to be investigated. The strategy was similar to that used for the light scattering measurements, that is to evaluate the sensitivity of the flash induced change upon the amount of R* formed and upon the presence of GTP. Oriented suspension samples of large area and low optical density were obtained, which, in the neutron beam of the I.L.L. (Grenoble) High Flux Reactor, allowed diffraction patterns to be collected very quickly : with 4 seconds of accumulation, an accuracy of ~ 0.1 % was obtained for the diffraction peak position, which are related to the disc spacing, and ~ 0.5 % for the relative intensity of the peak, which is related to the mass distribution within the disc lattice.

The strategy was the same as for the light scattering measurements : to investigate the sensitivity of the flash-induced changes to the amount of R* produced and the presence of GTP.

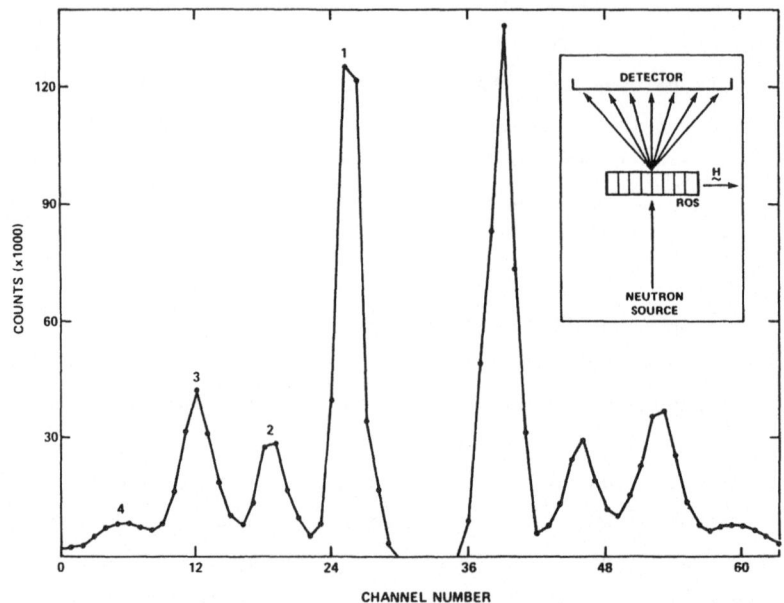

FIG. 5. Diffraction pattern recorded in 4 seconds.

As shown in figure 5, 4 orders of diffraction are recorded. In a
preliminary experiment where the time resolution was 1 min/spectrum, we
ascertained that the light-dependent change in the intensity of order 2
displayed all the characteristics of the light scattering binding and
dissociation signals. Namely, in the absence of GTP, changes in order 2
intensity can only be observed when a substantial amount of R* is
produced (i.e. 5 %–10 %). More significantly, this signal saturates in
amplitude at around the 10 % level of photolysis. With 1 mM GTP present,
order 2 intensity changes can be observed with a 0.1 % R* level and
saturates well below the 1 % level.

More information was obtained later with a time resolution of
4 sec./spectrum. It was known from earlier light scattering experiments
that at 6°C, the rise time of the release signal increased and moved
into the second range. From these two considerations, all neutron
diffraction experiments were carried out at 6°C. Again 4 orders of
diffraction were observed, although we shall choose not to use the 4th
order as its statistics are insufficient due to its weak magnitude. Each
measurement is made up of 64 time points and the photolysis flash
occured at the 15th point. Whenever possible, averages from several
comparable experiments are computed to enhance the signal-to-noise
ratio.

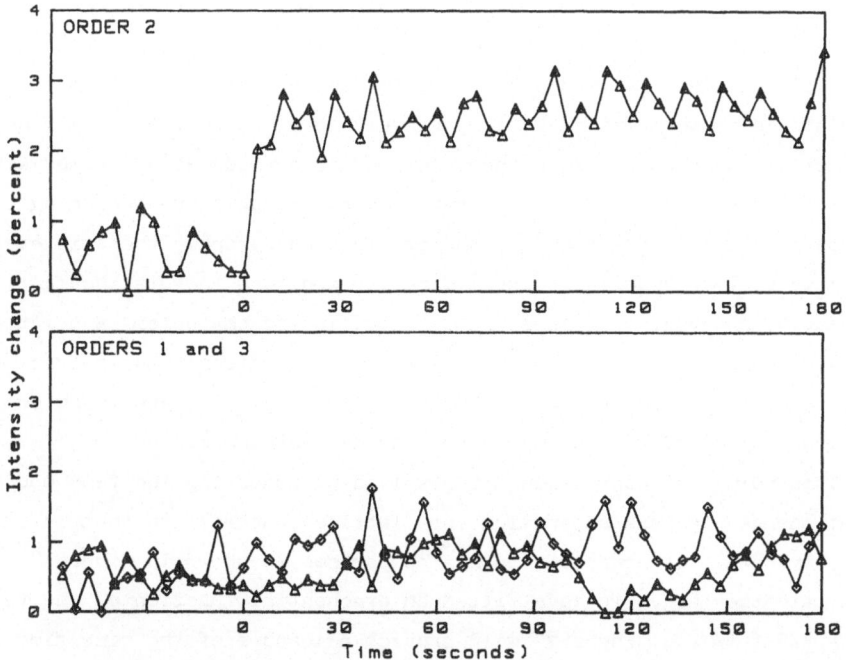

FIG. 6. Intensity changes observed for order 1, 2 and 3 upon a flash
photoexciting 0.5 % of the rhodopsin in the presence of 1 mM GTP.

Figure 6 displays the intensity changes observed for orders 1, 2 and 3. The light flash yielded 0.3 % R* and 1 mM GTP was added. From the lower panel, one can conclude that orders 1 and 3 do not change in amplitude upon the light flash. In contrast, order 2 as exhibited in the upper panel shows a light-induced increase in amplitude of about 2 %. This result should be understood in the crystalline context of the rod outer segment. Looking at the ROS as a 1-D crystal, the basic repeating unit, or unit cell, is a disc plus half the space between adjacent discs. This fundamental unit repeats itself every 300 A. The thickness of each disc is about 150 A, that is to say half of the repeat distance. The 2nd order in any 1-D diffraction pattern provides information about structural features that occur at twice the fundamental frequency (or conversely with a period that is half of the basic repeat period). In our case then, order 2 monitors changes at the surface of the discs as these surfaces occur every 150 A, that is twice as often as the basic unit cells which occur every 300 A.

It is now easy to accept why, of the 3 orders recorded only the 2nd one displays any change upon the light flash. As a signal that reflects the displacement of protein mass from the disc surface to the interdiscal space, the release signal is a change that affects structural features occuring at twice the fundamental frequency. That this phenomenon should manifest itself exclusively as a change in the amplitude of the 2nd order of diffraction is then of little surprise.

A diffraction pattern also contains another kind of information : the distance between its peaks is inversely proportional to the repeat distance of the unit cells in the object under consideration. Armed with this knowledge, we can now ask a very important question concerning the release signal. As described in the previous paragraphs, there is every reason to believe that this signal directly monitors the displacement of transducin, an event that is strictly axial in its character. A point of confusion now arises : a movement of the discs themselves (a rapprochement or moving apart) is also a strictly axial event and could conceivably give rise to a signal such as the release signal.

This point of contention is resolved by comparing the time course of the changes in peak separation (or lattice changes) to that of the intensity change of order 2. Such a comparison is shown in figure 7. To facilitate the task, we have fitted an exponential function of the form A (1 - EXP (-t/τ)) where τ will provide a measure of the rise time of the signal in question. In the lower panel, the distance between the two

2nd order peaks is seen to increase by about 0.3 % upon the light flash. (Recall from figure 5 that the diffraction pattern is symmetric and consists of pairs of peaks, one for each order observed.) This increase in peak spacing occurs slowly with a rise time of about 20 seconds. If the same fitting operation is performed on the order 2 intensity change, as shown in the upper panel of figure 7 , one obtains a value for significantly smaller than 20 seconds. From these kinetic considerations, we can now confidently conclude that the disc surface event as monitored by order 2 intensity changes (and by the light scattering release signal) reaches completion well ahead of the other event, that of the small amount of disc movement as reflected by the changes in peak spacings. The two effects are thus distinct, follow different time courses and can be differentiated kinetically.

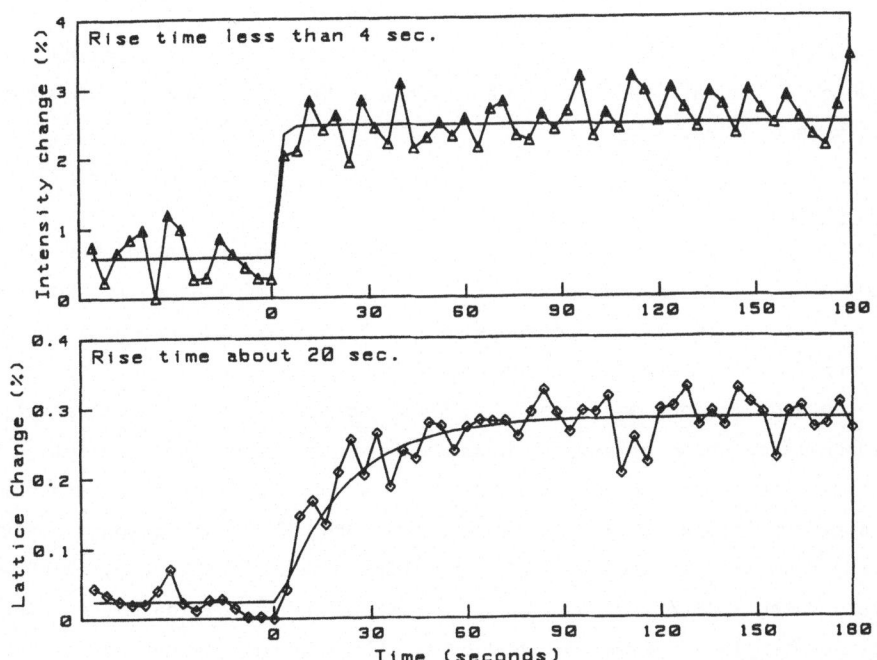

FIG. 7. Lattices changes observed for order 2. Sames conditions as for fig. 6.

Fast neutron diffraction has thus provided a finer structural understanding of the light scattering release signal. It provides a sounder basis for the interpretation of this signal as a direct measure of a strictly axial displacement of protein mass that occurs in the vicinity of the disc surface. It permits, on the basis of some simple kinetic comparison, the elimination of another strictly axial event, that of the movement of the discs themselves, as a possible origin for the release signal.

FAST TERMINATION : HOW AND WHERE TO BLOCK THE CASCADE

Not much is known yet on the cGMP gated channels which ultimately controls the response. They do not seem to possess a fast intrinsic deactivation mechanism to terminate the response. This indeed would not be satisfactory : if only this last stage was blocked and all the amplifying processes remained active, the response would indeed be terminated but the system could not be responsive again to a new stimulus. On the other hand blocking the first stage, that is the photoexcited receptors, is absolutely required but would not be sufficient, if the already activated transducin and phosphodiesterase are not switched off. The cascade needs to be blocked rapidly at its two main stages : at the receptor level, to stop the activation of Transducin, and at the level of Transducin-PDE coupling to stop the hydrolysis of cGMP.

R* blocking

An ATP dependent blocking of the cascade was first observed by Liebman[19]. The mechanism has now been fully elucidated by Kühn's group[20]. Its basis is the phosphorylation of photoexcited rhodopsin by a permanently active kinase, which does not require cyclic AMP, nor Ca^{++} and charged lipids as cofactors, but specifically and exclusively phosphorylates photoactivated rhodopsin. This important concept of sensitization of the activated receptor to a specific kinase appears now to be generalisable to other types of receptors. Phosphorylation is only a first step in the process, the blocking agent is the very soluble protein named "48 K" or Arrestin, which binds specifically to the multiphosphorylated site on rhodopsin, henceforth preventing further catalytic coupling to Tranducin.

The kinase, which phosphorylates very rapidly the receptors as soon as they are photoexcited, needs to be present only at a very low concentration. By contrast the arrestin molecules must remain bound stochiometrically and semi permanently to all the photoactivated and phosphorylated rhodopsins until slow metabolic processes (retinal hydrolysis, binding of a new 11-cis retinal and dephosphorylation) definitely inactivate and regenerate the rhodopsin molecules. A large number of Arrestin molecules are therefore required. Indeed, it was recognised recently that arrestin is a very abundant soluble protein, already identified by eye immunopathologists as the "retinal S Antigen" responsible for the retinal autoimmune disease known as uveitis[21]. Autoimmune uveitis may appear after a traumatism of the retina. It is caused by the release in the retina of large quantities of arrestin molecules upon the accidental breaking of the rod cell membrane.

Transducin-PDE coupling : mechanism of activation and termination

In the basal state, the cGMP phosphodiesterase activity is blocked by the inhibitory effect of a small subunit, PDEγ which hinders the hydrolytic activity of two large catalytic subunits PDEαβ. PDEγ is very sensitive to proteolysis and a short action of trypsin suffices for example to fully activate the PDE. Physiological activation amounts to the release of the inhibtion by PDEγ. It was recently demonstrated[22] that TαGTP, separated from Tβγ, binds to PDEγ and dissociates the inhibitory subunit from the catalytic ones. Under cytoplasmic ionic conditions, although TαGTP is soluble, it remains membrane bound when associated to PDEγ. The use of the non hydrolysable analog GTPγS allows to stabilise the TαGTPγS-PDEγ complex, which may then be extracted from the membrane by low ionic strength washing, together with the active PDEαβ complex. Both complexes are then easily isolated as separate entities. Activation results therefore from the high affinity of TαGTP for PDEγ. It will last as long as Tα remains in this high affinity state. Deactivation requires first that Tα reverses to the conformation observed with bound GDP, having a high affinity for binding to Tβγ and a low affinity for PDEγ. The hydrolysis step of GTP to GDP is a prerequisite to the termination of PDE activation. But the reversion of TαGTP to TαGDP may not be sufficient to terminate the PDE activity : full blocking requires the total release of PDEγ from Tα. This probably necessitates the intervention of Tβγ. There are evidences that TαGDP still binds PDEγ, although with a lower affinity than does TαGTP. In the absence of Tβγ, the complex TαGDP-PDEγ can be observed. An active role of GαGDP, isolated from the βγ subunits, is indeed implicit in the classical interpretation[23] of the inhibitory effects of Gβγ on hormone activated cyclases. The βγ subunit induces the reversion of PDE activity by competing with PDEγ, for binding with a high affinity to TαGDP, probably on the same site as that used by PDEγ. In both parts of the activation-deactivation cycle, Tβγ and PDEγ compete for binding to Tα : in the GTP bound conformation, Tα has a high affinity for PDEγ and a low one for Tβγ, the reverse being true when Tα has regained the GDP bound conformation :

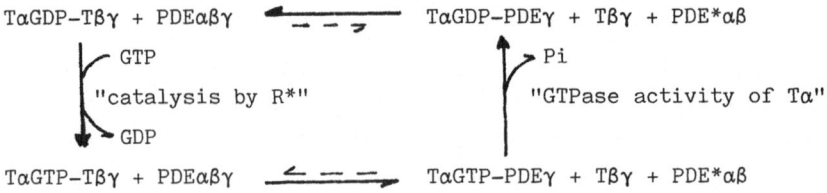

TαGDP-Tβγ + PDEαβγ ⟵ ⟶ TαGDP-PDEγ + Tβγ + PDE*αβ

GTP
"catalysis by R*"
GDP

Pi
"GTPase activity of Tα"

TαGTP-Tβγ + PDEαβγ ⟵ ⟶ TαGTP-PDEγ + Tβγ + PDE*αβ

Here lies a major problem : the slow rate of GTPase activity of Tα. The apparent lifetime of TαGTP seems much too long to account for a fast termination. But what is classicaly measured in a GTPase assay is the rate of hydrolysis of GTP in repetitive cycles of activation-deactivation of transducin. The maximum turn over rate measured, at 20° for cattle rods, never exceeds 3Pi/min/Transducin. As the rhodopsin catalysed GDP/GTP exchange step is very fast, and the recombination of TαGDP with Tβγ seems not to be rate limiting[18], this implies a lifetime of 20 sec. for TαGTP. But this 20 seconds cycle time might be divided in a short active time followed by a long "deadtime" during which TαGTP has become inactive or alternatively Tβγ has been modified in such a way that the already recombined TαGDP-T*γ complex cannot be immediatly reactivated. Phosphorylation of α subunits of G proteins have been reported in other systems. These phosphorylations are kinase C dependent and result from interactions between parrallel transduction cascades, possibly triggered by the same receptor. Similar processes may play a role in the visual process, but probably only for background adaptation. It seems more plausible that the fast termination of the response, independant of the previous state of illumination, is intrinsic to the amplification cascade.

CONCLUSION

As most biologists would agree, the sense of sight is of crucial importance to the survival of many higher animals. Man derives a huge amount of information about his surroundings via the visual input channel.

With so many recent findings pointing to substantial similiarities, both structural and functional between signal transduction in vision and in many hormonal systems, vision research has suddenly achieved a hitherto unknown level of relevance. Endowed with many desirable biochemical and physical features, the retinal rod outer segment has taken on the important role of a model system. In this chapter, our attempt has been to provide an overview of current vision research. We hope the reader has come to appreciate the wide range of techniques which have been brought to bear on the problem. These techniques vary widely but most are applied to take advantage of the very unique molecular arrangement of the rod outer segment.

*T. Minh Vuong is a post-doctoral fellow of the Helen Hay Whitney Foundation (New-York).

REFERENCES

1. M. Chabre. Trigger and amplification mechanisms in visual phototransduction, Ann. Rev. Biophys. Biophys.Chem. 14, 331-360 (1985).
2. L. Stryer. Cyclic GMP cascade of vision, Ann. Rev. Neurosci. 9, 87-119 (1986).
3. M. Chabre, C. Pfister, P. Deterre and H. Kühn. The mechanism of control of cGMP phosphodiesterase by photoexcited rhodopsin. Analogies with hormone controlled systems. In "Hormone and Cell regulation". Vol. 8, p. 87-98. J. Dumont and J. Nunez ed. Elsevier (1984).
4. R.A. Dixon et al. Cloning of the gene and cDNA for mammalian β adrenergic receptor and homology with rhodopsin. Nature 321, 75-79 (1986).
5. Tikubo et al. Cloning, sequencing and expression of complementary DNA encoding the muscarinic acetylcholine receptor. Nature 323, 411-416 (1986).
6. J.L. Benovic, R.A. Strasser, M.G. Caron and R. Lefkowitz. β adrenergic receptor kinase : identification of a novel protein kinase that phosphorylates the agonist occupied form of the receptor. Proc. Natl. Acad. Sc. US 83, 2797-2801 (1986).
7. E.A. Dratz and P.A. Hargrave. The structure of rhodopsin and the rod outer segment disk membrane. TIBS 8, 128-131 (1983).

8. H. Kühn. Interactions between photoexcited rhodopsin and light-activated enzymes in rods. In" Progress in Retinal Research", eds. N. Osborne and J. Chader, vol. 3, pp. 123-156. Oxford : Pergamon Press (1984).

9. C. Pfister, M. Chabre, J. Plouet, V.V. Tuyen, Y. De Kozak, J.P. Faure and H. Kühn. Retinal S antigen identified as the 48K protein regulating light dependent phosphodiesterase in rods. Science, 228, 891-893 (1985).

10. B. Honig, T. Ebrey, R.H. Callender, U. Dinur and M. Ottolenghi. Photoisomerisation, energy storage, and charge separation. A model for light energy transduction in visual pigment and bacteriorhodopsin. Proc. Natl. Acad. Sci., USA, 76, 2503-7 (1979).

11. A. Cooper. Energy uptake in the first step of visual excitation". Nature, London, 282, 531-3 (1979).

12. M. Chabre. Conformational and functional change induced in vertebrate rhodopsin by photon capture. In "The Biology of photoreception" eds D.J. Cosens and D. Vince Prue. S.E.B. Symposia XXXVI. Cambridge University Press (1983).

13. N. Bennett, M. Michel-Villaz, H. Kühn. Light-induced interaction between rhodopsin and the GTP-binding protein. Metarhodopsin II is the major photoproduct involved. Eur. J. Biochem. 127, 97-103 (1982).

14. D. Emeis, H. Kühn, J. Reichert and K.P. Hofmann. Complex formation between metarhodopsin II and GTP-binding protein in bovine photoreceptor membranes leads to a shift of the photoproduct equilibrium. FEBS Lett. 143, 29-34 (1982).

15. H. Kühn. Light- and GTP-regulated interaction of GTPase and other proteins with bovine photoreceptor membranes. Nature 283, 587-589 (1981).

16. U.A. Wilden and H. Kühn. Light-dependent phosphorylation of rhodopsin : number of phosphorylation sites. Biochemistry 21, 3014-3022 (1982).

17. N. Bennett and Y. Dupont. The G protein of retinal rod outer segments (Transducin) : Mechanism of interaction with rhodopsin and nucleotides. J. Biol. Chem. 260, 4156-4168 (1985).

18. B.K.K. Fung and L. Stryer. Photolyzed rhodopsin catalyses the exchange of GTP for bound GDP in retinal rod outer segments. Proc. Natl. Acad. Sci. USA 77, 2500-2504 (1980).

19. P.A. Liebman and E.N. Pugh. ATP mediates rapid reversal of cyclic GMP phosphodiesterase activation in visual receptor membranes. Nature 287, 734-736 (1980).

20. U. Wilden, S.W. Hall, H. Kühn. Phosphodiesterase activation by photoexcited rhodopsin is quenched when rhodopsin is phosphorylated and binds the 48KD protein of rod outer segments. Proc. Acad. Sc. USA, 83, 1174-1178 (1986).

21. C. Pfister, M. Chabre, J. Plouet, V. Tuyen, Y. De Kozak, J.P. Faure and H. Kühn. Retinal S antigen identified as the 48K protein regulating light dependent phosphodiesterase in rods. Science 228, 891-893 (1985).

22. P. Deterre, J. Bigay, M. Robert, C. Pfister, H. Kühn and M. Chabre. Activation of retinal rod cyclic GMP phosphodiesterase by Transducin. Characterization of the complex formed by phosphodiesterase inhibitor and transducin. Proteins Struct. Funct. Gen. 1, n° 2 (1986).

23. A.G. Gilman. G proteins and dual control of adenylate cyclase". Cell 36, 577-579 (1984).

24. H. Kühn, N. Bennett, M. Michel-Villaz and M. Chabre. Interactions between photoexcited rhodopsin and GTP-binding protein : kinetic and stoichiometric analysis from light-scattering changes. Proc. Natl. Acad. Sci. USA 18, 6873-6877 (1981).

25. T.M. Vuong, M. Chabre and L. Stryer. Millisecond activation of transducin in the cyclic nucleotide cascade of vision. Nature 311, 659-661 (1984).

26. H. Saibil, M. Chabre and D.L. Worcester. Neutron diffraction studies of retinal rod outer segments membranes. Nature 262, 266-270 (1976).

INOSITOL LIPID METABOLISM: GENERATION OF SECOND MESSENGERS

John R. Williamson and Carl A. Hansen

Department of Biochemistry and Biophysics
University of Pennsylvania School of Medicine
Philadelphia, Pennsylvania 19104

INTRODUCTION

It is generally accepted that the regulation of characteristic functions specific for different cells is brought about by the binding of agonists and hormones to receptor proteins located in the plasma membrane (1). Transfer of information from chemicals in the extracellular environment to the regulation of many intracellular enzymes and proteins is achieved by a number of different receptor-dependent signalling mechanisms. The best understood example of such a process is the β-adrenergic activation of adenylate cyclase, which causes an increased production of cAMP as a second messenger (2). The subsequent activation of cAMP-dependent protein kinase results in the phosphorylation and modulation of the activity of a variety of target enzymes and regulatory proteins within the cell with consequent alterations of specific cell functions (3).

Cell function can also be modulated by receptor-activated mechanisms that do not involve cAMP but rather are mediated by increases of the free Ca^{2+} concentration in the cytosol (4, 5). Calcium causes activity changes of a variety of proteins, including protein kinases, either directly or after binding to calmodulin or other Ca^{2+}-binding proteins. However, until recently the source of the Ca^{2+}, the amount and kinetics of the Ca^{2+} changes, and the mechanism of hormone-stimulated cellular Ca^{2+} mobilization were largely unknown.

Many studies have now established that a wide range of compounds including hormones, secretagogues, neurotransmitters, chemoattractants and other cell activating subtances that involve Ca^{2+} mobilization in the expression of the biological response cause an activation of a phosphodiesterase (phospholipase C), which breaks down inositol lipids in the plasma membrane (6-10). However, unlike receptor-mediated activation of adenylate cyclase, which produces cAMP as the only second messenger, receptor-mediated inositol lipid breakdown serves a dual-signalling role with production of two second messengers having different functions. One of these compounds, namely inositol 1,4,5-trisphosphate (IP_3) is responsible for eliciting intracellular Ca^{2+} mobilization (7, 10), while the second compound, 1,2-diacylglycerol (DAG) has as its primary signalling role the activation of a phospholipid-dependent protein kinase in the plasma membrane termed protein kinase C (11). Hence in principle, agents that interact with inositol lipid metabolism not only cause Ca^{2+} release with phosphorylation of proteins by Ca^{2+}-dependent protein kinases but also phosphorylation of a different set of proteins by activation of protein kinase C.

A challenge for future research is to elucidate how the different signalling systems interrelate with each other both at the level of generation of the second messengers through receptor coupling to GTP-binding proteins and at the level of expression of specific tissue functional effects through phosphorylation of regulatory proteins by the different types of protein kinases. The present article will summarize recent developments relating to the role of GTP-binding proteins in receptor-mediated activation of phospholipase C, the regulation of 1,2-diacylglycerol and IP_3 production as signal generators and the formation and metabolism of novel inositol phosphates that may have presently undiscovered functional roles.

GENERATION OF INTRACELLULAR SIGNALS

Figure 1 provides a schematic representation of the major metabolic events that occur following addition of a Ca^{2+} mobilizing agonist to a target cell. Receptor proteins in the plasma membrane span the phospholipid bilayer and have outwardly facing domains responsible for agonist binding and inwardly facing domains that interact with other proteins or receptor subunits on the inner leaflet of the membrane bilayer. Conformational changes within the receptor protein initiated by binding of specific agonists are thought to be responsible for information transfer from the outer to the inner domain. This is mediated by special-ized heterotrimeric GTP-binding proteins that selectively interact with specific target enzymes (12, 13). As discussed later, there is only circumstantial evidence indicating the involvement of a GTP-binding protein in receptor coupling to phospholipase C, and since it has not been isolated and characterized, it is denoted by Gx in Fig. 1. By analogy with other well-characterized GTP-binding proteins that interact with adenylate cyclase, Gx is assumed to become functionally active through dissociation of its GTP-activated α-subunit. Phospholipase C activity is found in both the plasma membrane and soluble fractions of the cell. Whether the GTP-bound α-subunit of the putative GTP-binding protein causes activation of only the membrane-bound phospholipase C or also promotes the binding and activation of soluble enzyme has not yet been elucidated.

The inositol lipid substrate that is responsible for Ca^{2+} signalling is phosphatidylinositol 4,5-bisphosphate (PIP_2), where the fatty acid side chains attached to glycerol are primarily stearyl (R_1) and arachidonyl (R_2). This lipid accounts for only about 0.1% of the total lipid com-position of the plasma membrane. Hydrolysis of the phosphodiester bond between the 3-position of glycerol and the 1-position of the inositol ring liberates 1,2-diacylglycerol in the membrane and the water soluble product *myo*inositol 1,4,5-trisphosphate or IP_3, which releases calcium sequestered in the endoplasmic reticulum. This mechanism accounts for a rapid increase of the cytosolic free Ca^{2+} following agonist stimulation. IP_3 is metabolized by two routes, one by a 5-phosphatase to *myo*-inositol 1,4-bisphosphate (IP_2) and the second by a specific ATP-dependent 3-kinase with the formation of *myo*inositol 1,3,4,5-tetrakisphosphate (IP_4). Degradation of IP_4 takes place also by the 5-phosphatase to a second IP_3 isomer, namely *myo*inositol 1,3,4-trisphosphate. In many cells the 5-phosphatase is found largely bound to the plasma membrane while the 3-kinase is located primarily in the soluble fraction. Of the inositol phosphate derivatives formed during hormone stimulation of cells, the 1,4,5-P_3 isomer is by far the most active in causing a release of calcium from intracellular organelles. The function of IP_4 is presently unknown.

28

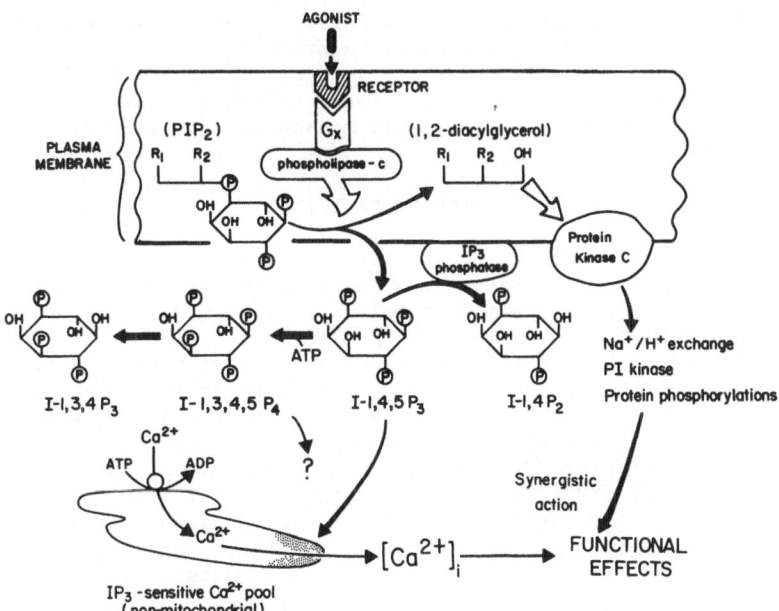

Fig. 1. Schematic representation of reactions involved in signal
generation from inositol lipid metabolism. PIP_2 = phosphatidylinositol 4,5-bisphosphate; G_x = postulated unidentified GTP-
binding protein; $I-1,4,5P_3$ (IP_3) = myoinositol 1,4,5-
trisphosphate; $I-1,4P_2$ (IP_2) = myoinositol 1,4-bisphosphate; I-
1,3,4,5P_4 (IP_4) = myoinositol 1,3,4,5-tetrakisphosphate; I-
1,3,4P_3 = myoinositol 1,3,4-trisphosphate; PI = phosphatidyl-
phosphatidylinositol; R_1 = fatty acid side chain (mainly stearyl);
R_2 = fatty acid side chain (mainly arachidonyl).

The other second messenger, 1,2-diacylglycerol, is the physiological
activator of protein kinase C. This protein kinase is also activated by
tumor-promoting agents such as phorbol myristate acetate (PMA), which like
diacylglycerol have the interesting effect of increasing the binding of
soluble enzyme to the plasma membrane and decreasing its requirements for
Ca^{2+} to resting cellular Ca^{2+} levels (11, 14-16). The physiological
target proteins for this kinase are still largely unknown, but in many
secretory cells it has the important functional effect of sustaining the
physiological response initiated by Ca^{2+} release (synergistic action with
Ca^{2+}). As depicted in Fig. 1, two processes that are enhanced through
protein kinase C activation, at least in some cells, are Na^+-H^+ exchange
across the plasma membrane leading to alkalinization of the cell interior,
a prerequisite for cell proliferation (17), and PI kinase activity (see
later discussion).

As previously mentioned, the inositol lipids contain a high proportion
of arachidonic acid in the 2-position of glycerol, and formation of
arachidonic acid, either through sequential degradation by diglyceride and
monoglyceride lipases or by a secondary activation of phospholipase A_2 by
Ca^{2+}, acts as a further signalling system. In many cells, notably platelets, neutrophils, and endothelial cells, arachidonic acid is metabolized
to a variety of eicosanoid mediators (prostaglandins, thromboxanes, and
leukotrienes). These metabolites can function as _inter_cellular mediators
to initiate responses in tissues other than those in which they are
produced or may combine with plasma membrane receptors of cells in the
same tissue. Prostaglandin H_2 and thromboxane A_2, for instance, interact

with receptors in platelets that are coupled to phospholipase C and hence produce a cascade-signalling effect by reinforcing the Ca^{2+}-mobilizing response to an original weak stimulatory signal (18-20). On the other hand, prostaglandin I_2 activates adenylate cyclase in platelets, which by causing an inhibition of inositol lipid metabolism, terminates the activational response (21).

RECEPTOR COUPLING THROUGH GTP-BINDING PROTEINS

Receptor-mediated effects on adenylate cyclase activity, either a stimulation (e.g. by β-adrenergic agonists) or an inhibition (e.g. by α_2-adrenergic agonists or muscarinic agonists acting on M_2 receptors) are mediated by two different GTP-binding proteins Gs (stimulatory) and Gi (inhibitory), which belong to a family of structurally and functionally related proteins (13, 22, 23). In non-stimulated cells, these proteins are thought to be present in the membrane as inactive $\alpha\beta\gamma$ oligomeric complexes with GDP bound to the α-subunit. In the case of Gs and Gi, the α-subunits are different (Mr = 52 kDa and 45 kDa for α_s and 41 kDa for α_i) while the β subunit (Mr = 35/36 kDa) and the γ subunit (Mr = 5-10 kDa) are similar. Receptor activation causes dissociation of the α-subunit accompanied by a replacement of bound GDP by GTP. Activation of adenylate cyclase occurs when α_s-GTP binds to the enzyme. This stimulatory signal is terminated by the intrinsic GTPase activity of the subunit, followed by reassociation of α_s-GDP with the $\beta\gamma$-subunits. Inhibition of activated adenylate cyclase by Gi on the other hand is thought to occur either by a direct inhibitory effect of the α_i-subunit (23, 24) or by a decrease in the amount of the free α_s-subunit as a consequence of an increased availability of $\beta\gamma$-subunits upon dissociation of Gi (25).

A number of approaches and tools have been used to identify and study the functional roles of these GTP-binding proteins. Specific bacterial toxins catalyse an ADP-ribosylation at particular sites on the α-subunits of the GTP-binding proteins by transfer of the ADP-ribose moiety from NAD^+ to specific amino acid sites in the proteins (13, 26). Cholera toxin causes ADP-ribosylation of Gs and produces a permanent activation of the α_s-subunit by inhibiting its GTPase activity, so that there is a persistent activation of adenylate cyclase. *Bordetella pertussis* toxin causes ADP-ribosylation of Gi and prevents dissociation of the α-subunit, with the result that the inhibitory effects of receptor coupling on adenylate cyclase are abolished. A second approach is provided by the use of nonhydrolyzable analogues of GTP, which stabilize the α-subunits in the active form since GTPase activity is prevented. An interesting property associated with the interaction between agonists, receptors, and GTP-binding proteins, is a shift of agonist binding from a high affinity to a low affinity state, which can be induced by activation of the GTP-binding coupling protein with GTP or its analogues (27). High affinity receptor binding sites consist of a ternary complex of agonist, receptor and GTP-binding protein, while low affinity sites correspond to a binary complex of agonist and receptor after dissociation of the GTP-binding protein (1).

The first indications that a GTP-binding protein may function as a signal transduction mechanism for Ca^{2+} mobilizing agonsits were reports that GTP or its nonhydrolysable analogues decreased the affinity of various receptors for binding to their agonists. Some examples of this effect are norepinephrine binding to α_1-receptors (28-30), carbachol interactions with muscarinic receptors (31), chemotactic peptide binding to neutrophil membranes (32), and vasopressin (33) and angiotensin (34) binding to liver plasma membranes. More conclusive evidence indicating that a GTP-binding protein may be specifically involved in inositol lipid metabolism has been obtained recently from a number of laboratories. After introduction of non-hydrolysable GTP analogues into mast cells, addition of extracellular Ca^{2+} caused an increased secretion of histamine without any change of cAMP levels (35). Similarly, addition of GTP or its analogues to permeabilized platelets decreased the Ca^{2+} requirements for serotonin release and promoted the formation of diacylglycerol, indicating a stimulation of phospholipase C (36, 37). In more recent work with homogenates and plasma membrane preparations from neutrophils (38), blowfly salivary glands (39), polymorphonuclear leukocytes (40), hepatocytes (41, 42), GH_3 pituitary cells (43), arterial smooth muscle (44), and cerebral cortical membranes (45), it has been shown that addition of GTP analogues alone or together with an agonist causes an increased breakdown of polyphosphoinositides and formation of inositol phosphates. The role of the stabilized (GTP-bound) α-subunit of the GTP-binding protein may be to decrease the Ca^{2+} requirement for activation of phospholipase C (46).

The identity of the GTP-binding protein that apparently interacts with phospholipase C has not been ascertained. In a number of cells, including polymorphonuclear leukocytes and neutrophils (40, 47-53), mast cells (54), human leukemic HL60 cells (55, 56) and the hybrid cell line WBC-264-9C (57), chemotactic peptide-induced interactions with inositol lipid metabolism, arachidonic acid release, O_2^- generation, Ca^{2+} mobilization and other responses were greatly inhibited by pretreatment of the cells with pertussis toxin. In these cells it was shown that the effects induced by the chemotactic peptide fMet-Lue-Phe were not mediated by changes of cAMP, but rather by a receptor-coupled activation of phospholipase C. The attenuation of all of these chemotactic peptide-induced effects by pertussis toxin was associated with the ADP-ribosylation of a 41 kDa membrane bound protein similar to G_i. The involvement of a G_i-like protein in coupling with the chemotactic peptide receptor in neutrophils (58) and HL-60 cells (59) was supported by further studies showing that addition of G_i (isolated from brain) to pertussis toxin pretreated membranes was able to restore fMet-Leu-Phe binding affinity, fMet-Leu-Phe-induced GTPase activity and inositol phosphate formation that had been inhibited by the pertussis toxin treatment. Addition of G_o (another GTP-binding protein isolated from brain having a 39 kDa α-subunit, which like G_i interacts with muscarinic receptors and is ADP-ribosylated by pertussis toxin [31]) was also able to reconsitute increased inositol lipid hydrolysis in pertussis toxin pretreated membranes from HL-60 cells (59). The conclusion reached from these studies, therefore, was that chemotactic peptide receptors in neutrophils and similar cells are coupled to a pertussis toxin-sensitive GTP-binding protein which mediates an activation of inositol lipid metabolism. The fact that the reconstitution experiments noted above were equally effective both with extrinsic G_i and G_o raises questions of specificity of GTP-binding protein functions. Alternatively, it cannot be excluded that the effects were produced by a different, unrecognized contaminating GTP-binding protein in the preparations of G_i and G_o used.

In other cell types, pertussis toxin treatment apparently has no effect in inhibiting the actions of Ca^{2+} mobilizing hormones (60). Thus, no effect of pertussis toxin pretreatment was observed on the binding of α_1-adrenergic ligands to rat kidney cortex membranes (61) or muscarinic ligands to receptors of astrocytoma cells (62) in the absence or presence of guanine nucleotides. More directly it has been shown (with intact or permeabilized cells) that treatment of astrocytoma and chick heart cells (63), 3T3 fibroblasts (64), pituitary GH_3 cells (65, 66), hepatocytes (67, 68) and pancreatic acinar cells (69) with concentrations of pertussis toxin that apparently cause a complete ADP-ribosylation of G_i and prevention of its inhibitory effects on adenylate cyclase, do not prevent agonist- or GTP analogue-induced increases of polyphosphoinositide breakdown or Ca^{2+} mobilization. In contrast, the stimulatory effects of angiotensin II on PIP_2 breakdown, $^{45}Ca^{2+}$ influx, cytosolic free Ca^{2+}, and prostaglandin E_2 synthesis in rat renal mesangial cells were inhibited by pertussis toxin pretreatment in the absence of changes of cAMP levels (70). In this study a protein band of Mr 42 kDa was ADP-ribosylated by pertussis toxin.

On the basis of presently available data, it is clear that receptor coupling to phospholipase C is mediated by a GTP-binding protein. The evidence is strong in neutrophils and similar cells that the G-protein is ADP-ribosylated and inactivated by pertussis toxin and that the ribosyl-ated subunit has a molecular weight very similar to that of the α-subunit of G_i. In other cells, however, the G-protein involved in Ca^{2+} mobili-zation appears not to be affected by pertussis toxin, suggesting that it is distinct from G_i. Thus, at present it is not clear whether there are fundamental differences between different cell types in the nature of the G-protein that couples receptors to phospholipase C. An alternative possibility is that there is a similar, as yet unidentified, G-binding protein that differs from G_i in its susceptibility to ribosylation and inactivation by pertussis toxin in different cells, but which uniquely couples to phospholipase C in all cells. One possibility that might explain such a behavior is association of the α-subunit with a $\beta\gamma$-subunit complex having a greater degree of hydrophobicity (due to a different γ-subunit) so that the G-protein heterotrimer is inserted to a greater extent into the phospholipid bilayer thereby hindering ADP-ribosylation of the α-subunit (71, 72). Probably each receptor type is complexed pri-marily with a particular G-protein in resting cells. Cell activation by physiological concentrations of agonists would then cause a unique channeling of information to effector enzymes (adenylate cyclase or phospholipase C) by release of an α-subunit containing amino acid sequences specific for recognition by the particular effector system. Whether some receptors, particularly at high occupancy with agonist, can cause an activation and dissociation of more than one G-protein is presently not clear. Characterization of G-proteins by sensitivity to pertussis toxin appears to represent a poor criteria for specificity of functional effects because of the heterogeneity of G-protein α-subunits in the molecular weight region of 40-41 kDa as revealed by immunological studies (72, 73) and ADP-ribosylation by cholera toxin (74, 75).

METABOLISM OF INOSITOL LIPIDS

The inositol lipids in the cell are in a continuous state of turnover as revealed by labeling studies with ^{3}H-inositol and ^{32}P-phosphate. The recognized pathways of phosphoinositide metabolism are depicted in Fig. 2. Two different types of enzymes are involved in phosphoinositide hydro-lysis, namely phosphomonoesterases as well as the phosphodiesterase, phospholipase C (76). The phosphomonoesterases selectively remove phos-phate from the 4 and 5 positions of the inositol ring and convert PIP_2 to phosphatidylinositol-4-phosphate (PIP) and the latter to phosphatidyl-inositol (PI), which represents more than 95% of the total inositol lipid pool. Together with PI and PIP kinases these enzymes are responsible for the turnover of the phosphate in the 4 and 5 positions of the inositol ring, while leaving the mass of these inositol lipids constant in the steady state. Phospholipase C causes hydrolysis of all three inositol lipids, with the production of the common product 1,2-diacylglycerol and the liberation of IP_1, IP_2, and IP_3, from PI, PIP, and PIP_2, respectively (see Fig. 2). Completion of the "PI cycle" is achieved by metabolism of diacylglycerol to phosphatidic acid (PA) by diacylglycerol kinase and subsequently to cytidine 5'-diphosphate-diacylglycerol, which condenses with myoinositol to reform PI in the endoplasmic reticulum. The incorporation of ^{32}P from ^{32}P-γATP into PA and subsequently into PI provided the basis for measurements of PI turnover in the earlier literature (See Ref. 9 for review).

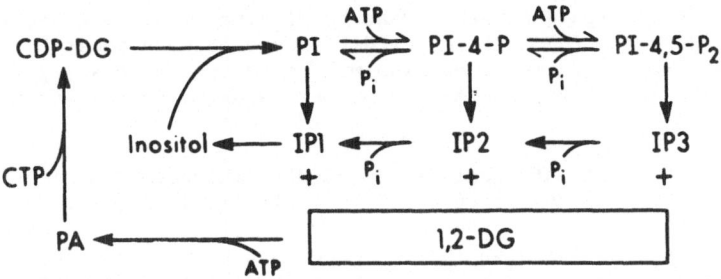

Fig. 2. Pathways of phosphoinositide metabolism. PI-4-P = phosphatidylinositol 4-phosphate; PI-4,5-P$_2$ = phosphatidylinositol 4,5-
bisphosphate; DG = diacylglycerol; PA = phosphatidic acid; CTP =
cytidine 5'-triphosphate; CDP = citidine 5'-diphosphate. Other
abbreviations are explained in the legend to Fig. 1.

Since diacylglycerol and IP$_3$ have separate and distinct roles as
signal generators, the relative availability of the inositol lipids to the
enzyme and the regulation of the substrate specificity of phospholipase C
are factors that will determine the amounts of these second messengers
produced during hormonal stimulation. Studies reviewed elsewhere (8, 9)
have shown that the inositol lipids are heterogeneously distributed in the
different cellular membranes and that only a small fraction of the total
inositol lipid content is available for metabolism by phospholipase C.
Isotopic turnover and cell fractionation studies have revealed distinct
hormone-sensitive and insensitive inositol lipid pools, particularly of
PI, which serves as the precursor for hormone-sensitive PIP and PIP$_2$ via
PI and PIP kinases (77-80). Presumably the hormone-sensitive inositol
lipid pools are located in the inner leaflet of the plasma membrane where
their rapid translational mobility in the bilayer makes them readily
accessible to plasma membrane-bound phospholipase C and inositol phosphate
kinases, despite their low content relative to other phospholipids. Since
movement of phospholipids between the bilayers (flip-flop) is very slow,
any inositol lipids in the outer leaflet will be in the hormone-
insensitive pool (80).

Many studies with intact cells, where the inositol lipids have been
labeled to uniform specific activity with [^{3}H]inositol or ^{32}P inorganic
phosphate, have shown that the breakdown of PIP and PIP$_2$ precedes that of
PI during hormone stimulation and is independent of an increase of the
cytosolic free Ca^{2+} concentration (7, 8). Preferential hydrolysis of the
polyphosphoinositides by hormonally activated phospholipase C is also
indicated by the faster rates of accumulation of IP$_3$ and IP$_2$ than of IP$_1$
after agonist addition (81, 82). Studies with a large number of different
cell types and agonists (e.g. 82-84) support the conclusion that the
initial effect of agonist stimulation is to increase the breakdown of
polyphosphoinositides with the formation of IP$_3$ and consequent mobilization of intracellular Ca^{2+}. There appear to be exceptions to this
general conclusion, however, since stimulation of phospholipase C in
pancreatic islets by nutrient secretagogues is probably secondary to an
activation of Ca^{2+} entry into the islet (85). Studies with purified
phospholipase C reconstituted along with radiolabeled inositol lipid
substrates into phospholipid vesicles have shown that whereas Ca^{2+}
stimulated the breakdown of all three inositol lipids, only the polyphosphoinositides were hydrolyzed in the presence of EGTA (86). Hydrolysis of
PI was completely dependent on Ca^{2+}, with a Michaelis constant of 1 to 2
μM (87). Majerus *et al.* (87) concluded that the delayed depletion of PI
levels may be secondary to the increase of cytosolic free Ca^{2+}, and

mediated by phospholipase C rather than by PI kinase, as originally suggested by Berridge (6). The major difference between these two proposals is that the former mechanism will produce more diacylglycerol relative to IP_3 rather than equal amounts as mandated by the second mechanism.

Direct measurements of agonist-induced diacylglycerol and inositol phosphate formation in hepatocytes (88-90), platelets (90, 91) and vascular smooth muscle (92) have shown that diacylglycerol increased rapidly, in accordance with an initial breakdown of polyphosphoinositides, but accumulated at later times in much greater stoichiometric amounts than IP_3, suggesting a delayed formation from PI. It is also evident that the phosphoinositides may not be the only precursors for diacylglcyerol production, as revealed by measurement of the composition of the fatty acid side chains of the diacylglcyerol (89). In a recent study with 3T3-L1 cells and HL-60 cells, Besterman et al. (93) showed that phorbol esters and platelet-derived growth factor (an agonist coupled to inositol lipid metabolism) caused rapid increases of both diacylglycerol and phosphocholine, suggesting a protein kinase C-mediated activation of a phospholipase C enzyme that hydrolyzed phosphatidylcholine. This mechanism may explain the later phase of the agonist-induced biphasic diacylglycerol production observed in vascular smooth muscle (92). The precise mechanisms regulating metabolism of PI and diacylglycerol production clearly require further investigation regarding possible heterogeneities of membrane-bound phospholipase C enzymes with different specificities to PI or other phospholipids compared with PIP and PIP_2. An unexplored possibility is that a number of different receptor-linked GTP-binding proteins are involved in activating PI-specific and PIP/PIP_2-specific phospholipase C enzymes. The relative accumulations of IP_3 and diacylglycerol will also be a function of their respective rates of metabolism as well as of their formation. Hence regulation can be exerted at the level of removal of either of the second messengers. Nevertheless, it is evident that a continuous agonist-mediated stimulation of PIP and PIP_2 breakdown would rapidly lead to depletion of the hormone-sensitive polyphosphoinositide pool and, consequently, to a cessation of IP_3 production and intracellular Ca^{2+} signalling without replenishment from PI by PI kinase. Since the levels of PIP and PIP_2 gradually increase after an initial fall, an enhanced flux through PI and PIP kinases is indicated. Studies using [^{32}P] phosphate labeling with platelets have demonstrated an increase flux through PIP kinase in conjunction with a thrombin-stimulated breakdown of PIP_2 (94).

One possible mechanism to account for increased flux from PI to PIP_2 by the inositol lipid kinases after hormonal stimulation is by a simple mass action effect, since product inhibition has been suggested for the PIP kinase (95). The properties of purified PIP kinase from rat brain have been investigated recently by Cochet and Chambaz (96). The enzyme was shown to be Ca^{2+}-independent and its activity was unaffected by protein kinase C (96), even though in intact platelets (97-99) and thymocytes (100) it has been shown that addition of phorbol esters to activate protein kinase C causes an increase of PIP and PIP_2 contents. Interestingly, an activation by phosphatidylserine has been observed, suggesting that the enzyme is sensitive to its lipid environment (96). Other studies (101) have suggested that the activity of brain PIP kinase is regulated by phosphorylation of a 48 kDa protein located exclusively in presynaptic nerve terminals. Whether some aspect of receptor-mediated signal generation causes a subcellular redistribution of PI or PIP kinase activity from the cytosol to the plasma membrane where the kinases would be accessible to the hormone-stimulated PI and PIP pools, as with activation of protein kinase C, requires investigation.

The physiological significance of the different inositol lipid substrates for phospholipase C is that it provides a means of separately regulating the production of the two second messengers diacylglcyerol and IP_3 and hence allows Ca^{2+} mobilization to be temporally dissociated from activation of protein kinase C. Thus, while the initiation of the cellular response requires Ca^{2+} mobilization, a sustained cellular response requires in addition an activation of protein kinase C (5, 11, 102-104). The sensitivity of the IP_3-mediated Ca^{2+} mobilization system seems to be greater than that for the diacylglycerol-mediated protein kinase C system since activation of phospholipase C by weak agonists (e.g. α_1-adrenergic stimulation of hepatocytes) can produce enough IP_3 to elicit maximal intracellular Ca^{2+} mobilization but only a partial activation of protein kinase C relative to a strong agonist such as vasopressin (8).

FORMATION AND METABOLISM OF INOSITOL PHOSPHATES: RELATIONSHIP TO Ca^{2+} RELEASE

By use of high performance liquid chromatography (HPLC) techniques for the separation and quantitation of radiolabeled inositol phosphates, a number of novel inositol phosphate products and isomers have been identified in hormonally stimulated cells. This has led to the realization that the production and metabolism of inositol phosphates is much more complicated than originally supposed (105). Nevertheless, these recent developments have strengthened rather than detracted from the proposal that $Ins(1,4,5)P_3$ (or its 1,2-cyclic form, see later) is the inositol phosphate metabolite directly involved in intracellular Ca^{2+} mobilization.

Irvine *et al.* (106) discovered that the inositol trisphosphate pool produced by carbachol stimulation of rat parotid glands for 15 min was heterogeneous and contained a large proportion of $Ins(1,3,4)P_3$ in addition to $Ins(1,4,5)P_3$. Further studies with the parotid glands (107) and with angiotensin II-stimulated guinea pig hepatocytes and human HL-60 leukemia cells (108) showed that this second IP_3 isomer accumulated after a short delay, whereas the formation of $Ins(1,4,5)P_3$ was immediate. The predominance of the $Ins(1,4,5)P_3$ isomer at early times after hormonal stimulation accounts for the fact that in a number of studies, *e.g.* permeabilized pancreatic acinar cells, which retain the ability of muscarinic agonists to activate inositol lipid metabolism, a good correlation between total IP_3 production and the amount of Ca^{2+} released from intracellular stores was seen under a variety of conditions (109). Likewise, studies with isolated hepatocytes stimulated with vasopressin (82), and PC12 pheochromocytoma cells stimulated with the muscarinic receptor agonist carbachol (110) showed an excellent correlation between the half-maximal agonist concentrations required for total IP_3 production and the increase of cytosolic free Ca^{2+} when measurements were made at early times after stimulation or initial rates of change were calculated. On the other hand, vasopressin concentration-dependency curves showed a poor correlation between peak IP_3 production and a maximum increase of cytosolic free Ca^{2+} (82, 111). This is due partly to an accumulation of the $Ins(1,3,4)P_3$ isomer at later times, but also to a lack of correspondence between receptor-stimulated phospholipase C activity and the sensitivity of the IP_3-mediated Ca^{2+} release system. Thus, a comparison of the effects of vasopressin, angiotensin II and norepinephrine in hepatocytes (112) showed that the maximum capacity of all three agonists to generate total IP_3 correlated well with maximum receptor binding capacity but not at all with maximum increases of the cytosolic free Ca^{2+} and phosphorylase activation, which were the same for all three agonists despite large differences of IP_3 generation.

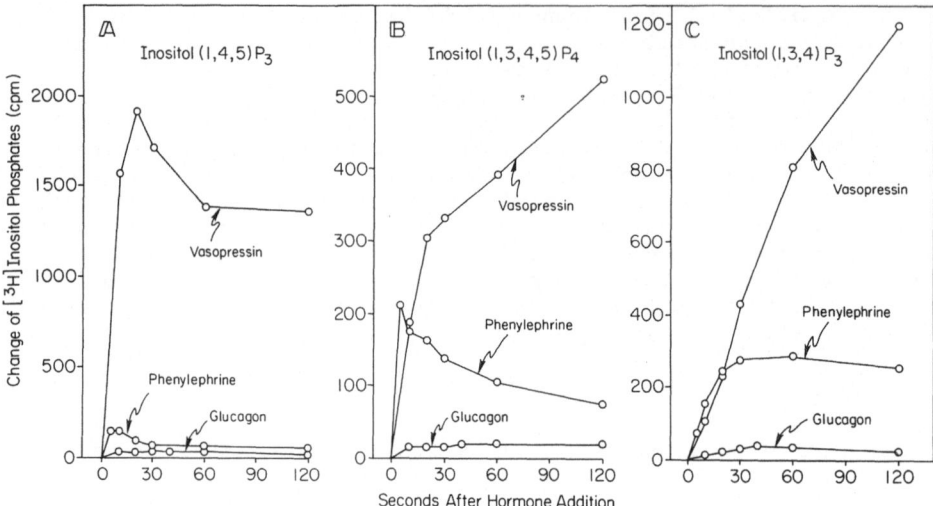

Fig. 3. Time courses for the relative increases of [³H] inositol phosphates above control levels after addition of 20 nM vasopressin,
10 μM phenylephrine or 10 nM glucagon to rat hepatocytes. The
experimental conditions and methods of analysis were the same as
those reported by Hansen *et al.* (115). Each point represents the mean value from 4 to 8 experiments. Typical values
for controls were: Ins(1,4,5)P₃ = 288 ± 26 cpm; Ins(1,3,4,5)P₄ =
54 ± 8 cpm; Ins(1,3,4)P₃ = 42 ± 4 cpm.

Time courses of the relative accumulations of Ins(1,4,5)P₃,
Ins(1,3,4,5,)P₄ and Ins(1,3,4)P₃ in [³H]-inositol prelabeled hepatocytes
maximally stimulated by vasopressin, phenylephrine (an α₁-adrenergic
agent) and glucagon are illustrated in Fig. 3. Strikingly different
increases of Ins(1,4,5)P₃ above background were observed for the three
agonists despite the fact that they increased the cytosolic free Ca²⁺ to
about the same maximal extent, as first reported by Charest *et al.* (113).
With vasopressin and phenylephrine at maximum effective concentrations,
the peak Ca²⁺ is reached after 5-10 s, while with glucagon a slower
increase peaking after 30-40 s is obtained. Clearly there is a very poor
correspondence between the amount of Ins(1,4,5)P₃ produced by the
different agonists and their ability to mobilize intracellular Ca²⁺.
Because of its large receptor reserve, vasopressin compared with phenylephrine generates an overabundance of Ins(1,4,5)P₃ which will rapidly
saturate the IP₃-binding sites of hepatic microsomes associated with the
Ca²⁺ release mechanism (114). These direct measurements of vasopressin-
induced Ins(1,4,5)P₃ formation in liver (Fig. 3A and see also Ref. 115)
support our earlier suggestion (82) that with this agonist sufficient
Ins(1,4,5)P₃ is produced in the first few seconds to elicit a maximal Ca²⁺
release. Similar observations have recently been made by Merritt *et al.*
(116) with pancreatic acinar cells, who showed that caerulein (a
cholecystokinin receptor agonist) caused much greater accumulations of
Ins(1,4,5)P₃ and Ins(1,3,4)P₃ than carbachol (a muscarinic receptor
agonist), even though these two agents were equipotent in increasing the
cytosolic free Ca²⁺.

Preliminary measurements of the mass of Ins(1,4,5)P₃ produced during
vasopressin stimulation (B.E. Corkey, personal communication) indicate a
value of about 2 μM in the steady state, which on the basis of a similar
specific activity corresponds to a peak Ins(1,4,5)P₃ increase of about 0.2
μM with phenylephrine. This concentration of Ins(1,4,5)P₃ is only sufficient to produce a submaximal release of Ca²⁺ in permeabilized hepatocytes

(117), suggesting that factors present in the intact cell may act cooperatively to sensitize the IP_3-mediated Ca^{2+} release system. Alternatively,
a localized higher than average concentration of $Ins(1,4,5)P_3$ may be
generated in the vicinity of a population of the endoplasmic reticulum
(e.g. adjacent to the plasma membrane) that contain the IP_3-sensitive Ca^{2+}
release sites (118).

The mechanism responsible for glucagon-induced Ca^{2+} release is
presently uncertain. The maximum increase of $Ins(1,4,5)P_3$ observed after
glucagon addition to hepatocytes was only about 20% of that produced by
phenylephrine (Fig. 3A), and corresponded to an average increase of $5 \pm 1\%$
above control levels. Cyclic AMP and its derivatives also elicit an
increase of the cytosolic free Ca^{2+} in hepatocytes (119) and mobilize
calcium from the same intracellular pool as that affected by vasopressin
(120, 121). Recent studies by Wakelam *et al*. (122), however, have
indicated that the Ca^{2+}-mobilizing effect of glucagon may be independent
of cAMP and induced by binding of glucagon to specific receptors that are
coupled directly to activation of phospholipase C and inositol lipid
metabolism. The glucagon-induced Ca^{2+} transient is kinetically and
quantitatively different from that observed with vasopressin (see Fig. 4),
but at present it is not clear whether the very small increase of Ins-
$(1,4,5)P_3$ produced by glucagon is sufficient to induce Ca^{2+} mobilization
or whether cAMP-dependent processes act cooperatively. The mechanism of
IP_3-induced Ca^{2+} release is considered to involve an activation of a
ligand-operated Ca^{2+} channel (123), but whether other factors regulate its
activity in the intact cell has not yet been clarified.

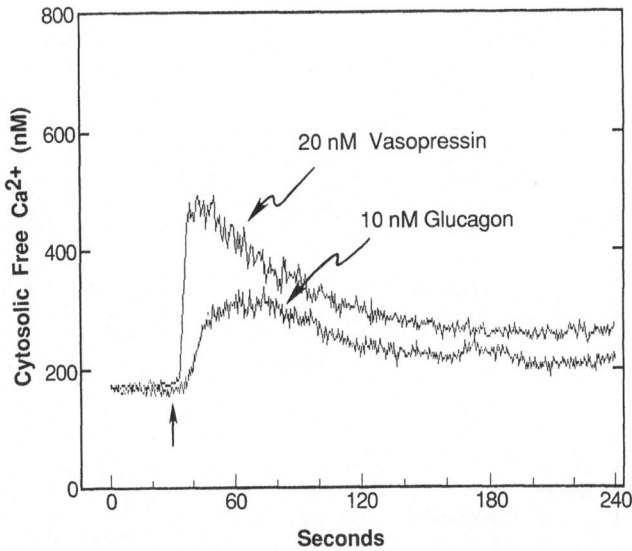

Fig. 4. Comparison of vasopressin and glucagon-induced Ca^{2+} transients in
        isolated hepatocytes. The cells were loaded with Quin 2 and the
        changes of cytosolic free Ca^{2+} were measured as previously
        described (82) except that the traces shown have been corrected
        for simultaneous changes of the NADH fluorescence. This correc-
        tion of the Quin 2-Ca^{2+} fluorescence change was proportionally
        greater after glucagon than after vasopressin addition and
        accounts for a lower maximal increase of cytosolic free Ca^{2+} with
        glucagon than with vasopressin (*c.f.* 113).

Irvine *et al.* (124) first reported the presence of an ATP-dependent enzyme in animal tissues that converted Ins(1,4,5)P_3 to Ins(1,3,4,5)P_4. This observation was confirmed and extended by Hansen *et al.* (115). The product produced by phosphorylation of Ins-(1,4,5)P_3 by a partially purified Ins(1,4,5)P_3 3-kinase from brain has now been identified and characterized by ^1H and ^{31}P nuclear magnetic resonance spectroscopic analysis (125), and the structure of Ins(1,3,4)P_3 has also been confirmed by NMR techniques (C. A. Hansen and J. R. Williamson, unpublished data). Identification of an Ins(1,4,5)P_3-kinase established the pathway for formation of Ins(1,3,4,5)P_4, an intermediate which had previously been detected following muscarinic receptor stimulation of brain slices (126) and 5-hydroxytryptamine stimulation of blowfly salivary glands (127).

As shown in Fig. 3B, Ins(1,3,4,5)P_4 accumulates in hepatocytes after additions of vasopressin, phenylephrine, or glucagon in relative amounts roughly in proportion to the effectiveness of these agonists to generate Ins(1,4,5)P_3. Although the early kinetics of formation of these inositol phosphates is not fully resolved, it is appparent that Ins(1,3,4,5)P_4 accumulates coincidentally with Ins(1,4,5)P_3. However, with vasopressin stimulation, Ins(1,3,4,5)P_4 continued to accumulate after Ins(1,4,5)P_3 declined from a peak at 20s to its steady state value after 1 min, whereas with phenylephrine, Ins(1,3,4,5)P_4 levels fell in parallel with those of Ins(1,4,5)P_3. These results suggest that the Ins(1,4,5)P_3 levels in vasopressin-stimulated hepatocytes were well above the Km for the Ins(1,4,5)P_3-kinase, while those after phenylephrine stimulation were within the Km region. A partially purified kinase preparation from brain gave a Km value of 0.2 μM (R. Johanson and J. R. Williamson, unpublished data), which is in line with the estimates of Ins(1,4,5)P_3 concentrations in liver reported above. In contrast, glucagon produced a small steady state increase of Ins(1,3,4,5)P_4 in accordance with the Ins(1,4,5)P_3 levels being well below the Km for the kinase. The accumulation of Ins(1,3,4)P_3 in hepatocytes after hormonal stimulation (Fig. 3C) was slower than that of Ins(1,3,4,5)P_4, consistent with it being the hydrolysis product formed by action of a 5-phosphatase (115, 124, 126).

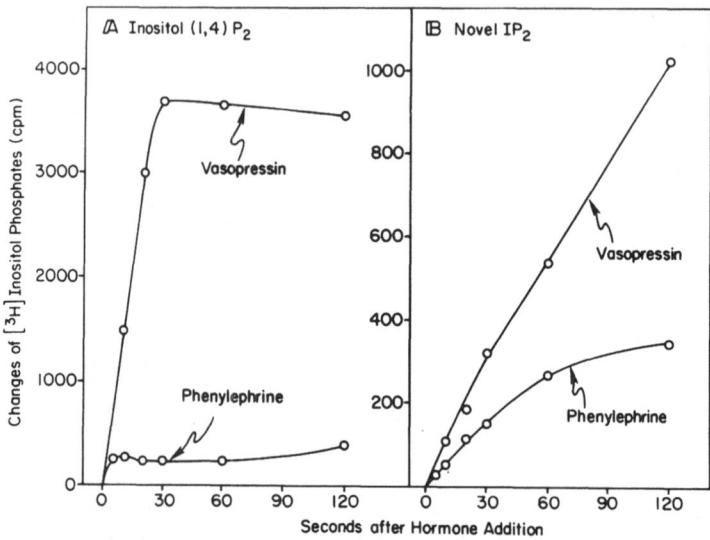

Fig. 5. Time courses for the relative increases of [^3H]inositol bisphosphates after addition of 20 nM vasopressin or 10 μM phenylephrine to rat hepatocytes. The experimental conditions were the same as in Fig. 3. Typical values for controls were: Ins(1,4)P_2 = 475 $\pm$ 61 cpm; novel IP$_2$ = 120 $\pm$ 21 cpm.

Fig. 5A shows that Ins(1,4)P$_2$ was also rapidly formed and accumulated to a greater extent than Ins(1,4,5)P$_3$, reaching steady state levels 5 s and 30 s, respectively, after addition of phenylephrine and vasopressin. Although Ins(1,4)P$_2$ is the product formed by 5-phosphatase hydrolysis of Ins(1,4,5)P$_3$, its large and rapid formation relative to Ins(1,4,5)P$_3$ suggests that it may be produced mainly by the direct action of phospholipase C on PIP in the plasma membrane. Fig. 5B shows that a second, as yet uncharacterized, IP$_2$ isomer also accumulates in hormone-stimulated hepatocytes. Its rate of accumulation is slower than that of Ins(1,3,4)P$_3$ (Fig. 3C) and it comigrates on HPLC columns with the IP$_2$ isomer formed by incubation of [^{3}H] inositol-labeled Ins(1,3,4,5)P$_4$ with a liver homogenate, suggesting that it is a product of Ins(1,3,4)P$_3$ hydrolysis (115). Glucagon also caused increases of Ins(1,4)P$_2$ and the novel IP$_2$, which in the steady state were 13 ± 1% and 57 ± 5%, respectively above control levels (data not shown).

A rapid formation of Ins(1,3,4,5)$_4$, which is kinetically, but not necessarily quantitatively similar to the formation of Ins(1,4,5)P$_3$ and which is accompanied by a slower rate of accumulation of Ins(1,3,4)P$_3$, has now been observed in pancreatic islets stimulated with glucose (128), muscarinic receptor-stimulated rat parotid slices (129) and insulin secreting RIN m5F cells (130), bradykinin-stimulated human A431 epidermoid carcinoma cells (131) and angiotensin II-stimulated rat adrenal glomerulosa cells (132). In other studies, the kinetics of formation of Ins-(1,3,4)P$_3$ in hormonally stimulated RlN m5F cells (133), rat pituitary tumor GH$_3$ cells (134), and rat pancreatic acinar cells (116) was shown to be slow relative to that of Ins(1,4,5)P$_3$, which slightly preceded the increase of cytosolic free Ca^{2+} (133). Meek (135) has described an interesting variant of the normal HPLC analysis for inositol phosphates by employing detection of inorganic phosphate after on-line enzymatic hydrolysis of the separated inositol phosphates. With this approach the *in vivo* concentration of Ins(1,4,5)P$_3$ in brain and salivary gland was estimated to be about 1 μM.

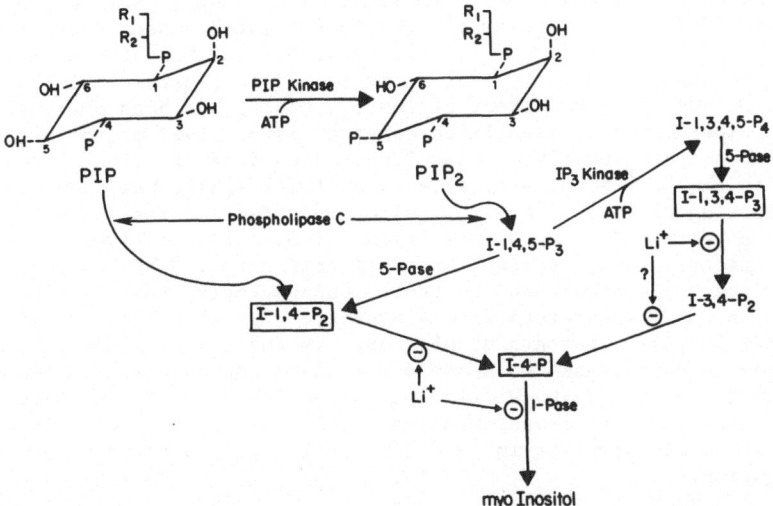

Fig. 6. Schematic diagram for the metabolism of inositol phosphates. The degradation pathway of I-1,3,4-P$_3$ to I-3,4-P$_2$ and I-4-P is speculative.

A schematic diagram of the enzymic steps involved in the formation and metabolism of inositol phosphates in mammalian cells based on the above studies is shown in Fig. 6. As discussed earlier, receptor-mediated activation of phospholipase C (via a specific GTP-binding coupling protein) induces hydrolysis of PIP and PIP_2 in the plasma membrane, with the simultaneous formation of $Ins(1,4)P_2$ and $Ins(1,4,5)P_3$. Metabolism of $Ins(1,4,5)P_3$ may proceed either to $Ins(1,4)P_2$ by a 5-phosphatase (136-138) or to $Ins(1,3,4,5)P_4$ by a specific ATP-dependent 3-kinase (115, 124). The same 5-phosphatase enzyme is probably involved in the conversion of Ins-$(1,3,4,5)P_4$ to $Ins(1,3,4)P_3$. Studies with a highly purified enzyme from brain cytosol have provided values of 3.6 μM and 0.4 μM for the Kms for $Ins(1,4,5)P_3$ and $Ins(1,3,4,5)P_4$ hydrolysis, respectively (139). Further metabolism of $Ins(1,4)P_2$ and $Ins(1,3,4)P_3$ has not been studied in detail and the intermediates shown are currently conjectural. The product of $Ins(1,3,4)P_3$ hydrolysis (novel IP_2 of Fig. 5B) is clearly a different IP_2 isomer from $Ins(1,4)P_2$. Whether it is $Ins(1,3)P_2$, the product of a 4-phosphatase activity as presumed by Hansen et al. (115), or Ins-$(3,4)P_2$, the product of a 1-phosphatase activity, is presently under investigation.

Several recent studies have indicated that $Ins(1,4)P_2$ is metabolized to Ins-4-P (132, 134, 137, 140, 141), which accumulates in hormonally stimulated cells prior to Ins-1-P (132, 134). The latter metabolite along with Ins 1, 2(cyclic)P (142), may be formed entirely from hydrolysis of PI as a post polyphosphoinositide hydrolysis event. With Li^+ pretreated cells, an accumulation of metabolites shown in boxes in Fig. 6, namely $Ins(1,3,4)P_3$, $Ins(1,4)P_2$ and Ins-4-P is observed (108, 115, 132, 140). These data suggest that an inositol polyphosphate 1-phosphatase is Li^+ sensitive in addition to the IP_1 phosphatase characterized by Hallcher and Sherman (143). Whether a 4-phosphatase enzyme activity with substrate specificity for $Ins(1,3,4)P_3$, $Ins(3,4)P_2$, or $Ins(1,4)P_2$ is also involved in the metabolism of these intermediates awaits further investigation. Possible ligand modulating effects of inositol phosphates on the various phosphatase activities have also not yet been sufficiently well studied.

The most intriguing aspect arising from elucidation of the inositol tris/tetrakisphosphate pathway for $Ins(1,4,5)P_3$ metabolism relates to its possible biological purpose. The most obvious one is that $Ins(1,3,4,5)P_4$ serves a unique signalling role for cellular hormonal effects. Preliminary studies from the author's laboratory suggested that it might have been involved in the mechanism of receptor-mediated enhanced permeability of the plasma membrane to Ca^{2+} (115). However, further studies with liver plasma membranes and brain synaptosomes failed to substantiate this hypothesis. In addition, no effect of $Ins(1,3,4,5)P_4$ has been obtained either on Ca^{2+} release from permeabilized hepatocytes or liver microsomes incubated with polyethyleneglycol plus GTP, and no sensitization of these systems to the Ca^{2+}-mobilizing effects of $Ins(1,4,5)P_3$ has been observed after addition of $Ins(1,3,4,5)P_4$. Also, no effect of $Ins(1,3,4,5)P_4$ was obtained on either Ca^{2+} uptake or release from isolated liver mitochondria and no effects on other systems or of effects produced by its injection into various cells have been reported. Consequently, its potential role as a new second messenger remains a mystery. Specific binding sites for [^{32}P]$Ins(1,3,4,5)P_4$ that are displaceable by $Ins(1,4,5)P_3$ have been identified in both plasma membrane particulate and soluble fractions in liver (R. Yasuda and J. R. Williamson, unpublished data). Whether or not these binding sites represent binding solely to inositol polyphosphate 5-phosphatase or to sites having a different function is presently under investigation.

An alternative possibility is that of the inositol phosphate metabolites produced during hormonal stimulation, only $Ins(1,4,5)P_3$ has a unique second messenger role, and that the $Ins(1,4,5)P_3$ 3-kinase pathway serves to modulate the intracellular $Ins(1,4,5)P_3$ concentration. The relative flux through the phosphorylation and dephosphorylation pathways for $Ins(1,4,5)P_3$ metabolism in intact cells during hormone stimulation has not been measured. A careful assessment of the relative amounts of 5-phosphatase and 3-kinase activities in the same tissue has also not been made, but since the ratio of the Km values for the two enzymes is 15-20 (for partially purified brain enzymes, 5-phosphatase $Km = 3.6\ \mu M$; 3-kinase $Km = 0.22\ \mu M$), it is apparent that at intracellular $Ins(1,4,5)P_3$ concentrations of 0.1 to 1 μM, flux through the kinase will be favoured over that through the phosphatase. In liver the total 5-phosphatase activity is considerably greater than the 3-kinase activity, hence flux through the two enzymes may be approximately the same, with the phosphorylation pathway being an effective means of disposal of $Ins(1,4,5)P_3$. A similar conclusion was reached by Biden and Wollheim (130) for the RIN m5F cell.

An unusual feature of the kinetics of $Ins(1,4,5)P_3$ accumulation after hormonal stimulation in many cells is its transience, with an initial peak after a few seconds followed by a fall to a steady state value slightly above resting levels (*e.g.* the response to phenylephrine with hepatocytes shown in Fig. 3A). If the rate of production of $Ins(1,4,5)P_3$ remains constant over the first few minutes, this phenomenon can only be accounted for by an activation of the 5-phosphatase or the 3-kinase. In this respect it is of interest that Biden and Wollheim (130) have reported that the $Ins(1,4,5)P_3$ kinase activity in a crude supernatant fraction from RIN m5F cells was activated by Ca^{2+} over the range from 0.2 to 10 μM with a change of $Vmax$ but not of Km, while the 5-phosphatase activity was unaffected by Ca^{2+}. Furthermore, in these cells after stimulation by carbamycholine and in human leukemic HL-60 cells stimulated with fMet-Leu-Phe (144), calcium depletion caused relative increases of $Ins(1,4,5)P_3$ accumulation and decreases of $Ins(1,3,4)P_3$ accumulation. A stimulation of $Ins(1,4,5)P_3$ kinase by Ca^{2+} is also suggested from the study by Rossier et al. (145) who investigated the metabolism of $Ins(1,4,5)P_3$ in permeabilized adrenal glomerulosa cells as a function of the medium Ca^{2+} concentration. Studies from the authors' laboratory with purified $Ins(1,4,5)P_3$ kinase have shown a 2-fold activation by Ca^{2+} over the physiological Ca^{2+} concentration range from 0.1 to 0.5 μM. Thus, the increase of cytosolic free Ca^{2+} induced by $Ins(1,4,5)P_3$ may provide a negative feedback signal to curtail the accumulation of $Ins(1,4,5)P_3$ by increasing its rate of disposal to $Ins(1,3,4,5)P_4$ and thence to $Ins(1,3,4)P_3$. With purified brain 5-phosphatase the Km for $Ins(1,3,4,5)P_4$ is about 10-fold lower than that for $Ins(1,4,5)P_3$ and if these relative values apply to enzyme from other tissues, they may explain the predominance of the $Ins(1,3,4)P_3$ isomer in the IP_3 pool at later times after hormonal stimulation, even in the absence of Li^+. It is also evident that the possible functional effects of $Ins(1,3,4)P_3$ have not been sufficiently well investigated. Preliminary studies from the author's laboratory show that it is effective in causing Ca^{2+} release from permeabilized hepatocytes and microsomes prepared from Syrian hamster insulinomas, but at much higher concentrations than those required for $Ins(1,4,5)P_3$-mediated Ca^{2+} release. However, at 10 μM, $Ins(1,3,4)P_3$ enhanced the effects of $Ins(1,4,5)P_3$ on Ca^{2+} release, suggesting that its accumulation in the cell may have physiological consequences.

The activity of a soluble inositol polyphosphate 5-phosphatase from human platelets (146) has been shown to be phosphorylated by protein kinase C with a substantial increase of activity (147). In accordance with the studies showing activation of the 5-phosphatase by protein kinase C *in vitro*, pretreatment of human platelets with diacylglycerol to activate protein kinase C *in situ* enhanced the rate of hydrolysis of $Ins(1,4,5)P_3$ when added to subsequently permeabilized cells (148). The purified brain enzyme is also phosphorylated by protein kinase C with an increase of V_{max}, and is inhibited by cAMP-dependent protein kinase with an increase of its K_m (139). The activity of the 5-phosphatase, therefore, seems to be modulated by covalent modification rather than by Ca^{2+} and may represent an additional mechanism for agonist-induced negative feedback signalling.

At present there is still some question as to whether the normal products of phospholipase C hydrolysis of inositol lipids are in the 1,2-cyclic form and whether these cyclic inositol phosphates represent the physiologically relevant species in intact cells. Most studies in which inositol phosphate accumulation has been measured have used acid extraction conditions, which opens the 1,2-cyclic bond and leaves the phosphate in the 1-position of the inositol ring. Early studies showed that the product of PI metabolism by phospholipase C consisted of a mixture of Ins-1-P and Ins-1,2(cyclic)-P, with the latter compound being converted to Ins-1-P by a tissue hydrolase before further hydrolysis to myoinositol (149). More recently, Wilson *et al.* (150, 151) found that hydrolysis of PI, PIP, and PIP_2 by purified phospholipase C was associated with the formation of both the 1,2-cyclic and the noncyclic phosphate esters of Ins-1-P, $Ins(1,4)P_2$ and $Ins(1,4,5)P_3$. Inositol 1,2-(cyclic) $4,5-P_3$ has also been shown to be the predominant IP_3 isomer formed in thrombin stimulated platelets after 10 s (152). The 5-phosphatase hydrolyses Ins 1,2-(cyclic) $4,5-P_3$ to Ins 1,2-(cyclic) $4P_2$, with a 10-fold higher K_m than for $Ins(1,4,5)P_3$, which is followed by further hydrolysis to Ins 1,2-(cyclic)-P and Ins-1-P by other cellular phosphatases (153). It has also been reported that Ins-1,2-(cyclic)$4,5-P_3$ is a substrate for the brain $Ins(1,4,5)P_3$ 3-kinase (124). In the few studies available, the biological effects of Ins-1,2-(cyclic) $4,5-P_3$ are similar to those of $Ins(1,4,5)P_3$ (151). Further studies with more carefully controlled extraction conditions are required to determine whether the 1,2-cyclic inositol phosphates represent the major products of inositol lipid hydrolysis in agonist-activated cells other than platelets. In stimulated GH_3 cells, for instance, no cyclic inositol polyphosphate accumulation was detected (134).

SUMMARY

Over the past few years considerable advances have been made towards understanding how agonist-mediated activation of inositol lipid metabolism results in the generation of second messengers responsible for the mediation of Ca^{2+}-dependent tissue functional responses. Indeed, the primary role of myoinositol 1,4,5-trisphosphate as a second messenger responsible for release of sequestered calcium from nonmitochondrial intracellular storate sites is now firmly established. However, many questions concerning the molecular mechanisms involved in signal generation remain unanswered. Much circumstantial evidence indicates that agonist-mediated receptor coupling to a phosphatidylinositol-specific phospholipase C is accomplished through a GTP-binding protein, but its identity and whether more than one GTP-binding protein is involved has not been established. The relationship and cellular location of hormone-sensitive and insensitive pools of the individual inositol lipids and their turnover requires further investigation, as do the characteristics of phospholipid phosphodiesterases with regard to their substrate specificities for the individual inositol lipids as well as for other phospholipids. Regulation of the activity and substrate specificities of these phosphodiesterases, whether by Ca^{2+} or other effectors, assumes an added significance in relation to the relative proportions of diacylglycerol and $Ins(1,4,5)P_3$ produced by a particular agonist. Thus, it is becoming apparent that diacylglycerol production with a consequent activation of protein kinase C may not be obligatorily linked to a simul-

taneous formation of Ins(1,4,5)P$_3$ and Ca^{2+} mobilization, while conversely
a Ca^{2+} signalling role through Ins(1,4,5)P$_3$ production can be achieved
with a minimal diacylglycerol production and activation of protein kinase
C.

The physiological role of protein kinase C activation in different
cells is currently a subject of much conjecture. Agonists that induce a
weak activation of phospholipase C and, therefore, a small increase of
diacylglycerol are likely to cause only a minimal activation of protein
kinase C, with phosphorylation of relatively few target proteins compared
with effects produced by a maximal activation of the enzyme by a strong
agonist or by phorbol esters. Consequently, effects produced by addition
of phorbol esters may not reflect the normal physiological response of the
cell to agonist stimulation (154). Activation of protein kinase C is
probably a requirement for sustained secretory responses and long term
hormonal effects on cell growth, but it also appears to have an important
negative feedback role in curtailing the initial agonist-induced response
resulting in acute desensitization (10, 11).

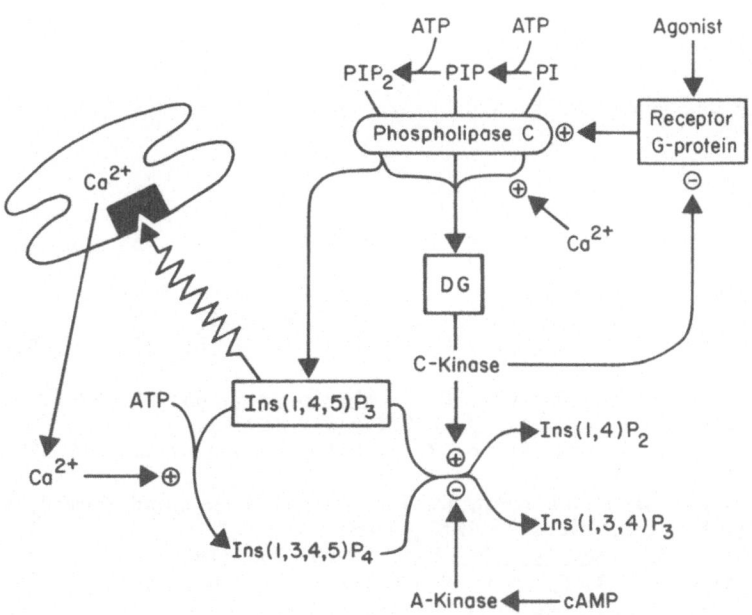

Fig. 7. Schematic representation of possible negative feedback effects of
        Ca^{2+} and protein kinase C activation on agonist-induced intra-
        cellular signalling mechanisms.

Potential mechanisms for negative feedback are illustrated in Fig. 7.
One mechanism whereby activated protein kinase C may negatively modulate
the agonist response is at the level of receptor G-protein coupling to
phospholipase C, either by phosphorylation of the receptor or possibly of
the G-protein itself (155). Effects at this level may result in homologous or heterologous desensitization, with a lack of tissue respon-
siveness to a variety of Ca^{2+} mobilizing agonists. On the other hand, a
protein kinase C-mediated activation of the inositol polyphosphate 5-
phosphatase, by increasing the removal of Ins(1,4,5)P$_3$, will tend to

diacylglycerol production and maintenance of protein kinase C activity. In contrast, inhibition of the inositol polyphosphate 5-phosphatase by cAMP-dependent protein kinase would have the opposite effect of maintaining $Ins(1,4,5)P_3$ levels. This interaction may be a component of the overall effects of glucagon as a Ca^{2+} mobilizing agonist in liver.

Fig. 7 also depicts how Ca^{2+} itself may interact with second messenger generation. As a positive effector of $Ins(1,4,5)P_3$ kinase, it will diminish the concentration of $Ins(1,4,5)P_3$ by promoting its conversion to $Ins(1,3,4,5)P_4$ and subsequently to $Ins(1,3,4)P_3$. This pathway may represent a major alternative route for removal of $Ins(1,4,5)P_3$ rather than by degradation to $Ins(1,4)P_2$ by the 5-phosphatase. At present it has not been ascertained whether $Ins(1,3,4,5)P_4$ or $Ins(1,3,4)P_3$ have any unique signalling roles in the cell. Also depicted in Fig. 7 is the possibility that the agonist-stimulated increase of cytosolic free Ca^{2+} may promote the breakdown of PI by a G-protein-activated phospholipase C, with an augmentation of diacylglcyerol production as a secondary event to the primary breakdown of the polyphosphoinositides. It is difficult to judge the significance of these interactions as general phenomena since they are based on limited experimental evidence. However, they suggest that many complexities in the regulation of the generation and metabolism of intracellular second messengers involved in Ca^{2+} mobilization and protein kinase C activation remain to be elucidated.

ACKNOWLEDGEMENTS

This work was supported by NIH Grants DK-15120, AA-05662 and HL-14461.

REFERENCES

1. Lefkowitz, R. J., Caron, M. G., and Stiles, G. L. New Engl. J. Med. 310: 1570-1579 (1984).
2. Schramm, M. and Selinger, Z. Science 225:1350-1356 (1984).
3. Cohen, P. Eur. J. Biochem. 151:439-448 (1985).
4. Williamson, J. R., Cooper, R. H. and Hoek, J. B. Biochim. Biophys. Acta 639:243-295 (1981).
5. Rasmussen, H. and Barrett, P. Q. Physiol Rev. 64:938-984 (1984).
6. Berridge, M. J. Biochem. J. 220:345-360 (1984).
7. Berridge, M. J. and Irvine, R. F. Nature (London) 312:315-321 (1984).
8. Williamson, J. R., Cooper, R. H., Joseph, S. K., and Thomas, A. P. Am. J. Physiol. 248:C203-C216 (1985).
9. Hokin, L. E. Ann. Rev. Biochem. 54:205-235 (1985).
10. Williamson, J. R. Hypertension 8:II-140-II-156 (1986).
11. Nishizuka, Y. Science 233:305-312 (1986).
12. Houslay, M. D. Trends Biochem. Sci. 9:39-40 (1984).
13. Gilman, A. G. Cell 36:577-579 (1984).
14. Hirasawa, K. and Nishizuka, Y. Ann. Rev. Pharmacol. Toxicol. 25: 147-170 (1985).
15. Ashendel, C. L. Biochem. Biophys. Acta 822:219-242 (1985).
16. Anderson, W. B., Estivol, A., Taprovaara, H., and Gopalakrishna, R. Adv. Cyclic Nucl. Prot. Phosph. Res. 19: 287-306 (1985).
17. Pouyssegur, J. Trends Biochem. Sci. 10:453-455 (1985).
18. Siess, W., Siegel, F. L. and Lapetina, E. G. J. Biol. Chem. 258: 11236-11242 (1983).
19. Pollock, K., Armstrong, R. A., Brydon, L. J., Jones, R. L. and MacIntyre, D. E. Biochem. J. 219:833-842 (1984).
20. Rittenhouse, S. E. Biochem. J. 222:103-110 (1984).
21. Watson, S. P., McConnell, R. T., and Lapetina, E. G. J. Biol. Chem. 259:13199-13203 (1984).

22. Smigel, M. D., Ross, E. M., and Gilman, A. G. In: "Cell Membranes: Methods and Reviews", Elson, E. L., Frazier, W. A. and Glaser, L., eds., Vol. 2, Plenum Press, New York, pp. 247-294 (1984).

23. Jakobs, K. H., Aktories, K., Minuth, M., and Schultz, G. Adv. Cyclic Nucl. Prot. Phosph. Res. 19:137-150 (1985).

24. Codina, J., Hildebrandt, J., Sunyer, T., Sekura, R. D., Manclark, C. R., Iyengar, R. and Birnbaumer, L. Adv. Cyclic Nucl. Prot. Phosph. Res. 17:111-125 (1984).

25. Smigel, M., Katada, T., Northup, J. K., Bokoch, G. M., Ui, M. and Gilman, A. G. Adv. Cyclic Nucl. Prot. Phosph. Res. 17:1-18 (1984).

26. Ui, M. Trends Pharmacol. Sci. 5:277-279 (1984).

27. Limbird, L. E. Biochem. J. 195:1-13 (1981).

28. Snavely, M. D. and Insel, P. A. Mol. Pharmacol. 22:532-546 (1982).

29. Goodhardt, M., Ferry, N., Geynet, P. and Hanoune, J. J. Biol. Chem. 257:11577-11583 (1982).

30. Lynch, C. J., Charest, R., Blackmore, P. F., and Exton, J. H. J. Biol. Chem. 260:1593-1600 (1985).

31. Florio, V. A. and Sternweiss, P. C. J. Biol. Chem. 260:3477-3483 (1985).

32. Koo, C., Lefkowitz, R. J. and Snyderman, R. J. Clin. Invest. 72: 748-753 (1983).

33. Cantau, B., Keppens, S., DeWulf, H. and Jard, S. J. Receptor Res. 1:137-168 (1980).

34. Crane, J. K., Campanile, C. P. and Garrison, J. C. J. Biol. Chem. 257:4959-4965 (1982).

35. Gomperts, B. D. Nature 306:64-66 (1983).

36. Haslam, R. J. and Davidson, M. M. L. FEBS Lett. 174:90-95 (1984).

37. Haslam, R. J. and Davidson, M. M. L. J. Receptor Res. 4: 605-629 (1984).

38. Cockcroft, S. and Gomperts, B. D. Nature (London) 314:534-536 (1985).

39. Litosch, I., Wallis, C. and Fain, J. N. J. Biol. Chem. 260:5464-5471 (1985).

40. Smith, C. D., Lane, B. C., Kusaka, I., Verghese, M. W. and Snyderman, R. J. Biol. Chem. 260:5875-5878 (1985).

41. Wallace, M. A. and Fain, J. N. J. Biol. Chem. 260:9527-9530 (1985).

42. Uhing, R. J., Jiang, H., Prpic, V. and Exton, J. H. FEBS Lett. 188: 317-320 (1985).

43. Lucas, D. O., Bajjalieh, S. M., Kowalchyk, J. A., and Martin, T. F. J. Biochem. Biophys. Res. Commun. 132:721-728 (1985).

44. Sasaguri, T., Hirata, M. and Kuriyama, H. Biochem. J. 231:497-503 (1985).

45. Gonzales, R. A. and Crews, F. T. Biochem. J. 232:799-804 (1985).

46. Smith, C. D., Cox, C. C. and Snyderman, R. Science 232:97-100 (1986).

47. Molski, T. F. P., Naccache, P. H., Marsh, M. L., Kermode, J., Becker, E. L. and Sha'afi, R. I. Biochem. Biophys. Res. Commun. 124:644-650 (1984).

48. Volpi, M., Naccache, P. H., Molski, T. F. P., Shefcyk, J., Huang, C.-K., Marsh, M. L., Munoz, J., Becker, E. L. and Sha'afi, R. I. Proc. Natl. Acad. Sci. USA 82:2708-2712 (1985).

49. Okajima, F. and Ui, M. J. Biol. Chem. 259:13863-13871 (1984).

50. Bokoch, G. M. and Gilman, A. G. Cell 39:301-308 (1984).

51. Bradford, G. P. and Rubin, R. P. FEBS Lett. 183:317-320 (1985).

52. Verghese, M. W., Smith, C. D., and Snyderman, R. Biochem. Biophys. Res. Commun, 127:450-457 (1985).

53. Lad, P. M., Olson, C. V., and Smiley, P. A. Proc. Natl. Acad. Sci. USA 82:869-873 (1985).

54. Nakamura, T. and Ui, M. J. Biol. Chem. 260:3584-3593 (1985).

55. Brandt, S. J., Dougherty, R. W., Lapetina, E. G. and Niedel, J. E. Proc. Natl. Acad. Sci. USA 82:3277-3280 (1985).

56. Krause, K. H., Schlegel, W., Wollheim, C. B., Andersson, T., Waldvogel, F. A., and Lew, P. D. J. Clin. Invest. 76:1348-1354 (1985).

57. Backlund, P. S. Jr., Meade, B. D., Manclark, C. R., Cantoni, G. L. and Aksamit, R. R. Proc. Natl. Acad. Sci. USA 82:2637-2641 (1985).

58. Okajima, F., Katada, T. and Ui, M. J. Biol. Chem. 260:6761-6765 (1985).

59. Kikuchi, A., Kozawa, O., Kaibuchi, K., Katada, T., Ui, M. and Takai, Y. J. Biol. Chem. 261:11558-11562 (1986).

60. Litosch, I. and Fain, J. N. Life Sci. 39:187-194 (1986).

61. Boyer, J. L., Garcia, A., Posadas, C. and Garcia-Sainz, J. A. J. Biol. Chem. 259:8076-8079 (1984).

62. Evans, T., Martin, M. W., Hughes, A. R. and Harden, T. K. Mol. Pharmacol. 27:32-37 (1985).

63. Masters, S. B., Martin, M. W., Harden, T. K. and Brown, J. H. Biochem. J. 227:933-937 (1985).

64. Murayama, T. and Ui, M. J. Biol. Chem. 260:7226-7233 (1985).

65. Schlegel, W., Wuarin, F., Zbaren, C., Wollheim, C. B. and Zahnd, G. R. FEBS Lett. 189: 27-32 (1985).

66. Martin, T. F. J., Lucas, D. O., Bajjaluh, S. M., and Kowalchyk, J.A. J. Biol. Chem. 261:2918-2927 (1986).

67. Pobiner, B. F., Hewlett, E. L. and Garrison, J. C. J. Biol. Chem. 260:16200-16209 (1985).

68. Uhing, R. J., Prpic, V., Jiang, H. and Exton, J. H. J. Biol. Chem. 261:2140-2146 (1986).

69. Merritt, J. E., Taylor, C. W., Rubin, R. P., and Putney, J. W. Jr. Biochem. J. 236:337-343 (1986).

70. Pfeilschifter, J. and Bauer, C. Biochem. J. 236:289-294 (1986).

71. Hilderbrandt, J. D., Codina, J., Rosenthal, W., Birnbaumer, L., Neer, E. J., Yamazaki, A and Bitensky, M. W. J. Biol. Chem. 260: 14867-14872 (1985).

72. Huff, R. M., Axton, J. M. and Neer, E. J. J. Biol. Chem. 260:10864-10871 (1985).

73. Gierschik, P., Falloon, J., Milligan, G., Pines, M., Gallin, J. I., and Spiegel, A. J. Biol. Chem. 261:8058-8062 (1986).

74. Owens, J. R., Frame, L. T., Ui, M. and Cooper, D. M. F. J. Biol. Chem. 260:15946-15952 (1985).

75. Verghese, M, Uhing, R. J. and Snyderman, R. Biochem. Biophys. Res. Commun. 138:887-894 (1986).

76. Irvine, R. F., Letcher, A. J., Lander, D. J. and Dawson, R. M. C. In: "Inositol and Phosphoinositides," Bleasdale, J. E., Eichberg, J. and Hauser, G., eds., pp. 123-135, Humana Press, New Jersey, (1985).

77. Seyfred, M. A. and Wells, W. W. J. Biol. Chem. 259:7659-7665 (1984).

78. Seyfred, M. A. and Wells, W. W. J. Biol. Chem. 259:7666-7672 (1984).

79. Koreh, K. and Monaco, M. E. J. Biol. Chem. 261:88-91 (1986).

80. Vickers, J. D. and Mustard, J. F. Biochem. J. 238:411-417 (1986).

81. Berridge, M. J. Biochem. J. 212:849-858 (1983).

82. Thomas, A. P., Alexander, J., and Williamson, J. R. J. Biol. Chem. 259:5574-5584 (1984).

83. Vicentini, L. M., Ambrosini, A., Di Virgilio, F., Pozzan, T. and Meldolesi, J. J. Cell. Biol. 100:1330-1333 (1985).

84. Taylor, C. W., Merritt, J. E., Putney, J. W. Jr., and Rubin, R. P. Biochem. J. 238:765-772 (1986).

85. Best, L. Biochem. J. 238:773-779 (1986).

86. Wilson, D. B., Bross, T. E., Hofmann, S. L., and Majerus, P. W. J. Biol. Chem. 259:11718-11724 (1984).

87. Majerus, P. W., Wilson, D. B., Connolly, T. M., Bross, T.E. and Neufeld, E. J. Trends Biochem. Sci. 10:168-171 (1985).

88. Thomas, A. P., Marks, J. S., Coll, K. E. and Williamson, J. R. J. Biol. Chem. 258:5716-5725 (1983).

89. Bocckino, S. B., Blackmore, P. F. and Exton, J. H. J. Biol. Chem. 260:14201-14207 (1985).

90. Preiss, J., Loomis, C. R., Bishop, W. R., Stein, R., Niedel, J. E. and Bell, R. M. J. Biol. Chem. 261:8597-8600 (1986).

91. Rittenhouse, S. E. and Sasson, J. P. J. Biol. Chem. 260:8657-8660 (1985).

92. Griendling, K. K., Rittenhouse, S. E., Brock, T.A., Ekstein, L. S., Gimbrone, M. A., Jr., and Alexander, R. W. J. Biol. Chem. 261:5900-5906 (1986).

93. Besterman, J. M., Duronio, V., and Cuatrecasas, P. Proc. Natl. Acad. Sci. USA 83:6785-6789 (1986).

94. Wilson, D. B., Neufeld, E. J. and Majerus, P. W. J. Biol. Chem. 260:1046-1051 (1985).

95. Van Rooijen, L. A. A., Rossowska, M. and Bazan, N. G. Biochem. Biophys. Res. Commun.126:150-155 (1985).

96. Cochet, C. and Chambaz, E. M. Biochem. J. 237:25-31 (1986).

97. Watson, S. P. and Lapetina, E. G. Proc. Natl. Acad. Sci. USA 82: 2623-2626 (1985).

98. deChaffoy de Courcelles, D., Roeveus, P. and van Belle, H. FEBS Lett. 173:389-393 (1984)

99. Halenda, S. P. and Feinstein, M. B. Biochem. Biophys. Res. Commun. 124:507-513 (1984).

100.Taylor, M. V., Metcalfe, J. C., Hesketh, T. R., Smith, G. A. and Moore, J. P. Nature 312:462-465 (1984).

101.Gispen, W. H., van Dongen, C.J., de Graan, P. N. E., Oestreicher, A. B. and Zwiers, H. In: Inositol and phosphoinositides: Metabolism and Regulation, J. E. Bleasdale, J. Eichberg, and G. Hauser, eds., Humana Press, Clifton, N.J., (1985).

102.Kojima, I., Kojima, K. and Rasmussen, H. J. Biol. Chem. 260:9177-9184 (1985).

103.Barrett, P. Q., Kojima, I., Kojima, K., Zawalich, K., Isales, C. M. and Rasmussen, H. Biochem. J. 238:893-903 (1986).

104.Barrett, P. Q., Kojima, I., Kojima, K., Zawalich, K., Isales, C. M. and Rasmussen, H. Biochem. J. 238:905-912 (1986).

105.Michell, R. H. Nature 319:176-177 (1986).

106.Irvine, R. F., Letcher, A. J., Lander, D. J.and Downes, C. P. Biochem. J. 223:237-243 (1984).

107.Irvine, R. F., Anggard, E. E., Letcher, A. J., and Downes, C. P. Biochem. J. 229:505-511 (1985).

108.Burgess, G. M., McKinney, J. S., Irvine, R. F., and Putney, J. W. Biochem. J. 232:237-248 (1985).

109.Streb, H., Heslop, J. P., Irvine, R. F., Schulz, I. and Berridge, M. J. J. Biol. Chem. 260:7309-7315 (1985)

110.Vicentini, L. M., Ambrosini, A., DiVirgilio, F., Meldolesi, J. and Pozzan, T. Biochem. J. 234:555-562 (1986).

111.Charest, R., Prpic, V., Exton, J. H. and Blackmore, P. F. Biochem. J. 227:79-90 (1985).

112.Lynch, C. J., Blackmore, P. F., Charest, R. and Exton, J. H. Mol. Pharmacol. 28:93-99 (1985).

113.Charest, R., Blackmore, P. F., Berthon, B. and Exton, J. H. J. Biol. Chem. 258:8769-8773 (1983).

114.Spat, A., Bradford, P. G., McKinney, J. S., Rubin, R. P. and Putney, J. W., Jr. Nature (London) in press (1986).

115.Hansen, C. A., Mah, S., and Williamson, J. R. J. Biol. Chem. 261: 8100-8103 (1986).

116.Merritt, J. E., Taylor, C. W., Rubin, R. P. and Putney, J. W. Jr. Biochem. J. 238:825-829 (1986).

117.Joseph, S. K., Thomas, A. P., Williams, R. J., Irvine, R. F., and Williamson, J. R. J. Biol. Chem. 259:3077-3081 (1984).

118.Putney, J. W. Jr. Cell Calcium 7:1-12 (1986).

119. Sistare, F. D., Picking, R. A. and Haynes, R. C. Jr. <u>J. Biol. Chem.</u> 260:12744-12747 (1985).
120. Combettes, L., Berthon, B., Binet, A. and Claret, M. <u>Biochem. J.</u> 237:675-683 (1986).
121. Blackmore, P.F. and Exton, J. H. <u>J. Biol.Chem.</u> 261:11056-11063 (1986).
122. Wakelam, M. J. O., Murphy, G. J., Hruby, V. J., and Houslay, M. D. <u>Nature</u> 323:68-71 (1986).
123. Joseph, S. K. and Williamson, J. R. <u>J. Biol. Chem.</u> 261:in press (1986).
124. Irvine, R. F., Letcher, A. J., Heslop, J. P. and Berridge, M. J. <u>Nature</u> 320:631-634 (1986).
125. Cerdan, S., Hansen, C. A., Johanson, R., Inubushi, T. and Williamson, J. R. <u>J. Biol. Chem.</u> 261:in press (1986).
126. Batty, I. R., Nahorski, S. R., and Irvine, R. F. <u>Biochem. J</u>. 232:211-215 (1985).
127. Heslop, J. P., Irvine, R. F., Tashjian, A. H., and Berridge, M. J. <u>J. Exp. Biol.</u> 119:395-401 (1985).
128. Turk, J., Wolf, B. A. and McDaniel, M. L. <u>Biochem. J.</u> 237:259-263 (1986).
129. Hawkins, P. T., Stephens, L. and Downes, C. P. <u>Biochem. J.</u> 238:507-516 (1986).
130. Biden, T. J. and Wollheim, C. B. <u>J. Biol. Chem.</u> 261:11931-11934 (1986).
131. Tilly, B. C., vanParidon, P. A., Wirtz, K. W. A., deLaat, S. W. and Moolenaar, W. H. <u>Biochem. J.</u> in press (1986).
132. Balla, T., Baukal, A. J., Guillemette, G., Morgan, R. O., and Catt, K. J. <u>Proc. Natl. Acad. Sci.</u> USA in press (1986).
133. Wollheim, C. B. and Biden, T. J. <u>J. Biol. Chem.</u> 261:8314-8319 (1986).
134. Dean, N. M. and Moyer, J. D. <u>Biochem. J.</u> in press (1986).
135. Meek, J. L. <u>Proc. Nat. Acad. Sci.</u> USA 83:4162-4166 (1986).
136. Seyfred, M. A., Farrell, L. E., and Wells, W. W. <u>J. Biol. Chem.</u> 259:13203-13208 (1984).
137. Storey, D. J., Shears, S. B., Kirk, C. J. and Michell, R. H. <u>Nature</u> (London) 312:374-376 (1984).
138. Joseph, S. K. and Williams, R. J. <u>FEBS Lett.</u> 180:150-154 (1985).
139. Hansen, C. A., Johanson, R., Williamson, M. T., Filburn, C., and Williamson, J. R. manuscript in preparation (1986).
140. Siess, W. <u>FEBS Lett.</u> 185:151-156 (1985).
141. Sherman, W. R., Munsell, L. Y., Gish, B. G., and Honchar, M. P. <u>J. Neurochem.</u> 44:798-807 (1985).
142. Dixon, J. F. and Hokin, L. E. <u>J. Biol. Chem.</u> 260:16068-16071 (1985).
143. Hallcher, L. M. and Sherman, W. R. <u>J. Biol. Chem.</u> 255:10896-10901 (1980).
144. Lew, P. D., Morad, A., Krause, K.-H., Waldvogel, F. A., Biden, T. J. and Schlegel, W. <u>J. Biol. Chem.</u> in press (1986).
145. Rossier, M. F., Dentand, I. A., Lew, P. D., Capponi, A.M. and Vallotton, M. B. <u>Biochem. Biophys. Res. Commun.</u> 139:259-265 (1986).
146. Connolly, T. M., Bross, T. E. and Majerus, P. W. <u>J. Biol. Chem.</u> 260:7868-7874 (1985).
147. Connolly, T. M., Lawing, W. J. Jr., and Majerus, P. W. <u>Cell</u>, 46: 951-958 (1986).
148. Molina y Vedia, L. M. and Lapetina, E. G. <u>J. Biol. Chem.</u> 261:10493-10495 (1986).
149. Dawson, R. M. C., Freinkel, N., Jungalwala, F. B. and Clarke, N. <u>Biochem. J.</u> 122:605-607 (1971).
150. Wilson, D. B., Bross, T. E., Sherman, W. R., Berger, R. A., and Majerus, P. W. <u>Proc. Natl. Acad. Sci.</u> USA 82:4013-4017 (1985).
151. Wilson, D. B., Connolly, T. M., Bross, T. E., Majerus, P. W.,

Sherman, W. R., Tyler, A. N., Rubin, L. J., and Brown, J. E.
J. Biol. Chem. 260:13496-13501 (1985).

152. Ishi, H., Connolly, T. M., Bross, T. E. and Majerus, P. W. Proc.
Natl. Acad. Sci. 83:6397-6401 (1986).

153. Connolly, T. M., Wilson, D. B., Bross, T. E. and Majerus, P. W.
J. Biol. Chem. 261:122-126 (1986).

154. Pandol, S. J. and Schoeffield, M. S. J. Biol. Chem. 261:4438-4444
(1986).

155. Sibley, D. R. and Lefkowitz, R. J. Nature 317:124-129 (1985).

BRADYKININ-INDUCED INOSITOL PHOSPHATE METABOLISM IN HUMAN A431 EPIDERMOID

CARCINOMA CELLS

B.C. Tilly[1], P.A. van Paridon[2], I. Verlaan[1], K.W.A. Wirtz[2],
S.W. de Laat[1] and W.H. Moolenaar[1]

[1] Hubrecht Laboratory
Uppsalalaan 8
3584 CT Utrecht
The Netherlands

[2] Department of Biochemistry
State University of Utrecht
Padualaan 8
3584 CH Utrecht
The Netherlands

SUMMARY

Addition of the nonapeptide bradykinin to human A431 epidermoid
carcinoma cells causes an immediate release of inositol phosphates and a
rapid rise in $[Ca^{2+}]_i$. Half-maximal stimulation occurs at a bradykinin
concentration of 4 nM. In the continuous presence of a saturating hormone
concentration (1 µM) the inositol phosphate accumulation levels off within
approximately 2 min; however, subsequent stimulation with histamine,
another activator of phospholipase C, gives an additional increase in
inositol phosphate production. Separation of the inositol phosphates by
HPLC-anion exchange chromatography, reveals a rapid but transient
accumulation of $Ins(1,4,5)P_3$, followed by an increase in $Ins(1,3,4,5)P_4$
and $Ins(1,3,4)P_3$. Our data support a precursor/product-relationship
between $Ins(1,4,5)P_3$ and $Ins(1,3,4)P_3$ with $Ins(1,3,4,5)P_4$ as intermediate
and argue against a messenger role of $Ins(1,3,4,5)P_4$ and $Ins(1,3,4)P_3$ in
Ca^{2+} signalling.

INTRODUCTION

A number of calcium-mobilizing hormones, neurotransmitters and growth
factors induce a rapid breakdown of phosphoinositides after binding to
their receptor (Michell et al., 1981; Berridge, 1984). This receptor-
mediated hydrolysis results in the formation of at least two second
messengers: diacylglycerol, a stimulator of protein phosphorylation via
protein kinase C (Nishizuka, 1984) and inositol(1,4,5)trisphosphate, which
act by mobilizing intracellular calcium (Berridge and Irvine, 1984).
Recently it has been shown that in addition to $Ins(1,4,5)P_3$, receptor
activation results in the formation of $Ins(1,3,4)P_3$ and $Ins(1,3,4,5)P_4$
(Batty et al., 1985; Heslop et al., 1985; Hansen et al., 1986). The
biological function of these inositol phosphates is still unclear.

Bradykinin, a nonapeptide (Arg-Pro-Pro-Gly-Phe-Ser-Pro-Phe-Arg) involved in many biological processes (Regoli and Barabé, 1980), is known
to activate a rapid hydrolysis of PtdInsP$_2$ and to increase intracellular
free calcium in several neuronal and epithelial cell systems (Pidikiti et
al., 1985; Yano et al., 1985; Francel and Dawson, 1986; Derian and
Moskowitz, 1986). We studied the effects of bradykinin on the kinetics of
inositol phosphate formation in human A431 epidermoid carcinoma cells, a
cell line widely used for EGF receptor studies. Experimental details are
described elsewhere (Tilly et al., in press).

INOSITOL FORMATION IN BRADYKININ-STIMULATED A431 CELLS

Stimulation of near confluent of A431 cells, pre-labelled with [^{3}H]-
myo-inositol, with bradykinin in the presence of 10 mM LiCl causes an
immediate accumulation of inositol phosphates (Figure 1). This stimulation
is dose-dependent, with a half-maximal stimulation at a concentration of
4 nM, which is close to the reported K$_d$ for [^{3}H]-bradykinin-receptor
binding (Innes et al., 1983). A saturating respons is observed at approximately 1 µM, raising the inositol phosphate level to 3-4 times the basal
level. The bradykinin-induced formation of inositol phosphates is a very
rapid process; within 15 sec after hormone addition a two to three-fold
increase in inositol phosphate levels is observed (Figure 1B). Despite the
continuous presence of a saturating concentration of bradykinin (1 µM),
the accumulation of inositol phosphates levels off within 2 min. Addition
of histamine (100 µM), an inducer of polyphosphoinositide-turnover in

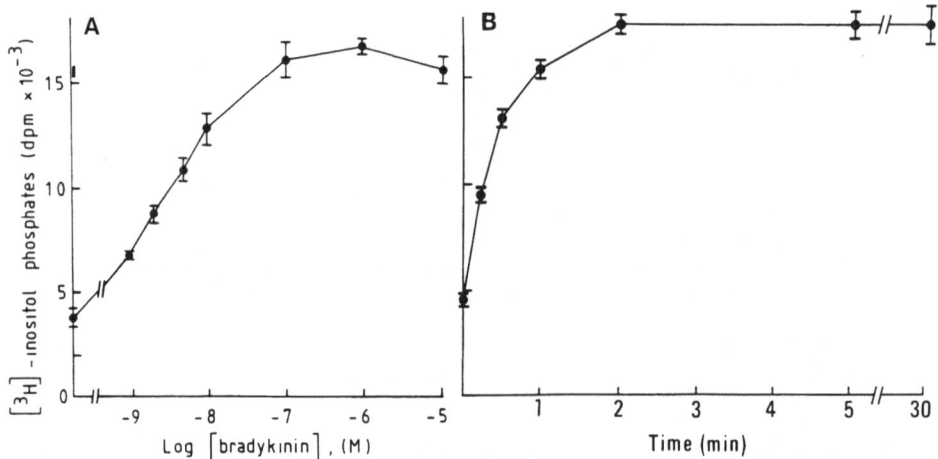

Fig. 1. Dose-dependency (A) and time-course (B) of bradykinin-induced
inositol phosphate formation. Near confluent cultures of A431
cells, pre-labelled with [^{3}H]-myo-inositol for 24 hr, were stimulated with bradykinin in the presence of 10 mM LiCl. The incubations
were terminated by replacing the medium for TCA and an inositol
phosphate fraction (InsP$_N$-fraction, containing InsP, InsP$_2$ and
InsP$_3$) was prepared as described elsewhere (Tilly et al., in press).
A: a dose-response curve was obtained by incubating the cells with
various concentrations of bradykinin for 10 min. B: Cultures were
incubated in a medium containing 10 mM LiCl for 30 min. Bradykinin
(1 µM) was added at the indicated times prior to termination. Data
are expressed as mean ± S.E. for 3 determinations.

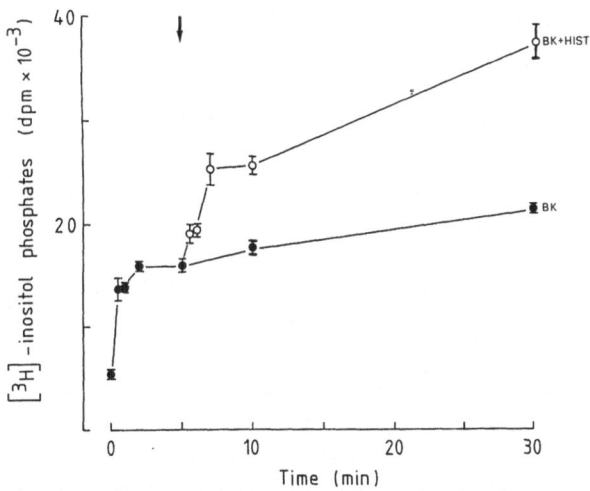

Fig. 2. <u>Additivity of histamine- and bradykinin-induced inositol phosphate accumulation.</u> Histamine (100 μM) or bradykinin (1 μM) was added to cultures 5 min after stimulation with bradykinin (1 μM). Arrow indicates time-point of addition of histamine or bradykinin.

neuronal tissue (Daum et al., 1984), to cells pre-stimulated with bradykinin gives an additional increase in inositol phosphate levels, whereas a second addition of bradykinin (final concentration 2 μM) does not (Figure 2). A possible explanation for this phenomenon is the existence of different agonist-sensitive pools of phosphoinositides. The bradykinin-sensitive PtdInsP$_2$-pool is rapidly depleted after addition of the hormone and prevents further increases in Ins(1,4,5)P$_3$ levels. On the other hand, the histamine sensitive pool is not affected and receptor activation leads to a further increase in inositol phosphate levels.

A second possibility is that the receptor-mediated phosphoinositide hydrolysis is under inhibitory control, perhaps through diacylglycerol-activated protein kinase C (cf. Brock et al., 1985; Orellana et al., 1985; Leeb-Lundberg et al., 1985). Activation of protein kinase C by treating the cells with the phorbol ester TPA, however, gives only a slight reduction in the bradykinin-induced inositol phosphate levels (Tilly et al., in preparation), suggesting that protein kinase C is not involved in the observed desensitization.

KINETICS OF THE BRADYKININ-INDUCED INOSITOL PHOSPHATE FORMATION

The time course of formation and degradation of [^{3}H]-inositol phosphates is studied by means of HPLC-anion exchange chromatography, using a 0-1.5 M ammonium formate/phosphoric acid gradient (Tilly et al., in press). The elution profile (Figure 3) shows that, in addition to the identified Ins(1)P, Ins(1,4)P$_2$ and Ins(1,4,5)P$_3$, several other putative inositol phosphates are present in unstimulated A431 cells. From their ionic mobilities and from previous analyses in other cell systems (Irvine et al., 1985; Heslop et al., 1985) we tentatively identified them as Ins(1,3,4)P$_3$, InsP$_4$, InsP$_5$ and InsP$_6$.

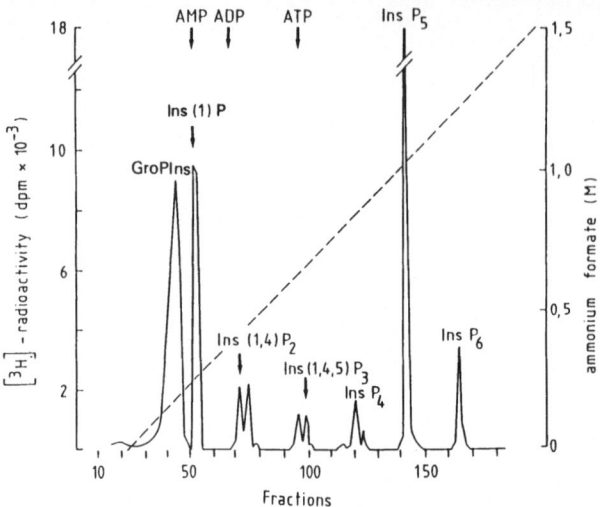

Fig. 3. <u>Elution profile of inositol phosphates by HPLC</u>. Inositol phosphates
were separated by HPLC-anion exchange chromatography using a
Nucleosil NH$_2$ column (Machery-Nagel) eluted with a 45 ml linear
gradient to 1.5 M ammonium formate/phosphoric acid (pH = 3.7). The
elution profile from the column represents the ^{3}H-radioactivity in
the fractions (fraction size: 250 μl). The ammonium formate gradient
is indicated by the dashed line. For experimental details see Tilly
et al. (in press). The peak co-eluting with ATP and running close
to Ins(1,4,5)P$_3$ is Ins(1,3,4)P$_3$.

The kinetics of the formation of the various inositol phosphates are
monitored between 0 and 10 min after addition of 1 μM bradykinin (Figure
4). In the presence of the hormone there is an acute, approximately ten-
fold increase in Ins(1,4,5)P$_3$ within 15 sec. This stimulation of PtdIns-
(4,5)P$_2$ hydrolysis is very transient, since Ins(1,4,5)P$_3$ returns to pre-
stimulatory levels after 1-2 min. The bradykinin-induced rise in Ins-
(1,4,5)P$_3$ is followed by a subsequent, but delayed increase in InsP$_4$ and
Ins(1,3,4)P$_3$, both reaching maximal levels respectively 30 and 60 sec
after addition of the hormone. Unlike Ins(1,4,5)P$_3$ the accumulation of
both InsP$_4$ and Ins(1,3,4)P$_3$ are much less transient, an approximately
6-fold increase is still measured 10 min after bradykinin addition. This
clear separation in time between the formation of Ins(1,4,5)P$_3$, InsP$_4$ and
Ins(1,3,4)P$_3$ supports a precursor/product relationship between Ins(1,4,5)P$_3$
and Ins(1,3,4)P$_3$, with InsP$_4$ as intermediate. These findings are entirely
consistent with the view proposed by Batty et al. (1985) supported by the
observation of a Ins(1,4,5)P$_3$-kinase in various mammalian tissues (Hansen
et al., 1986; Irvine et al., 1986).

Bradykinin also evokes a dramatic increase in the levels of InsP and
InsP$_2$. The accumulation of InsP$_2$ starts a few seconds after Ins(1,4,5)P$_3$
and reaches its maximal level within 30 sec (Tilly et al., in press). We
conclude that part of the Ins(1,4,5)P$_3$ formed is immediately converted into
Ins(1,4)P$_2$ by a 5-phosphatase, leaving the resulting Ins(1,4,5)P$_3$ to be
phosphorylated to Ins(1,3,4,5)P$_4$.

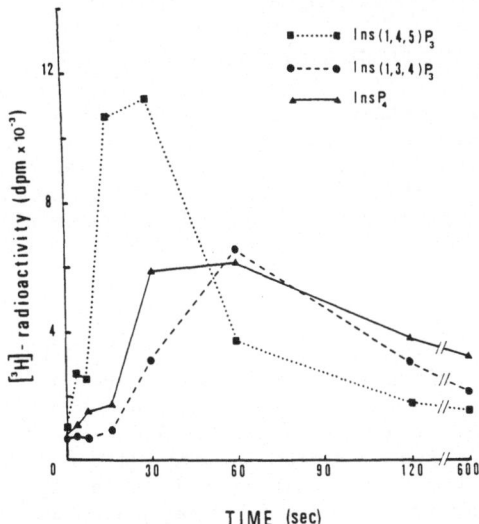

Fig. 4. Time-course of bradykinin-induced inositol phosphate metabolism.
[³H]-myo-inositol (15 µCi/ml) pre-labelled cultures were stimulated for the indicated times with bradykinin (1 µM). Incubations
were terminated by replacing the medium for TCA and the inositol
phosphates were analyzed by HPLC. Data are all derived from one
representative experiment (N = 3).

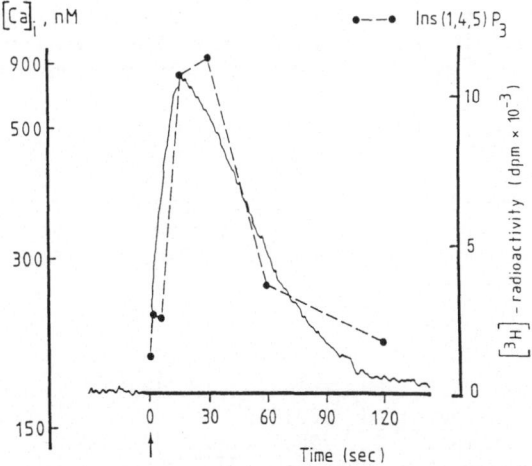

Fig. 5. Time-course of bradykinin-induced Ca^{2+}-signal. Quin-2 load cells
were stimulated with 1 µM bradykinin (arrow). The continuous line
represents the Ca^{2+}-signal, the dashed line the formation of
Ins(1,4,5)P$_3$ as measured by HPLC anion exchange chromatography.
For experimental details and calibration procedures see Moolenaar
et al. (1984, 1985).

BRADYKININ-INDUCED INTRACELLULAR CALCIUM MOBILIZATION

Stimulation of A431 cells with bradykinin evokes a rapid but transient rise in cytoplasmic free Ca^{2+}, as measured by quin-2 fluorescence (Figure 5). The increase in $[Ca^{2+}]_i$ starts without a detectable lag-period (< 1 sec), reaching a peak value within 10-20 sec, and returns to the pre-stimulatory level within 2 min. The bradykinin-induced calcium mobilization is only slightly attenuated (< 20%) in the absence of extracellular calcium, indicating that the Ca^{2+}-signal is due to the release of Ca^{2+} from intracellular stores.

The alterations in the levels of $Ins(1,4,5)P_3$ after stimulation with bradykinin coincide perfectly with the changes in $[Ca^{2+}]_i$ (Figure 5). This close correlation between $Ins(1,4,5)P_3$ accumulation and calcium mobilization confirms the hypothesis that binding of $Ins(1,4,5)P_3$ to its receptor on the endoplasmic reticulum causes calcium release (Berridge and Irvine, 1984; Spät et al., 1986). The rapid formation of both $Ins(1,3,4)P_3$ and $InsP_4$ make them suitable candidates for a function as second messenger. The kinetics of their formation, however, argues against a role in calcium signalling, since both $Ins(1,3,4)P_3$ and $InsP_4$ accumulate only after a lag period of 15 sec and reach peak levels when the Ca^{2+}-signal is already declining.

CONCLUSIONS

Bradykinin rapidly evokes the formation of inositol phosphates and a rapid but transient rise in cytoplasmic free calcium in human A431 cells. The $Ins(1,4,5)P_3$ formed is partly dephosphorylated to $Ins(1,4)P_2$ by a 5-phosphatase and partly converted into $Ins(1,3,4)P_3$ by phosphorylation/dephosphorylation via $Ins(1,3,4,5)P_4$. Analysis of the kinetics of inositol phosphate formation reveals that, unlike $Ins(1,4,5)P_3$, neither $Ins(1,3,4)P_3$ nor $InsP_4$ are involved in calcium signalling.

Human A431 epidermal carcinoma cells are widely used for studying the biological activities of epidermal growth factor (EGF). Since an EGF-induced phosphoinositide turnover has previously been described (Sawyer and Cohen, 1981), we will investigate the action of EGF on inositol phosphate metabolism and compare it with that of bradykinin. These studies are in progress and will be reported elsewhere (Tilly et al., in preparation).

ACKNOWLEDGEMENTS

We thank Leen Boom and Carmen Kroon for preparing the figures, and Eveline Hak for typing the manuscript. This work was supported in part by the Netherlands Cancer Foundation (Koningin Wilhelmina Fonds).

REFERENCES

Batty, I.R., Nahorshi, S.R., and Irvine, R.F., 1985, Biochem. J. 232:211.
Berridge, M.J., 1984, Biochem. J. 220:345.
Berridge, M.J., and Irvine, R.F., 1984, Nature 312:315.
Brock, T.A., Rittenhouse, S.E., Powers, C.W., Ekstein, L.S., Gimbrone, M.A., and Alexander, R.W., 1985, J. Biol. Chem. 260:14158.
Daum, P.R., Downes, C.P., and Young, J.M., 1984, J. Neurochem. 43:25.
Derian, C.K., and Moskowitz, M.A., 1986, J. Biol. Chem. 261:3831.
Francel, P.C., and Dawson, G., 1986, Biochem. Biophys. Res. Comm. 135:507.
Hansen, C.A., Mah, S., and Williamson, J.R., 1986, J. Biol. Chem. 261:8100.

Heslop, J.P., Irvine, R.F., Tashjian, A.T., and Berridge, M.J., 1985, J. exp. Biol. 119:395.

Innis, R.B., Manning, D.C., Stewart, J.M., and Snyder, S.H., 1981, Proc. Natl. Acad. Sci. USA 78:2630.

Irvine, R.F., Änggård, E.E., Letcher, A.J., and Downes, P.C., 1985, Biochem. J. 229:505.

Irvine, R.F., Letcher, A.J., Heslop, J.P., and Berridge, M.J., 1986, Nature, 320:631.

Leeb-Lundberg, L.M.F., Cotecchia, S., Lomasney, J.W., DeBernardis, J.F., Lefkowitz, R.J., and Caron, M.G., 1985, Proc. Natl. Acad. Sci. USA 82:5651.

Michell, R.H., Kirk, C.J., Jones, L.M., Downes, C.P., and Creba, J.A., 1981, Philos. Trans. R. Soc. London Ser. B. 296:123.

Nishizuka, Y., 1984, Nature 308:693.

Orellana, S.A., Solski, P.A., and Brown, J.H., 1985, J. Biol. Chem. 260:5236.

Pidikiti, N., Gamero, D., Gamero, J., and Hassid, A., 1985, Biochem. Biophys. Res. Comm. 130:807.

Regoli, D., and Barabé, J., 1980, Pharmacol. Rev. 32:1.

Sawyer, S.T., and Cohen, S., 1981, Biochemistry 20:6280.

Spät, A., Fabiota, A., and Rubin, P., 1986, Biochem. J. 233:929.

Yano, K., Higashida, H., Hattori, H., and Nozawa, Y., 1985, FEBS Lett. 181:403.

MECHANISMS OF SIGNAL TRANSDUCTION BY GROWTH FACTOR RECEPTORS

W.H. Moolenaar, L.H.K. Defize, B.C. Tilly, A.J. Bierman
and S.W. de Laat

Hubrecht Laboratory
Netherlands Institute for Developmental Biology
Uppsalalaan 8
3584 CT Utrecht
The Netherlands

ABSTRACT

Immediate consequences of growth factor-receptor interaction include tyrosine-specific protein phosphorylations, inositol lipid breakdown and changes in cytoplasmic pH (pH_i) and in the level of free Ca^{2+}. The rise in pH_i has a permissive effect on DNA synthesis and is mediated by Na^+/H^+ exchanger in the plasma membrane, which is turned on by protein kinase C. The Ca^{2+} signal is generated through the inositol lipid pathway and may contribute to the expression of certain proto-oncogenes. Monoclonal antibodies to the epidermal growth factor receptor can act as partial agonists in that they can induce tyrosine kinase activity without inducing inositol lipid breakdown and ionic changes. These antibodies fail to induce DNA synthesis, suggesting that tyrosine kinase activation is not sufficient for stimulation of cell proliferation.

KEY WORDS

Growth factors, ionic changes, inositol lipid breakdown, tyrosine kinase, monoclonal antibodies, cell proliferation.

INTRODUCTION

The molecular mechanism by which polypeptide growth factors regulate cell proliferation are not understood. Insight into these complex mechanisms may be obtained by studying the earliest detectable events occurring in growth factor-stimulated cells. Growth factors like epidermal growth factor (EGF) or platelet-derived growth factor (PDGF) initiate their action by binding to specific cell surface receptors. The activated receptor mediates a cascade of rapid biochemical changes in the target cell, which ultimately lead to stimulation of DNA synthesis and cell division.

One of the immediate consequences of growth factor binding is the activation of a protein kinase specific for tyrosyl residues, which is intrinsic to the receptor molecule. Activation of this kinase results in receptor autophosphorylation as well as in the phosphorylation of various substrate proteins[1]. To date, it has not been possible to relate increased tyrosine kinase activity to specific physiological changes in stimulated cells. Other early consequences of receptor activation include the breakdown of inositol phospholipids, a transient rise in free Ca^{2+} ($[Ca^{2+}]_i$) and the activation of Na^+/H^+ exchange in the plasma membrane resulting in a sustained increase in cytoplasmic pH (pH_i).

Most cells maintain their pH_i at 7.0–7.4. This is well above the electrochemical equilibrium value of 6.0–6.4 that is predicted by the Nernst equation from a transmembrane potential of approximately –60 mV. In vertebrate cells, the specific H^+-extruding mechanism which raises pH_i appears to be Na^+/H^+ exchange[2,3]. The functioning of the Na^+/H^+ exchanger in the plasma membrane and its role in pH_i homeostasis is most easily assessed by continuously monitoring the rapid recovery of pH_i to its resting level after a sudden acidification of the cytoplasm, as induced by a NH_4^+-prepulse or by weak acids. In most cells, this pH_i recovery process follows an exponential time course and is entirely due to net H^+ extrusion via the Na^+/H^+ exchanger, which utilizes the energy stored in the transmembrane Na^+ gradient. The major determinant of the rate of the Na^+/H^+ exchanger is pH_i. At normal pH_i values (near 7.0) the exchanger is relatively inactive, although the steep transmembrane Na^+ gradient could theoretically raise pH_i about one unit more alkaline.

As pH_i falls below a certain "threshold", the Na^+/H^+ exchanger is increasingly stimulated. Aronson and co-workers[4] were the first to point out that the Na^+/H^+ exchanger is apparently set in motion through allosteric activation by cytoplasmic H^+ at a regulatory site which is distinct from the internal H^+ transport binding site. The relatively strong pH_i sensitivity of the exchanger is, of course, a crucial property for an H^+ extruding system to maintain pH_i at a critical level. Furthermore, a change in the apparent affinity of the exchanger for internal H^+ could provide a powerful mechanism by which external stimuli regulate the physiological state of the exchanger.

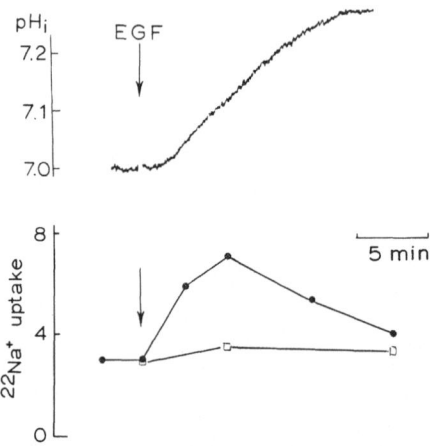

Fig. 1. Activation of Na^+/H^+ exchange and rise in pH_i induced by EGF in human fibroblasts. Shift in pH_i was measured by BCECF fluorescence. Initial rates of $^{22}Na^+$ uptake were measured over 3 min intervals in the presence (□) or absence (●) of 1 mM amiloride.

Figure 1 shows a typical example of an alkaline pH_i shift after addition of EGF to quiescent cells loaded with the pH-sensitive dye bis(carboxyethyl)carboxyfluorescein (BCECF). The shift in pH_i is initiated within 20-30 sec and is complete by 10-15 min. The elevated pH persists for as long as the growth factor is present. In general, the induced alkalinizations range from 0.1-0.3 pH unit; they are inhibited by amiloride and by Na^+ removal and are accompanied by a transient increase in amiloride-sensitive $^{22}Na^+$ uptake (for review see ref. 5). Furthermore, the rise in pH_i is converted into a fall in pH_i when the direction of the transmembrane Na^+ gradient is reversed[6]. When taken together, these data convincingly demonstrate that the mitogen-induced pH_i rise is mediated by the Na^+/H^+ exchanger.

How does receptor occupancy lead to activation of the Na^+/H^+ exchanger? Recent studies have shown that the activation is attributable to an alkaline shift in the pH_i sensitivity of the exchanger[5,7]. As mentioned above, this pH_i sensitivity is determined by an allosteric H^+ binding site on the cytoplasmic face of the exchanger. It thus seems plausible to assume that the altered pH_i sensitivity of the exchanger is due to some conformational change resulting in an increased pK_a of the regulatory binding site. Thus, the physiological effect of growth factors on the Na^+/H^+ exchanger is to increase its pH_i threshold, that is the level to which pH_i must rise before the exchanger virtually shuts off. Indeed, the Na^+/H^+ exchanger is only transiently stimulated by external stimuli and its activity returns to the control level once pH_i has attained its new stable value (Fig. 1).

The obvious next question is by which route growth factors modify the pH_i sensitivity of the exchanger to raise pH_i. An important key comes from studies using the tumour promoter 12-0-tetradecanoyl-13-acetate (TPA), which binds to and directly activates protein kinase C. Under normal conditions, kinase C is activated by endogenous diacylglycerol (DG) derived from the breakdown of inositol phospholipids. Both TPA and synthetic DG are capable of mimicking the effects of growth factors on Na^+/H^+ exchange and pH_i in various cell types[5]. The simplest explanation for these findings is that kinase C directly phosphorylates the exchanger, but it cannot, of course, be excluded that kinase C acts in a more indirect way to activate Na^+/H^+ exchange. Information concerning the molecular structure of the exchange carrier may become available in the near future, and this should greatly facilitate further study of the role of kinase C and other kinases in the activation process.

The apparent involvement of kinase C strongly suggests that an alkaline shift in pH_i is not uniquely induced by growth factors and phorbol esters but is a common cellular response in the action of those hormones and neurotransmitters that trigger the hydrolysis of inositol phospholipids and thereby generate diacylglycerol. It should be emphasized, however, that there is evidence that pathways other than kinase C may be involved in Na^+/H^+ exchange activation[8], but the biochemical nature of these pathways remains to be elucidated.

Of critical importance is the question of whether the observed rises in pH_i are essential for the initiation of DNA synthesis. A convincing demonstration of a signalling role for Na^+/H^+ exchange and pH_i in mitogenesis has been made with fertilized sea urchin eggs, in which pH_i must rise to permit DNA synthesis to begin[9]. A similar result has been obtained by Pouysségur et al.[10] who used mutant fibroblasts lacking a functional Na^+/H^+ exchanger to show that below a certain threshold value (around 7.2) pH_i becomes limiting for cell proliferation. Other studies seem to confirm that a mitogen-induced rise in pH_i is a permissive rather than a triggering event which is necessary for progression through S-phase[11,12].

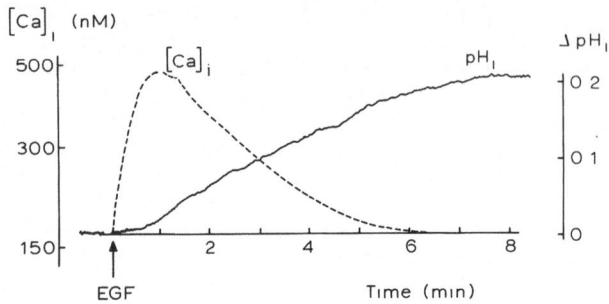

Fig. 2. Changes in $[Ca^{2+}]_i$ and pH_i in human A431 cells in response to EGF (100 ng/ml).

Ca^{2+} MOBILIZATION BY GROWTH FACTORS

In addition to activating Na^+/H^+ exchange, mitogens like EGF and PDGF induce a rapid but transient rise in $[Ca^{2+}]_i$ in their target cells, as measured by quin-2 fluorescence[13-15]. Subsequent to growth factor binding, the $[Ca^{2+}]_i$ rise is initiated without a detectable lag period (< 1 sec). It usually peaks within 30-60 sec and then returns to its resting level during the next 5-10 min. Figure 2 illustrates the time courses of both the $[Ca^{2+}]_i$ transient and the rise in pH_i as induced by EGF in human A431 cells.

In general, the rapid $[Ca^{2+}]_i$ signals in response to extracellular stimuli are mediated by the second messenger inositol 1,4,5-trisphosphate (IP_3), which triggers the release of Ca^{2+} from the endoplasmic reticulum (see next section). Mitogens like PDGF, serum and thrombin indeed appear to mobilize Ca^{2+} from internal stores, as demonstrated by studies with the fluorescent Ca^{2+} indicator quin-2 and by IP_3 measurements. In contrast, the Ca^{2+} signal evoked by EGF in quin-2 loaded A431 and 3T3 cells does not seem to originate from internal stores and has been attributed to the activation of a Ca^{2+} influx pathway ("channel" or carrier) in the plasma membrane[14,15]. This interpretation is based mainly on the finding that the transient increase in $[Ca^{2+}]_i$ in response to EGF does not occur in Ca^{2+}-free media and is abolished by Ca^{2+} entry blockers such as La^{3+} and Mn^{2+}. These results are intriguing because they suggest that there is a fundamental difference between the receptors for EGF and PDGF in terms of their $[Ca^{2+}]_i$-raising mechanisms; however, some caution is needed in interpreting the disappearance of the Ca^{2+}-quin-2 response to EGF when external Ca^{2+} is removed, since it is conceivable that intracellular quin-2 (a Ca^{2+} chelator) somehow interferes with the proper functioning of the EGF receptor, particularly in the absence of extracellular Ca^{2+}.

Regardless of the distinct mechanisms by which growth factors raise $[Ca^{2+}]_i$ (influx or intracellular release), it is generally accepted that Ca^{2+} may play an important role as a second messenger which regulates numerous cellular activities. Since the increase in $[Ca^{2+}]_i$ is short-lived, lasting only for 5-10 min after receptor stimulation, it obviously cannot directly mediate such late events as the initiation of protein synthesis and DNA synthesis, which begin only after many hours. Instead, the transient increase in $[Ca^{2+}]_i$ is more likely to trigger a sequence of early cellular changes occurring within minutes of growth factor binding. In this context it is noteworthy that artificial elevation of $[Ca^{2+}]_i$ by means of an ionophore mimics EGF and PDGF in rapidly inducing the expression of the c-*fos* and c-*myc* proto-oncogenes[16] (Kruijer et al., in preparation). Thus, Ca^{2+} may have a key role in mediating, either directly or indirectly, the early transcriptional effects of growth factors.

Table 1. ^{3}H-inositol Phosphate Levels in A431 Cells

| | d.p.m. ($\times 10^{-2}$) per Sample | |
	Control	EGF
IP$_3$ (1,4,5)	11.0	18.0
IP$_3$ (1,3,4)	3.5	11.0
IP$_4$	5.3	9.0
IP$_5$ + IP$_6$	117.0	83.0

Levels of inositol phosphates were measured in nearly confluent cultures pre-labelled to isotopic equilibrium with ^{3}H-inositol (2 µCi/ml). Treatment with EGF (100 ng/ml) lasted for 60 sec. Cell extracts were processed for analysis of ^{3}H-inositol phosphates by an HPLC anion-exchange system (Tilly et al., in prep.).

INOSITOL LIPID BREAKDOWN

Much attention has recently been focussed on the role of inositol 1,4,5-trisphosphate (IP$_3$(1,3,5)) as a specific releaser of intracellular Ca^{2+} following receptor stimulation[17]. It is now becoming increasingly apparent that several additional inositol polyphosphates, with as yet unknown functions, are produced in stimulated cells. In particular, an IP$_3$-(1,3,4)-isomer has been detected which seems to be formed by dephosphorylation of inositol-1,3,4,5-tetrakisphosphate (IP$_4$; ref. 18). Furthermore, the evidence of IP$_5$ and IP$_6$ has been described[19]. Although it seems likely that IP$_3$(1,3,4) and IP$_4$ may have second messenger functions, this idea remains to be tested.

We have separated ^{3}H-inositol phosphates from EGF-treated A431 cells using an HPLC anion-exchange system. Growing A431 cells were found to contain two IP$_3$ isomers at roughly equal concentrations, while three prominent peaks with decreasing ionophoretic mobilities were tentatively identified as IP$_4$, IP$_5$ and IP$_6$ respectively[20]. Addition of EGF causes a small but significant increase in the level of IP$_3$(1,4,5), while there is a ∿3-fold increase in IP$_3$(1,3,4) concentration within 1 min after a nearly two-fold increase in IP$_4$ (Table 1). These results, although preliminary, strengthen the view that IP$_3$(1,3,4) and/or IP$_4$ may have an as yet unknown second messenger role. Time-course studies and micro-injection experiments should help to gain further insight into the metabolism and physiological functions of the individual inositol phosphates in growth factor action.

DISSOCIATION OF SIGNAL PATHWAYS BY ANTI-RECEPTOR MONOCLONAL ANTIBODIES

A potentially powerful tool for the dissociation of molecular events in the signalling cascade is provided by the availability of monoclonal antibodies to the EGF receptor. We have used three different anti-EGF receptor monoclonal IgG's, directed against distinct epitopes of the extracellular domain of the human EGF receptor[21], to test their ability to act as partial or full agonists of the EGF receptor. All three antibodies (named 2E9, 2D11 and 2G5, respectively) are able to immunoprecipitate a functional EGF receptor showing EGF dependent tyrosine kinase activity. Monoclonal 2E9 is unique in that it recognizes a peptide determinant at or close to the EGF binding domain of the receptor. As a consequence, 2E9 competitively inhibits EGF receptor binding. In contrast, the other monoclonals (2D11 and 2G5) are

Table 2. Comparison of the Biological Effects of EGF
and Anti-EGF Receptor Monoclonal Antibodies
on Human A431 Cells and Fibroblasts

	EGF	2E9	2D11
Precipitation of EGF Receptor		+	+
EGF Binding Competition	+	+	-
Stimulation of Tyrosine Kinase	+	+	+
Morphological Changes	+	-	+
Inositol Phosphate Formation	+	-	-
Rise in $[Ca^{2+}]_i$	+	-	-
Rise in pH_i	+	-	-
Stimulation of DNA Synthesis	+	-	-

Stimulation of DNA synthesis was tested on quiescent
fibroblasts; all other effects on A431 cells. For
further detail see ref. 21.

directed to bloodgroup A-specific carbohydrate structures on the EGF
receptor and fail to affect EGF binding. We have tested these antibodies
for their EGF-like properties in stimulating the receptor-mediated tyrosine
phosphorylations, cytoplasmic alkalinization, Ca^{2+} mobilization and DNA
synthesis. As summarized in Table 2, all three monoclonal IgG's can stimulate the tyrosine-specific autophosphorylation of the 170 kD EGF receptor
both in isolated A431 membranes and in intact cells[21]. Interestingly, none
of these antibodies is capable of triggering the formation of inositol
phosphates or the induction of ionic changes (Table 2), even after addition
of a second cross-linking anti-IgG (not shown). Finally, the anti-receptor
antibodies fail to stimulate DNA synthesis in quiescent human fibroblasts.
Stimulation of the receptor's intrinsic tyrosine kinase is apparently not
sufficient, by itself, to elicit a mitogenic response (for further details
see ref. 21). Another important conclusion from those results is that
stimulation of the EGF receptor kinase does not necessarily activate the
post-receptor pathways that leads to inositol lipid breakdown and to an
increase in $[Ca^{2+}]_i$ and in pH_i.

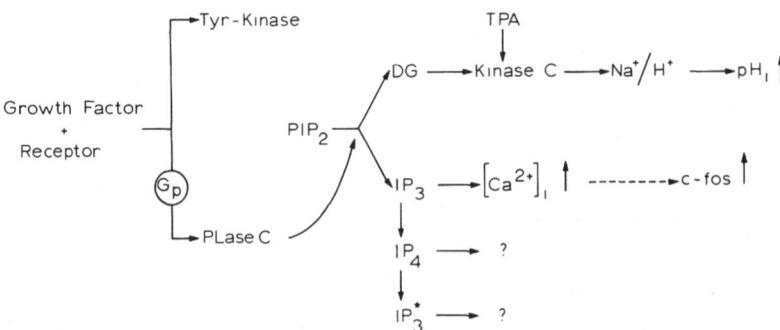

Fig. 3. Proposed sequence of events after growth factor binding. G_p is the
putative GTP-binding protein; PLase C = phospholipase C; PIP_2 =
phosphatidylinositol 4,5-bisphosphate; DG = diacylglycerol; IP_3 =
inositol 1,4,5-trisphosphate; IP_3^* = inositol 1,3,4-trisphosphate.

A schematic representation of the receptor-linked signal pathways is shown in Figure 3. In this simplified scheme, activation of Na^+/H^+ exchange and Ca^{2+} mobilization occur in parallel but independently of each other through the action of phospholipase C, whereas the tyrosine-specific protein kinase initiates a distinct pathway.

Although uncertainty still exists about the biological significance of each of these "early" steps in the eventual initiation of DNA synthesis, occurring many hours later, it seems likely that the various events evoked by inositol lipid breakdown act in concert with tyrosine-specific protein phosphorylations to stimulate cell proliferation.

ACKNOWLEDGEMENTS

We thank Daniëlle Steggink for preparing the manuscript. This work was supported by the Netherlands Cancer Foundation (Koningin Wilhelmina Fonds) and the Organisation for the Advancement of Pure Research (ZWO).

REFERENCES

1. T. Hunter and J.A. Cooper, Protein-tyrosine kinase, Ann. Rev. Biochem. 54:897 (1985).
2. A. Roos and W. Boron, Intracellular pH, Physiol. Rev. 61:296 (1981).
3. W.H. Moolenaar, Regulation of cytoplasmic pH by Na^+/H^+ exchange, Trends in Bioch. Sci. 11:141 (1986).
4. P.S. Aronson, J. Nee and M.A. Suhm, Modifier role of internal H^+ in activating the Na-H exchanger in renal microvillus membrane vesicles, Nature 299:161 (1982).
5. W.H. Moolenaar, Effects of growth factors on intracellular pH regulation, Ann. Rev. Physiol. 48:363 (1986).
6. W.H. Moolenaar, R.Y. Tsien, P.T. van der Saag and S.W. de Laat, Na^+/H^+ exchange and cytoplasmic pH in the action of growth factors in human fibroblasts, Nature 304:645 (1983).
7. J. Pouysségur, The growth factor-activatable Na^+/H^+ antiporter, Trends in Bioch. Sci. 10:453 (1985).
8. F. Vara and E. Rozengurt, Stimulation of Na^+/H^+ antiport activity by EGF and insulin occurs without activation of protein kinase C. Biochem. Biophys. Res. Comm. 130:646 (1985).
9. M.J. Whitaker and R.A. Steinhardt, Ionic regulation of egg activation. Quart. Rev. Biophys. 15:593 (1982).
10. J. Pouysségur, C. Sardet, A. Franchi, G. l'Allemain and S. Paris, A specific mutation abolishing Na^+/H^+ antiport activity in hamster fibroblasts precludes growth at neutral and acidic pH. Proc. Natl. Acad. Sci. USA 81:4833 (1984).
11. R. Bravo and H. MacDonald-Bravo, Effect of pH on progression to S-phase in mouse fibroblasts. FEBS Lett. 195:309 (1986).
12. W.H. Moolenaar, L.H.K. Defize and S.W. de Laat, Ionic signalling by growth factor receptors. J. exp. Biol. 124:359 (1986).
13. W.H. Moolenaar, L.G.J. Tertoolen and S.W. de Laat, Growth factors immediately raise cytoplasmic free Ca^{2+} in human fibroblasts. J. Biol. Chem. 259:8066 (1984).
14. W.H. Moolenaar, R.J. Aerts, L.G.J. Tertoolen and S.W. de Laat, The epidermal growth factor-induced calcium signal in A431 cells. J. Biol. Chem. 261:279 (1986).
15. T.R. Hesketh, J.P. Moore, J.D.H. Morris, M.V. Taylor, J. Rogers, G.A. Smith and J.C. Metcalfe, A common sequence of calcium and pH signals in the mitogenic stimulation of eukaryotic cells. Nature 313:481 (1985).

16. R. Bravo, J. Burckhardt, T. Curran and R. Müller, Stimulation and inhibition of growth by EGF in different A431 cell clones is accompanied by the rapid induction of c-*fos* and c-*myc* proto-oncogenes. EMBO J. 4:1193 (1985).

17. M.J. Berridge and R.F. Irvine, Inositol trisphosphate, a novel second messenger in cellular signal transduction. Nature 312:315 (1984).

18. I.R. Batty, S.R. Nahorski and R.F. Irvine, Rapid formation of inositol-tetrakisphosphate following muscarinic receptor stimulation of rat cerebral cortical slices. Biochem. J. 232:211 (1985).

19. J.P. Heslop, R.F. Irvine, A.T. Tashjian and M.J. Berridge, Inositol tetrakis- and pentakisphosphates in GH$_4$ cells. J. exp. Biol. 119: 396 (1985).

20. B.C. Tilly, P. van Paridon, I. Verlaan, K.W.A. Wirtz, S.W. de Laat and W.H. Moolenaar, Inositol phosphate metabolism in bradykinin-stimulated human A431 carcinoma cells. Relationship to calcium signalling. Biochem. J., in press.

21. L.H.K. Defize, W.H. Moolenaar, P.T. van der Saag and S.W. de Laat, Dissociation of cellular responses to epidermal growth factor using monoclonal antibodies. EMBO J. 5:1187 (1986).

STUDY OF THE MECHANISM OF HORMONE INDUCED DESENSITIZATION

AND INTERNALIZATION OF BETA-ADRENERGIC RECEPTORS

Rainer Joachim Box and Matthys Staehelin

Departement Forschung
Kantonsspital
Hebelstrasse 20
CH-4031 Basel
Switzerland

The hormone epinephrine exerts its effects by binding to specific plasmamembrane receptors. The receptors interact with the membrane-bound enzyme complex of adenylate cyclase which catalyzes the formation of adenosine-3', 5'-cyclic monophosphate (cyclic AMP). Cultured S49 mouse lymphoma cells become refractory or desensitized to hormone after stimulation for relatively short periods (20-30 minutes), i.e. no further cyclic AMP is synthetized when the cells are subsequently exposed to a second hormone dose. Desensitization during the first 30 minutes after hormone addition to intact cells is characterized by a rapid decline in the rate of cyclic AMP production (Figure 1) and uncoupling of cell surface receptors followed by internalization of the receptors (1,2,3). In S49 cells, at 37°C, 50-60 % of the receptors originally present on the plasma membrane are internalized 2-4 minutes after administration of agonist (e.g. 1 micromolar isoproterenol). About 40 % of the receptors, however, remain on the cell surface, even after stimulation with isoproterenol for several hours (2).

QUANTITATIVE RELATIONSHIP BETWEEN RECEPTOR NUMBERS AND CYCLIC AMP YIELD

In order to see if internalization of receptors could account for the decrease in cyclic AMP production observed during desensitization, we investigated the quantitative relationship between receptor numbers and cyclic AMP yield after stimulation of intact cells with hormone. For this purpose, surface receptors were made inaccessible to hormone by chemical modification, thus simulating 'desensitized' cells at various stages of receptor internalization.

Receptors were blocked irreversibly in a dose-dependent manner by covalent reaction with the antagonist bromoacetylalprenololmenthane (BAAM) described by Pitha (4,5). Treated cells were washed excessively to remove unreacted antagonist. The adopted washing procedure sufficed to remove also similar concentrations of the reversibly binding antagonist alprenolol. Available receptor numbers were then assessed by binding studies with the hydrophilic radioligand ^{3}H-CGP-12177 which specifically recognizes receptors at the cell surface alone (6).

The blockade of cell surface receptors was irreversible since Scatchard plots from saturation binding experiments using tritiated CGP-12177 were

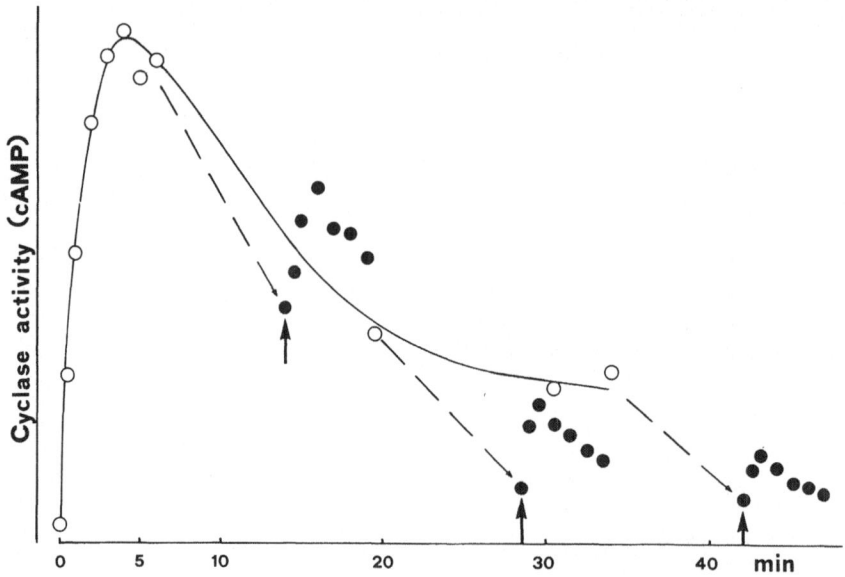

Figure 1. Desensitization of adenylate cyclase activity. S49 cells
were incubated with 1 micromolar (-)-isoproterenol (IPR)
at 37°C in Dulbecco-modified Eagle medium (DMEM) containing
no phosphodiesterase inhibitor. Aliquots were taken and
assayed for cyclase activity (o). At the times indicated
(dashed lines) further aliquots were washed with hormone-
free medium twice, and, after the final centrifugation, were
resuspended in an identical volume of DMEM. The cells were
then rechallenged with 1 micromolar IPR at the times denoted by the vertical arrows (●).

linear and parallel for treated and untreated cells, revealing differences
in the maximal number of binding sites. Dose-response curves of isoproterenol dependent cyclic AMP accumulation were similar in shape, and halfmaximal
cyclic AMP values were reached at similar hormone concentrations, only that
less cyclic AMP accumulated in treated cells compared to non-treated cells.
In contrast, when using the displaceable antagonist alprenolol, Scatchard
plots were linear but not parallel to the plot of control data. For both,
extrapolation of the maximal number of binding sites gave similar values.
The dose-response curve of alprenolol treated cells was identical to the one
obtained for the untreated control.

    Cells treated with different BAAM concentrations were challenged with
isoproterenol, and the amount of accumulated cyclic AMP was measured as described elsewhere (7). The cyclic AMP response decreased almost proportionally
with the decreasing number of available receptors (8). It was concluded that
nearly all receptors have to be accessible to the hormone to give a full response and that no spare receptors exist (cf. 9).
    Since in fact the decrease in hormone-accessible receptors is parallel-
led by a decrease in cyclic AMP synthesis, the removal of receptors from the

cell surface during desensitization may possibly explain the decline in the rate of cyclic AMP production observed (e.g. Figure 1).

THE EFFECT OF FORSKOLIN ON DESENSITIZATION

It has been reported that the diterpene forskolin (7-acetoxy-8,13-epoxy-1a,6,9a-trihydroxy-labd-14-ene-11-one), derived from the roots of Coleus forskohlii, reverses desensitization (10). After addition of forskolin desensitized cells showed equal levels of cellular cyclic AMP as control cells treated with hormone and forskolin simultaneously. Forskolin is currently thought to act directly on the catalytic subunit of adenylate cyclase (11). At present, however, the mode and precise site of forskolin action are still unclear. There is some evidence that forskolin interacts also with the nucleotide regulatory protein (10,12, references in 13). In S49 cells low doses of forskolin potentiate the effect of the beta-adrenergic agonist isoproterenol in terms of cellular cyclic AMP accumulation while also reducing slightly, by a factor of 2-3, the isoproterenol concentration required to give a halfmaximal response.

We have investigated how forskolin achieves the reported reversal of desensitization. Using BAAM treated cells and desensitized cells, we found that forskolin does not affect any specific step of desensitization, but that the apparent reversal of desensitization is the consequence of several effects: a) a slight reduction in the requirement of receptors to obtain a full cyclic AMP response, and b) the prolongation of cyclic AMP accumulation in the presence of forskolin, and c) the fact that desensitized cells contain initially higher amounts of cyclic AMP than control cells (8).

A model for the activation of adenylate cyclase by forskolin has been proposed (14) which explains the forskolin effect with the dissociation of the beta-subunit from the guanine nucleotide binding protein N_s. Forskolin is supposed to impede the reassociation of the beta-subunit with the alpha-subunit of N_s by stabilizing the activated enzyme complex consisting of the alpha-subunit and the catalytic subunit of adenylate cyclase. The stabilizing effect of forskolin on the activated enzyme complex has been demonstrated experimentally (15) and has been made use of in studies of the turnover of adenylate cyclase (Bockaert, personal communication). Our finding that the time course of cyclic AMP accumulation is prolonged in the presence of forskolin (8) is also in accordance with the proposed scheme. There is, however, conflicting data regarding the destabilizing influence of the beta-subunit or the beta-subunit of transducin, respectively, on the active enzyme complex in the presence of forskolin (15,16,17) which may require further modification of the dissociation model.

RECEPTOR INTERNALIZATION AND DESENSITIZATION: A MUTUAL MECHANISM?

It has recently been shown that in S49 cells the beta-adrenergic receptors are phosphorylated by a hormone stimulated but cyclic AMP-independent protein kinase (18,19). It is conceivable that receptor phosphorylation is involved in the desensitization of the adenylate cyclase system. Phosphorylation may cause uncoupling of the receptors from the guanine nucleotide regulatory protein N_s, or, alternatively, it may provide the signal to initiate receptor internalization.
Uncoupling has recently been demonstrated to precede receptor internalization in S49 cells (1). Since receptor phosphorylation is maximal before the maximal desensitization is reached (18), it is possible that receptor phosphorylation initially uncouples the receptors and then leads to their internalization and to desensitization.

This sequence would perhaps adequately describe the events of desensitization in simple terms, provided that receptor uncoupling and internalization result directly in the desensitization of the enzyme. However, there is evidence that receptor internalization is not required for functional desensitization, i.e. that internalization of receptors and desensitization occur separately after being triggered off by the binding of the hormone and uncoupling of the receptors. The internalization of the receptors, namely, can be blocked by phenylarsine oxide (20) or by reducing the cellular ATP content (21), by concanavalin A (22), and by treatment with BAAM (unpublished observation). Under such conditions, agonist-induced uncoupling is not diminished and the decline in adenylate cyclase activity can still be observed.

It remains to be seen if future findings prove these results to be compatible with the suggested mutual regulatory mechanism.

ACKNOWLEDGEMENT

We thank the Swiss National Science Foundation (Grant No. 3.645-0.84) for the funding of this work.

REFERENCES

(1) Hertel,Portenier,Staehelin, Evidence for the appearance of an uncoupled form of the beta-adrenergic receptor distinct from the internalized receptor. - J.Cell.Biochem. 30: 219-225 (1986).
(2) Portenier,Staehelin, (^{3}H)CGP-12177: Its use in determining changes in cell surface receptors. - In:'Adrenergic Receptors: Molecular Properties and Therapeutic Implications.' Symposia Medica Hoechst 19: 481-496. Schattauer Verlag, Stuttgart-New York (1985).
(3) Hertel,Müller,Portenier,Staehelin, Determination of the desensitization of beta-adrenergic receptors by (^{3}H)CGP-12177. - Biochem.J. 216: 669-674 (1983).
(4) Pitha,Zjawiony,Nasrin,Lefkowitz,Caron, Potent beta-adrenergic antagonist possessing chemically reactive group. - Life Sci. 27: 1791-1798 (1980).
(5) Pitha,Hughes,Kusiak,Dax,Baker, Regeneration of beta-adrenergic receptors in senescent rats: A study using an irreversible binding agonist . - Proc.Natl.Acad.Sci.USA 79: 4424-4427 (1982).
(6) Staehelin,Hertel, (^{3}H)CGP-12177, a beta-adrenergic ligand suitable for measuring cell surface receptors. - J.Recept.Res. 3: 35-43 (1983).
(7) Staehelin,Simons, Rapid and reversible disappearance of beta-adrenergic cell surface receptors. - EMBO J. 1: 187-190 (1982).
(8) Box,Portenier,Staehelin, submitted.
(9) Insel,Stoolman, Radioligand binding to beta-adrenergic receptors of intact cultured S49 cells. - Mol.Pharmacol. 14: 549-564 (1978).
(10) Darfler,Mahan,Koachman,Insel, Stimulation by forskolin of intact S49 lymphoma cells involves the nucleotide regulatory protein of adenylate cyclase. - J.Biol.Chem. 257: 11901-11907 (1982).
(11) Seamon,Daly, Activation of adenylate cyclase by the diterpene forskolin does not require the guanine nucleotide regulatory protein. - J.Biol. Chem. 256: 9799-9801 (1981).
(12) Morris,Bilezikian, Evidence that forskolin activates turkey erythrocyte adenylate cyclase through a noncatalytic site. - Arch.Biochem.Biophys. 220: 628-636 (1983).
(13) Daly, Forskolin, adenylate cyclase and cell physiology: An overview. - Adv,Cyclic Nucl.Prot.Phosp.Res. 17: 81-90. Raven Press, New York (1984).
(14) Barber,Goka, Adenylate cyclase activity as a function of forskolin concentration. - J.Cyclic Nucl.Prot.Phosph.Res. 10: 23-29 (1985).
(15) Yamashita,Kurokawa,Higashi,Danura,Ishibashi, Forskolin stabilizes a

functionally coupled state between activated guanine nucleotide binding stimulatory protein N_s, and catalytic protein of the adenylate
cyclase system in rat erythrocytes. - Biochem.Biophys.Res.Comm. 137:
190-194 (1986).

(16) Bockaert,Deterre,Pfister,,Guillon,Chabre, Inhibition of hormonally
regulated adenylate cyclase by the beta-gamma subunit of transducin. -
EMBO J. 4: 1413-1417 (1985).

(17) Enomoto,Asakawa, Inhibition of catalytic unit of adenylate cyclase and
activation of GTPase of N. protein by beta-gamma subunits of GTP-
binding proteins. - FEBS Lett. 202: 63-68 (1986).

(18) Strasser,Sibley,Lefkowitz, A novel catecholamine-activated adenosine
cyclic 3',5'-phosphate independent pathway for beta-adrenergic receptor phosphorylation in wild-type and mutant S49 lymphoma cells:
Mechanism of homologous desensitization of adenylate cyclase. -
Biochemistry 25: 1371-1377 (1986).

(19) Benovic,Strasser,Caron,Lefkowitz, Beta-adrenergic receptor kinase:
Identification of a novel protein kinase that phosphorylates the ago-
nist-occupied form of the receptor. - Proc.Natl.Acad.Sci.USA 83: 2797-
2801 (1986).

(20) Hertel,Coulter,Perkins, A comparison of catecholamine-induced internalization of beta-adrenergic receptors and receptor-mediated endocytosis of epidermal growth factor in human astrocytoma cells. - J.Biol.
Chem. 260: 12547-12553 (1985).

(21) Hertel,Coulter,Perkins, The involvement of cellular ATP in receptor-
mediated internalization of epidermal growth factor and hormone-
induced internalization of beta-adrenergic receptors. - J.Biol.Chem.
261: 5974-5980 (1986).

(22) Waldo,Northup,Perkins,Harden, Characterization of an altered membrane
form of the beta-adrenergic receptor produced during agonist-induced
desensitization. - J.Biol.Chem. 258: 13900-13908 (1983).

IMMUNOLOGICAL STIMULATION OF MAST CELLS DEGRANULATION:

ROLE OF CYTOSOLIC pH, Na^+ AND Ca^{+2} IONS

I. Pecht, R. Schweitzer-Stenner, R. Gertler, M. Wolf,
Y. Zisman and B. Reck
Department of Chemical Immunology, The Weizmann Institute of
Science, Rehovot, 76100, Israel

Transient changes in the free cytosolic concentrations of essential cations such as protons, sodium or calcium are emerging as cellular signalling elements. Evidence for their involvement in vital processes such as cell-proliferation and differentiation and the coupling between excitation and contraction or stimulus and secretion in respective muscle or nerve cells are well documented[1-6]. Changes in cytosolic steady-state concentrations of these cations may be caused by net influx into the cytosol from the extracellular medium or recruitement from intracellular stores. Depending on concentration gradients existing for specific ions, exchange or net influx processes prevail. Thus, for example, stimulation of fibroblasts by growth factors was shown to activate a Na^+/H^+ exchanger leading to a net uptake of sodium ions and protons efflux[1-3,7]. For a wide range of cells, e.g. B or T-lymphocytes, one of the earliest events following stimulation is a transient rise in cytosolic free Ca^{+2} which is due either to release from cellular stores or to influx via specific ion channels opened in their plasma membranes.[5,6,8]

Mast-cells are activated to secrete preformed and newly synthesized mediators of the immediate-type hypersensitivity by specific antigen induced crosslinking of IgE-class antibodies which are bound at their membranes[8,9]. Interest in resolving the biochemical processes which couple the antigen induced stimulus to the eventual secretion is of both fundamental nature as well as applied one. This is because (a) the nature and sequence of biochemical steps leading to degranulation are still largely unknown and (b) these processes provide the molecular basis of allergy[9].

The crucial role of Ca^{+2} ions in coupling stimulus to secretion from mast cells is well established. The process is strictly dependent on the presence of extracellular calcium[10]. Furthermore, a transient rise in cytosolic free calcium ($[Ca^{+2}]_i$) following antigen stimulation has been clearly demonstrated: This problem has originally been investigated using $^{45}Ca^{+2}$ as tracer, yielding evidence for a net uptake of these ions by the cells[11]. More recently, the fluorescent chelating agent quin-2 was employed to directly monitor changes in the cytosol free Ca^{+2}[12]. This reagent enabled detailed time resolution of the actual transient rise in $[Ca^{+2}]_i$[13-15]. Still, a quantitative correlation between experiments monitoring both the radioactive tracer and the fluorescent chelator is clearly required in order to unambiguously define the source(s) of the calcium ions arising in the cytosol. It is therefore rather interesting that we have recently observed that antigen-induced secretion from mast cells may be achieved without any detectable rise in $[Ca^{+2}]_i$[15]. This can be attained when the cells are stimulated in the presence of tumor promoting phorbol esters (e.g. phorbol 12-myristate 13-acetate - PMA)[14,15]. Depending on the concentration employed, PMA was shown to have both, a synergistic effect with antigen (at low PMA $\leq$ 10 nM) and an inhibitory effect at higher concentration. However, the calcium signal was found to be suppressed over the whole range of PMA added. These observations led us to explore the potential role of other cations primarily protons and sodium in the coupling antigen stimulus to secretion in these cells. The particular consideration of these cations is founded on the above mentioned reports[1-4,7] describing activation of the Na^+/H^+ exchanger by growth factors known to trigger the formation of ciacylglycerol for which PMA serves as an effective analog. Indeed, modulation of the free cytosolic concentrations of these ions was found to markedly affect mediators secretion from RBL-2H3 cells.

MATERIALS AND METHODS

Reagents

2',7-bis-(2-carboxyethyl)-5, or 6-)-carboxyfluorescein (BCECF) was purchased from Molecular Probes, OR, USA. Monensin and amiloride from Sigma. Quin2-AM and ouabain from Calbiochem. The DNP specific monoclonal A-2 IgE secreting line was kindly given by Dr. V.T. Oi, Department of Genetics, Stanford University School of Medicine. Antigen was bovine serum albumin (BSA) derivatized with an average of 8 DNP residues per molecule.

Cells

The secreting subline RBL-2H3 [16] of the rat basophilic leukemia was regularly grown in 250 ml tissue culture flasks (Nunc, Denmark or Falcon, Oxnard, CA) in Eagle's minimal essential medium (Gibco, Grand Island, NY) supplemented with 15% heat inactivated fetal calf serum (Biolab, Jerusalem, Israel), 4 mM Glutamine and antibiotics. Cells were kept in a humidified 5% CO_2 incubator and were subcultured every 48 to 56 hours. The cells were collected by scraping with a mechanical rubber scraper (policeman), washed and transferred at $2X10^6$ cells in 10 ml of fresh medium.

Degranulation experiments

RBL-2H3 cells were loaded with labelled serotonin by an overnight incubation with 5 μCi/ml [^{3}H]-5-hydroxytryptamine creatinine sulfate (11 Ci/mmol, Radiochemical Centre, Amersham). During the following morning, the cells were suspended, washed and after counting and adjusting their density, they were resuspended ($1x10^6$ cells/ml) in Tyrode medium containing 137 mM NaCl, 2.7 mM KCl, 0.4 mM NaH_2PO_4, 2 mM $CaCl_2$, 1 mM $MgCl_2$, 20 mM HEPES and 5.6 mM glucose, pH 7.4 and an appropriate amount of monoclonal, DNP-specific IgE antibody. After incubation for 90 minutes to allow binding of the IgE, the cells were washed again with Tyrode buffer and degranulation was assayed in a total volume of 150 μl cell suspension aliquots each in a 96 flat-bottom-wells microtiter plate (Nunclon) containing $1 \cdot 10^5$ cells and the indicated reagents per well. At the end of

35 min incubation at 37 °C, the cells were spun down and 100 μl aliquots of the supernatants counted. The total radiolabeled 5-hydroxytryptamine incorporated into the cells was determined by dissolving the cells in 50 μl of 1 N NaOH and counting the radioactivity of the whole sample. Release is expressed as the percentage of the total content of labeled 5-hydroxytryptamine.

The extent of mediators release caused by cell lysis was checked, following the above procedure, by monitoring the activity of the cytosolic enzyme lactate dehydrogenase (LDH) in the cells' medium. This was done by measuring the change in concentration of NADH added to the medium according to the routine assay procedure for the enzyme [17]. The dependence of monensin induced mediators-release on metabolic energy was examined in experiments where each of the inhibitors deoxyglucose (10 mM) or NaCN (100 μM) or NaN$_3$ (100 μM) were added to the reaction medium. No glucose was present in these experiments.

Fluorometric monitoring of $[Ca^{+2}]_i$ and $[pH]_i$

a) Adherent cells were scraped with fresh medium and incubated with DNP-specific monoclonal IgE antibody under saturating conditions for 1 hour at 37°C. The cell suspension was then centrifuged, supernatant discarded, and the cells washed twice with Tyrode buffer, pH 7.4 to remove the excess of IgE. Cell density was adjusted to 2×10^6 cells/ml and cells incubated with quin2-AM (final concentration 20 μM) for 30 minutes at 37°C. The cell suspension was again centrifuged, the supernatant discarded as completely as possible to remove untrapped dye. Finally, cells were resuspended in Tyrode buffer at 10^6 cells/ml for fluorescence measurements. These were performed at 37°C with 1 ml aliquots of the cell suspension in a 7x7 mm cuvette using a Perkin Elmer MPF 44A spectrofluorometer. During the measurement, cells were kept in suspension by a stirrer which was inserted into the cuvette without obstructing the optical measurements. After measuring the emission spectrum of the cell suspension and ascertaining that the quin2-AM had been hydrolyzed, i.e., the maximum emission had shifted to 492 nm, the different stimulating agents were introduced into the cuvette. The calibration procedure for the calculation of cytosolic-free Ca^{+2} concentrations ($[Ca^{+2}]_i$) in RBL-2H3 cells was done according to Tsien et al.[18] using parameters obtained as follows: At the end of recording the reaction, the cells were lysed by adding Triton X100 (final concentration 0.1% v/v) to set free the trapped quin2 and saturate it with Ca^{+2} from the extracellular medium. This leads to the maximal fluorescence intensity F_{max}, corresponding to the total quin2 concentration of the cell suspension. By the subsequent addition of MnCl$_2$ (final concentration 0.5 mM) Ca - quin2 is replaced nearly quantitatively by Mn - quin2 complex, which has a negligibly small fluorescence yielding the value of F_{min}. If MnCl$_2$ (final concentration 0.5 mM) is added to the intact cells prior to lysis, a slight drop in fluorescence intensity, ΔF, is observed. This corresponds to extracellular quin2 which may sometimes leak from the cells.

b) Adherent RBL-2H3 cells were scraped, washed with Tyrode buffer, pH 7.4 and set to 10^6 cells/ml. The cells were then incubated in suspension with the acetoxymethylester BCECF/AM (final concentration 1 μM) for 1 hour at 37°C, then centrifuged, the supernatant discarded, and the cells resuspended in Tyrode buffer, pH 7.4 at 10^6 cells/ml. Fluorescence measurements were performed in the same way as described above for quin2, except that BCECF fluorescence was monitored at 526 nm and excited at 505 nm[19].

RESULTS AND DISCUSSION

Crosslinking of $Fc_\varepsilon R$ by specific IgE-antigen interaction leads to a transient rise in free cytosolic Ca^{+2} concentration in RBL-2H3 cells.[13,15] The strict dependence of this rise in cytosolic free calcium on maintaining crosslinked IgE on the cell's membrane is illustrated in Figure 1. In these experiments an excess of a monovalent hapten, 2,4,6, trinitro-toluene (TNT), is competing with the antigen (BSA - DNP_8) for the binding to IgE and causes the dissociation of crosslinked antibody-antigen aggregates. This dose-dependent competition is monitored in parallel either via the decrease in calcium signal or the reduction of secretion. The physico-chemical definition of the signalling element of a crosslinked oligomeric membrane-bound IgE is still not satisfactorily understood. Studies employing covalently linked oligomers of IgE have shown that already dimers have the capacity of triggering secretion, though the larger oligomers are significantly more effective[20,21]. Still, these results fail to provide clear insight into the physical proximity requirements for two or more Fc_ε receptors in order to initiate the biochemical cascade culminating in degranulation. Thus, employing the same DNP-specific monoclonal IgE (as in experiments illustrated in figure 1) and chemically synthesized divalent haptens with spacers of different length and rigidity we are trying to resolve the above issue[22]. For example, the flexible symetric and divalent hapten (α-N-DNP $(Ala)_3$-NH_2-$(CH_2)_7$-NH_2-$(Ala)_3$-α-N-DNP was found to be a very poor secretion inducing agent. This was the case although it binds well to the IgE and induces its dimerization as evidenced by titration in homogeneous solutions (Figure 2). Moreover, this divalent hapten is also an effective

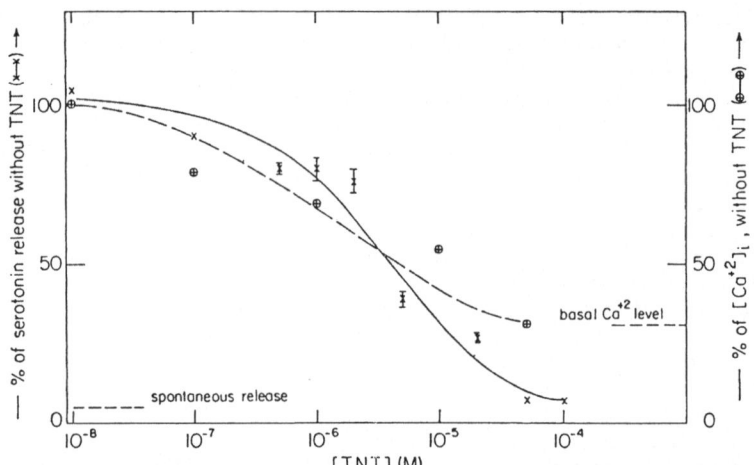

Fig. 1. Dose response curves demonstrating the inhibitory effect of the monovalent hapten: 2,4,6 - trinitrotoluene (TNT) on antigen (5 ng BSA-DNP_8 per 10^6 cells/ml) induced degranulation and on the rise in $[Ca^{+2}]_i$; (the affinities of DNP and TNT to the IgE antibody are nearly the same, $K_{DNP-lys} = 5.3 \times 10^5 M^{-1}$ $K_{TNT} = 6.3 \times 10^5 M^{-1}$ at $37^\circ C$; (TNT was chosen as a competing hapten because of its less intefering optical properties for the quin 2 experiments). Degranulation is presented as percent of ^{3}H-serotonin released from preloaded cells, release without TNT is taken as to 100% (x--x). $[Ca^{+2}]_i$ was calculated from Ca-quin 2 signals (λ em = 492 nm, λ ex = 339 nm) and is presented as percent of $[Ca^{+2}]_i$ obtained by antigen triggering without TNT (o---o).

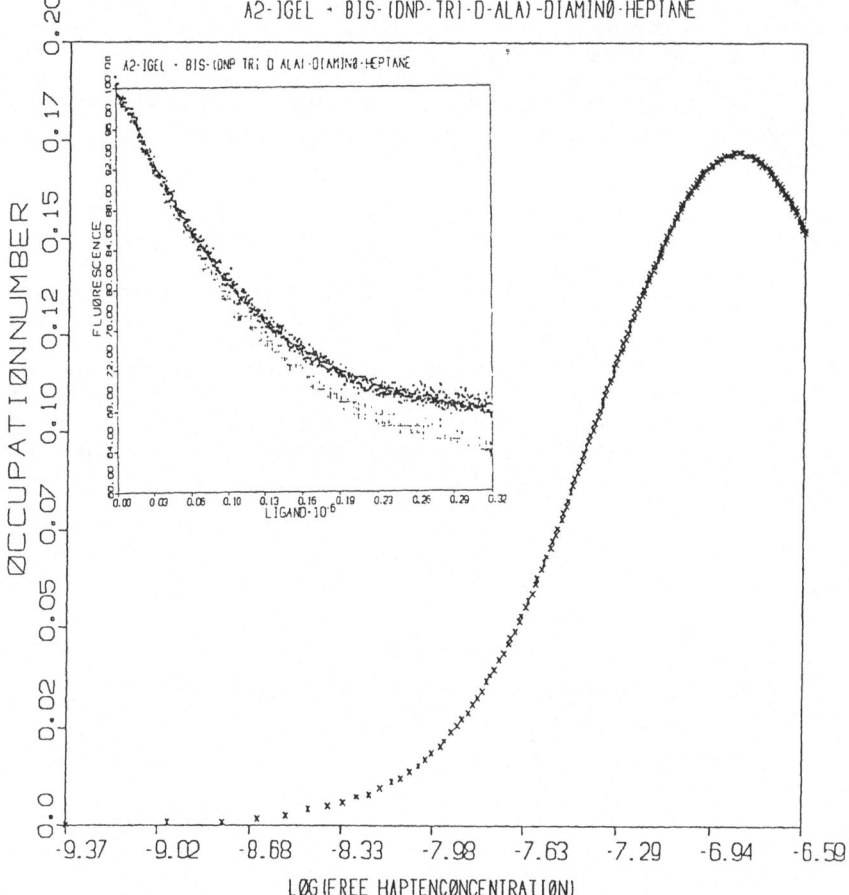

Fig. 2. Fluorescence titration (insert) in aqueous 0.15 M NaCl 0.1 M
sodium borate buffer pH 7.4 solution of a monoclonal IgE class
antibody (A 2) specific to 2,4 dinitrophenyl hapten by a divalent
hapten: bis(α-N-dinitrophenyl-tri-D-alanyl)-N,N, diaminoheptane.
Excitation was at 280 nm, emission was monitored at 340 nm.
Titrant-hapten solution was added continuously from a motor-
driven syringe. The lower set of data (grey) represents the
actual titration, while the upper set of points are after
correction for non-specific quenching (e.g. inner filter effect)
and dilution. The fitted curve yielded by the analysis done
according to a model involving oligomer formation is presented by
a solid line drawn through the data.

The values derived for the respective equilibrium constants are:
$K_{int} = 3.3 \cdot 10^{-6}$ M, $K_{agg} = 3.4 \cdot 10^{5}$ M and $J_2 = 26$. Where K_{int} is
the intrinsic binding constant, K_{agg} is for the initial step of
aggregation, i.e. binding of a second IgE to the complex of the
first IgE with one divalent hapten and J_2 is the ring closure
equilibrium constant.

The results of the fitting procedure presented in the form of the
distribution in mole fraction of the antibodies transformed into
dimers as a function of added divalent hapten. Further details
are provided elsewhere[22].

inhibitor of antigen induced secretion (Figure 3) supporting further its effective binding to the cell associated IgE. Still it fails to induce secretion on its own (bottom of figure 3). One attractive rational for this apparent discrepancy would be that crosslinking by such a flexible divalent hapten fails to provide the steric constraints imposed by globular polyvalent antigen. Obviously, this interesting feature of ligand-dependent aggregation as a cellular stimulus requires further investigation.

Returning to the antigen-induced transient rise in cytosolic free calcium, as illustrated above (Figure 1), the level of antigen induced increase in $[Ca^{+2}]_i$ is a sensitive function of the extent of $Fc_\epsilon R$ aggregation state. Since several cellular processes are involved in regulating $[Ca^{+2}]_i$, the dynamic nature of this parameter is not surprising. Unexpected, however was the observation that certain biochemical modulation can cause the RBL-2H3 cells to undergo degranulation under conditions where no increase in $[Ca^{+2}]_i$ is detectable. Tumor promoting phorbol esters, being structural analogs of diacyl glycerol, were shown to be effective activators of protein kinase C (PKC) and are therefore widely employed in studies of the role of this enzyme in different cellular processes[23]. While rat peritoneal mast cells[24] were shown to be directly activated to secretion by PMA (phorbol 12 myristate - 13 acetate), their tumor analog the RBL-2H3 cells are not[14]. Still, PMA in conjunction with antigen were found to be effective degranulating

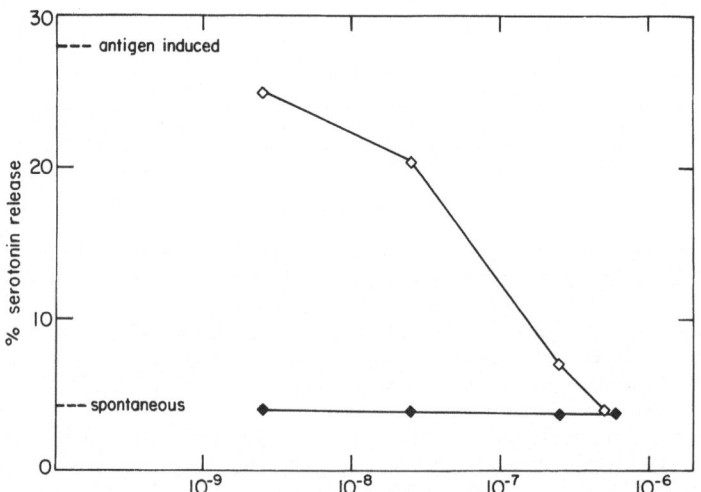

Fig. 3.   Inhibition of antigen (BSA - DNP_8) induced degranulation of the RBL-2H3 cells by the divalent hapten bis-α-N-DNP $(Ala)_3$,N,N' diamino heptane. A DNP specific IgE was used to prime the cells and the divalent hapten was added 2 minutes before the antigen (10 ng BSA-DNP_{15} per 10^6 cells/ml tyrode buffer, pH 7.4) (open symbols). Degranulation is expressed as percentage of secreted ^{3}H-serotonin, which the cells had incorporated during an overnight incubation with 5-Hydroxy (6-^{3}H) tryptamine creatinine sulfate.

Full points indicate release experiments with the divalent hapten itself at the given concentrations. Additionally, spontaneous and maximal ^{3}H-serotonin release from RBL-2H3 cells triggered with antigen alone are indicated.

agents. As illustrated in Figure 4, titration of the RBL-2H3 cells with increasing amounts of PMA, at constant antigen concentration leads first to a synergistic increase in secretion followed by a later range of inhibition. Noteworthy however is the observation that almost half of the amount of mediators secretion induced by antigen alone is maintained throughout, even at the highest PMA concentrations employed. Moreover, when the rise in $[Ca^{+2}]_i$ is monitored in parallel and under identical conditions to those employed for measurements of secretion, a very dramatic decrease in the transient rise in $[Ca^{+2}]_i$ (so-called calcium signal) is observed already in the PMA range where synergism is still maintained. Moreover, although no detectable rise in $[Ca^{+2}]_i$ is observed at the higher range of PMA concentrations employed, the secretion is maintained. PKC activation causes the suppression of the transient increase in $[Ca^{+2}]_i$ probably by more than one mechanism. This is evidenced by experiments where the relative points in time of introducing the perturbants are varied. For example, preincubating the cells with PMA causes a time dependent suppression of the calcium signal while addition after antigen stimulus causes an accelerated decay of the $[Ca^{+2}]_i$ basal level (Figure 5). Evidently, both calcium extrusion mechanisms as well as

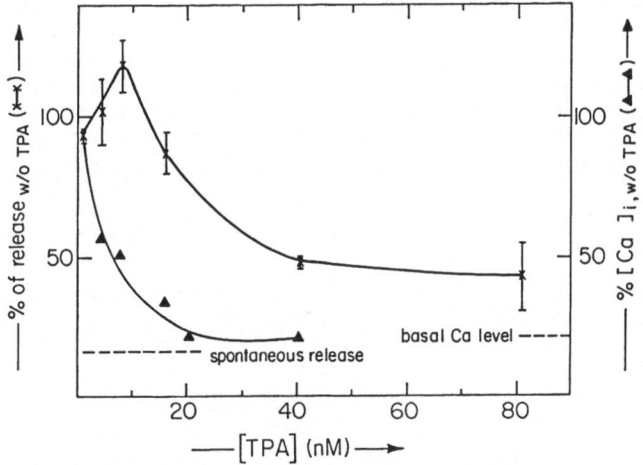

Fig. 4. Dose response curves demonstrating the effect of PMA on the antigen induced rise in $[Ca^{+2}]_i$ (Δ - Δ) and on the degranulation (x - x) of RBL-2H3 cells. Degranulation is expressed by the percent of serotonin released (100% = release without PMA). Rise in $[Ca^{+2}]_i$ was calculated from changes in the emission of quin 2 at 492 nm (excitation at 339 nm) and is presented as percentage of the rise in cytosolic free Ca^{+2} (obtained from the Ca-quin 2 signal observed without PMA taken as 100%). PMA was added 1 minute before antigen (5 ng BSA-DNP$_8$) to 1 ml cell suspension (10^6 cells). Serotonin release without PMA = 26.6 $\pm$ 0.8% of total ^{3}H-serotonin, spontaneous release = 4.3 $\pm$ 0.9%; without PMA $[Ca^{+2}]_i$ = 765 nM, basal $[Ca^{+2}]_i$ = 145 $\pm$ 19 nM.

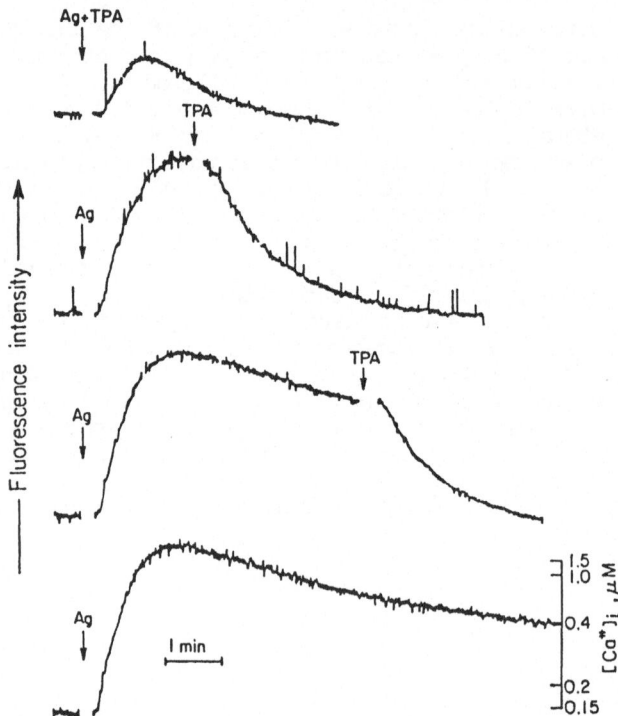

Fig. 5.   Effect of PMA on the antigen induced rise in $[Ca^{+2}]_i$ of RBL-2H3
cells as monitored by quin 2 emission at 492 nm, excited at 339
nm.   81 nM PMA were added at different times after the addition
of antigen (5 ng BSA-DNP$_8$/ml) to a stirred cell suspension (10^6
cells/ml tyrode buffer pH 7.4) thermostated at 37°C.   In the
experiments without PMA (lowest trace) $[Ca^{+2}]_i$ was calculated by
a calibration procedure which employs the maximal emission
obtained for total quin 2 concentration (after cell lysis with
0.5% Triton X-100) and minimal emission (after addition of Mn^{+2}
to a final concentration of 0.5 mM to the lysed cells).

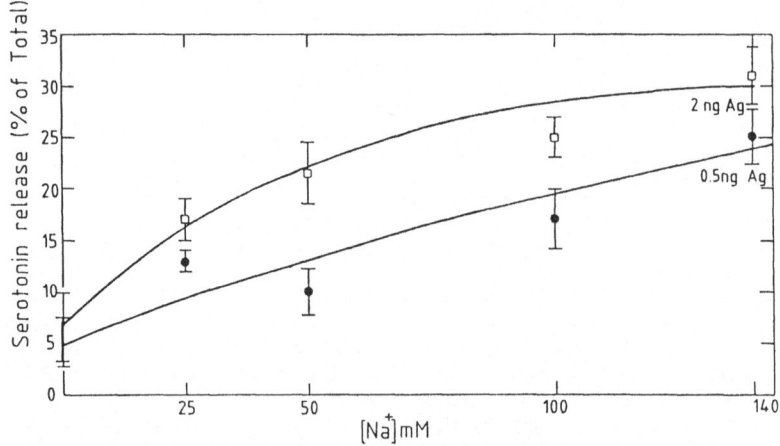

Fig. 6.   Dependence of antigen-induced serotonin secretion from RBL-2H3
cells on Na$^+$ ions concentration in their medium.   Assay
procedures are as described in the text.   Osmolarity was
maintained constant by the addition of a compensating
concentration of tetramethyl ammonium chloride.

inhibition of its influx from both extra- and intracellular sources are activated by PKC mediated pathways. These processes maintaining the homeostatis of $[Ca^{+2}]_i$ are lines to be further pursued.

Another intriguing and fundamental facet emerging from the above observations is the nature of coupling mechanisms that the RBL-2H3 cells employ in maintaining a pronounced extent of antigen mediated secretion even though no rise in $[Ca^{+2}]_i$ is detected. One specific activity which has been assigned to PKC activation is that of the cellular Na^+/H^+ exchanger. This system originally identified and characterized in brush-border membrane vesicles[25] was later also identified in the plasma membrane of a wide range of animal cells; from erythrocytes, epithelial cells and fibroblasts to skeletal muscle and neuronal ones[26]. It employs the Na^+ concentration gradient between the medium and cytosol to actively extrude intracellular protons[27]. Hence, it has a major role in maintaining intracellular pH homeostasis[28].

A significant shift in cytosolic Na^+/H^+ balance has been reported to be induced by various growth factors and mitogens. Thus, Na^+ influx and cytosolic alkalinization due to the parallel protons efflux were proposed to provide intracellular second messengers for the initial stimulus at the plasma membrane[1-4]. The Na^+/H^+ exchanger was postulated to be the component performing this process. Recently modulation of the exchanger by PKC has been directly implicated at least in one cell type, rat thymocytes[29]. PKC activation of Na^+/H^+ exchanger was shown to cause a shift in its pH dependence and therefore enhance its activity[29]. Identification of the sodium-proton exchanger as a target of activation by PKC prompted the investigation of the possibility that the exchanger would also be affected by PKC in the RBL-2H3 cells. This could provide the alternative coupling element between antigen stimulus and the secretion operating when the calcium signal is suppressed by PMA.

The dependence of antigen induced secretion on sodium ions concentration in the medium of the RBL-2H3 cells is illustrated in Figure 6. Evidently, a rather limited extent of secretion can be attained when this cation is eliminated. Several other different experimental protocols which affect the cytosolic Na^+/H^+ balance were therefore employed and their correlation with the extent of mediator secretion was examined. First, the effect of PMA itself on cytosolic pH of RBL-2H3 cells was examined by monitoring the fluorescence of BCECF. No change could however be resolved throughout the range of PMA employed (5 - 80 nM). By contrast, addition of specific antigen caused a small but consistant acidification of the RBL-2H3 cytosol. This could be promptly reversed upon addition of PMA (Figure 7). A quantitative examination of this process is now being pursued.

An artificial perturbation by the proton/sodium exchanging ionophore monensin was shown to effectively induce non-cytotoxic serotonin release. As illustrated in figure 8, it causes release of more than 40% of these cells' contents at 1 μM concentration, with a half-maximal dose of 0.2 μM at pH 7.4. This release is independent of extracellular calcium, since an essentially identical dose response curve to the one shown in figure 8 (pH 7.4) is obtained in the presence of 3 mM EGTA. The monensin-induced release is, in contrast, sensitive to both pH and sodium ion concentration of the release medium (figure 8). It markedly increases with a rise in external pH and is suppressed by lowering external sodium concentrations. The extent of LDH activity in the supernatants of the cells exposed to doses of monensin up to 5 μM was similar to that of the control cells' supernatants which were not exposed to this ionophore or were triggered by antigen.

In experiments designed to check the dependence of the monensin-induced release on maintained glycolysis or oxidative phosphorylation, unequivocal results were obtained. In several experiments substitution of glucose by deoxyglucose suppressed secretion, also, cyanide or azide were not effective inhibitors. In fact, these caused increased cell mortality to the extent that their effect on the monensin induced release is inconclusive.

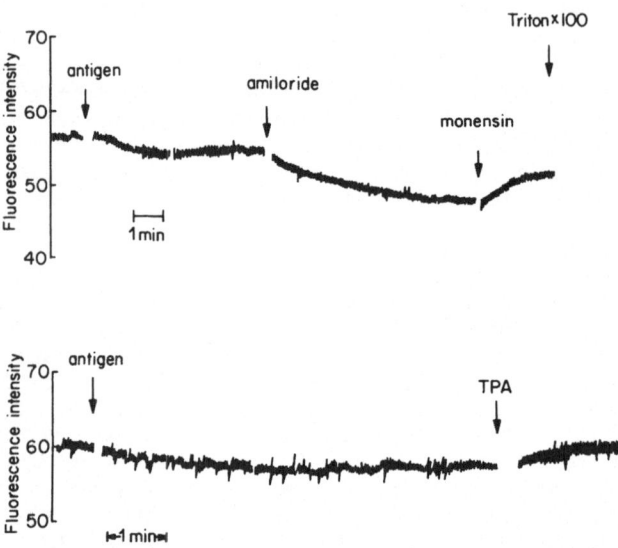

Fig. 7. Traces of the BCECF fluorescence intensities at λ_{ex} = 505 nm, λ_{em} = 526 nm, demonstrating intracellular pH changes of RBL-2H3 cells. Upper trace: cells were triggered sequentially by the addition of antigen (5 ng BSA-DNP$_{15}$ per 10^6 cells/ml), amiloride (final concentration 100 μM) and monensin (final concentration 0.5 μM). Lower trace: cells were triggered with antigen (2 ng BSA-DNP$_{15}$ per 10^6 cells/ml) and subsequently by the addition of PMA (50 nM final concentration).

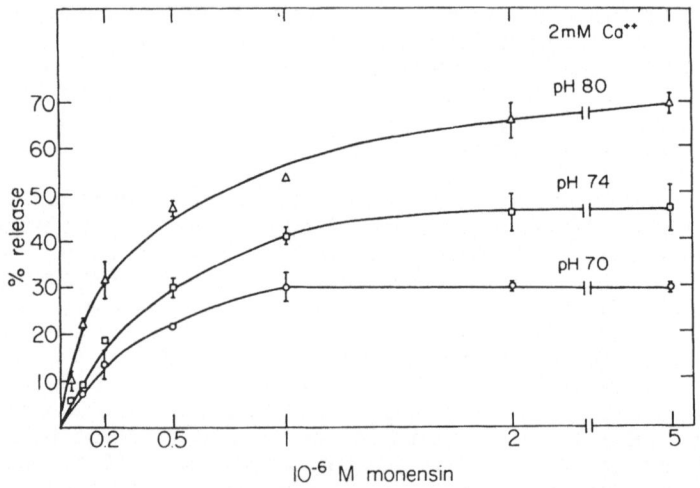

Fig. 8. Monensin induced serotonin release from RBL-2H3 cells; dose and pH dependence of the process.

The change in pH_i caused by monensin is illustrated in figure 9 where a trace of fluorescence intensity of the trapped indicator BCECF in a cell suspension is shown. The alkalinization brought about by monensin is evident. The concomitant changes in $[Ca^{+2}]_i$ were monitored in separate experiments via the emission intensity of quin2. A transient rise in $[Ca^{+2}]_i$ is indeed observed upon addition of monensin, however when external free calcium is practically eliminated by addition of excess EGTA (3 mM), monensin does **not** cause any significant change in $[Ca^{+2}]_i$.

Several other ionophores carrying preferentially sodium ions were examined for their capacity to cause serotonine release. Significantly none of these examined (4-Crown-12; Gramicidin; Eniatin B and Beauvericin) were effective. The most probable reason for this behavior lies in the Na^+/H^+ exchange capacity of monensin which is not shared by the other ionophores examined.

To differentiate between the observed monensin induced serotonin release and mediators release by degranulation we examined, in parallel experiments, both: (1) activity of β-hexoseaminidase, a granule-stored enzyme, and (2) amounts of ^{3}H-serotonin released. Significantly, only in antigen induced release we observed the concomitant presence, in the RBL-2H3 supernatants, of both the macromolecular enzyme, β-hexoseaminidase and of serotonin. Monensin failed in releasing β-hexasoaminidase activity into the cell's medium. Thus, the serotonin is released by monensin in a different mechanism; most probably by a non-physiological process allowing the efflux of the low molecular weight mediator-serotonin while maintaining unperturbed the macromolecular ones, like β-hexoseaminidase. Monensin probably induces a sodium-proton exchange also across the granular membrane. Hence, it would perturb the protonation equilibrium of serotonin and allow efflux of its deprotonated non-charged form.

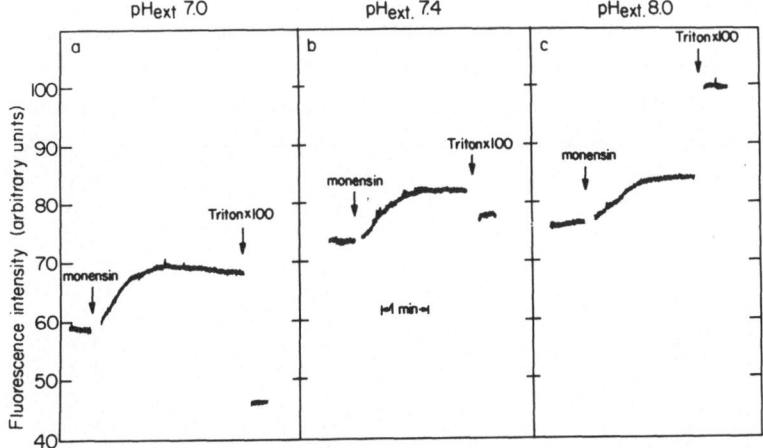

Fig. 9. Monitoring changes in cytosolic pH of RBL-2H3 cells suspended in buffers (HEPES) at increasing external pH and treated with monensin (1 μM final concentration). Emission intensities of the trapped fluorescent dye 2',7'-bis-(2-carboxyethyl)-5 (and -6)-carboxyfluorescein (BCECF, λ_{ex} = 505 nm, λ_{em} = 526 nm) are expressed in arbitrary units.

Amiloride and its analogs are well established inhibitors of the cellular Na^+/H^+ antiporter[30]. We have examined whether it affects the antigen or monensin induced release. As shown in Table 1, it causes a pronounced inhibition of both types of induced secretion. Amiloride by itself, even at the highest concentration employed (100 μM), did not cause a significant elevation of the basal level of release ($\approx$ 3%). The inhibition of either antigen or monensin-induced release exhibited some dependence on the length of incubation of the cells with amiloride. Clearly, however, it is an effective inhibitor when employed in the higher end of the reported concentration range for its activity in other cell types[30,31].

Finally, a different cellular control mechanism of intracellular sodium concentration is attained by the Na^+/K^+ ATPase pump[31]. A potent inhibitor of this ATPase is ouabain. Application of this reagent to the RBL-2H3 cells caused already by itself a significant amount of secretion e.g. about 10% of the cell's contents were secreted by treatment with 0.1 mM ouabain and a two-fold increased secretion was attained by 1.0 mM.

Experiments described in this report show that in antigen stimulated mast cells of the type illustrated by the RBL-2H3 line several transient changes in cytosol ion composition occur: The transient rise in free calcium shown to couple stimulus to function in a wider range of cells is also observed in mast cells. This calcium signal is dramatically suppressed by PMA, most probably because the activated PKC causes blockage of Ca^{+2} influx and its enhanced extrusion. In pursuing the possibility that the Na^+/H^+ exchanger of the RBL-2H3 cells is activated by PKC and serves as an alternative coupling mechanism following PMA treatment, we observed that antigen stimulation causes by itself a slight decrease in pH_i of these cells. However, PMA does not induce any change. Still the sensitivity of antigen induced degranulation to changes in the cytosol Na^+/H^+ was evidenced by the inhibition exerted with amiloride and secretion enhanced by ouabain. In this context it is of interest that the Na^+/H^+ ionophore monensin which causes release of low molecular weight

Table 1. Inhibition by amiloride of serotonin release from RBL-2H3 cells induced by antigen or by monensin

Amiloride (uM)	Antigen	Monensin	% Inhibition
1	+	−	4.1
5	+	−	12.0
10	+	−	16.6
20	+	−	20.8
50	+	−	41.6
1	−	+	3.8
2	−	+	34.6
5	−	+	50.0
10	−	+	61.5
20	−	+	65.0
50	−	+	65.3

Release with antigen (1 ng DNP_8-BSA) was 24%, release induced by monensin (0.1 uM) was 26% of the cell's serotonin contents. All results are averages of triplicate experiments.

mediators like serotonin, does not induce real degranulation. It should however be also noticed that opposite changes in pH_i are caused by antigen and monensin: The former induces acidification while the latter alkalinization. Hence, one can exclude the monensin increase in pH_i per se as a cause for degranulation and it has now to be critically examined whether acidification has a messenger role or is an outcome of antigen induced degranulation. Thus, the mechanism by which antigen stimulated secretion is partly maintained although the calcium signal is eliminated by PMA is still unresolved. The mutual relation among the different potential second messengers require further investigation.

Acknowledgement

This work was supported by grants from the Council for Tobacco Research USA Inc. and from The Hermann and Lilly Schilling Foundation for Medical Research, W. Germany.

REFERENCES

1. E. Rozengurt, Adv. Enzymol. Regul. 19:61 (1981).
2. W. Moolenaar, R. Tsien, P. Van der Saag and S. de Laat, Nature 304:645 (1983).
3. S. Shuldiner and E. Rozengurt, Proc. Natl. Acad. Sci. USA 79:7778 (1982).
4. T.M. Connolly and L.E. Limbird, Proc. Natl. Acad. Sci. USA 80:5320 (1983).
5. A. Weiss, J. Imboden, D. Shoback and J. Stobo, Proc. Natl. Acad. Sci. USA 81:4169 (1984).
6. A. Alcover, M.J. Weiss, F.J. Daley and E.L. Reinherz, Proc. Natl. Acad. Sci. USA 83:2614.
7. G. L'Alleimain, A. Franchi, E. Cragoe Jr. and J. Pouyssegur, J. Biol. Chem. 259:4313 (1984).
8. A. Corcia, R. Schweitzer-Stenner, I. Pecht and B. Rivnay, EMBO J. 5:849 (1986).
9. T. Ishizaka and K. Ishizaka, Prog. Allergy 34:188 (1984).
10. J.L. Mongar and H.O. Schild, J. Physiol. 140:272 (1958).
11. J.C. Foreman, M.B. Hallet and J.L. Mongar, J. Physiol. 271:193 (1977).
12. R.Y. Tsien, Biochemistry 19:2396 (1980).
13. M.A. Beaven, J. Rogers, J.P. Moore, T.R. Hesketh, G.A. Smith and J. Metcalfe, J. Biol. Chem. 259:7129 (1984).
14. R. Sagi-Eisenberg and I. Pecht, Immunol. Lett. 8:237 (1984).
15. R. Sagi-Eisenberg and I. Pecht, Nature 313:59 (1985).
16. E.L. Barsumian, C. Isersky, M.G. Petrino and R.P. Siraganian, Europ. J. Immunol. 11:317 (1981).
17. J.D. Taurog, C. Fewtrell and E. Becker, J. Immunol. 122:2150 (1979).
18. R.Y. Tsien, T. Pozzan and T.J. Rink, J. Cell. Biol. 94:325 (1982).
19. T.J. Rink, R.Y. Tsien and T.J. Pozzan, J. Cell. Biol. 95:189 (1982).
20. D.M. Segal, J.D. Taurog and H. Metzger, Proc. Natl. Acad. Sci. USA 74:2993 (1977).
21. A.K. Menon, D. Holowka and B. Baird, J. Cell. Biol. 98:577 (1984).
22. R. Schweitzer-Stenner, E. Luescher, A. Licht and I. Pecht, (1986), submitted.
23. Y. Nishizuka, Science 233:305 (1986).
24. Y. Katakami, K. Kaibuchi, M. Sawamura, Y. Takai and Y. Nishizuka, Biochem. Biophys. Res. Commun. 121:573 (1984).
25. H. Murer, U. Hopfer and R. Kinne, Biochem. J. 154:597 (1976).
26. P. Cala, J. Gen. Physiol. 76:683 (1980).
27. S. Paris and J. Pouyssigur, J. Biol. Chem. 258:3503 (1983).

28. A. Roos and W.F. Boron, Physiol. Rev. 61:296 (1981).
29. S. Grinstein, S. Cohen, J.D. Goetz, A. Rothstein and E.A. Gelfand, Proc. Natl. Acad. Sci. USA 82:1429 (1985).
30. D.J. Benos, Am. J. Physiol. 242:C131 (1982).
31. S. Grinstein, S. Cohen and A. Rothstein, J. Gen. Physiol. 83:341 (1984).
32. L.C. Cantley, Curr. Top. Bioenerg. 11:201 (1980).

NICOTINIC ACETYLCHOLINE RECEPTOR AND MYASTHENIA GRAVIS

STUDIED BY ANTI-RECEPTOR MONOCLONAL ANTIBODIES

Socrates J. Tzartos

Dep. Biochem.
Hellenic Pasteur Institute
127, Vas. Sofias Ave.
Athens 115 21, Greece

INTRODUCTION

Antibodies to the nicotinic acetylcholine receptor (AChR) cause the disease myasthenia gravis (MG). Both the molecule and the disease have been extensively studied and serve as models for similar systems. Monoclonal antibodies (mAbs) to the AChR have been used for several years as probes for the study of the AChR and the MG. In the introduction, some general information will be presented on the AChR, the disease and the mAbs.

1. Acetylcholine receptor (AChR)

The AChR mediates neurotransmission at the neuromuscular junctions and at the fish electric organs. This molecule in addition to being the best studied neurotransmitter receptor, it is also a model system for other transmitter and hormone receptors (reviewed in ref. 1-4). The AChR is available in large quantities from the electric organs of the marine elasmatobranch Torpedo (about 1 nmole/g wet weight) and from the eel Electrophorus electricus. Mammalian muscles however contain about 1,000 times less AChR than Torpedo electric organs thereby making biochemical studies with mammalian AChR rather difficult. The AChR is an integral protein of the postsynaptic membrane. Detergent-solubilized Torpedo AChR monomers have a molecular weight of about 290,000[3] and contain at least two acetylcholine binding sites[1-3]. SDS-denaturation of the molecule yields four different subunits with apparent molecular weights of about 40K (α) 48K (β) 58K (γ) and 64K (δ) in a molar ratio $\alpha_2\beta\gamma\delta$[1-3]. All four subunits are glycosylated. The DNA for all four AChR subunits has been cloned and their complete amino acid sequence has been determined[3,5,6].

Acetylcholine released from the nerve terminal (upon the arrival of the nerve impulse) binds on the AChR and causes opening of the cation channel which allows Na^+ to enter the muscle cell. The cation channel is an integral part of the AChR as has been clearly shown by two kinds of studies. First, preparations of affinity-purified Torpedo AChR comprising of only the four kinds of subunits, when reconstituted into artificial membranes showed agonist-dependent cation flux[2,3]. Second, more recently, injection of mRNA derived from the cDNA encoding the four kinds of AChR subunits into Xenopus oocytes directed the synthesis of a functional AChR molecule[7]. All four kinds of subunits are required for the assembly of a fully functional AChR[7] and, interestingly, some interspecies hybrid molecules (comprizing of a mixture of Torpedo and calf or mouse AChR subunits) are also at least equally functional[8,9].

A large part of the biochemical studies on AChR have been made possible because some snake toxins (e.g. α-bungarotoxin) bind very specifically near the acetylcholine binding sites which are located on the α-subunits[1-3]. These toxins have been used as tools to localize, quantitate and purify the AChR and to inhibit its function. In addition, local anesthetics block the response to agonists in a non-competitive way[3]. However, detailed studies of a large molecule like the AChR cannot be accomplished by using probes directed towards only two classes of sites. Many more probes for the various sites of the AChR molecule were necessary and such are now the mAbs.

2. Myasthenia gravis (MG)

The modern era in MG research started in 1973 with two complementary reports. Patrick and Lindstrom[10] reported that rabbits injected with purified AChR from electric organs of the Electrophorus electricus produced anti-AChR antibodies and showed severe symptoms of muscular weakness, similar to those of human MG. The experimental disease which is caused by injecting animals with AChR (or directly with anti-AChR antibodies) is called experimental autoimmune myasthenia gravis (EAMG). The same year Fambrough et al[11] studied the numbers of toxin binding sites in muscles of healthy and myasthenic humans and reported that MG patients had only 11-30% as much AChR as did healthy controls.

Today MG is thought to be mainly due to a loss of AChR (caused by antibodies) which impairs neuromuscular transmission and results in muscular weakness and fatigability (reviewed in ref. 12-14). Antibody-mediated AChR loss is caused by at least two mechanisms: (a) bivalent and polyvalent antibodies crosslink membrane-bound AChRs resulting in an increase of the AChR internalization and degradation rate (antigenic modulation)[15,16] and (b) complement binds on those antibodies which are bound on the membrane AChRs and mediates lysis of the AChR-containing membranes[13]. Direct blockage of AChR by antibodies to the acetylcholine binding site might be an additional cause of impairement of the neuro-muscular transmission[17].

Antibodies to human muscle AChR are detected in about 85-90% of MG patients' sera whereas they are essentially non-existent in healthy humans[12,14]. Relative decrease in anti-AChR antibody titer in a particular patient's sera correlates well with clinical improvement. However, studies of populations of MG patients show only a weak correlation between absolute anti-AChR concentration and severity of the disease[18]. Whether or not only certain antibody specificities cause impairment of neuromuscular transmission is yet questionable. Synthesis of anti-AChR antibodies is in part regulated by AChR-specific helper T-lymphocytes which have recently been isolated from several MG patients[19]. It is not yet known what triggers the autoimmune responce in MG[12-14]. Obviously, homogeneous antibodies directed to specific sites of the AChR are an indispensable group of probes for the disease.

3. Monoclonal antibodies (mAbs)

Monoclonal antibodies are an excellent tool for studying many biological systems[20,21]. mAbs can be obtained against various sites of a macromolecule, thus permitting a detailed study of its entire surface. They are homogeneous antibody preparations each one usually directed against a single site (epitope) on the corresponding antigen. Two of their most attractive features are that they can be' obtained in unlimited quantities at any time and that specific mAbs can be produced against non-purified and even unknown antigens[20,21].

Contrary to the polyclonal sera which are produced in vivo, mAbs are manufactured in the laboratory by culturing antibody-producing cells immortalized by fusion with myeloma cells. Injection of an immunogen (e.g. a macromolecule) into an animal stimulates the production of a large number of lymphocytic clones which secrete heterogeneous antibodies against the immunogen. However, since each lymphocyte and its progeny produce a single antibody species, culture of isolated lymphocytes from immunized animals would result in the production of homogeneous antibody preparations i.e. mAbs. Because lymphocytes do not normaly survive for a long period in culture, Kohler and Milstein made them immortal by fusing them with immortal myeloma cells[20]. The principle for the production of mAbs is the following: Spleen cells from an immunized animal are fused with a myeloma cell line, usually by the addition of polyethylene glycol. The cells are then dispensed into a large number of wells and are cultured in selective medium (hypoxanthine + aminopterine + thymidine, HAT) which kills the unfused myeloma cells while unfused spleen cells die spontaneously. Thus, only the hybrid cells survive. Subsequently, the small fraction of the specific antibody-producing hybrids have to be identified, isolated and cultured to produce large amounts of mAbs.

We devised a modification of the classical fusion technique which has facilitated and accelerated the production of a large number of mAbs[22,23]. The main modification of this technique, which is called "direct cloning", is that the polyethylene glycol-treated spleen-myeloma hybrids are, after the necessary washings, mixed immediately with agar. They are then dispensed into 24-well plates and when the agar is solidified it is overlayed with culture medium. The overlaying medium allows for subsequent screening assays in order to find the antibody-containing wells whereas the agar keeps the various clones separate into distinct colonies. Each colony of the antibody-positive wells is then individually transfered by pasteur pipets into new wells. From each original positive well usually only one of the transfered colonies is found to be positive. This technique saves considerable time and effort[23].

Receptors, which are usually present in very small quantities, have largely profited from the mAb technology. Examples are the receptors for adrenaline, insulin, EGF, TSH, the T cell antigen receptor (all reviewed in ref 24, 25) and the AChR. Several laboratories have produced mAbs to the AChR from various species which are also successfully used by several other laboratories for the study of the molecule and the disease MG. By using the "direct cloning" technique we have produced about 150 mAbs to AChRs from fish electric organs and mammalian muscles. These mAbs have been well characterized and have been proved excellent tools. In the following chapters will be presented some information obtained mainly by our and also by other groups, using the anti-AChR mAbs. First, some of the techniques we used will be described .

CHARACTERIZATION OF THE ANTI-AChR mAbs   LOCALIZATION OF THE CORRESPONDING EPITOPES

In order to make the best use of a mAb an extensive characterization of its various parameters has first to be performed. This includes validation of its homogeneity, determination of its immunoglobulin class and subclass, affinity for the immunogen, crossreactivity with antigens related or not with the immunogen and localization of the corresponding epitope on the immunogen. Among 137 of our anti-AChR mAbs, all being from rat origin, 44 are IgG1, 55 IgG2a, 21 IgG2b, 1 IgG2c and 16 IgM[22,26-28]. The binding affinities for the immunogen of 40 mAbs varied from K_D 0.17 nM to 92 nM[22]. About 60 of our mAbs are absolutely species-specific, 18 crossreact with muscle and electric organ AChR from almost all species tested (usually with lower affinity), whereas the rest crossreact with AChRs from some only of the tested species[22,26-29]. Some of our mAbs also crossreact to a certain extent with apparently irrelevant proteins[30].

Perhaps the most difficult and the most valuable kind of mAb characterization is the localization of their binding sites (epitopes) on the antigen. We have used several different techniques by which a considerable progress has been achieved towards this goal. These techniques fall within two groups: The first group provides the means for localizing the binding of mAbs on the denatured subunits and their segments. The second group aims at localizing mAb binding on the three-dimensional structure of the intact AChR; this involves mainly competition experiments among mAbs for binding on the AChR. Since the detailed three-dimensional structure of the AChR is not yet known, both approaches are required. Undoubtedly each approach benefits from and supplements the other.

1. Determination of mAb Binding Sites by the Use of Isolated Subunits, their Fragments and Synthetic Peptides

First we determined the subunit specificity of the mAbs. The AChR pentamer was dissociated into its subunits with the denaturing detergent SDS. The subunits were then radiolabeled with ^{125}I, mixed and immunoprecipitated with the test mAbs and the immunoprecipitates were analysed by polyacrylamide gel electrophoresis (PAGE) followed by autoradiography to identify the specifically immunoprecipitated ^{125}I-labeled subunits. In this way we characterized all mAbs derived from animals immunized with denatured AChR subunits and more than half of the mAbs derived from rats immunized with intact AChR. Several mAbs of the latter group do not bind detectably to denatured subunits[22,26-28]. Several mAbs crossreact with more than one subunits due to the similarities which exist among the subunits. However, even in this case their binding to a certain subunit is usually of much higher affinity than to other subunits thus permiting the determination of their actual specificity to a single subunit[22,26,33]. Use of immunoblots gave similar results[31] although occasionally immunoblots were less specific.

Since the AChR subunits are large polypeptides with 437-501 aminoacids each[5], a more precise localization of the binding sites for the mAbs had subsequently to be performed. Thus the ^{125}I-labeled subunits were proteolysed and binding of the mAbs to the ^{125}I-labeled peptides was determined by immunoprecipitation as with the intact subunits[32,33]. With this technique the mAbs were divided into 28 different groups. Recently Ratnam et al[34] by the use of an interesting peptide mapping approach, mapped the approximate location of several mAbs on the amino acid sequence of the α, β and δ subunits. Using antibodies specific for the C-terminus of these subunits they could identify all the proteolyticaly derived overlapping peptides of each subunit containing the C-terminal end. The segment of a subunit in which a mAb binding site is located was determined from the difference between the size of the smallest C-terminal peptide that bound this mAb and of the largest C-terminal peptide that did not bind it.

When the amino acid sequence of the AChR subunits was revealed by cDNA techniques[5] a powerful approach for the almost ultimate characterization of the AChR epitopes became available, i.e. the use of synthetic peptides[35,36]. Small synthetic peptides, mimicking parts of the subunits, with amino acids as few as 8-10 are sufficient to bind to anti-AChR mAbs[36]. Actually, in solid phase systems even tri- and di-peptides have been reported to having specificity for mAbs to certain antigens[37]. With the use of synthetic peptides Lindstrom and collaborators determined almost the exact sites of the epitopes for 14 anti-α-subunit mAbs[36].

Finally, expression into E. coli of large quantities of recombinant fusion proteins, containing segments of the AChR α-subunit, allowed us to localize an important region on the α-subunit, namely the main immunogenic region (MIR)[38].

2. Determination of mAb Binding Sites by the Use of Intact AChR

All the above techniques have been proved valuable for determining the anti-AChR mAb binding sites. Especialy by the use of synthetic peptides it is expected to obtain a thorough mapping of the mAbs which bind to the denatured subunits. However, this kind of mapping will not reveal the location of the epitopes on the intact AChR molecule. Furthermore, several mAbs to the intact AChR do not bind to the denatured subunits or their peptides. For these reasons, in parallel with the previous techniques, we have devised techniques involving competition among mAbs (or mAbs and antisera) for binding to the intact AChR. Although various techniques are used, all are based on the fact that when an antibody binds to the antigen it covers and so protects the region surrounding the specific epitope against binding of other antibody molecules. The arm of an antibody molecule bound on its antigen covers about 10 nm², i.e. an area much larger than its epitope which is composed of approximately 5-6 amino acids (~2 nm²). Thus two antibodies with distinct but neighbouring epitopes may exclude completely each other from binding to the antigen. However, this apparent disadvantage is at the same time profitable since only by effective competitions between antibodies to distinct but neighboring sites their relative position can be determined. For example, if mAb A does not compete with mAb C, but B competes with both A and C we can conclude that mAbs A, B and C, bind to distinct epitopes and that the epitope for B is between those for A and C.

In recent experiments, while mapping the binding of mAbs to the cytoplasmic side of the AChR, we were able to determine the sensitivity of the antibody-competition techniques[39,40]. Based on the knowledge of the exact sites of the epitopes for some of these mAbs on the amino acid sequence of the AChR subunits[34,36] we showed that the antibody competition technique is indeed very sensitive since no significant competition was observed between mAbs to epitopes separated by only 4-7 amino acids. Furthermore we could even differentiate between mAbs to partially overlapping epitopes since although they completely competed each-other they exhibited different degrees of competition with other mAbs[39,40].

The various antibody-competition techniques which we have used include:
a. Immobilization of the AChR on Sepharose beads[22,26] or on plastic plates[40]; then protection of a region of the AChR by an unlabeled protecting mAb and addition of an unlabeled[22,26] or ^{125}I-labeled test mAb[40]; binding of the unlabeled test mAb was determined by testing for their presence (or absence) in the supernatants (by an additional radio-immunoassay), whereas binding of the labeled test mAbs was determined directly by measuring the radioactivity bound to the insoluble matrix. Using this technique we mapped the extracellular side of the AChR and identified the MIR on its α-subunits[22,26].
b. Preincubation of the soluble protecting mAb with ^{125}I-toxin-labeled AChR followed by incubation with the immobilized on Sepharose beads test

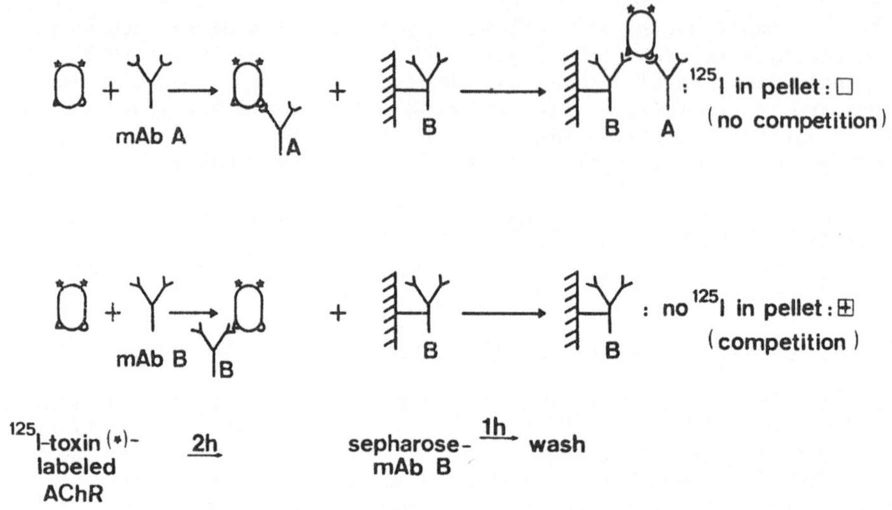

Fig. 1. Competition between mAbs for binding to the intact toxin-labeled AChR. ^{125}I-toxin-labeled AChR was preincubated with a saturating amount of the soluble mAb and then the mixture was incubated with a Sepharose-bound mAb to determine whether or not the "protected" AChR could bind on it. This technique was used to map mAbs binding to the cytoplasmic side of the AChR[39,40].

mAb. Thus competition between two mAbs resulted in low radioactivity bound on the Sepharose whereas absence of competition resulted in highly labeled Sepharose beads. The outline of this technique is shown in Fig. 1. Using this technique we mapped parts of the cytoplasmic side of the AChR and we found that the sequences α339-378 and β336-469 are extended over large distances on the surface of the molecule[39,40].

c. In competition experiments between the rat mAbs and human myasthenic antibodies all three reagents were in solution (^{125}I-toxin labeled AChR, protecting mAb and test myasthenic serum). Then the human antibodies were selectively immunoprecipitated with anti-human gamma-globulin, coprecipitating any bound to them labeled AChR[41]. Using this technique we mapped the antigenic specificities of the antibodies from myasthenic patients and found that the majority of them are directed against the MIR[41]. Similar anti-AChR mAb competition techniques have been also used by other groups[42,43,62].

PROGRESS ON AChR RESEARCH BY THE USE OF mAbs

Among the first uses of the anti-AChR mAbs was the demonstration that the Torpedo AChR subunits present similarities and we suggested that they may be derived from a common ancestral gene, since single mAbs crossreacted with more than one subunit[26]. Subsequently the AChR subunit similarities were confirmed by amino acid and cDNA sequencing studies[5,44]. The extent of similarities among AChRs and their subunits from various species was also investigated by determining the crossreactivities of the various anti-subunit mAbs with AChR from electric organs and muscles from different species[22,26-29]. These also were later confirmed by amino acid and cDNA sequencing studies (reviewed in ref. 4).

Some anti-AChR mAbs recognize certain presumably unrelated proteins including uncharacterized bacterial[45] and protozoan proteins[30] or certain specific proteins e.g. phosvitin[33,46], actin and tubulin[47]. Whether these crossreactions reflect amino acid sequence and/or conformational similarities with functional or evolutionary significance, or simply reflect non-significant accidental small similarities remains to be seen.

Renaturation of certain sites of the SDS-denatured cytoplasmic AChR α-subunit has been monitored and enhanced by anti-AChR mAbs. Two distinct conditions were found (low SDS- or high lipid-dependent) under which a high affinity for α-bungarotoxin was recovered (K_D ~3 and 0.5 nM respectively)[48,49]. Many mAbs which bind mainly to the intact AChR and very weakly to the denatured subunits bound very significantly to the renatured α-subunit. Binding of some mAbs to the renatured subunit further improved its affinity for α-bungarotoxin[49].

Interesting information on the conformation of the AChR has also been obtained by the mAbs. Anti-AChR mAbs have been used to modify existing models and to propose new models on AChR subunit conformation. From the hydrophobicity profile of the amino acid sequence of the α-subunit it was proposed that each subunit spans four times the membrane[3,5,6] through four hydrophobic α-helices while a fifth transmembrane domain was proposed to be formed by an amphipathic α-helix which might contribute to the assembly of the ion channel[4] (Fig. 2). By the use of mAbs, specific amino acid sequences were localized on the extracellular and cytoplasmic sides of the AChR. By this kind of information the above model was first supported[4,50] then supplemented[51] and subsequently modified[36]. Thus Lindstrom and collaborators showed that mAbs to sites which were earlier proposed to be extracellular (α152-159) actually bound to the cytoplasmic surface of the AChR a fact which prompted them to suggest two more transmembrane domains for the α-subunit[51] (model C in Fig. 2). They also showed that mAbs to the regions thought to form the amphipathic α-helix (the supposed part of the channel) actually bind on the cytoplasmic surface of the AChR and they suggested that this region, as well as another one that was also previously proposed to be transembranous, are actually located on the cytoplasmic side of the AChR[36].

By competition experiments among mAbs directed against the cytoplasmic side of the AChR we confirmed that at least part of the amphipathic segment is indeed not transmembranous and we showed that it must be extended over a large distance on the surface of the cytoplasmic side of the AChR probably not forming large loops[39,40]. With the same kind of experiments we also found some putative points of vicinity between γ and δ subunits as inferred by effective competition between mAbs to different subunits.

Efforts have been made to study AChR function with the mAbs. We expected to find mAbs to specific subunits and sites (binding away from the acetylcholine binding site) to interfere with the function of the channel. However, the great majority of our anti-AChR mAbs did not seem to have any such effect. Four mAbs were found to inhibit carbamylcholine-induced channel opening efficiently. This was tested on Torpedo AChR reconstituted into artificial lipid vesicles, by measuring $^{22}Na^+$ flux into the vesicles[52,53]. Unexpectedly, all four mAbs seem to bind to the cytoplasmic surface of the AChR. The two mAbs are anti-β, one is anti-α, and one is anti-γ. When Fab fragments of two of them were tested, no blockage of the channels was observed which suggests that at least these two mAbs do not exert their effect by binding to sites of special significance for the channel but probably they distort AChR conformation by crosslinking. Therefore, contrary to the bulk of information obtained by the mAbs on AChR structure, their contribution to research on AChR function has not reached the expected level.

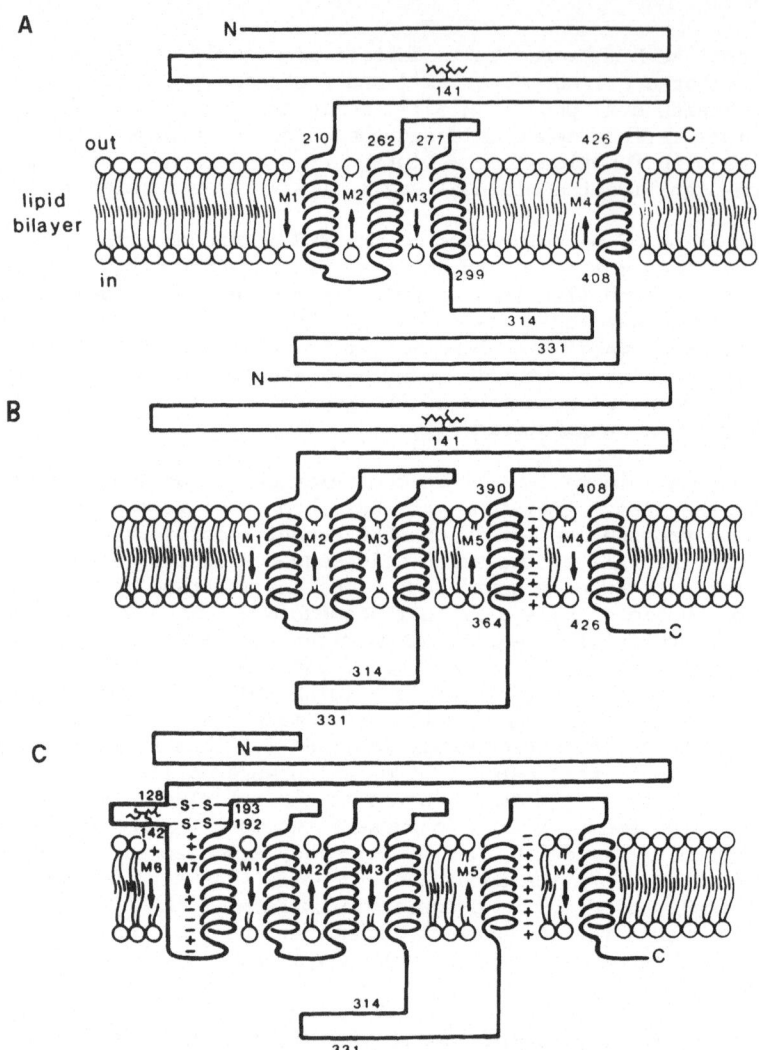

Fig. 2. Three alternative models of the transmembrane organization of the AChR α-subunit. A. The four hydrophobic transmembranous domain (M1-M4) model[3,5,6]. B. The five transmembrane domain model[4]; this introduces the amphipathic transmembrane domain (M5) believed to be part of the ion channel(4). C. The seven transmembrane domain model resulted from immunochemical experiments by Lindstrom and collaborators[51]. Later the same group proposed that M4 and M5 are not transmembranous but rather cytoplasmic[36]. (From Criado et al[51]).

Anti-AChR mAbs have also been proved valuable in neuronal nicotinic AChR research. These AChRs seem to fall within two main groups, characterized by their ability to bind[54] or not[55] α-bungarotoxin. Biochemical studies of the latter group were slow to develop because of the absence of appropriate probes. Fortunately some mAbs to the AChR from fish electric organs and mammalian muscles bind to both kinds of neuronal AChRs. This is especially true for mAbs to the MIR suggesting that this region is well preserved between neuronal and muscle AChR. Such mAbs have been used to localise, purify and study chick (56,57) and fish brain AChR[58].

Other successfull applications of the anti-AChR mAbs which will not be described in this review include their use for the purification of electric organ and muscle AChR[59], in the analysis of acetylcholine and toxin binding sites[43,60-62], in the study of AChR[64-66], in the selection of specific cDNA clones[67,68], in the study of AChR derived from the cDNA technology[7,68] and for the elucidation of the T-B cell interactions involved in the immune responce to AChR[69-71].

PROGRESS ON MYASTHENIA GRAVIS BY the USE OF mAbs

1. The Main Immunogenic Region (MIR)

The first valuable piece of information obtained by the mAb competition experiments was the discovery of the MIR on the α-subunit of the Torpedo AChR[26]. We subsequently found that there is a homologous MIR on all AChRs tested as far as rat and human anti-AChR antibodies are concerned[22,23,27,29,41]. About two-thirds of the anti-AChR antibodies from immunized rats and from human myasthenic patients are directed to this region.

Localization of the MIR has been the subject of considerable investigation. It is mainly located on the α-subunit since many mAbs to this region have been found to bind detectably to the denatured α-subunit. The MIR is located on the extracellular surface of the AChR as has been clearly shown by the localization of anti-MIR mAb binding sites using immunoelectron microscopy[29,36] and biochemical techniques[29], as well as by the capacity of these mAbs to cause AChR loss on cell cultures[73-75] and on experimental rats[26,72]. The MIR is located away from the acetylcholine and α-bungarotoxin binding sites since mAbs against it do not compete with α-bungarotoxin for binding to the AChR[22,26,27,52,53]. It is apparently a relatively small region since the mAb-competition techniques by which it was determined[22,26] define relatively small regions[39,40]; moreover all anti-MIR mAbs which were tested for binding to proteolytic peptides of the Torpedo α-subunit bound to the same group of overlapping peptides[33]. Recently Barkas et al[38,68] and Ratnam et al[34] localized the MIR near the N-terminal end of the α-subunit; with the use of recombinant bacterial fusion proteins containing segments of the α-subunit we showed that 6 anti-MIR mAbs bind to a fusion protein containing the amino acid sequence α 6-85; at least three of them bind to the α 37-85 part[38,68]. Lindstrom and collaborators[34] using proteolytic peptides showed that another anti-MIR mAb binds to the sequence α46-120. Thus the 40 amino acid long segment extending between α46-85 seems to essentially contain the MIR. The MIR does not seem to be glycosylated since the specific proteolytic peptides to which the anti-MIR mAbs bound did not contain any sugar moieties[32]. Moreover, the only residue of the α-subunit with a potential to bind carbohydrate, namely Asp-141, is located away from the MIR-containing part of the α-subunit. Further studies with synthetic peptides are needed for the localization of the exact site of the MIR.

The MIR probably consists of more than one epitopes. This is inferred from several pieces of information. Thus the binding characteristics of the anti-MIR mAbs are heterogeneous in terms of species cross-reactivity and of their mode of binding to the AChR (some crosslink two AChRs, others do not)[22,26,27,29]. Idiotypic analysis[76] as well as their analysis by isoelectric focusing[77] gave also patterns of heterogeneity. Further, as was mentioned above, three of the six tested anti-MIR mAbs could not bind

to the amino acids α37-85 of the fusion proteins[38,68].   Finally, the
different anti-MIR mAbs used as protectors of the human AChR against human
MG serum antibodies caused similar but clearly not identical degrees of
inhibition of binding of human antibodies to the AChR[41].

The MIR is a phylogenetically conserved region because anti-MIR mAbs
are in general much more crossreactive among species than mAbs to other
regions and because the α-subunit segment containing the amino acids 37-120
is very well conserved (there is only one amino acid difference between
human and calf AChR on this part of the molecule). The fact that the MIR ,
although highly immunogenic, is a phylogenetically conserved region, may
mean that it protrudes from the remaining molecule and it may form a mobile
segment of the AChR such as to be more accessible to the immunocompetent
cells[78].   Use of the mAbs in sophisticated electron microscopy studies[4,79] is expected to determine the exact location of the MIR on the
intact AChR.

In terms of antibody binding capacity and immunogenicity the MIR is
very sensitive to denaturation by SDS[22,26,48,49] but very resistant to
proteolysis[80] (and Kordossi and Tzartos, unpublished). Another important
extracellular site, the α-bungarotoxin binding site, is also similarly
sensitive to SDS-denaturation and resistant to proteolysis[48,80]. On the
contrary, many epitopes on the cytoplasmic side are resistant to SDS-
denaturation and very sensitive to proteolysis[22,26,28,80]. This observation may reflect a more generalized difference between the two sides of
the AChR.

The pathogenic role of the anti-MIR antibodies has been studied both
in vivo, by the use of experimental animals, and in vitro, by the use of
cell cultures. We have concluded, as it is analysed in the following
chapters, that the anti-MIR antibodies play a very important role in MG.

2.   Passive Transfer of Experimental Autoimmune Myasthenia Gravis (EAMG) in Rats by Anti-AChR mAbs

Several groups have succeded in rendering experimental animals
myasthenic by injecting anti-AChR mAbs[26,72,81,82]. mAb preparations in
quantities larger than the total AChR content of the used animals caused
muscular weakness and loss of roughly half of the muscle AChR within 1-2
days. In addition to the EAMG caused by mAbs directed to unspecified sites
on the AChR, EAMG was also achieved by mAbs to two defined regions, namely
the toxin binding site and the MIR. Gomez and Richman injected mAbs
directed to the α-bungarotoxin binding site, to chicken and obtained
muscular weakness within only an hour from the injection. The AChR content
of the chicken was not affected and apparently muscular weakness was
induced by blocking AChR function[17].

We have injected thirteen different mAbs to experimental rats. Five of
them were anti-MIR, one bound to the extracellular side of the α-subunit,
four to the cytoplasmic side of various subunits and three to AChRs from
species other than rats[26,72]. All five anti-MIR mAbs (three IgG1 and two
IgG2a) caused, within two days, severe symptoms of EAMG in the rats and
reduced their AChR content to 47-60%. None of the other mAbs caused any
symptoms of muscular weakness nor did they reduce the AChR content of the
animals although the mAb to the extracellular side of the α-subunit bound
to the majority of the AChR molecules from the sacrificed rats[72]. These
results suggest that the anti-MIR antibodies are very potent in passively
transfering EAMG independently of their IgG subclass specificity. Since
only a single mAb directed to non-MIR sites of the extracellular surface of
the rat AChR could be tested we cannot draw conclusions on the effect of
mAbs to other than the MIR sites.

3. Use of AChR-bearing Cell Cultures

Accelerated internalization and degradation rate of AChR by antibody-crosslinking (antigenic modulation) is considered to be one of the two main mechanisms of AChR loss in MG[12,16]. Antigenic modulation of AChR can be analysed in tissue culture; this has been used to study the modulating potency of animal and human antibodies on AChR of animal tissue cultures[15,16]. Such antibodies were shown to cause a two- to three-fold enhancement of AChR internalization and degradation rate. mAbs to the AChR also caused antigenic modulation of the AChR in animal muscle cell cultures[73,74,83]. We have used both animal and human muscle cell cultures to study the effect of anti-AChR mAbs and also the effect of groups of antibodies from human patients directed to certain regions of the AChR, especially those directed to the MIR[73,75,63].

The technique which is usually used to study antigenic modulation of AChR in animal muscle cell cultures is in principle the following: The AChR-bearing cells are incubated with ^{125}I-α-bungarotoxin which labels practically irreversibly the AChR molecules present on the surface of the cells. After eliminating the excess toxin the test antibodies are added and incubated for several hours at 37°C. Internalized AChR (normaly or induced by antibody-crosslinking) carries with it the bound labeled toxin, thus both are degraded. Eventually, free ^{125}I-tyrosine is released in the medium which is proportional to the amount of the internalized toxin-labeled AChRs[15]. Measurement at time intervals of the ^{125}I content of aliquots from the culture media gives an indication of the AChR internalization rate.

We have used the above technique to study the effect of anti-AChR mAbs on the internalization rate of the AChR from calf[73] and mouse[74] muscle cell cultures. Anti-MIR mAbs were very efficient in increasing internalization rate up to threefold. Univalent Fab fragments of the mAbs, as expected, lost their capacity to accelerate AChR internalization apparently because they do not crosslink AChRs[12]. When these fragments were subsequently crosslinked with anti-rat gamma-globulin they exerted again their activity. The anti-β-subunit mAb 73 which was previously found unable to passively transfer EAMG could only slightly, if at all, accelerate AChR internalization rate in the mouse cell cultures. It became however very efficient when it was crosslinked with anti-rat gamma-globulin. We can therefore conclude that its inefficiency when used alone was due to its low, or inexistent, AChR crosslinking capacity. In any case the anti-MIR mAbs were shown also in this case to be especially potent.

Although we clearly showed that anti-MIR antibodies can cause EAMG and can accelerate internalization of AChR in cell cultures, the relative contribution to the disease of the anti-MIR and the antibodies to other AChR regions in human MG sera was unknown. Because satisfactory isolation of the antibody groups with specificity for the MIR or for other regions of the intact human AChR is yet very difficult, we used an indirect way with the help of the mAbs: The MIR of the AChR on cultured mouse muscle cells was "protected" by excess Fab fragments from an anti-MIR mAb (mAb 35) which do not induce AChR degradation. Subsequently the test MG serum was added. Its effect was thus due only to the antibodies which bind away from the MIR. The capacity of the anti-MIR antibodies of the test serum to cause internalization and degradation of the AChR was estimated by subtracting the effect of the serum to the "mAb-protected" AChR from that to the non-protected AChR. Similarly, the effect of the MG antibodies to the region corresponding to the anti-β-subunit mAb 73 was also estimated; here the intact mAb was used as protector since this mAb did not cause antigenic modulation. Fig. 3 shows an extreme example of an MG serum where almost

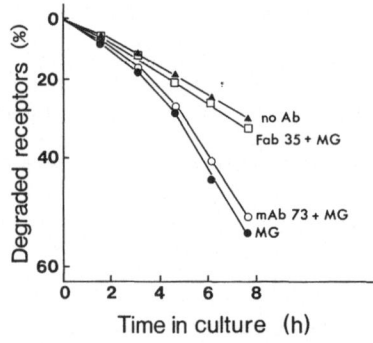

Fig. 3. mAb-mediated protection of mouse muscle cell surface AChR against
antigenic modulation (i.e. increased AChR internalization and
degradation rate) by a serum from a selected myasthenic patient
(MG). ^{125}I-toxin-labeled surface AChR was shielded, or not, with
the protecting mAb. Then the test MG serum was added and incubated
at 37°C. The amount of internalized and degraded toxin-AChR
complexes was estimated by the amount of released radioactivity in
the medium. Protection by the Fab fragments of the anti-MIR mAb 35
is nearly complete (From Tzartos et al[74]).

all its effect was due to its anti-MIR antibody fraction. Fig. 4 shows the
application of the described technique to the study of 27 MG sera. It shows
that 68+16% of the antigenic modulation capacity of the sera is due to
their anti-MIR antibody fraction in the same sera. This fraction formed the
69% of the total antibodies in the MG sera. In conclusion, as far as it
concerns the one of the pathogenic mechanisms in MG, i.e. the antigenic
modulation, the MIR seems to be the main effective region[74].

A major disadvantage of all the studies on antigenic modulation of
AChR by human sera performed until recently, including the above described,
is that they have used animal instead of human cell cultures. The MG
serum titers for rat and mouse AChR are only ~2-20% of those for human
AChR[84]. Moreover, this crossreactivity is not representative because
certain AChR regions tend to induce more crossreactive antibodies than
others[41]. This could give rise to misleading conclusions. We have
recently overcome these disadvantages by using human embryonic myotube
cultures[75,63]. These cells produce a lower amount of surface AChR whose
binding affinity for α-bungarotoxin is significantly lower than to the
animal cell AChR, resulting in fast dissociation of the bound labeled
α-bungarotoxin. Furthermore some MG sera replaced bound ^{125}I-α-burgaro-
toxin even in the presence of excess ^{125}I-α-bungarotoxin. Because of its
low affinity for toxin, AChR antigenic modulation could not be measured

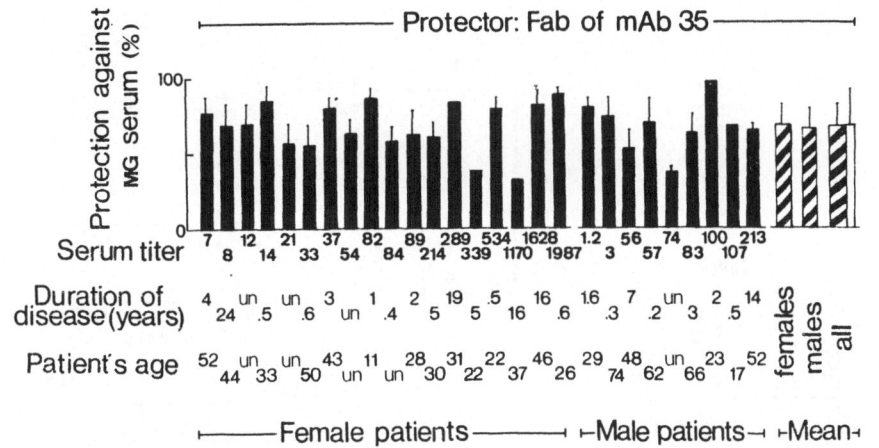

Fig. 4. mAb-mediated protection of mouse muscle cell surface AChR against
antigenic modulation by various sera. Toxin-labeled surface AChR
was shielded, or not, with the protecting Fab fragments of the
anti-MIR mAb 35. Then the test MG sera were added and incubated
for 7 h at 37°C. Large bars mean that protection was very
efficient. The empty bar at the right end shows the mean inhibition, by the Fab, of binding of these sera to the detergent-
solubilized human muscle AChR. (Modified from Tzartos et al[74]).

with the previously described system which involves the use of prelabeled
AChR. Instead, the myotubes were first incubated with the antibodies at
37°C or 15°C for several hours, followed by the addition of [125]I-
α-bungarotoxin and incubation for another hour to estimate the remaining
toxin binding sites. At 15°C no internalization of AChR occurs[15] thus, a
decrease in bound radioactivity was interpreted as blockage of the
existing toxin-binding sites by the antibodies. This effect was taken into
account for estimation of the actual amount of the remaining surface AChR
at 37°C. The remaining AChRs were the result of the continuous synthesis
of new AChRs and the increased internalization rate of all the AChRs. Fig.
5 shows the outline of this technique.

We have used the human myotube cultures with the above technique in
two kinds of experiments. First we studied the effect of human MG sera
from patients in various states of the disease and with different anti-AChR
antibody titers. Because there is not good correlation between disease
severity and patients' anti-AChR antibody titers, we aimed to investigate
whether antigenic modulation correlates better with disease severity or
with antibody titer. Study of 29 MG sera showed clearly that their antigenic modulation capacity depended mainly on the final concentration of the
antibody present, independently of the severity of the patients' state of
the disease[63].

We then studied the effect of several mAbs on antigenic modulation of
the human AChR[75]. More mAbs could be tested on these cells, including some
to extracellular sites other than the MIR, because we have available more
such mAbs binding to the human than to the mouse AChR. We observed that
when synthesis of new AChRs was blocked by cycloheximide the system became

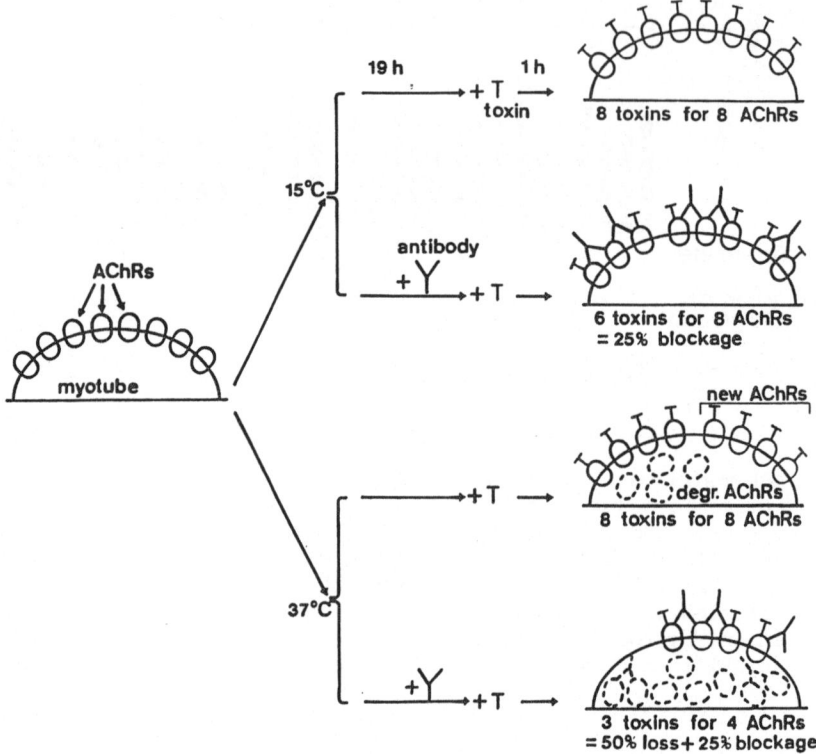

Fig. 5. Determination of antigenic modulation of AChR on human myotube
cultures by mAbs. Embryonic human myotubes were first incubated at
15°C or 37°C for several hours with the test antibodies. Then
^{125}I toxin was added to measure the remaining free surface AChRs.
Decrease of toxin binding at 15°C was due to blockage of existing
sites, whereas decrease of binding at 37°C was due to both blockage
and accelerated internalization of AChRs. Actual remaining surface
AChR due to antibody effect was estimated from the ratio: %
detected toxin sites at 37°C over % detected sites at 15°C.

more sensitive, possibly because newly inserted AChRs into the membrane may
be more resistant to internalization. The effect of some mAbs to various
AChR sites and subunits is shown in Fig. 6. Binding of the
anti-MIR mAb 35 to myotubes decreased surface AChR content down to less
than one-third of the control. However, at high concentrations most of its
effect disappeared apparently because at high antibody excess it did not
crosslink AChRs anymore, but formed only complexes comprizing of 2 mAbs
with one AChR molecule. Two mAbs against subunits other than the α (an
anti-β and an anti-γ) also induced AChR loss in this system (Fig. 6). This
is especially interesting because it was not known whether the formation of
large antibody-AChR complexes is required, or simple crosslinking of two
AChR molecules by one bivalent antibody is sufficient to induce AChR internalization. Addition of single IgG mAbs against unique epitopes on the β -
or γ-subunit should be unable to form complexes larger than two AChRs to
one antibody molecule. In fact we have seen that the anti-γ mAb 66 forms
only such small complexes with the AChR in solution[28]. Therefore the
results in Fig. 6 suggest that small antibody-antigen complexes are
sufficient to cause AChR antigenic modulation.

In conclusion the anti-AChR mAbs have already been proved valuable
tools for the progress in understanding AChR and MG. The new powerful

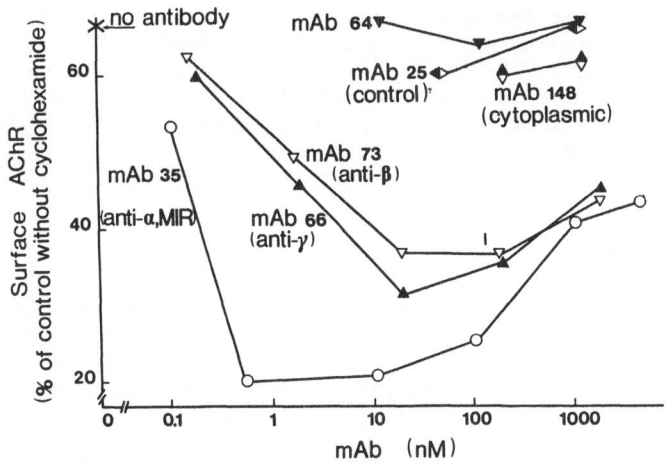

Fig. 6. Effect of mAbs on the internalization rate of AChR on human muscle
cell cultures. Synthesis of new AChRs was blocked with 20 μg/ml
cycloheximide. Incubation with the mAbs was for 7 h at 37°C. At
15°C these mAbs exhibited ≤10% inhibition of toxin binding (not
shown). (From Tzartos and Starzinski-Powitz[75]).

techniques recently introduced in AChR research, especialy the DNA
recombinant technology, have not undermined the role of the mAbs but have
rather boosted their use, each technology having benefited from the other.

SUMMARY

The nicotinic acetylcholine receptor (AChR) is a 290,000 Dalton
protein formed by five transmembrane subunits ($\alpha_2\beta\gamma\delta$) which has been
extensively studied. Autoantibodies to the AChR cause the disease
myasthenia gravis (MG) in human patients and in experimental animals.
Anti-AChR monoclonal antibodies (mAbs) have been available since 1979. We
have, and are still performing a thorough characterization of our 150 mAbs,
especially by localizing the sites of the corresponding epitopes on the
AChR molecule. These mAbs, as well as anti-AChR mAbs of other investigators
have significantly contributed towards understanding AChR and MG. One of
their interesting contributions has been the discovery and characterization
of the main immunogenic region (MIR) of AChR to which are directed about
two-thirds of the anti-AChR antibodies from human patients and from rats
immunized with intact AChR. The MIR is located on the extracellular side

of the α-subunit, near its N-terminus. It is distinct from the acetylcholine and toxin binding sites and is not involved in the regulation of
the ion channel. Anti-MIR mAbs can passively transfer experimental MG into
rats and can cause AChR loss in muscle cell cultures. The anti-MIR antibody
fraction of human patients' sera is mainly responsible for causing AChR
loss in cell cultures.

ACKNOWLEDGEMENTS

I am indepted to my colleagues who participated in some of the studies
described here, including T. Barkas, A. Starzinski-Powitz, J. Lindstrom ,
A. Kordossi, D. Sophianos, K. Soteriadou, A. Efthimiadis, A. Kokla and A.
Tzinia. Supported by the Greek General Secretariat of Research and
Technoogy.

REFERENCES

1. B. Conti-Tronconi, and M. Raftery, The nicotinic cholinergic receptor,
    Ann. Rev. Biochem. 51:491 (1982).
2. A. Karlin, The anatomy of a receptor, Neurosci. Comm. 1:111 (1983).
3. J.-P. Changeux, A. Devillers-Thiery, and P. Chemouilli, Acetylcholine
    receptor: an allosteric protein, Science, 225:1335 (1984).
4. R.M. Stroud, and J. Finer-Moore, Acetylcholine receptor structure,
    function, and evolution, Ann. Rev. Cell Biol. 1:317 (1985).
5. M. Noda, et al, Structural homology of Torpedo Californica
    acetylcholine receptor subunits, Nature 302:528 (1983).
6. T. Claudio, M. Ballivet, J. Patrick, and S. Heinemann, Nucleotide and
    deduced amino acid sequences of Torpedo Californica acetylcholine
    receptor gamma-subunit, Proc. Natl. Acad. Sci. USA, 80:1111 (1983).
7. M. Mishina, et al, Expression of functional acetylcholine receptor from
    cloned cDNAs, Nature, 307:604 (1984).
8. T. Takai, et al, Primary structure of gamma-subunit precursor of
    calf-muscle acetylcholine receptor deduced from the cDNA sequense,
    Eur. J. Biochem. 143:109 (1984).
9. M. M. White, K.M. Mayne, H.A. Lester, and N. Davidson, Mouse-Torpedo
    hybrid acetylcholine receptors: Functional homology does not equal
    sequence homology, Proc. Natl. Acad. Sci. USA, 82:4852 (1985).
10. J. Patrick, and J. Lindstrom, Autoimmune responce to acetylcholine
    receptors, Science, 180:871 (1973).
11. D. M. Fambrough, D.B. Drachman, and S. Satiamurti, Neuromuscular
    junction in myasthenia gravis: Decreased acetylcholine receptors,
    Science, 182:293 (1973).
12. D. Drachman, Myasthenia gravis: Immunobiology of a receptor disorder,
    Trends Neurosci. 6:446 (1983).
13. A. G. Engel, Myasthenia gravis and myasthenic syndromes, Ann. Neurol.
    16:519 (1984).
14. J. Lindstrom, Immunobiology of myasthenia gravis, experimental
    autoimmune myasthenia gravis, and Lambert-Eaton syndrome, Ann. Rev.
    Immunol. 3:109 (1985).
15. S. Heinemann, S. Bevan, R. Kullberg, J. Lindstrom, and J. Rice,
    Modulation of the acetylcholine receptor by anti-receptor
    antibodies, Proc. Natl. Acad. Sci. USA, 74:3090 (1977).
16. I. Kao, and D.P. Drachman, Myasthenic immunoglobulin accelerates
    acetylcholine receptor degradation, Science, 197:527 (1977).
17 C. M. Gomez, and D.P. Richman, Anti-acetylcholine receptor antibodies
    directed against the alpha-bungarotoxin binding site induce a unique

form of experimental myasthenia, Proc. Natl. Acad. Sci. USA, 80:4089 (1983).

18. J. M. Lindstrom, M.E. Seybold, V.A. Lennon, S. Whittingham, and D. Duane, Antibody to acetylcholine receptor in myasthenia gravis: Prevalence, clinical correlates and diagnostic value, Neurology, 26:1054 (1976).

19. R. Hohlfeld, K.V. Toyka, K. Heininger, H. Grosse-Wilde, and I. Kalies, Autoimmune human T lymphocytes specific for acetylcholine receptor, Nature, 310:244 (1984).

20. G. Kohler, and C. Milstein, Continuous cultures of fused cells secreting antibody of predefined specificity, Nature, 256:495 (1975).

21. C. Milstein, From antibody structure to immunological diversification of immune response, Science, 231:1261 (1986).

22. S. J. Tzartos, D.E. Rand, B.E. Einarson, and J.M. Lindstrom, Mapping of surface structures of Electrophorus acetylcholine receptor using monoclonal antibodies, J. Biol. Chem. 256:8635 (1981).

23. S. J. Tzartos, Monoclonal antibodies as probes of the acetylcholine receptor and myasthenia gravis, Tr. Biochem. Sci. 9:63 (1984).

24. M. F. Greaves, ed. "Monoclonal antibodies to receptors. Probes for receptor structure and function," in the series "Receptor and recognition" vol. B17, Chapman and Hall, London, (1984).

25. J. C. Venter, C.M. Fraser, and J. Lindstrom, eds. "Monoclonal and anti-idiotypic antibodies. Probes for receptor structure and function" in Receptor Biochemistry and methodology" Vol.4, Alan R. Liss, N.Y (1984).

26. S. J. Tzartos, and J.M. Lindstrom, Monoclonal antibodies to probe acetylcholine receptor structure: Localization of the main immunogenic region and detection of similarities between subunits, Proc. Natl. Acad. Sci. USA, 77:755 (1980).

27. S. Tzartos, L. Langeberg, S. Hochschwender, and J. Lindstrom, Demonstration of a main immunogenic region on acetylcholine receptors from human muscle using monoclonal antibodies to human receptor, FEBS Lett. 158:116 (1983).

28. S. Tzartos, L. Langeberg, S. Hochschwender, L. Swanson, J. Lindstrom, Characteristics of monoclonal antibodies to denatured Torpedo and to native calf acetylcholine receptors: species, subunit and region specificity, J. Neuroimmunol. 10:235 (1986).

29. P. Sargent, B. Hedges, L. Tsavaler, L. Clemmons, S.J. Tzartos and J. Lindstrom, The structure and transmembrane nature of the acetylcholine receptor in amphibian skeletal muscle as revealed by cross-recting monoclonal antibodies, J. Cell Biol. 98:609 (1984).

30. K. Soteriadou, A. Tzinia, and S. Tzartos, Common antigenic determinants between acetylcholine receptor and protozoan membranes, in: "Mechanism of Action of the Nicotinic Acetylcholine Receptor," A. Maelicke, ed. Springer-Verlag, H (1986).

31. W. J. Gullick, and J.M. Lindstrom, Structural similarities between acetylcholine receptors from fish electric organs and mammalian muscle, Biochemistry, 21:4563 (1982).

32. W. J. Gullick, S.J. Tzartos, and J. Lindstrom, Monoclonal antibodies as probes of acetylcholine receptor structure. I. Peptide mapping, Biochemistry, 20:2173 (1981).

33. W. Gullick, and J. Lindstrom, Mapping the binding of monoclonal antibodies to the acetylcholine receptor from Torpedo Californica, Biochemistry, 22:3312 (1983).

34. M. Ratnam, P. Sargent, V. Sarin, J.L. Fox, D. Le Nguyen, J. Rivier, M. Criado, and J. Lindstrom, Location of antigenic determinants on primary sequences of the subunits of the nicotinic acetylcholine receptor by peptide mapping. Biochemistry, 25:2621 (1986).

35. M. A. Juillerat, T. Barkas, and S.J. Tzartos, Antigenic sites of the nicotinic acetylcholine receptor cannot be predicted from the hydrophilicity profile, FEBS Lett. 168:143 (1984).

36. M. Ratnam, D. Le Nguyen, J. Rivier, P. Sargent, and J. Lindstrom, Transmembrane topography of the nicotinic acetylcholine receptor: immunochemical tests contradict theoretical predictions based on hydrophobicity profile. Biochemistry, 25:2633 (1986).

37. H. M. Geysen, Antigen-antibody interactions at the molecular level: adventures in peptide synthesis, Imm. Today, 6:364 (1985).

38. T. Barkas, A. Mauron, B. Roth, J.M. Gabriel, S.J. Tzartos, M. Juillerat, C. Alliod, and M. Ballivet, Localization of the main immunogenic region and toxin binding site of the nicotinic acetylcholine receptor, Ann. N.Y. Acad. Sci. in press (1986).

39. S. J. Tzartos, and A. Kordossi, acetylcholine receptor conformation probed by subunit-specific monoclonal antibodies, in "Mechanism of Action of the nicotinic acetylcholine receptor," A. Maelicke, ed. Springer-Verlag, Heidelberg, in press (1986).

40. A. Kordossi, and S.J. Tzartos, Mapping the cytoplasmic side of the Torpedo acetylcholine receptor by monoclonal antibodies, in preparation, (1986).

41. S. J. Tzartos, M. Seybold, and J. Lindstrom, Specificities of antibodies to acetylcholine receptors in sera from myasthenia gravis patients measured by monoclonal antibodies, Proc. Natl. Acad. Sci. USA, 879:188 (1982).

42. B. Garabedian, and E. Morel, Monoclonal antibodies against the human acetylcholine receptor, Biochem. Biophys. Res. Commun. 113:1 (1983).

43. P. Whiting, A. Vincent, and J. Newsom-Davis, Monoclonal antibodies to Torpedo acetylcholine receptor. Characterization of antigenic determinants within the cholinergic binding site. Eur. J. Biochem. 150:533 (1985).

44. M. Raftery, M. Hunkapiller, C. Strader, and, L. Hood, Acetylcholine receptor: Complex of homologous subunits, Science, 208:1454 (1980).

45. K. Stefansson, M.E. Dieperink, D.P. Richman, C.M. Gomez, and L.S. Marton, Sharing of antigenic determinants between the nicotinic acetylcholine receptor and proteins in Escherichia coli, Proteus Vulgaris and Klebsiella Pneumoniae, N. Engl. J. Med., 312:221 (1985).

46. S. Pizzighella, A.S. Gordon, M.C. Souroujon, D. Mochly-Rosen, A. Sharp, and S. Fuchs, An anti-acetylcholine receptor monoclonal antibody cross-reacts with phosvitin, FEBS Lett. 159:246 (1983).

47. K. Soteriadou, A. Tzinia, and S.J. Tzartos, in preparation, (1986).

48. S. J. Tzartos, and J.-P. Changeux, High affinity binding of α-bungarotoxin to the purified α subunit and its 27K proteolytic peptide from Torpedo acetylcholine receptor. Requirement for sodium dodesyl sulfate, EMBO J. 2:381 (1983).

49. S. J. Tzartos, and J.-P. Changeux, Lipid-dependent recovery of α-bungarotoxin and monoclonal antibody binding to the purified α-subunit from Torpedo marmorata acetylcholine receptor, J. Biol. Chem. 259:11512 (1984).

50. J. Lindstrom, M. Criado, S. Hochschwender, J. Fox, and V. Sarin, Immunochemical tests of acetylcholine receptor subunit models, Nature, 311:573 (1984).

51. M. Criado, S. Hochschwender, V. Sarin, J. Fox, and J. Lindstrom, Evidence for unpredicted transmembrane domains in acetylcholine receptor subunits, Proc. Natl. Acad. Sci. USA, 82:2004 (1985).

52. J. Lindstrom, S.J. Tzartos, and W. Gullick, Structure and function of the acetylcholine receptor molecule studied using monoclonal antibodies, Ann. N.Y. Acad. Sci. 377:1 (1981).

53. K. Wan, and J. Lindstrom, Effects of monoclonal antibodies on the function of purified acetylcholine receptor from Torpedo Californica reconstituted into liposomes, Biochemistry, 24:1212 (1985).

54. B. M. Conti-Tronconi, S.M. Dunn, E.A. Barnard, J.O. Dolly, F.A. Lay, N. Ray, and M.A. Raftery, Brain and muscle nicotinic acetylcholine receptors are different but homologous proteins, Proc. Natl. Acad. Sci. USA, 82:5208 (1985).

55. J. Boulter, K. Evans, D. Goldman, G. Martin, D. Treco, S. Heinemann and J. Patrick, Isolation of a cDNA clone coding for a possible neural nicotinic acetylcholine receptor alpha-subunit, Nature, 319:368 (1986).

56. L. Swanson, J. Lindstrom, S.J. Tzartos, L. Schmued, D.D. O'Leary, and W.M. Cowan, Immunohistochemical localization of monoclonal antibodies to the nicotinic acetylcholine receptor in the middbrain of the chich, Proc. Natl. Acad. Sci. USA, 80:4532 (1983).

57. P. Whiting, and J. Lindstrom, Purification and characterization of a nicotinic acetylcholine receptor from chich brain, Biochemistry, 25:2082 (1986).

58. J. M, Henley, M. Mynlieff, J. Lindstrom, and R.E. Oswald, Interaction of monoclonal antibodies to electroplaque acetylcholine receptors with the alpha-bungarotoxin binding site of goldfish brain, Brain Res. 364:405 (1986).

59. M. Y. Momoy, and V.A. Lennon, Purification and biochemical characterization of nicotinic acetylcholine receptors of human muscle, J. Biol. Chem. 257:12,757 (1982).

60. C. Mochly-Rosen, and S. Fuchs, Monoclonal anti-acetylcholine receptor antibodies directed against the cholimergic binding site, Biochemistry, 20:5920 (1981).

61. D. Watters, and A. Maelicke, Organization of ligand binding sites at the acetylcholine receptor: A study with monoclonal antibodies, Biochemistry, 22:1811 (1983).

62. M. Mihovilovic, and D.P. Richman, Modification of alpha-bungarotoxin and cholinergic ligand-binding properties of Torpedo acetylcholine receptor by an anti-acetylcholine receptor monoclonal antibody, J. Biol. Chem. 259:15051 (1984).

63. S. J. Tzartos, D. Sophianos, K. Zimmermann, and A. Starzinski-Powitz, Antigenic modulation of human muscle acetylcholine receptor by myasthenic sera. Serum titer determines receptor internalization, J. Immunol. 136:3231 (1985).

64. J. P. Merlie, R. Sebbane, S.J. Tzartos, and J. Lindstrom, Inhibition of glycosylation with tunicamycin blocks assembly of newly synthesized acetylcholine receptor subunits in muscle cells, J. Biol. Chem. 257:2694 (1982).

65. R. Sebbane, G. Clokey, J.P. Merlie, S.J. Tzartos, and J. Lindstrom, Characterization of the mRNA for mouse muscle acetylcholine receptor alpha-subunit by quantitative translation in vitro, J. Biol. Chem. 258:3294 (1983).

66. J. P. Merlie, Biogenesis of the acetylcholine receptor, a multisubunit integral membrane protein, Cell, 36:573 (1984).

67. J. Giraudat, A. Devillers-Thiery, C. Auffray, F. Rougeon, and J.P. Changeux, Identification of a cDNA clone coding for the acetylcholine binding subunit of Torpedo marmorata acetylcholine receptor, EMBO J. 1:713 (1982).

68. T. Barkas, A. Mauron, B. Roth, C. Alliod, S.J. Tzartos, and M. Ballivet, Expression cloning and fusion proteins as tools to study receptor structure, in "Mechanism of Action of the nicotinic acetylcholine receptor," A. Maelicke, ed. Springer-Verlag Heidelberg, in press (1986).

69. B. C.G. Schalke, W.E.F. Klinkert, H. Wekerle, and D.S. Dwyer, Enhanced activation of a T cell line specific for acetylcholine receptor by using anti-acetylcholine receptor monoclonal antibodies plus receptors, J. Immunol. 134:3643 (1985).

70. Y. Zhang, S.J. Tzartos, B. Schalke, A. Melms, and H. Wekerle, Interaction between acetylcholine receptor-specific T and B

lymphocytes: Antigen presentation by B hybridoma cells and enhancing effect of monoclonal antibodies on T cell activation, <u>Ann. N.Y.</u> Acad. Sci. in press (1986).

71. R. Hohlfeld, K. Toyka, M. Michels, K. Heininger, B. Conti-Tronconi, and S.J. Tzartos, Acetylcholine receptor-specific human T-lymphocyte lines, <u>Ann. N.Y. Acad. Sci.</u> in press (1986).

72. S. J. Tzartos, S. Hochschwender, and J. Lindstrom, Passive transfer of experimental autoimmune myasthenia gravis by monoclonal antibodies to the main immunogenic region of the acetylcholine receptor, in preparation (1986).

73. B. Conti-Tronconi, S.J. Tzartos, and J. Lindstrom, Monoclonal antibodies as probes of acetylcholine receptor structure. II. Binding to native receptor, <u>Biochemistry,</u> 20:2181 (1981).

74. S. J. Tzartos, D. Sophianos, and A. Efthimiadis, Role of the main immunogenic region of acetylcholine receptor in myasthenia gravis. An Fab monoclonal antibody protects against antigenic modulation by human sera. <u>J. Immunol.</u> 134:2343 (1985).

75. S. J. Tzartos, and A. Starzinski-Powitz, Decrease in acetylcholine receptor content of human myotube cultures mediated by monoclonal antibodies to α, β and γ subunits, <u>FEBS Lett.</u> 196:91 (1986).

76. J. Killen, S. Hochschwender, and J. Lindstrom, The main immunogenic region of acetylcholine receptors does not provoke the formation of antibodies to a predominant idiotype, <u>J. Neuroimmunol.</u> 9:229 (1984).

77. A. Bionda, M. De Baets, S.J. Tzartos, J. Lindstrom, and W.O. Weigle, Spectrotypic analysis of antibodies to acetylcholine receptors in experimental autoimmune myasthenia gravis, <u>Clin. and Exp. Immunol.</u> 57:41 (1984).

78. J. Tainer, E. Getzoff, Y. Patterson, A. Olson, and R.A. Lerner, The atomic mobility component of protein antigenicity, <u>Ann. Rev. Immunol.</u> 3:501 (1985).

79. A. Brisson, and P. Unwin, Quaternary structure of the acetylcholine receptor, <u>Nature,</u> 315:474 (1985).

80. B. Einarson, B. Gullick, B. Conti-Tronconi, and J. Lindstrom, Subunit composition of fetal calf muscle nicotinic acetylcholine receptor, <u>Biochemistry,</u> 21:5295 (1982).

81. D. P. Richman, C. Gomez, P. Berman, S. Burres, and B.G.W. Arnason, Monoclonal anti-acetylcholine receptor antibodies can cause experimental myasthenia, <u>Nature,</u> 286:738 (1980).

82. V. A. Lennon, and E.H. Lambert, Myasthenia gravis induced by monoclonal antibodies to acetylcholine receptors, <u>Nature,</u> 285:238 (1980).

83. M. Souroujon, D. Mochly-Rosen, A. Gordon, and S. Fuchs, Interaction of monoclonal antibodies to Torpedo acetylcholine receptor with the receptor of skeletal muscle, <u>Muscle and Nerve,</u> 6:303 (1983).

84. J. Lindstrom, M. Campbell, and B. Nave, Specificities of antibodies to acetylcholine receptors, <u>Muscle Nerve,</u> 1:140 (1978).

MUTANT RECEPTORS FOR LOW DENSITY LIPOPROTEIN IN FAMILIAL

HYPERCHOLESTEROLEMIA

Wolfgang J. Schneider

Department of Biochemistry
University of Alberta
Edmonton, Alberta, Canada

INTRODUCTION

The concept of receptor-mediated metabolism of lipoproteins emerged from studies initiated in 1973 on human skin fibroblasts grown in culture[1]. These experiments were designed to elucidate the normal function of low density lipoprotein (LDL), about which little was known at that time. Biochemical studies showed that a specific cell surface receptor, the LDL receptor, mediates the binding, uptake and degradation of LDL, thus supplying almost all cells in the body with cholesterol. Detailed insight into the molecular mechanisms underlying this complex process was obtained from studies with fibroblasts derived from patients with the phenotype of homozygous familial hypercholesterolemia (FH). As in many other biological systems, the expression of a disease state in a defined cellular system was essential to the discovery of the causal factor: FH is now one of the best characterized genetic diseases at the molecular level. As will be outlined below, in FH several groups of mutations occur naturally in the structural gene for the LDL receptor which disrupt its normal function and lead to severe hypercholesterolemia, producing myocardial infarctions and premature atherosclerosis. Thus, the important role of lipoprotein receptors in normal physiology is underscored by the dramatic consequences of their functional absence.

THE LDL RECEPTOR - PHYSIOLOGICAL ASPECTS

The LDL Receptor Pathway

This pathway, depicted in Fig. 1, describes how cells, through receptor-mediated endocytosis of LDL, control their internal cholesterol concentration. The cholesterol derived from the lysosomal hydrolysis of LDL cholesterol esters mediates a complex array of feedback control mechanisms that protects the cell from overaccumulation of cholesterol. First, the LDL-derived cholesterol suppresses the activity of the enzyme 3-hydroxy-3-methylglutaryl CoA reductase (HMG-CoA reductase), the rate controlling enzyme in cholesterol biosynthesis, thereby turning off cholesterol synthesis in the cell[2]. Second, the cholesterol activates a cholesterol-esterifying enzyme called acyl-CoA:cholesterol acyltransferase (ACAT) so that excess cholesterol can be stored as droplets of cholesteryl esters[3].

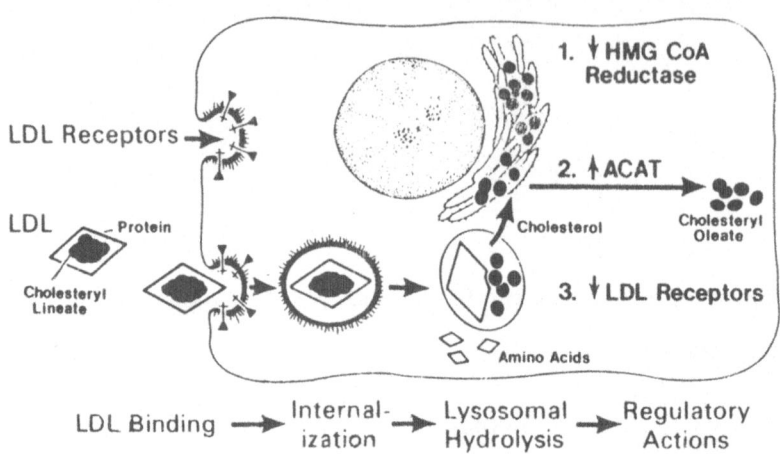

Fig. 1. Sequential steps in the LDL pathway in cultured mammalian cells. LDL denotes low density lipoprotein; HMG CoA reductase denotes 3-hydroxy-3-methylglutaryl CoA reductase; and ACAT denotes acyl-CoA: cholesterol acyltransferase.

Third, the cholesterol suppresses the synthesis of new LDL receptors, preventing further cellular entry of LDL and thus overloading with cholesterol[4].

The overall effect of this regulatory system is to coordinate the intracellular and extracellular sources of cholesterol in order to maintain a constant level of cholesterol within the cell while the external supply in the form of lipoproteins undergoes fluctuation. Human fibroblasts and other mammalian cells are able to subsist in the absence of lipoproteins because they can synthesize cholesterol from acetyl CoA. In contrast, when LDL is accessible, the cells primarily use the LDL receptor to take up LDL and keep their own cholesterol synthesis suppressed[5,6,7].

Familial Hypercholesterolemia: Clinical Consequences of the Genetic Disruption of the LDL Receptor Pathway

The delineation of the LDL receptor pathway in fibroblasts was greatly facilitated by studies on fibroblasts from a large number of patients with the typical phenotype of homozygous FH. In 1974, it was recognized that mutations affecting the function of the LDL receptor are responsible for FH[1,8]. Now, 12 years later, at least 11 different alleles at the LDL receptor locus have been identified in 104 fibroblast lines from patients with the clinical features of FH homozygotes.

FH is characterized by three cardinal features: (i) a selective elevation in the plasma level of LDL; (ii) deposition of cholesterol in abnormal sites, in particular in tendons (formation of xanthomas) and in arteries (forming atheromas); and (iii) inheritance as an autosomal dominant trait with a gene dosage effect. This means that individuals who inherit two mutant alleles (FH homozygotes) are more severely affected than those with one mutant allele (FH heterozygotes)[9,10]. Heterozygotes occur at a frequency of about 1 in 500 persons, while homozygotes occur at a frequency of about 1 in one million persons among European and American populations. Through the delineation of the molecular defects in certain patients (see below), it is now clear that the severely affected offspring of a marriage between two heterozygotes can be either true homozygous or actually a heteroallelic genetic compound[9,11,12,13,14]. Thus, the term <u>homozygote</u> in most cases is not strictly correct in a genetic sense, yet is a convenient one and is generally applied to patients with two mutant alleles at the LDL receptor locus.

Familial Hypercholesterolemia is the outstanding example of a single-gene mutation that regularly produces atherosclerosis. The deposition of LDL-cholesterol within the intima of major arteries in homozygotes becomes rapidly progressive, and leads to remarkably uniform clinical findings. Myocardial infarction, angina pectoris, and sudden death usually occur in homozygotes before the age of 15[15].

The clinical features of heterozygous FH are more variable and less severe than those of the homozygous form. Heterozygous men experience a myocardial infarction before 60 years of age with a 75% probability; in normal men, this risk is 15%[16,17]. Despite the expression of the same genetic abnormality and similarly elevated plasma LDL levels, heterozygous women suffer from coronary heart disease less often and at a later age than do heterozygous men. The risk for female heterozygotes to develop coronary artery disease by age 60 is about 45% as compared with 10% in unaffected females[16,17].

THE LDL RECEPTOR - MOLECULAR ASPECTS

Structure of the Normal Human LDL Receptor

The complete amino acid sequence of the human LDL receptor was derived from the nucleotide sequence of the full-length receptor cDNA. The strategy for the isolation of this cDNA is outlined in Ref. 18. The amino acid sequence, in combination with biochemical experiments[19], revealed that the mature receptor can be divided into five domains. A schematic model of the LDL receptor structure is shown in Fig. 2. In the following, the distinguishing features of these domains will be highlighted.

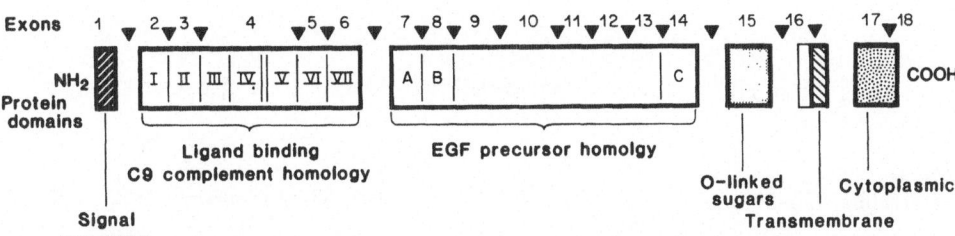

Fig. 2. Exon organization and protein domains in the human LDL receptor. The domains of the protein are delimited by thick black lines and are labeled in the lower portion of the figure. The 7 cysteine-rich, 40-amino acid repeats in the LDL binding domain are assigned roman numerals I-VII. Repeats IV and V are separated by 8 amino acids. The 3 cysteine-rich repeats in the EGF precursor homology domain are lettered A-C. The positions at which introns interrupt the coding region are indicated by arrow heads. Exon numbers are shown between the arrow heads (reprinted with permission from Ref. 20; Copyright 1985 by the AAAS).

First domain: ligand binding   The mature receptor (after cleavage of a 21-residue signal sequence) consists of 839 residues. The ligand binding domain (292 residues) is located at the NH_2-terminus of the receptor. These 292 amino acids, 42 of which are cysteines, exist in a tightly folded conformation held together by multiple disulfide bonds. Analysis of the amino acid sequence and exon/intron mapping data[20] suggest that the first domain is made up of seven repeats, each consisting of 40 amino acids. Within these 40 residues, 6 cysteines are spaced 4-7 amino acids apart and are located exactly in register for all repeats. At the COOH-terminus of each repeat unit, there is a cluster of negatively-charged amino acids. These sequences (Fig. 3) are complementary to positively charged sequences in the receptor's ligands, apo E[21] and apo B[22]. It has been speculated[22,23] that the negatively charged clusters of amino acids within the cysteine-rich repeat sequences constitute binding sites for apo E and apo B. Studies by Mahley and associates[24,25,26] and others[27,28] strongly support this notion. In addition, the disulfide bridges may be important in stabilizing the binding site structure, in particular when the receptor recycles via the endosome with its acidic environment[19].

Second domain: EGF precursor homology   This region, consisting of about 400 residues, is homologous to a portion of the extracellular domain of the precursor to epidermal growth factor (EGF)[18,19]. The growth factor is a peptide of 53 amino acids, which is derived by proteolytic cleavage from a large, membrane-bound precursor of 1217 residues[29,30]. The portion of the precursor facing the cytoplasm contains multiple homologous but not identical repeats of the EGF sequence, as well as unrelated spacer sequences. The EGF precursor homology region of the LDL receptor is flanked by 3 cysteine-rich repeats (A, B, and C in Fig. 2), which are highly homologous to four repeat sequences in the EGF precursor[20,29,31]. Repeats A, B, and C of the LDL receptor are also homologous to certain proteins of the blood clotting system, such as Factor IX, Factor X, and protein C[20,31].

All observations regarding the significance of the existence of these homologies seem to suggest that the homologous regions arose by a duplication of an ancestral gene. These genes might have differentiated further by, e.g., acquisition of exons from other genes providing specific functions, such as nutrient delivery (LDL) or signaling of cell proliferation through the secretion of a peptide (EGF).

Apo-E
(Residues 140-150)   -His-Leu-Arg-Lys-Leu-Arg-Lys-Arg-Leu-Leu-Arg-

Apo-B
(Residues 276-286)   -Thr-Thr-Arg-Leu-Thr-Arg-Lys-Arg-Gly-Leu-Lys-

Apo-B,E(LDL) Receptor
(Consensus Sequence)   -Cys-Asp-X-X-X-Asp-Cys-X-Asp-Gly-Ser-Asp-Glu-

Fig. 3. Structural similarities between the receptor binding domain of apo-E
         and a region of apo-B. The homologous positively charged residues
         lysine and arginine are boxed in the sequence of apo-E and apo-B.
         The consensus sequence of the postulated ligand binding site of the
         apo-B, E(LDL) receptor, which is enriched in the negatively charged
         residues aspartic and glutamic acids (underlined), is shown for
         comparison.

Third Domain: 0-linked sugars   There is a sequence of 58 amino acid residues that contains 18 serines or threonines immediately external to the membrane[18]. This region contains a large number of carbohydrate chains attached to the hydroxyl groups of Ser and Thr[19,32]. The elongation of these sugar chains during posttranslational processing of the receptor brings about the dramatic increase in apparent molecular weight that is observed in biosynthetic studies of radiolabeled receptor (see below). Carbohydrate chains in 0-linkage are also present on other parts of the receptor; these chains do not appear to be clustered as they are in the 0-linked sugar domain[33]. The role of 0-linked carbohydrate in the function of the LDL receptor, if any, is unknown.

Fourth domain: membrane-spanning region   Twenty-two uncharged and hydrophobic amino acids constitute the only membrane-spanning portion of the receptor[19]. There is a single intramembraneous cysteine residue in the human receptor, which is replaced by an alanine in the bovine receptor[19]. Since the two receptors function similarly, it appears likely that this cysteine residue is not involved in a disulfide bridge, but rather exists in a reduced state.

Fifth domain: cytoplasmic region   At the COOH-terminus of the LDL receptor, 50 amino acid residues project into the cytoplasm. The intracellular location of this tail region was demonstrated in proteolysis experiments by the use of an anti-peptide antibody directed against the COOH-terminus sequence[19]. As outlined below, the cytoplasmic region is crucial for clustering of LDL receptors in coated pits, a prerequisite for subsequent internalization of receptor/LDL complexes. Localization to coated pits may occur through interaction of the cytoplasmic tail with clathrin itself or with some clathrin-associated protein on the intracellular side of the plasma membrane. In search of a mediator site on the LDL receptor, the intracellular regions of other coated pit receptors have been investigated for common features, but no obvious structural element has been identified[23]. Thus, while the function of the cytoplasmic tail of the LDL receptor is established, the mechanism of its action has yet to be elucidated.

The Biosynthetic Pathway of LDL Receptor in Normal Human Fibroblasts

The biosynthesis of LDL receptors can be conveniently studied by metabolic labeling of cultured cells with ^{35}S-methionine, followed by immunoprecipitation of receptor protein with a monoclonal anti-receptor antibody[34]. Experimental procedures are detailed in Refs. 13 and 35. The normal human LDL receptor is initially synthesized as a precursor with an apparent M_r of 120,000, as estimated by SDS polyacrylamide gel electrophoresis (SDS-PAGE). It contains asparagine-linked (N-linked) high mannose oligosaccharide chains and N-acetylgalactosamine (GalNAc) residues attached to serine and threonines by 0-linkage[32]. Between 30 and 60 min after synthesis, the LDL receptor precursor undergoes a sudden shift in apparent M_r from 120,000 to 160,000 in SDS-PAGE[13]. This change in apparent M_r is due to the elongation of the 0-linked chains[32] which temporally coincides with the maturation of the N-linked carbohydrate moiety[32]. The calculated molecular weight of the protein portion of the receptor is 93,102, and including the mature carbohydrate its M_r is 115,000, yet the 0-linked carbohydrate which forms a cluster in a Ser/Thr rich region of the receptor[19] causes it to migrate to the position of a 160 kDa-protein in SDS-PAGE[32].

In other studies, it was shown that the 120 kDa precursor is localized intracellularly[14], and that the addition of N-linked carbohydrate chains is required neither for transport to the surface nor for the receptor's ability to bind LDL and to recycle. The role of O-linked carbohydrate in the receptor's cellular itinerary is now under investigation. A deletion of the clustered O-linked carbohydrates does not appear to impair the function of the LDL receptor[33]. However, there is a small number of isolated chains of O-linked carbohydrates in addition to the clustered chains, and their significance for receptor function needs to be delineated[33].

Molecular Defects in LDL Receptors of Patients with Familial Hypercholesterolemia

At least 11 different mutant alleles at the LDL receptor locus have been described[14,15,36,37]. Fibroblasts from 104 FH homozygotes were analyzed in terms of LDL receptor structure and function; these studies revealed that the mutations could be divided into four classes (for review, see Ref. 23). The mutant alleles are summarized in Table 1 and described in the following.

Class 1 mutations: no detectable precursor These alleles, designated R-0, fail to express receptor proteins as measured by functional assays (binding of ^{125}I-LDL) or attempts to immunologically identify proteins with a variety of monoclonal and polyclonal antibodies directed against the LDL receptor. R-0 alleles are the most frequent of the mutant alleles and probably account for about 40% of all mutations at the LDL receptor locus. This class most likely includes nonsense mutations that introduce termination codons early in the protein coding region, but point mutations in the promoter that block transcription, mutations in intron/exon junctions that lead to abnormal splicing of the mRNA, and large deletions may also be included in this class of mutant alleles. Examples are described in Refs. 13 and 14.

Class 2 mutations: precursor not processed These alleles specify receptor precursors that are produced in normal or reduced amounts, have normal (120 kDa) or abnormal (100 kDa, 135 kDa) molecular weight, but do not undergo any apparent increase in molecular weight after synthesis. Receptors encoded by these alleles never reach the cell surface[14].

Most of the mutant receptor precursors in this class have an apparent M_r of 120,000 (allele designation, R-120) but precursors with apparent molecular weights of 100,000 (R-100 allele) and 135,000 (R-135 allele) have also been identified as members of this class[14]. These M_r abnormalities are likely due to alterations in the length of the protein chain and do not result from changes in the carbohydrate moieties[14].

Variants of the class 2 mutations were observed in a black American family, in Afrikaners, and in WHHL rabbits[36]. In these variants the receptor is produced as a precursor of apparently normal M_r that is processed to the mature form, but much slower than the normal precursor. Eventually, about 5-10% of the receptors reach the cell surface as 160 kDa mature proteins. Once on the cell surface, these receptors display a reduced ability to bind LDL; that means that the mutation has disrupted both transport and binding capacity of the receptor[36]. Very recently, a mutation in the WHHL rabbit receptor gene has been identified that causes the deletion of four amino acids in the LDL binding domain (T. Yamamoto, M.S. Brown, J.L. Goldstein et al., unpublished observations). While this might explain the inability of the mutant receptors to bind LDL normally, the reason for their delayed processing is less obvious. In any event, expression of the mutant gene in transformed cells will provide the answer as to whether this mutation is sufficient to bring about the WHHL phenotype.

The molecular basis of other defects in the class 2 mutations and their

Table 1. Mutations at the LDL receptor locus that produce familial hypercholesterolemia (FH)

| Class of mutation | Allele designation[1] | Apparent receptor mass on SDS gels (kDa) | | Receptor location | | | | LDL binding to intact cells | Frequency in FH patients |
| | | Precursor | Mature | Plasma membrane | | | Extracellular | | |
				Intracellular	Coated pits	Noncoated regions			
Class 1: No detectable precursor	R-0	None	None					None	Common
Class 2: Precursor not processed	R-100	100	100	+				None	Rare; found in Lebanese
	R-120	120	120	+				None	Common
	R-135	135	135	+				None	Rare
Class 2 variant: Precursor processed slowly, mature receptor binds LDL poorly	R-120 $\xrightarrow{\text{slow}}$ 160 b⁻	120	160	+	(+)[2]			Reduced	Rare; found in Afrikaners and in WHHL rabbits
Class 3: Precursor processed normally, mature receptor binds LDL poorly	R-140 b⁻	100	140	+				Reduced	Rare
	R-160 b⁻	120	160	+				Reduced	Common
	R-210 b⁻	170	210	+				Reduced	Rare
Class 4: Precursor processed normally, mature receptor binds LDL normally, but does not enter coated pits	R-150 i⁻, sec	110	150			(+)	+	Normal binding; defective internalization	Rare
	R-160 i⁻	120	160			+		Normal binding; defective internalization	Rare
	R-155 i⁻	115	155			+		Normal binding; defective internalization	

[1]Allele designations are based on the apparent molecular weight (in kilodaltons) on SDS polyacrylamide gel electrophoresis of the mature form of the receptor, i.e. the predominant form observed after a 2 hr pulse followed by a 2 hr chase. b⁻ denotes defective LDL binding; i⁻ denotes defective LDL binding; sec denotes secretion from the cell.
[2]Symbols in parentheses refer to minor populations.

Reproduced, with permission, from the Annual Review of Cell Biology, Vol. 1. Copyright 1985 by Annual Reviews Inc.

variants are not known. It seems likely that the failure of transport arises from some subtle changes in structure, similar to the one occurring in the WHHL rabbit.

Class 3 mutations: precursor processed, abnormal binding   Receptors in this class of mutants reach the cell surface at normal rates, but are able to bind only less than 15% of the normal amount of LDL[14],[15],[34]. Most of class 3 receptors have normal M_r in SDS-PAGE (allele designation, R-160 b⁻). Receptors with apparent M_r of 140,000 (R-140 b⁻)[14] and 210,000 (R-210 b⁻)[13] have been described as well. All mature receptor proteins in this class originate as precursors with apparent molecular weights that are 40,000 less, i.e., 120,000, 100,000, and 170,000, respectively. It appears that the abnormal molecular weights are due to alterations in the amino acid sequence and not to changes in carbohydrate content[32].

The structure of the LDL binding domain - 7 repeats of a 40 amino acid sequence - suggests an explanation for the abnormally sized, binding deficient receptors in class 3. The DNA encoding the repeat structure might be susceptible to deletion or duplication following mispairing and recombination of homologous regions during meiosis. Such deletions or duplications would change size and binding capacity of the receptors.

Class 4 mutations: precursor processed, receptor binds LDL but does not cluster in coated pits   These mutations have been termed internalization-defective[38]. From the description of the process of receptor-mediated endocytosis (see Introduction) it is apparent that clustering of receptors in coated pits is a prerequisite for subsequent internalization of ligand; clustering of LDL receptors is disrupted in class 4 mutations. To date, the molecular defects in five patients with this phenotype have been elucidated. The original example was patient J.D.[38]. While he inherited a R-0 (class 1) allele from his mother, he inherited from his father a gene that produces a cell surface receptor of normal size and normal LDL binding capacity, but which is unable to carry the bound LDL into the cell (R-160 i⁻ allele)[39]. Thus J.D. is a compound heterozygote with regards to his genotype at the LDL receptor locus.

It was suspected early that the R-160 i⁻ allele is producing a receptor altered in its intracellular portion of 50 amino acid residues, because this region of the receptor is believed to mediate the clustering of receptors into coated pits through interaction with cytoplasmic coated pit specific proteins[39]. The molecular defect, indeed, was found to reside in the intracellular domain[40]. A point mutation has converted the codon for tyrosine 807 into one for cysteine. Subsequently, four other patients with internalization defects have been identiifed and their genes analyzed. In one of them, the internalization-defective allele produces a receptor that lacks the intracellular domain and the membrane-spanning region, and the truncated protein is secreted from these cells[41]. The secreted receptor has an apparent M_r of 150,000 (allele designation, R-150 i⁻,sec). Another is a patient from Japan[42], the offspring of consanguineous parents, apparently homozygous for a R-150 i⁻,sec allele. Two more internalization-defective alleles were identified and the mutant genes analyzed[37]. In one (FH patients 682 and 683), the cytoplasmic domain consists of only two residues due to a point mutation that changes a tryptophan codon into a termination codon[37]. The other mutant gene (found in patient FH 763) contains a four-base duplication, producing a frameshift that alters the reading frame. The cytoplasmic tail of this mutant receptor is made up of six of the normal plus eight additional amino acids instead of the normal 50[37].

In conclusion, the analysis of class 4 mutations at the LDL receptor locus strongly suggests that the signal for movement of the receptor into coated pits resides in its cytoplasmic domain. In addition, the function of the cysteine-rich region at the aminoterminus as the ligand binding site is now well established. Future analysis of mutant receptor genes that affect posttranslational processing, or that lead to abnormally large LDL receptors (e.g., R-210 b⁻) can be expected to shed light on the function of other LDL receptor domains.

REFERENCES

1. M.S. Brown, and J.L. Goldstein, Proc. Natl. Acad. Sci. U.S.A. 71:788-792 (1974).
2. M.S. Brown, S.E. Dana, and J.L. Goldstein, J. Biol. Chem. 249:789-796 (1974).
3. J.L. Goldstein, S.E. Dana, and M.S. Brown, Proc. Natl. Acad. Sci. U.S.A. 71:4288-4292 (1974).
4. M.S. Brown, and J.L. Goldstein, Cell 6:307-316 (1975).
5. M.S. Brown, and J.L. Goldstein, Science 191:150-154 (1976).
6. J.L. Goldstein, and M.S. Brown, Curr. Topics Cell. Reg. 11:147-181 (1976).
7. J.L. Goldstein, and M.S. Brown, Ann. Rev. Biochem. 46: 897-930 (1977).
8. J.L. Goldstein, and M.S. Brown, J. Biol. Chem. 249:5153-5162 (1974).
9. D.S. Fredrickson, J.L. Goldstein, and M.S. Brown, in: "The Metabolic Basis of Inherited Disease," J.B. Wyngaarden, and D.S. Fredrickson, eds., McGraw-Hill, New York (1978).
10. J.L. Goldstein, and M.S. Brown, John Hopkins Med. J. 143:8-16 (1978).
11. A.K. Khachadurian, Am. J. Med. 37:402-407 (1964).
12. J.L. Goldstein, S.E. Dana, G.Y. Brunschede, and M.S. Brown, Proc. Natl. Acad. Sci. U.S.A. 72:1092-1096 (1975).
13. H. Tolleshaug, J.L. Goldstein, W.J. Schneider, and M.S. Brown, Cell 30:715-724 (1982).
14. H. Tolleshaug, K.K. Hobgood, M.S. Brown and J.L. Goldstein, Cell 32:941-951 (1983).
15. J.L. Goldstein, and M.S. Brown, in: "The Metabolic Basis of Inherited Disease," J.B. Stanbury, J.B. Wyngaardon, and D.S. Fredrickson, et al., eds., McGraw-Hill, New York (1983).
16. J. Slack, and N.C. Nevin, J. Med. Genet. 5:4-8 (1968).
17. N.J. Stone, R.I. Levy, D.S. Fredrickson, and J. Verber, Circulation 49:476-488 (1974).
18. T. Yamamoto, C.G. Davis, M.S. Brown, W.J. Schneider, M.L. Casey, J.L. Goldstein, and D.W. Russell, Cell 39:27-38 (1984).
19. D.W. Russell, W.J. Schneider, T. Yamamoto, K.L. Luskey, M.S. Brown, and J.L. Goldstein, Cell 37:577-585 (1984).
20. T.C. Südhof, J.L. Goldstein, M.S. Brown, and D.W. Russell, Science 228:815-822 (1985).
21. T.L. Innerarity, K.H. Weisgraber, K.S. Arnold, S.C. Rall, Jr., and R.W. Mahley, J. Biol. Chem. 259:7261-7267 (1984).
22. T.J. Knott, S.C. Rall, Jr., T.L. Innerarity, S.F. Jacobson, M.S. Urdea, B. Levy-Wilson, L.M. Powell, R.J. Pease, R. Eddy, H. Nakai, M. Byers, L.M. Priestly, E. Robertson, L.B. Rall, C. Betsholtz, T.B. Shows, R.W. Mahley, and J. Scott, Science 230:37-43 (1985).
23. J.L. Goldstein, M.S. Brown, R.G.W. Anderson, D.W. Russell, and W.J. Schneider, Ann. Rev. Cell Biol. 1:1-39 (1985).
24. R.W. Mahley, T.L. Innerarity, Biochim. Biophys. Acta 737:197-222 (1983).
25. T.L. Innerarity, and R.W. Mahley, Biochemistry 17:1440-1447 (1978).

26. R.E. Pitas, T.L. Innerarity, and R.W. Mahley, J. Biol. Chem. 255:5454-5460 (1980)

27. S.K. Basu, J.L. Goldstein, R.G.W. Anderson, and M.S. Brown, Proc. Natl. Acad. Sci. U.S.A. 73:3178-3182 (1976).

28. W.J. Schneider, U. Beisiegel, J.L. Goldstein, and M.S. Brown, J. Biol. Chem. 257:2664-2673 (1982).

29. J. Scott, M. Urdea, M. Quiroga, R. Sanchez-Pascador, N. Fong, M. Selby, W.J. Rutter, and G.I. Bell, Science 221:236-240 (1983).

30. A. Gray, T.J. Dull, and A. Ullrich, Nature 303:722-725 (1983).

31. R.F. Doolittle, D.-F. Feng, and M.S. Johnson, Nature 307:558-566 (1984).

32. R.D. Cummings, S. Kornfeld, W.J. Schneider, K.K. Hobgood, H. Tolleshaug, M.S. Brown and J.L. Goldstein, J. Biol. Chem. 258:15261-15273 (1983).

33. C.G. Davis, A. Elhammer, D.W. Russell, W.J. Schneider, S. Kornfeld, M.S. Brown, and J.L. Goldstein, J. Biol. Chem., in press (1986).

34. U. Beisiegel, W.J. Schneider, J.L. Goldstein, R.G.W. Anderson, and M.S. Brown, J. Biol. Chem. 256:11923-11931 (1981).

35. W.J. Schneider, J.L. Goldstein, and M.S. Brown, Methods Enzymol. 109:405-417 (1985).

36. W.J. Schneider, M.S. Brown, and J.L. Goldstein, Mol. Biol. Med. 1:355-367 (1983).

37. M.A. Lehrman, J.L. Goldstein, M.S. Brown, D.W. Russell, and W.J. Schneider, Cell 41:735-743 (1985).

38. M.S. Brown, and J.L. Goldstein, Cell 9:663-674 (1976).

39. J.L. Goldstein, M.S. Brown, and N.J. Stone, Cell 12:629-641 (1977).

40. C.G. Davis, M.A. Lehrman, D.W. Russell, R.G.W. Anderson, M.S. Brown, and J.L. Goldstein, Cell 45:15-24 (1986).

41. M.A. Lehrman, W.J. Schneider, T. Südhof, M.S. Brown, J.L. Goldstein, and D.W. Russell, Science 227:140-146 (1985).

42. Y. Miyake, S. Tajima, T. Yamamura, and A. Yamamoto, Proc. Natl. Acad. Sci. U.S.A. 78:5151-5155 (1981).

CALCIUM AND GUANOSINE NUCLEOTIDE MODULATION OF MELANOTROPIN RECEPTOR

FUNCTION AND ADENYLATE CYCLASE IN THE M2R MELANOMA CELL LINE

Jeffrey E. Gerst and Yoram Salomon

Department of Hormone Research
Weizmann Institute of Science
Rehovot 76100, Israel

INTRODUCTION

Melanocytes are specialized melanin producing cells of neural embryonic origin, which function to regulate cutaneous pigmentation in vertebrate systems (Parker, 1981). Both α- and β-melanotropins (α-MSH, β-MSH) are thought to regulate melanin synthesis in these cells via cAMP-dependent stimulation of the enzyme, tyrosinase (monophenol monooxygenase), in direct correlation with adenylate cyclase (AC) activity (Johnson and Pastan, 1972; Varga et al., 1974: O'Keefe and Cuatrecasas, 1974). Calcium has been implicated in the actions of both melanotropic hormones (Vesely and Hadley, 1974; Van de Veerdonk and Brouwer, 1973), as well as for adrenocorticotropin (ACTH) (Bar and Hechter, 1969; Sayers et al, 1972; Perchellet and Sharma, 1979), on the stimulation of AC activity, though the specific step(s) in the cyclase pathway that require calcium have not been adequately elucidated. Recently, several reports have shown that millimolar levels of extracellular free calcium are important for both ACTH binding and subsequent corticosteroid secretion in rat adrenal cortical cells (Cheitlin et al., 1985), and for photoaffinity labeling and subsequent melanophore dispersion in Xenopus, using photoreactive α-MSH (De Graan et al., 1982a; De Graan et al., 1982b). However, in these studies the direct activation of hormone-sensitive adenylate cyclase, as a consequence of hormone binding, was not examined. Therefore, it was of importance to determine the specific role of calcium in both melanotropin binding and subsequent AC activation.

We have previously examined the specificity of the MSH receptor and hormone-sensitive AC in the M2R mouse melanoma cell line (Gerst et al., 1986a). In another report we demonstrate that melanotropin receptor regulation of AC in M2R mouse melanoma cells is under dual control by both calcium and guanosine nucleotides (Gerst et al., 1986b). Binding of β-MSH to its receptor and activation of membrane-bound AC can be demonstrated to depend on two classes of calcium-pertinent sites which saturate in the micromolar range. Post-receptor events leading to the activation of adenylate cyclase were found to be calcium-insensitive, thus the role of calcium in melanotropin regulation of AC is shown to be confined to the level of hormone binding.

Calcium and guanosine nucleotides modulate both melanotropin receptor affinity and receptor occupancy, thus signifying a potentially important mechanism for receptor fine-tuning. In addition, we report on the inhibition of melanotropin receptor function by certain antagonists of calmodulin. It may be that the calcium sensitivity involved in melanotropin receptor action is conferred to the receptor by either calmodulin or a calmodulin-like calcium binding protein, or is an inherent property of the receptor itself.

Calcium dependency of β-MSH action in intact M2R melanoma cells

β-MSH (0.1 μM) strongly stimulates the intracellular accumulation of [2-^{3}H] cAMP in intact M2R cell monolayers prelabeled with [2-^{3}H] adenine (Fig.1). This cellular response to β-MSH is seen to be absolutely dependent on the concentration of extracellular free calcium. In the absence of extracellular calcium (< 50 nM), achieved by the inclusion of EGTA (2 mM) in the culture medium, there is essentially no response of M2R cells to the addition of β-MSH. A dose-dependent increase in β-MSH-sensitive cAMP accumulation is observed upon a gradual increase in extracellular free calcium (achieved by a calculated reduction in added EGTA concentrations). This calcium sensitivity of β-MSH-stimulated cAMP accumulation is observed over two ranges of free calcium concentrations: one saturated below 10 μM and a second saturated above 100 μM. In contrast, no such calcium dependency is observed for cAMP accumulation stimulated by prostaglandin E$_1$ (PGE$_1$) or determined in the absence of stimulant (basal).

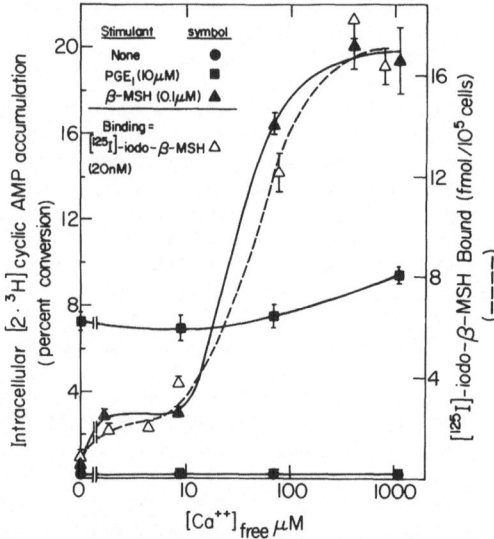

Fig. 1. Calcium dependency of β-MSH action in intact M2R melanoma cells. M2R cell monolayers, prelabeled with [2-^{3}H]adenine, were incubated in the absence (basal) or presence of β-MSH (0.1 μM), or PGE$_1$ (10 μM), and [2-^{3}H]cAMP accumulation was determined at various concentrations of extracellular free calcium. Binding of [^{125}I]-iodo-β-MSH to cell monolayers was determined under similar incubation conditions. Methods for the determination of [2-^{3}H]cAMP accumulation and hormone binding to intact cells were as described by Gerst et al., (1986a).

On the assumption that these two hormones operate through different receptor systems, but utilize a common GTP-regulatory protein and AC pool, it can be suggested that the difference in calcium dependency between β-MSH and PGE_1 must reside at a level preceeding G protein activation. To examine this hypothesis we tested for the binding of $[^{125}I]$-iodo-β-MSH to intact M2R cell monolayers, under conditions identical to those used for the cAMP accumulation studies. It was found that the binding of $[^{125}I]$-iodo-β-MSH (20 nM) to intact M2R cells is calcium concentration-dependent and this dependence correlates well with the calcium dependency observed for the stimulation of cAMP accumulation (Fig. 1). Similarly, two ranges of extracellular free calcium that facilitate hormone binding are observed: one saturated below 10 μM free calcium and one saturated above 100 μM free calcium. Thus the role of calcium in the regulation of β-MSH action is to facilitate the interaction between β-MSH and its receptor.

Calcium dependency of β-MSH activation of AC in M2R cell membranes.

In order to test the effects of calcium on adenylate cyclase under cell-free conditions, agonist-stimulated AC activity was measured in a plasma membrane-enriched fraction derived from M2R cells, prepared according to Gerst et al., (1986a). The assay was performed either in the presence (10 μM) or absence (< 5 nM) of added calcium, using Ca-EGTA buffers (Fig. 2). β-MSH-stimulated AC activity is increased by a factor of 3 when determined in the presence of added calcium, while in contrast, no stimulating effect on the response of AC to PGE_1 is evident. Similarly, the presence of calcium did not augment the response of AC to forskolin or guanosine-5'-O-(3-thiotriphosphate) (GTPγS), and did not change the activity of the unstimulated enzyme. However, a general non-specific inhibitory effect by calcium on AC activity can be observed.

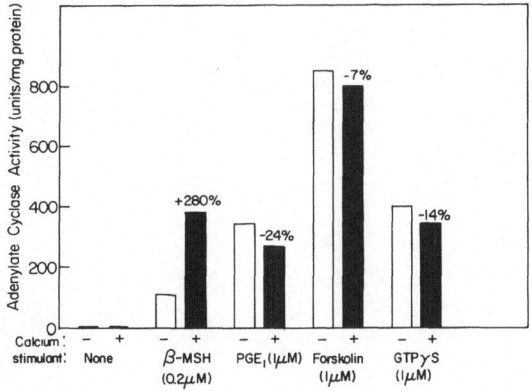

Fig. 2. Effect of calcium on adenylate cyclase activity in M2R plasma membranes. M2R plasma membranes were incubated alone or with β-MSH (0.2 μM), PGE_1 (1 μM), forskolin (1 μM), or GTPγS (1 μM), in the presence or absence of 10 μM calcium, under AC assay conditions. Determination of AC activity was as described by Gerst et al., (1986a). The results show the net increment in stimulated over unstimulated AC activity.

Next the dose-dependent stimulation of AC activity by β-MSH was determined in the presence or absence of added calcium. Calcium was found to shift the half-maximally effective concentration of β-MSH by a factor of 5, from 369±8 nM to 75±8 nM (n=3), as calculated from the net increment of hormone-stimulated AC activity over that of the unstimulated enzyme. Clearly, calcium appears to increase the apparent affinity of β-MSH for its receptor, by a factor of 5, as deduced from these studies.

Calcium- and Guanosine Nucleotide-dependent binding of β-MSH to M2R cell membranes.

To test the above hypothesis we examined the binding of [125I]-iodo-β-MSH (5-750 nM) in the presence or absence of 100 μM calcium (Fig. 3). The results of the dose-dependent binding of β-MSH under these conditions indicate that calcium shifts the apparent binding affinity of β-MSH by a factor of 20, from 418±40 (n=3) in the absence to 23±3 nM (n=7) in its presence. The increase in the binding affinity of β-MSH, seen in the presence of calcium, agrees well with observations shown above for β-MSH-stimulated AC activity. It should be noted, however, that the calcium-mediated shift in affinity is only 5-fold under these conditions, presumably due to the presence of GTP (10 μM) in the AC assay.

Guanosine nucleotides are known to decrease the affinity of certain hormones and neurotransmitters for their receptors, mediating their effects through the G-regulatory binding proteins involved in regulation of adenylate cyclase (Rodbell, 1980). GTP is essential for the stimulation of AC activity, however, when GTP (25 μM) is added to the binding assay the apparent binding affinity of β-MSH is shifted by a factor of 2, from 23±3 nM (n=7) to 49±4 nM (n=3). In addition, the maximal binding of β-MSH to its receptor is decreased by 63±4% (n=3). In comparison, the apparent affinities of β-MSH, as determined for AC activation or β-MSH

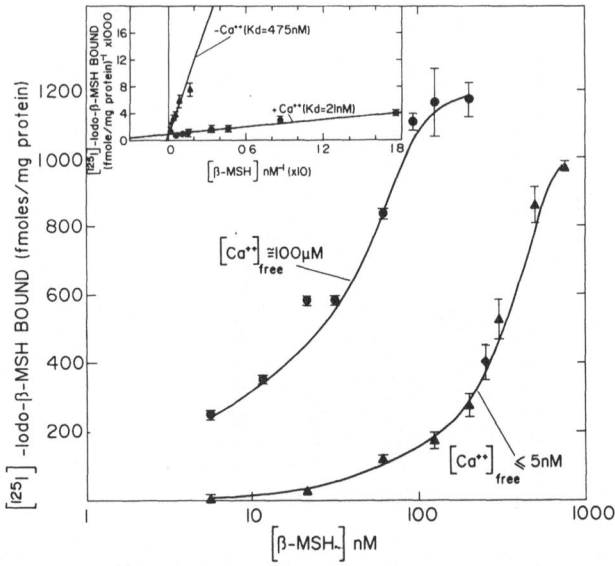

Fig. 3. Calcium-dependent binding of β-MSH of M2R plasma membranes. Binding of [125I]-iodo-β-MSH (5-750 nM) to M2R cell membranes was determined in the presence or absence of 100 μM calcium. Determination of hormone binding to cell membranes was as described by Gerst et al., (1986b).

binding, are similar, 75±6 (n=3) and 49±4 nM (n=3), respectively, in the presence of calcium and GTP.

The inhibitory effect of guanosine nucleotides on β-MSH binding was found to be concentration-dependent. The concentration of GTP required for half-maximal inhibition of hormone binding was found to be 0.9±0.2 μM (n=3). Non-hydrolysable analogs of GTP, such as guanosine-5'-(β,ɣ-imido)triphosphate (GppNHp), were similarly found to affect the affinity of β-MSH binding to its receptor. We also have examined the effects of guanosine nucleotides on the stability of pre-formed receptor-hormone complex. As can be seen (Fig. 4), GTP (25 μM) induces the dissociation of β-MSH pre-bound to M2R plasma membranes. The half-life of the receptor-hormone complex was found to be 3.5 minutes (4.0±0.3 min, n=4) under these conditions, although the extent of dissociation did not exceed 70% after 15 minutes of incubation with GTP. Similarly, both GTPɣS and guanosine 5'-O-(2-thiodiphosphate) (GDPβS) also facilitated the dissociation of bound β-MSH.

As shown above, β-MSH binding to its receptor at physiological hormone concentrations (≤ 5nM) is absolutely calcium-dependent, however at hormone concentrations in excess of 100 nM β-MSH binding becomes calcium-independent (Fig. 3). This indicates that calcium is not absolutely required for the association of β-MSH and its receptor. To test whether calcium is required for maintaining stability of the receptor-hormone complex β-MSH (25-175 nM) was pre-bound (40 min) to M2R membranes, under standard binding assay conditions, in the presence of calcium (1 mM) (Fig. 5). Excess EGTA (10 mM) was then added to reduce the free calcium concentration to ≤ 15 nM and the amount of residual bound hormone was determined. It can be seen that pre-bound β-MSH is rapidly dissociated after the addition of EGTA. At sub-saturating concentrations of hormone

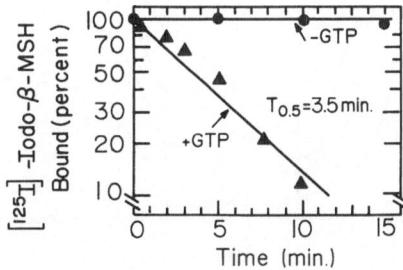

Fig. 4. GTP-induced dissociation of β-MSH bound to M2R plasma membranes. [^{125}I]-iodo-β-MSH pre-bound to M2R membranes, under standard assay conditions, was challenged with GTP (25 μM). Residual bound hormone was determined at various times after the addition of GTP or an equivalent aliquot of phosphate-buffered saline (-GTP). Method was as described by Gerst et al., (1986b).

(≤ 75nM) no apparent hormone re-binding is observed within 30 minutes of measurement. In contrast, at higher concentrations of β-MSH (> 75 nM) re-binding of the hormone in the absence of calcium is evident in a time and concentration-dependent manner. This receptor-hormone complex formed in the absence of calcium was insensitive to any further addition of EGTA (not shown). Upon re-addition of calcium to the assay a rapid increase in β-MSH binding is evident. Hormone binding upon the re-establishment of initial free calcium concentrations equilibrated in a time-dependent manner and reached the same levels to those seen prior to the addition of EGTA.

Experiments Figs. 3 and 5 demonstrate that two states of the receptor-hormone complex can be formed. The first represents a calcium-dependent high affinity state formed in the presence of calcium, while the second represents a low affinity calcium-independent state formed in the absence of calcium (see Fig. 6). It should be emphasized that the transition between the first state to the second requires prior dissociation of the receptor-hormone complex and may not be achieved solely by the simple dissociation of calcium.

To determine whether the GTP-induced dissociation of the receptor-hormone complex is also calcium-sensitive, we compared the ability of GTP to dissociate either state of bound β-MSH. Binding was done at hormone concentrations required to yield significant levels of both calcium-containing and calcium-depleted complexes. When these two forms of bound β-MSH were challenged with GTP (25 μM) their half-lives were found to be identical, 2.6±0.2 min and 2.5±0.4 min (n=3), respectively. The results clearly indicate that the effect of GTP on the dissociation of bound β-MSH is independent of the conditions under which the bound complex is formed, i.e., in the presence or absence of calcium.

The metal ion specificity of β-MSH binding to its receptor was tested using a variety of divalent metal cations substituting for

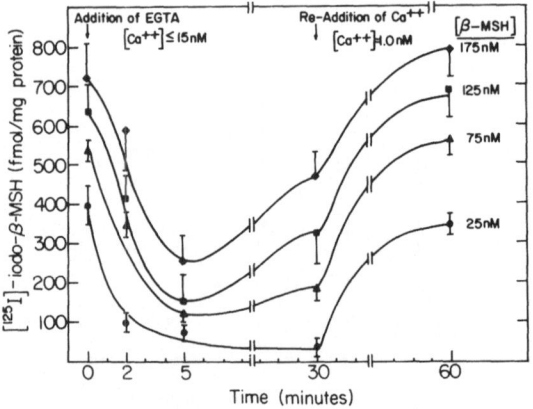

Fig. 5. Reversibility of β-MSH binding. [^{125}I]-iodo-β-MSH (25-175 nM) pre-bound to M2R plasma membranes was challenged with EGTA (10 mM) and residual bound hormone was determined for up to 30 min thereafter. Subsequently, CaCl$_2$ was added to restore the free calcium concentration to 1 mM and hormone rebinding was determined up to 30 min thereafter as described by Gerst et al., (1986b).

calcium. The order of effectivity of the various cations tested was $Ca^{++} >> Sr^{++} \gtrsim Ni^{++} > Ba^{++} > Mn^{++} > Cd^{++} > Co^{++} > Cu^{++} >> Mg^{++} >$ no ion added. Binding in the absence of added metal cations, in the presence of EGTA, was <5% of the calcium-facilitated binding. While both Sr^{++} and Ni^{++} are good substitutes for calcium, neither metal is physiologically present in significant concentrations to support hormone binding in vivo. These results indicate that calcium is likely to be the physiologically relevant metal cation involved in the regulation of the actions of β-MSH.

Effect of Calmodulin Antagonists on the actions of β-MSH

We have tested whether various calmodulin inhibitors, such as the phenolthiazine, fluphenazine, and the cytotoxic peptide, melittin, might have an effect on calcium-dependent β-MSH binding and/or stimulation of AC. It can be demonstrated that these substances inhibit both β-MSH binding and stimulation of AC in a dose-dependent manner and at concentrations similar to those described for the inhibition of calmodulin-dependent phosphodiesterase (Barnette et al., 1983) and myosin light chain kinase activity (Katoh et al., 1982). For example, half-maximal inhibition β-MSH-binding by either fluphenazine or melittin can be shown to occur at 14±1 μM (n=2) and 0.7±0.2 μM (n=3), respectively. The inhibitory effects of these substances on hormone binding strongly implicate a possible calmodulin involvement in β-MSH binding in the M2R cell line (see Fig. 6). Either calmodulin or a calmodulin-like calcium binding protein might therefore be considered to confer calcium sensitivity to the MSH receptor, in addition to the possibility that calcium binding may be an inherent property of the receptor itself.

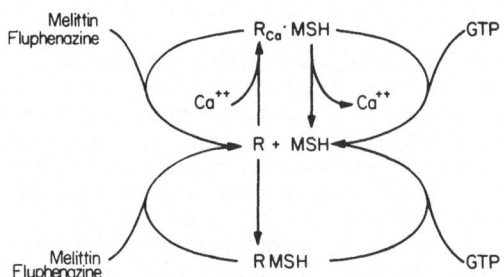

Fig. 6. A model for the regulation of β-MSH binding to its receptor. Binding of MSH to its receptor (R) may occur either in the presence or absence of calcium (Ca^{++}), leading to the formation of a high affinity EGTA-sensitive calcium-containing complex ($R_{Ca} \cdot MSH$) or a low affinity EGTA-insensitive calcium-depleted complex (R-MSH). Interconversion between $R_{Ca} \cdot MSH$ and R-MSH requires prior dissociation of the receptor-hormone complex and subsequent re-binding. Formation of either complex results in direct stimulation of AC. In addition, both complexes are equally sensitive to guanosine nucleotides, which decrease the stability of the complexes, leading to enhanced dissociation. Calmodulin antagonists, such as fluphenazine, or melittin, inhibit the formation of either complex in a dose-dependent manner.

DISCUSSION

We have demonstrated that β-MSH stimulation of adenylate cyclase, in cultured melanoma cells and in plasma membranes derived thereof, is calcium-dependent, and that the specific role of calcium is to facilitate the direct interaction between the hormone and its receptor. Calcium acts in a concentration-dependent manner, mediated by two separate classes of calcium binding sites, to regulate the affinity of the melanotropin receptor towards the hormone. At physiological hormone concentrations β-MSH regulation of AC activity is strictly calcium-dependent, and involves a freely reversible mechanism for the binding and/or dissociation of the hormone.

The MSH receptor appears to belong to a unique class of peptide hormone receptors, which also include the ACTH receptor, that specifically require calcium for their activity. Similarly, both receptor types belong to a large class of hormone receptors whose affinity is modulated by GTP. However, the regulatory effects of guanosine nucleotides appear to be calcium-insensitive, at least with respect to the MSH receptor.

The location(s) of the calcium-pertinent sites involved in MSH action have not been elucidated, however it may be assumed that these sites are exposed either to the extracellular milieu, the cytosol, or both. Since the concentration of extracellular free calcium is on the millimolar level it can be expected that extracellularly-oriented calcium binding domains would saturate under physiological conditions. Hence, under these conditions, calcium regulation of melanotropin action would be minimal unless conditions allowing for significant changes in extracellular calcium prevailed. If, on the other hand, the calcium binding domains are intracellularly-oriented, then changes in cytosolic free calcium levels (resting levels of calcium are approximately 100 nM) could be expected to confer a significant regulatory ability to these sites. Recently, Kojima et al., (1985) demonstrated the existence of ACTH-mediated calcium fluxes in isolated adrenal cortical cells. By analogy, melanotropins may also regulate calcium fluxes across the plasma membrane and thus associate hormone binding with the modulation of intracellular free calcium levels. The receptor-hormone complex may also be considered to function by itself as a calcium gating apparatus.

The embryology of melanocyte differentiation may provide a clue to the origin and nature of the calcium-pertinent sites involved in melanotropin action. Melanocytes differentiate from cells of the neural crest and then distribute themselves throughout the basal cell layer of the epidermis (Parker, 1981). However, melanocytes clearly differ from epidermal cells, both in morphology and in the metabolism of tyrosine and 3,4-dihydroxyphenylanine (DOPA), which is similar to that of nervous tissue (Parker, 1981). In addition to the distinct similarities between melanocytes and nervous tissue, melanotropins, corticotropins, and their common precursor, pro-opiomelanocortin (POMC), have been demonstrated to be present in various regions of the central nervous system (Krieger et al., 1980). Interestingly, both melanotropins and corticotropins have also been suggested to modulate the presynaptic release of neurotransmitter at the neuromuscular junction (Johnson et al., 1983). It is apparent then, that in addition to their actions on melanocytes, melanotropins themselves may be intimately involved in the activity of other cell types. Since the regulation of neuronal intracellular calcium concentration is of key importance for the presynaptic release of neurotransmitters into the synaptic cleft, then local changes in synaptic calcium may significantly affect both neurotransmitter release and synaptic transmission. If under certain physiological conditions, i.e., during calcium influx into the presynaptic terminal, calcium becomes limiting in the synaptic cleft, the activity of melanotropins may be expected to decline. Thus, melanotropin activity in the cleft could be modulated by synaptic transmission and vice versa. Calcium-limiting conditions in the synapse or neuromuscular junction may well have been a prerequisite for the development of a calcium-dependent mechanism for the regulation of melanotropin action. Thus, because of their neural origin, melanocytes may retain and utilize this calcium-dependent binding mechanism so intimately involved in the actions of MSH. Whether the calcium binding domains are localized intracellularly or extracellularly, their regulation of melanotropin binding activity appears to have strong physiological significance.

ACKNOWLEDGEMENT

This work was supported by a grant to Y.S. from the Hermann and Lilly Schilling Foundation for Medical Research, Federal Republic of Germany (Herman and Lilly Schilling stiftung fur medizinische Forshung im stiftenverband fur die Deutsche Wissenschaft). Y.S. is the Charles W. and Tillie Lubin Professor for Hormone Research. J.E.G.; in partial fulfillment of Ph.D. thesis work at the Feinberg Graduate School of the Weizmann Institute of Science. The authors gratefully acknowledge Rona Levin for her excellent secretarial assistance.

REFERENCES

Bar, H.P., and Hechter, O., 1969, Adenyl cyclase and hormone action. Proc. Natl. Acad. Sci. USA 63:350.

Barnette, M.S., Daly, R., and Weiss, B., 1983, Inhibition of calmodulin activity by insect venom peptides. Biochem. Pharmacol. 32:2929.

Cheitlin, R., Buckley, D.I., and Ramachadran, J., 1985, The role of extracellular calcium in corticotropin steroidogenesis. J. Biol. Chem. 260:5323.

De Graan, P.N.E., Everle, A.N., and Van de Veerdonk, F.C.G., 1982, Calcium requirement for α-MSH action on tail fin melanophores of Xenopus tadpoles. Mol. Cell. Endocrinol. 26:327.

De Graan, P.N.E., Van Dorp, C.J., and Van de Veerdonk, F.C.G., 1982, Calcium sites in MSH stimulation of Xenopus melanophores: Studies with photo-reactive α-MSH. Mol. Cell. Endocrinol. 26:315.

Gerst, J.E., Sole, J., Mather, J.P., and Salomon, Y., 1986a, Regulation of adenylate cyclase by β-melanotropin in the M2R melanoma cell line. Mol. Cell. Endocrinol. 46:137.

Gerst, J.E., Sole, J., and Salomon, Y., 1986b, Dual regulation of β-MSH receptor function and adenylate cyclase by calcium and guanosine nucleotides in the M2R. Melanoma cell line. Mol. Pharmacol. In press.

Johnson, G., and I. Pastan., 1972, Dibutyryl adenosine 3',5'-monophosphate induces pigment production in melanoma cells. Nature New Biol. 237:267.

Johnston, M.F., Kravitz, E.A., Meiri, H., and Rahaminoff, R., 1983, Adrenocorticotropic hormone causes long-lasting potentiation of transmitter release from frog motor nerve terminals. Science 220:1071.

Katoh, N., Raynor, R.L., Wise, B.C., Schatzman, R.C., Turner, R.S., Helfman, D.M., Fain, J.N., and Kuo, J., 1982, Inhibition by melittin of phospholipid-sensitive and calmodulin-sensitive Ca^{2+}-dependent protein kinases. Biochem. J. 202:217.

Kojima, I., Kojima, K., and Rasmussen, H., 1985, Role of calcium and cAMP in the action of adrenocorticotropin on aldosterone secretion. J. Biol. Chem. 260:428.

Krieger, D.T., Liotta, A.S., Brownstein, M.J., and Zimmerman, E.A., 1980, ACTH, β-lipotropin, and related peptides in brain pituitary, and blood. Rec. Prog. Horm. Res. 36:277.

O'Keefe, E., and Cuatrecasas, P., 1974, Cholera toxin mimics melanocyte stimulating hormone in inducing differentiation in melanoma cells. Proc. Natl. Acad. Sci. USA 71:2500.

Parker, F., 1981, Skin and hormones. in: "Textbook of Endocrinology, Sixth Edition. R.H. Williams ed., W.B. Saunders Publishers, Philadelphia. p. 1080.

Perchellet, J.P., and Sharma, R., 1979, Mediatory role of calcium and guanosine 3',5'-monophosphate in adrenocorticotropin induced steroidogenesis by adrenal cells. Science 203:1251.

Rodbell, M., 1980, The role of hormone receptors and GTP regulatory proteins in membrane transduction. Nature 284:17.

Sayers, G., Beall, R.J., and Seele, S., 1972, Isolated adrenal cells: Adrenal corticotropic hormone, calcium, steroidogenesis and cyclic adenosine monophosphate. Science 175:1131.

Van de Veerdonk, F.C.G., and Brouwer, E., 1973, Role of calcium and prostaglandin (PGE_1) in MSH-induced activation of adenylate cyclase in Xenopus laevis. Biochem. Biophys. Res. Commun. 52:130.

Varga, J.M., Dispasquale, A., Pawelek, J., McGuire, J.S., and Lerner, A.B., 1974, Regulation of melanocyte stimulating hormone action at the receptor level: Discontinuous binding of hormone to synchronized mouse melanoma cells during the cell cycle. Proc. Natl. Acad. Sci. USA 71:1590.

Vesely, D.L., and Hadley, M.E., 1971, Calcium requirement for melanophore-stimulating hormone action on melanophores. Science 173:923.

STRUCTURE-FUNCTION STUDIES OF GROWTH-FACTOR RECEPTORS

Mona Bajaj, Michael D. Waterfield and Thomas L. Blundell*

Ludwig Institute for Cancer Research at the Imperial Cancer
Research Fund, Lincoln's Inn Fields, London WC2A 3PX, U.K.
* Dept. of Crystallography, Birkbeck College, Malet St
London WC1 7HX, U.K.

SUMMARY

The extracellular regions of the human and drosphila EGF, c-erb-2 and
human insulin receptors each contain two large, homologous domains (L)
which are probably comprised of at least four α-helices followed by turns
of conserved lengths and β-strands. In the human and drosphila EGF and
c-erb-2 receptors these homologous domains are each followed by a series
of smaller cystine rich domains (S) to give a gene duplicated structure
$L_1S_{11}S_{12}S_{13}L_2S_{21}S_{22}S_{23}$. In the human insulin receptor the second series
of cystine domains is replaced by a different sequence. These duplicated
structures are probably organised as a pseudo-symmetrical dimer contained
within one chain with a "hypervariable" region at one end. This region
is a candidate for hormone or growth-factor binding.

INTRODUCTION

Growth-factors interact with specific cell-surface receptors to
induce a complex cascade of biochemical events which can result in DNA
synthesis and proliferation of target cells. Several growth-stimulating
peptides (epidermal growth-factor, EGF; platelet derived growth-factor,
PDGF; insulin-like growth factors, IGF's; and insulin) and their respective
receptors have now been identified and characterised. Although the mechanism by which these molecules regulate cell growth and proliferation is
as yet unknown, some insight into their function has come from the discovery
that they are structurally and functionally similar to several of the
retroviral oncogene products (Bishop, 1983; Hunter and Cooper, 1985;
Waterfield, 1985). Thus there is evidence that certain genes capable of
transforming cells encode growth factors (Doolittle et al., 1983; Waterfield
et al., 1983) or abnormal growth-factor receptors (Downward et al., 1984;
Ullrich et al., 1984; Sherr et al., 1985); the expression of these genes
allows cells to divide in a constitutive manner.

Recently the complete amino-acid sequences of the human EGF receptor
(Ullrich et al., 1984) and human insulin receptor(Ullrich et al., 1985)
have been deduced from cDNA clones while the sequence of a drosphila EGF
receptor homologue has been derived from the genomic DNA (Livneh et al.,
1985; Schejter et al., 1986). All these receptor sequences suggest the
presence of a single hydrophobic transmembrane strand. Such a topology
clearly distinguishes these receptors from the class of proteins exemplified

by the acetylcholine receptor, which is a multi-subunit protein and spans the membrane bilayer several times (Noda et al., 1983).

All the receptor molecules cited above comprise of an extracellular ligand binding domain, a single hydrophobic transmembrane region and a cytoplasmic tyrosine-specific kinase, which in the case of the EGF receptor molecule is homologous to the transforming protein of the avian erythroblastosis virus, v-erb (Downward et al., 1984). A high homology is also observed between the insulin receptor tyrosine kinase domain and the ros oncogene encoded tyrosine kinase (Ullrich et al., 1985).

In the human and drosphila EGF receptor, the extracellular regions which are responsible for binding the growth factors contain a gene-duplicated cystine rich sequence (Ullrich et al., 1984; Livneh et al., 1985). A similar arrangement of cystine rich domains is also observed in the neu oncogene; which is repeatedly activated in neuro- and glioblastomas. Structural analyses of the rat neu oncogene and its human cellular homologue c-erb-2, suggest these genes to encode an EGF-receptor related protein for an as yet unidentified growth factor (Bargmann et al., 1986; Yamamoto et al., 1986).

The insulin receptor comprises an α-subunit which binds the hormone and a β-subunit which contains the transmembrane an the kinase sequence, and these form a disulphide-linked $\alpha_2\beta_2$ quaternary structure (Massague et al., 1981). The primary structure of the insulin receptor (Ullrich et al.,1985) shows that the α-subunit is homologous to the cystine-rich region of the EGF receptor.

In this paper we review the evidence for gene divergent evolution of the four receptors. We show that the extracellular regions of both the human EGF and the c-erb-2 receptors comprise two large, homologous, probably α/β domains (L), each followed by several smaller cystine rich domains (S) to give an arrangement of L_1CCCL_2CCC. The drosphila receptor has a similar duplicated sequence of large α/β domains and cystine-rich domains, but the second set of cystine-rich domains is further duplicated. Finally, in the insulin receptor only the large α/β domain is duplicated and the second set of cystine rich domains is replaced by a quite different sequence, which contains few cystines. We consider the implications of this arrangement for receptor binding and the initiation of the biological response by these growth factors and hormones.

METHODS

We aligned first the two cystine rich domains of the human EGF receptor and then extended the alignment back into the preceding sequences. This revealed a pattern of conserved residues many of which were found to be invariant or conservatively varied in the drosphila EGF receptor. Finally, we aligned the human insulin and c-erb-2 receptor sequences.

The sequence alignments were then used to derive "fingerprints" which were used to search the Protein Information Resource data bank (Baker et al., 1984) using the soft-ware utilities of the Protein Sequence Query (PSQ) system and the template fitting proram, FITEM (Taylor, 1985).

Secondary structure prediction was carried out using the methods of Garnier et al., (1978) and Chou and Fasman (1974) and by searching for patterns of hydrophobic residues using methods similar to those of Lim (1974). We also identified flexible/hydrophilic regions using the method of Karplus and Schulz (1985). The most probable strand and turn regions were then used to investigate different topological arrangements that might contribute to domains.

For any individual pair of sequences the number of identities could be increased by manipulation of the alignments. However, our alignment places most insertions and deletions at what appear to surface loops and maximisès the conservatively varied residues,

The homologies show that the drosphila, human EGF and c-erb-2 receptor sequences are most closely related. However, the % homology between the duplicated sequences within the EGF receptors are comparable with the homology of the human EGF receptor to the human insulin receptor. The greatest divergence appears to be in the second domain of the insulin receptor and this may be related to the lack of the contiguous cystine rich repeats in this receptor. From this analysis it appears evident that the EGF receptors have diverged at a later stage than the human EGF and human insulin receptors. The internal duplication of the receptors may have taken place at approximately the same time if we assume equal rates of evolution for the two halves as for the two receptors, or slightly before if we assume that the restraints on the two halves are more equal, i.e. if they are involved functionally in a symmetrical way.

In Table 1 we have also indicated those positions where in the eight aligned domain sequences (two from each receptor), all/most residues are hydrophobic. This leads to a repeating pattern of four (and perhaps a fifth) motifs:

-8	-7	-6	-5	-4	-3	-2	-1	1	2	3	4
X	–	–	X	–	(X)	X	–	G	–	X	(X)	X

where X are usually hydrophobic, (X) are often hydrophobic, – is any residue but is usually hydrophilic, and G is glycine with only one exception(HIR II). We have aligned the sequences in Table 1 so that this repeat is evident. We suggest that it reflects similar supersecondary motifs that are repeated along the sequence of each domain. The conserved hydrophobics presumably reflect amino acids that contribute to the core of the protein. Glycines are most usually conserved in turns between secondary structures with tight loops (Sibanda and Thornton, 1985).

In order to identify the secondary structures we used the prediction algorithms of Garnier et al (1978), Chou and Fasman (1974) and Lim (1974). As sequence predictions using these methods are not likely to be more than 60% correct (Kabsch and Sander, 1983), we have treated the results with caution. We tended to use cumulative predictions not only by summing the individual probabilities suggested by each method, but also for homologous stretches of aligned sequences on the basis that secondary structure is more conserved in evolution than primary structure. The predictions are shown under the aligned sequences in Table 1.

RESULTS AND DISCUSSION

Table 1 shows our preferred alignment of the larger, non-cystine rich domains which occur twice in each of the extracellular domains of the four receptor sequences. Table 2 shows the number of identities between each pair of sequences for these domains and the percentage homologies.

The cummulative secondary structure prediction (shown at the bottom of Table 1) shows evidence of an alternating α-helix/β-sheet structure which would be characteristic of a parallel β-pleated sheet with α-helices packed against it. Noting the identical and conserved hydrophobic residues in Table 1, a clearly repeating pattern emerges which corresponds to those regions indicated as α-helical or β-strands. Those regions thought

Table 1. Sequence alignment of the non-cystine rich domains of the drosphila EGF receptor (DER), human EGF receptor (HER), c-erb-2 receptor (c-erb) and the human insulin receptor (HIR).

```
                                         G
DER I     43  G Y V D N G N M K V C I G T K S R L S V P S N K E H   68
HER I      1              L E E K K V C Q G T S N K L T Q L G T F E D   22
c-erb I                  A A S T Q V C T G T D M K L R L P A S P E T   41        MOTIF I
HIR I                  L H L Y P G E V C P G M D I R   14
DER II   353  T C P G V T V   359
HER II   312  V C N G I G I G E F K D S L S   326
c-erb II 341  V C Y G L G M E H L R E V R A   355
HIR II   311  V C H L L E G E K T I   321

DER I     69  H Y R N L R D R Y T N G T Y V D G N L K T W L P N Y T   97
HER I     23  H P L S L Q R M F N N C E V V L G N L E I T Y V Q R N Y   50
c-erb I   42  H L D M L R H L Y Q G C Q V V Q G N L E L T Y L P T N A S   70
HIR I     15  N N L T R L H E L E N C S V I D G N L I F L Q ... P E D F R   47
DER II   360  L H A G N I D S F R N C T V I S G D L H I L D Q T F S G F D V Y A N Y T M G P R Y I   402      MOTIF II
HER II   327  I N A T N I K H F K N C T S I S G D L H I L P V A F R G D S F T H T P   361
c-erb II 356  V T S A N I Q E F A G C K K I F G S L A F L P E S F D G D P A S N T A   390
HIR II   322  D S V V T S A Q E L R G C T V I N G S L T I N I R G G N N   349

DER I     98  D L S F L D N I R E V T G Y T L I S   115
HER I     51  D L S F L K T I Q E V A G Y V L I A L N T V E R   74
c-erb I   71  L S F L K - I L I   87
HIR I     48  L S F P K - I L I T D D Y L F R V Y   67
DER II   403  P L D P E R E V F S T V K E I I T G T H   428                         MOTIF III
HER II   362  P L D P Q E L D I F H K T V K E I I T G F L N I Q A W P   387
c-erb II 391  P L Q P E Q L Q V F E T L E E I I S G Y L Y I S A W P   416
HIR II   350  P L A E E L E A N L G L L E E I S G Y L K I R R S Y   374

DER I    116  H V D V K K V F P K L Q I I R G R T L F S L S V E E K Y A L F V T Y S   151
HER I     75  H N Q V R Q V P L Q R L R G T Q L F E D N Y A L A V L S N Y D A N K T   106
c-erb I   68  G L E S L K D L F P N L T V I F E M V   99
HIR I    429  P Q F R N L S Y F R N L E T I I H G R Q L M E S M F A A L A I V K S   461
HER II   388  E N H T D L H A F E N L E I I R G R T K Q H G Q F S L A V V S L   419       MOTIF IV
c-erb II 417  D S L S F L K F E Q N L Q V I R G R I L H N G A Y S L T L Q G L   448
HIR II   375  A L V S L S F E R R K L R I G E T L E I G N Y S F Y A L D N Q N   407

DER I    152  K M Y T L E I P D L R D L N G Q V G F   171
HER I    307  G L R E L P M R N L Q E I L H G A V R F   126
c-erb I  136  G L R E L G L R N L T E I I T R G S V R I   155
HIR I    100  H L K E L G L Y N L M N I T R G S V R I   119
DER II   462  S L Y S L E M R N L K Q I I S G V V I   481                          MOTIF V
HER II   420  N I T S L G L R S L K E I S A G S V I I   439
c-erb II 449  G I S W D W S K H N L T T Q G K L F   428
HIR II   408  L R Q L W D W S K H N L T T Q G K L F   428
```

α-helix —— β-strand-turn-βstrand
```
```

Table 2. Number of identities and percentage homologies (in brackets) between each pair of the non-cystine rich domains aligned in TABLE 1.

```
DER11   Cpk      Ches          Cthg        Cwgegpkn          Cqkfsklt        Cspq    Caggr    Cygpkpre
HER11   Cqk      Cdps          Cpngs       Cwgageen          Cqkltkii        Caqq    C Sgr    Crgkspsd
HIR11   Cgdi     Cpgtakgktn    Cpatv(8)    Cwths             Cqk    v        Cpti    Ckshg    Ctaegl
cERB11  Chp      Cspm          Ckgsr       Cwgessed          Cqsltrtv        Gagg    Car      Ckgplptd

DER21   Cekngti  Csdq          Cnedg       Cwgagtdq    Clt   Cknfnengt       Ciad    Cgyisnaykfdngt
HER21   Ckatgq   Chal          Cspeg       Cwgpeprd    Cvs   Crnvsrgre       Cvdk    Cnllegeprefvense
cERB21  Cvgagla  Chql          Cargh       Cwgpgtq     Cvn   Csqflrgqe       Cvee    Crylqglpreyvnarh

DER31   Crk      Chpl          Cel         Ctnygyheqv  Csk   Cthykrreq       Cete    Cpadhytdeeqre

DER12   C        Chlf          Cagg        Ctgptqkd    Cia   Cknffdqav       Skee    Cppm(23)
HER12   C        Chnq          Cagg        Ctqpresd    Clv   Crkfrdeat       Ckdt    Cppl(23)
HIR12   C        Cheq          Clgn        Csqpddptk   Cva   Crnfyldgr       Cvet    Cppp(8)
cERB12  C        Cheq          Cagg        Ctgpkhsd    Cla   Clhfnhsgi       Celh    Cpal(23)

DER22   Cki      Chpe          Crt         Cngagadh    Cqe   Cvhvrdgqh       Cvse    Cpknkyndrgv
HER22   Ciq      Chpe          Clpqamnit   Ctgrgpdn    Ciq   Cahyidgph       Cvkt    Cpag(18)
cERB22  Clp      Chpe          Cqngsvt     Cfgpeadq    Cva   Cahykdppf(9-12) Cvar    Cpsg(19)

DER32   Cfq      Rhpe          Cng         Ctgpgadd    Cks   Crnfk(17-20)    Ctsk    Cple(14)

DER13   Cvke     Cpghllrdnga-Cvrs          Cpqdkmdkgge Cvp   Cngp            Cpkt    C(16-19)
HER13   Cvkk     Cprnyvvtdhg-Cvra          Cgadsye(8)  Ckk   Cegp            Cegp
HIR13   Cvnfsf   Chqyvihnnk  Cipe          Cpsgytm(8)  Ctp   Clgp            Cpkv
cERB13  Cvta     Cpnylstdvg  Ctlv          Cplhnqe(9)  Cek   Cskp            Carv

DER23   Cre      Chat          Cdg         Ctpg(8)     Ctt   Cnla(13)        Cll(4)   C(25)
HER23   Chl      Chpn          Ctyg        Cgleg       C
cERB23  Cqp      Cpins         Cths        Cvdlddkg    C
```

Table 3. The alignment illustrates the occurence of a 3-fold repeat in each cystine rich domain, each repeating unit containing 8 cystine residues.

	DER1	HER1	c-erbB2I	H1R1	DER2	HER2	c-erbB2II	HIR2
DER1	0							
HER1	36(46)	0						
c-ERB1	48(62)	47(57)	0					
H1R1	25(44)	30(45)	19(37)	0				
DER2	21(40)	23(35)	22(35)	26(44)	0			
HER2	23(41)	27(39)	23(39)	28(44)	37(49)	0		
c-ERB2	22(36)	20(37)	24(37)	25(35)	31(46)	41(56)	0	
HIR2	20(37)	24(35)	24(40)	19(32)	23(42)	21(40)	30(43)	0

to be α-helical tend to have pairs of residues between the conserved hydrophobic positions whereas those regions which are predicted as β-sheet tend to have conserved hydrophobic residues at alternating positions in the sequences. Most intriguing of all is the tendency to find a conserved glycine at positions between the α-helical and β-sheet regions. This is consistent with a tight turn between the α-helix and β-strand. Interestingly, the repeating α-helix-G-β-sheet regions show some similarities in the pattern of hydrophobics on each side of the glycine, indicating that they may have rather similar conformations.

In total there are four repeating α/β motifs and perhaps a fifth in -NH$_2$ terminal residues which is poorly conserved. It is most likely that these form a parallel β-sheet flanked by α-helices. There are two possibilities. If the domains are not associated in pairs there are insufficient β-strands to form a β-barrel structure and it is likely that a twisted sheet is formed similar to that in many α/β -structures. However, it is possible that the domains associate in pairs with a local dyad axis. In this case the two structures could form an eight (or less likely ten) stranded β-barrel with surrounding helices. Alternatively the two β-sheets could hydrogen bond to form an extended β-sheet protein. In both cases all the glycine would be at one end of the sheet. This end shows far more evolutionary conservation both in terms of identities of residues and in terms of insertions and deletions. It can therefore be suggested that this region might have a common structural-functional role in the different receptors.

An examination of the proposed glycosylation site shows that they occur most frequently in the strands we have indicated as α-helical, occasionally in the turn regions and least commonly in the β-strand region.

Table 3 shows the alignment of the cystine rich regions. These occur twice in each of the EGF receptors immediately after the larger domains described above, but in the insulin receptor they occur only once between the two larger domains. Interestingly, in the drosphila EGF receptor sequence, the second cystine rich domain is immediately followed by a third cystine-rich domain. Each cystine-rich domain can be arranged as three (or two in the third region of the drosphila receptor) sub-domains with eight cystines. This is reminiscent of the EGF precursor (Gray et al., 1983; Scott et al., 1983) and the LDL receptor (Sudhof et al., 1985), but the repeated sequences show no homologies with these regions or indeed with any other repeated cystine rich repetitive sequences (for example those of neurotoxins). In the absence of cystine bridge information it is not possible to be sure that the cystine pattern is similar in each repeated sequence. The variation of both the number and the nature of amino acids in the loops between the conserved cystines indicates that these repeated structures probably have a structure in which the cystines are bunched in the core and the loops extend out into the solvent. In this way they would resemble the structures of the neuro-toxins which also show highly variable loop structures.

Figure 1 shows possible structures for the EGF and insulin receptors. The close proximity of the second cystine rich domains of the EGF receptors to the transmembrane sequences indicates that these domains must be close to the membrane. In our suggested arrangement the duplicated sequences form symmetrical dimeric structures. This is a common feature of internally duplicated sequences for example in the lens crystallins (Blundell et al., 1981) or the aspartic proteinases (Tang et al., 1978). This arrangement would place the larger domains away from the membrane.

Let us now consider arguments concerning the binding of the hormone. The wide distribution of cystine rich domains suggest that they play a crucial role in the function of the receptor molecules. If these regions are involved in ligand-binding it is possible that cystine exchange might occur- this has not been strictly excluded for either EGF or insulin. Alternatively the variable loops may have evolved to form complememetary regions. However, this idea is not very attractive as a cystine rich region would not be the most suitable structure for mediating the biological response as it would be less easily able to undergo a conformational exchange. Its position close to the membrane would also make it less available to the hormone or growth-factor.

A more attractive model would involve the pseudo-symmetrical pair of larger domains forming a region not unlike the heavy and light chain variable regions of an immunoglobulin domain. Indeed such a model would be consistent with most of the varaibility between the insulin and EGF receptors being at one end of the sheet, i.e. the variable residues between the more conserved α-helix, β-strand regions act as a sort of hypervariable region. In this way they form regions complememtary to the very different EGF and insulin molecules.

NEW APPROACHES TO THE UNDERSTANDING OF GROWTH FACTOR AND ITS RECEPTOR

Of course the ultimate description of growth-factor and receptor interaction will come from studies involving direct binding of ligand to purified receptors, which now seems possible. We are pursuing this attractive possibility by isolating and characterising complexes between the ligands and the receptor-binding domains, which appear to be largely extrinsic to the membrane. Such complexes may be studied by both spectroscopic and X-ray techniques. In the meantime other approaches may be helpful in further defining receptor interactions. Site-directed mutagenesis combined with recently devoloped computer-aided design appears to offer particular advantages and is one of the approaches being followed at present.

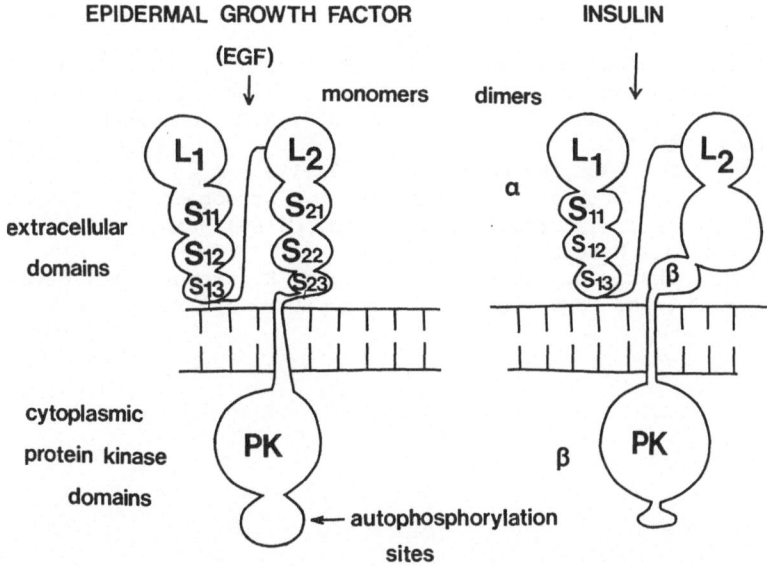

Figure 1.   Proposed domain structures for the human EGF and Insulin Receptors.

REFERENCES

Bargmann,C., Hung, H.C., and Weinberg, R.A., (1986) Nature 319:226.

Bishop, J.M. 1983 Rev. Biochem. 52:301.

Blundell, T.L., Lindley, P., Miller, L., Moss, D., Slingsby, C., Turnell, W.G. and Wistow, G., 1981 Nature 289:771.

Chou, P.Y. and Fasman, G.D., 1974 Biochemistry 13:222.

Doolittle, R.F., Hunkapiller, M.W., Hood, L., Devare, S.G., Robbins, K.C., Aaronson, S.A., and Antoniades, H.N., 1983 Science 221:275.

Downward, J., Yarden, Y., Mayes, E., Scrace, G., Totty, N., Stockwell, P., Ullrich, P., Schlessinger, J. and Waterfield, M.D., 1984 Nature 307:521.

Garnier, J., Osgusthorpe, D.J., and Robson, B. 1978 J.Mol. Biol. 120:97.

Gray, A., Dull, T.J., and Ullrich, A., 1983 Nature 303:722.

Hunter, T. and Cooper, J.A., 1985 Ann. Rev. Biochem.54:897.

Kabsch, W. and Sander, C.,(1983) FEBS Lett. 155:179.

Karplus, M.A., and Schulz, G.E., 1986 Naturwissenschaften 72:212.

Lim, V.I., 1974 J.Mol.Biol. 88:873.

Livneh, E., Glazer,L., Segal, D., Schlessinger, J., and Shilo, B.-Z., 1985 Cell 40:599.

Massague, J., Pilch, P.F., and Cassell., 1981 J.Biol. Chem. 256:3182.

Noda, M., Takahashi, H., Tanabe, T., Toyosato, M., Kikyotani, S., Furutani, Y., Hirose, T., Takashima, H., Inayama, S., Miyata, T., and Numa, s., 1983, Nature 302:528.

Schejter, E.D., Glazer, L., Segal., D., Ullrich, A., Schlessinger, J., and Shilo, B.-Z., 1986, submitted for publication.

Scott, J., Urdea, M., Quirago, H., Sandchez-Pescador, R., Fong, N., Selby, M., Rutter, W.J., and Bell, G.I., 1983, Science, 221:236.

Sherr, C.J., Rettenmier, C.W., Sacca, R., Roussel., M.F., Look, A.T., and Stanley, E.R., 1985, Cell 41:665.

Sibanda, B.L., and Thornton, J.M., 1985, Nature 316:170.

Sudhof, T.C., Russell, D.W., Goldstein, J.L., Brown, M.S., Sanchez-Pescador, R., and Bell, G.I., 1985, Science 228:893.

Taylor, W.R., 1985, J.Mol.Biol. 188:233.

Ullrich, A., Bell, J.R., Chen, E.Y., Hervera, R., Petruzelli, L.M., Dull, T.J., Gray, A., Coussens, L., Liao, Y.-C., Tsobokawa, M., Mason, A., Seeburg, P.H., Grunfeld, C., Rosen, O.M., and Ramachandran, J., 1985, Nature 313:756.

Ullrich, A., Coussens, L., Hayflick, J.S., Dull, T.J., Gray, A., Tam, A.W., Lee, J., Yarden, Y., Liberman, T.A., Schlessinger, J., Downward, J., Mayes, E.L.V., Whittle, N., Waterfield, M.D., and Seeburg, P.H., Nature 309:418.

Waterfield, M.D., 1985, Prog.Med.Virol. 32:129.

Waterfield, M.D., Scrace, G.T., Whittle, N., Stroobant, P., Johnsson, A., Wasteson, A., Westermark, B., Heldin, C.-H., Huang, J.S., and Deuel, T.F., 1983, Nature 304:35.

Yamamoto, T., Ikawa, S., Akiyama, T., Semba, K., Nomura, N., Miyajima, N., Saito, T., and Toyoshima, K., 1986, Nature 319:230.

MOLECULAR INTERACTIONS OF GANGLIOSIDE RECEPTORS WITH TETANOTOXIN ON SOLID SUPPORTS, AQUEOUS SOLUTIONS AND NATURAL MEMBRANES

Ephraim Yavin, Philip Lazarovici and Anetta Nathan

Department of Neurobiology
The Weizmann Institute of Science
Rehovot, Israel

INTRODUCTION

Homeostatic regulation of the mammalian cell is mediated, among other factors, by extrinsic signals in the form of hormones, growth substances or pharmacologically active agents which interact with specific cell surface determinants. The molecular details by which the message generated from these interactions is transduced through the lipid milieu to affect cell function are unclear.

Gangliosides are ubiquituous lipid constituents which interact with such specific ligands as toxins, glycoprotein hormones, viruses and interferon (1). The molecular diversity, the amphiphatic properties and the proximity of their binding determinants to the cell surface render the gangliosides as targets for specific membrane perturbations hence as potential candidates for signal transduction through the bilayer.

In our laboratory we are studying the role of gangliosides in binding and internalization of tetanotoxin (TeTo) using nerve cells in tissue culture. Specific polysialogangliosides which accumulate in the course of maturation of fetal cerebral neurons (2) or these which are artificially added to polysialoganglioside-deficient neural cells (3) seem to participate in the process of toxin translocation from an extracellular to an operationally defined intracellular milieu (4).

TeTo is a most powerful bacterial neurotoxin which is recognized by peripheral nerves, internalized, transported to the central nervous system and released at the synaptic junction of a second nerve where it blocks neurotransmitter release, thus exerting biotoxicity (5). The toxin is composed of a heavy chain (87 Kd) and a light chain (47 Kd) connected by a disulphide bridge (6). Papain digestion of the holotoxin cleaves the heavy chain to yield a fragment of 47 Kd (fragment IIc) and a larger fragment (fragment Ibc) which is composed of the remainder of the heavy chain with the attached light chain (7). It is now conceived (8) that these fragments contain two separate and distinct active sites: a) a ganglioside binding site located in fragment IIc, assumed to be responsible for toxin binding and internalization and b) a biotoxic site, located in the heavy chain of fragment Ibc assumed to be involved in the central inhibition of transmitter release.

TeTo is believed to enter the central nervous system by a process known as retrograde axonal transport (9-11). The toxin interacts with specific membrane components of a peripheral nerve, is internalized and transported to the central nervous system in association with defined vesicular structures as revealed by E.M. autoradiography. The toxin, which is bioactive during this process, crosses the postsynaptic membrane to exert its action. The first events in this process is the recognition of TeTo by surface membrane receptor(s) and the linkage between receptor recognition and neuroskeletal system action in moving bioactive, undegraded and vesicular bound polypeptide.

In previous studies using rat primary neuronal cultures enriched in polysialogangliosides (12) somatic neurohybrid cells (3) and human erythrocytes artificially supplemented with these substances (13,14) we defined a ganglioside-mediated three-step interaction of the toxin with the cells. The first step involved a reversible, low ionic strength and energy-independent binding process followed by a second, salt-insensitive and detergent nonextractable toxin-cell association, operationally defined as sequestration. Finally, in living cells, this process resulted in an energy-dependent internalization defined by the criteria of resistance of the toxin-membrane complex to neuraminidase treatment (4).

Despite these observations, little is known concerning the details of the molecular interaction between the toxin and the ganglioside receptors in the native membrane. Recently, using an affinity purified TeTo fraction (15), we have demonstrated that binding to guinea pig synaptic preparations depends on gangliosides but is facilitated by a protease-sensitive and phospholipase-sensitive component (16). Similarly, we have found that TeTo receptors on nerve cells contain a trypsin-sensitive component (17). The existence of this component was demonstrated upon cell treatment with trypsin and chymotrypsin. Each of the proteases reduced binding by about 40% compared to control values. Treatment of the trypsin insensitive receptors with sialidase caused an additional, and nearly total loss of binding activity. Loss of binding was also encountered after methanol extraction of cells indicating the lipid nature of the remaining receptor. Cross-linking agents such as formaldehyde and glutaraldehyde were also effective in abolishing binding activity. The data suggested that in addition to a sialic acid component, binding of TeTo to cerebral neurons was facilitated by a protease-removable and formaldehyde-inactivated component.

2. THE GANGLIOSIDIC NATURE OF THE TeTo RECEPTOR

A major question raised by these studies concerned the molecular specificity of the ganglioside for maximal binding. In one approach, we have adsorbed pure gangliosides to nitrocellulose paper and overlayed ^{125}I-labeled TeTo. Secondly, we have exposed cell lipid extracts after separation by TLC or cellular proteins after SDS gel electrophoresis to ^{125}I-labeled toxin to study binding. Thirdly, we have artificially supplemented a variety of cells with the appropriate gangliosides and obtained elevated levels of TeTo association with the cells (3,13). Alternatively, we applied sialidase to trim the cell surface or used sodium periodate to oxidize sialic acid residues (18). The latter modulations of the cell surface emphasized the unique properties of the ganglioside receptor and led to further studies designed to understand the biophysical properties of the toxin in its interaction with the ganglioside.

2.1. TeTo Interaction with Gangliosides on Nitrocellulose Paper

An overlay of ^{125}I-TeTo on nitrocellulose paper embedded with various gangliosides is depicted in Figure 1. It is evident that GT1b and GD1b were the most active substrates interacting with the toxin.

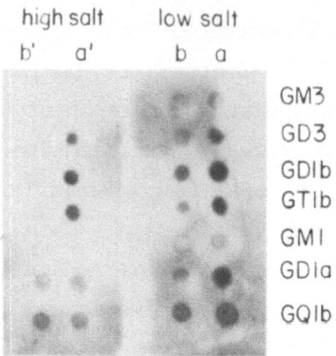

Figure 1. Binding of ^{125}I-TeTo to pure gangliosides dot-spotted on nitro-cellulose paper. For procedural details, see Ref. 18.

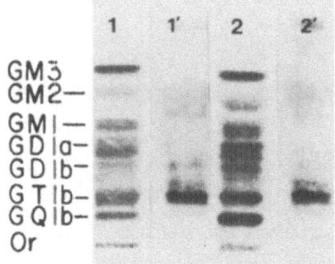

Figure 2. (A) Autoradiography of ^{14}C-glucosamine labeled gangliosides of PC-12 cells in the absence (1) or presence (2) of NGF. Overlay with ^{125}I-TeTo in the absence (1') or presence (2') of NGF. Abbrev: tetra (GQ1b); tri (GT1b); di (GD1a, GD1b); and mono (GM1, GM2, GM3) sialogang-liosides. For procedural details, see Ref. 12.

Binding to GM3 and GM1 gangliosides was lowest, but appeared substan-tially higher with GD3. Surprisingly enough, GQ1b which is a higher homolog of GT1b, bound relatively less TeTo. Essentially, the data pre-sented here are in accord with the findings of Holmgren et al. (19) sug-gesting GD1b and GT1b as the most effective receptors for the toxin.

2.2. TeTo Interaction with Gangliosides after Separation on Silica Gel G

A thin layer chromatoplate of ^{14}C labeled glucosamine lipid extracts from a clonal line of pheochromocytoma, PC-12 (20) is shown in Figure 2. Clearly, PC-12 cells contain the entire spectrum of complex gangliosides as usually seen in the normal nervous system (21,22). Furthermore, after treatment with NGF (see lane 2), an overall increase in the labeling of complex gangliosides was observed.

When the isolated lipids from the cells were subjected to ^{125}I-labeled TeTo (Lanes 1', 2'), the label appeared to be confined mainly to the area corresponding to GT1b and lesser to an area corresponding to GD1b gangliosides. Little if any label was present on GD1a or GM1 gangliosides. Western blots of the cellular proteins separated on SDS-PAGE gel exhibited little if any binding as shown in figure 3. At the front, binding corresponded to a region where gangliosides migrated on the gel. The data indicated that the receptor for the toxin is lipid soluble and that no associated protein component can substitute it.

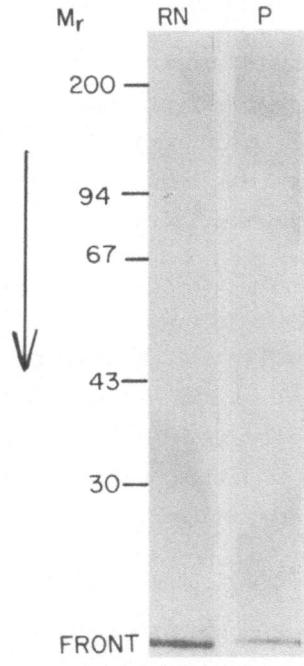

Figure 3. Binding of ^{125}I-TeTo to nitrocellulose paper after electroelution of SDS-acylamide electrophoresed proteins from rat neuron (RN) cultures and PC-12 (P) pheochromocytoma cells.

2.3. TeTo Interaction with Cells Artificially Supplemented with Gangliosides

Somatic neurohybrid SB21-B1 cells grown in serum exhibit limited capacity to bind ^{125}I-labeled tetanus toxin and cannot synthesize gangliosides higher than GM2. By 6 hr after supplementing the culture medium with pure or mixtures of brain gangliosides, binding of ^{125}I-labeled TeTo to cells increased approximately 8-fold (3). The uptake of added gangliosides was saturable and was facilitated by serum removal by 2.1-fold. Enhancement of TeTo binding to cells depended on the ganglioside species and concentration; GT1b (25 μg/ml) was respectively, two and three times as effective as GD1b and GM1 in increasing toxin binding. Reconstitution of ganglioside-mediated TeTo binding activity was a reversible phenomenon; removal of medium gangliosides caused a 3-fold drop in toxin binding by 24 h, after which an apparent plateau for at least 3 days above the basal level was established. As in cerebral cultures, binding of toxin to ganglioside-supplemented neurohybrid cells exhibited salt and sialidase sensitivity and was enhanced 2.6-fold at 37°C compared to 0-4°C. The resultant temperature-dependent toxin-cell association was found sialidase insensitive. Fixation of cells by formaldehyde or treatment of ganglioside-supplemented cells with trypsin had no substantial effect on ganglioside-mediated binding of the toxin while a methanol/chloroform treatment of cells caused a 91.4% loss of binding activity.

These studies demonstrated that the neural hybrid cells can insert gangliosides into their membranes resulting in binding of TeTo to the cells. The interaction was indistinguishable from that of primary cortical neurons suggesting that the interaction of the toxin with the cell membrane ganglioside is the most prominent and initial event in toxin translocation.

Primary cortical neurons were also able to incorporate GD1b from the culture medium. As seen in Table 1, binding of TeTo to control cultures increased over 5-fold over a period of one week in vitro. Addition of 25 μg/ml GD1b at day 1 or 4 increased binding of the toxin to near saturation levels. At day 7 or 21 little or no excess binding was seen in the presence of gangliosides suggesting that the cells reached saturation with respect to toxin binding sites.

Table 1. Effect of ganglioside (GD1b) supplement on TeTo binding to cultured nerve cells.

Days in culture	Non-supplemented	Supplemented
	---cpm/well---	
1	2215	8625
4	4820	9450
7	11080	12890
21	12340	11400

Cells were prepared and incubated with 25 μg/ml GD1b in serum free DMEM as described elsewhere (4).

Additional studies were carried out with human erythrocytes artificially supplemented with GD1b and GT1b gangliosides (13,14). Under these conditions a macromolecular complex of about 700,000 daltons containing toxin and gangliosides has been isolated and characterized by Sephacryl S-300 gel permeation chromatography, SDS-gel electrophoresis, immunoprecipitability and biotoxicity. This complex was obtained only in ganglioside-supplemented cells and not when free ^{3}H-GD1b was reacted with ^{125}I-labeled toxin in solution in the absence of cells. The hydrophobicity properties acquired as a result of ganglioside-toxin interaction, presumably at the cell surface suggested a conformation change of the toxin which is believed to facilitate its penetration into the bilayer.

2.4. TeTo Interaction with PC-12 Cells after Periodate Oxidation.

The ability of PC12 cells to form their own complex gangliosides such as GT1b and GD1b was already noted (21,22). In spite of that, these cells bind rather ineffectively TeTo in comparison to fetal cerebral neurons in monolayer cultures. The limited binding capacity may stem from the cryptic state of the ganglioside receptor at the cell surface. This prompted us to search for agents known to alter the cell surface sialic acid. Sodium periodate was found most effective in enhancing binding of TeTo as clearly demonstrated in Table 2.

Table 2 Effect of periodate oxidation on binding of ^{125}I-tetanotoxin to NGF-treated or untreated PC12 cells

Cell Treatment	Untreated		+NGF		Stimulation by NGF (X)
	cpm/µg DNA	fold change	cpm/µg DNA	fold change	
Control	41± 4		98± 12		2.4
Periodate	109± 9	2.7	252± 16	2.6	2.3

Cells in 35 mm culture dishes grown for 2 days in the absence or presence of 50 ng/ml NGF were treated with 2 mM periodate for 20 min at 4°C. Excess reagent was removed and cell monolayer rinsed twice with Pi/saline. ^{125}I-tetanotoxin (100,000 cpm) was added for 75 min at 37°C. Cell associated radioactivy was determined by the gamma-counter and DNA was determined on the pellet. Values represent triplicate cultures ±SEM.

Stimulation of binding by periodate was most pronounced in cells which contained GD1b and GT1b gangliosides. It was close to zero in L-8 fibroblasts and was very insignificant in astrocyte cultures. Periodate-stimulated binding was specific as it was inhibited by excess (500-fold) unlabeled toxin and was 15-fold greater when TeToB, an affinity purified toxin was used instead of TeToA. The latter is highly neurotoxic but has no ability to bind to gangliosides (15). When periodate treated cells were subjected to borohydride, binding activity was restored to normal levels. Periodate caused also a profound stimulation of binding when used on cells which have been artificially supplemented with G1b gangliosides.

Periodate, a well-known oxidant used at low concentrations (1-2mM), low temperature (4°C) and for short time periods (10-20 min), has been shown to cleave the ninth or eighth carbon of sialic acid residues on the cell surface. The aldehyde generated under these conditions would appear to serve as a better acceptor for TeTo than the hydroxyl group. Oxidation of the cell surface with periodate emphasized again the pivotal role of the lipid-linked sialic acid in binding of tetanotoxin.

<u>3</u>. TeTo UNDERGOES CHANGES AFTER INTERACTION WITH GANGLIOSIDES

In initial experiments we observed that Triton X-100 solubilized preparations of synaptosomes or ganglioside-enriched human erythrocytes after incubation with TeTo gave rise to high molecular weight complexes (300-700 Kd) when subjected to Sephacryl S-300 gel permeation chromatography (23). Studies utilizing gel electrophoresis, analytical ultracentrifuge circular dichroism (CD) and electron microscopy (EM) were initiated to monitor the biophysical properties of the toxin molecule after binding/sequestration.

Particularly, we have utilized the analytical ultracentrifuge to test the hypothesis that GT1b/GD1b (G1b) gangliosides interact specifically with TeTo to induce conformational changes and aggregation of the polypeptide under aqueous conditions. As shown in Figure 4, the interaction between GT1b/GD1b gangliosides and affinity purified TeToB gave rise to aggregates of a sedimentation coefficient of 12 and 24S.

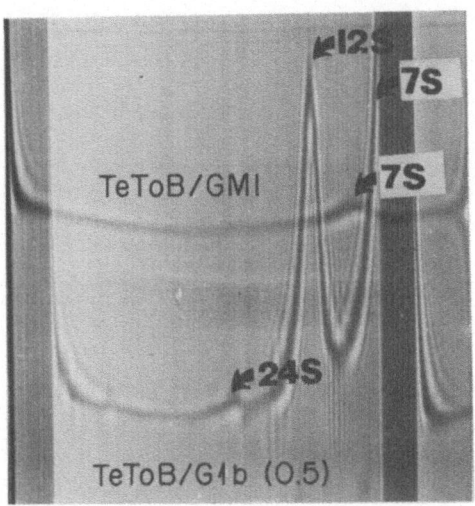

Figure 4. Separation of affinity purified TeToB/GM1 and TeToB/G1b complexes by analytical ultracentrifuge. For details, see Ref. 23.

There were no high molecular weight peaks when TeToB was sedimented in the presence of GM1 or in the presence of 10^{-10}M GT1b ganglioside. The latter observation suggested that the aggregates were formed only in the presence of micellar rather than monomeric gangliosides. The sedimentation pattern of TeToA in the presence of G1b ganglioside showed the vast majority (98%) of the toxin to reside in the 7S peak which is equivalent to an $Mr \simeq 150,000$.

The sedimentation values obtained after running various TeToB/GT1b complexes were used to estimate the number of toxin molecules associated with the ganglioside micelle (Table 3). At a molar ratio ranging between 3.1-5.8, the apparent size of the complex was Mr=450,000-550,000 while at ratios above 5.8, these values were Mr=600-750,000. The number of toxin molecules attached per micelle was estimated after subtracting the average Mr of the GT1b micelle. Using a value of Mr=120-150,000 for TeTo, it would appear that at a molar ratio of 26, there were three toxin molecules per micelle, while at a molar ratio below 15 only 1-2 monomers were present. These observations indicated for the first time polyvalent interactions between TeTo molecule and micellar gangliosides much like those observed for cholera toxin and GM1 gangliosides (24-26).

Table 3: TeToB concentration-dependent aggregation in the presence of GT1b gangliosides.

Molar ratio	Molecular Weight			
(toxin/ GT1b micelle)	Mr > 600,000		Mr < 600,000	
3.1	N.D.	(-)	500	(1-2)
5.8	N.D.	(-)	450	(1)
15.0	600-650	(2)	450-450	(1)
26.0	700-750	(3)	400-450	(1)

Molecular weights were calculated from sedimentation values. The amount of toxin monomers estimated per ganglioside micelle is given in parenthesis,

While aggregate formation was unequivocally achieved in the presence of G1b gangliosides, the CD spectra of the toxin was shifted from a random to a highly α-helical form albeit by nearly every lipid examined.

Based on this and other observations, we predicted that the changes in the optical properties as monitored by CD are due to a discrete hydrophobic region on the heavy chain of the toxin molecule which exhibited a low selectivity towards the lipid environment. The lack of specificity with respect to the secondary toxin structure was found in accord with the low specificity often encountered when TeTo was reacted with GM1 (7) or with other glycolipids as critically reviewed by Wiegandt (27). It may also point out the importance of bulk lipids in facilitating a high molecular weight polypeptide to transverse the energy barrier imposed by the lipid bilayer.

In addition to the conformational changes induced by bulk lipids, we have also recorded the molar ellipticity changes of the bulk gangliosides before and after interaction with TeTo. The negative values of pure gangliosides in solution appeared to reflect primarily the optical properties of the sialyl moieties but also additional domains on the ganglioside molecule as previously demonstrated (28). The shift in molar ellipticity from a negative to a positive value upon interaction with a highly competent (i.e. GDlb or GTlb) or an irrelevant (GMl or GD3) ganglioside could indicate a shift of the sialyl moiety of the ganglioside towards a more stable configuration. Indeed, it has been demonstrated that gangliosides may perturb the bilayer organization when present in excess concentration (>20%) with respect to total lipids (29). In some membrane bilayers, several integral membrane proteins have been selectively, laterally excluded by glycolipid domains (30). Therefore, we would like to postulate that upon interaction between TeToB and membrane bound polysialogangliosides, the lipid bilayer may undergo a reorganization leading to formation of macromolecular toxin-micelle aggregates. Lateral or vertical movement of such ganglioside-toxin complexes alone or in association with some membranous or cytoskeletal elements in conjunction with aggregate dissociation may provide the basis for understanding TeTo binding, sequestration and translocation through the neuronal membrane. At this point it seems to us that the role of the ganglioside is to sequester TeTo at the cell surface by modifying its hydrophobicity for further processing.

REFERENCES

1. S.I. Hakomori, Ann. Rev. Biochem 50:733-764 (1981).

2. E. Yavin, E. and Z. Yavin, Dev. Neurosci. 2:25-37 (1979).

3. E. Yavin, Arch. Biochem. Biophys. 230:129-137 (1984).

4. E. Yavin, Z. Yavin, and L.D. Kohn, J. Neurochem. 40:1212-1219 (1983).

5. E. Habermann, Infections of the nervous system, in: "Handbook of Clinical Neurology," P.J. Vinhen and G.W. Bruyh, eds., North Holland, Amsterdam (1979).

6. J.P. Robinson and J.H. Hash, Mol. Cell Biochem. 48:33-44 (1982).

7. T.B. Helting and O. Zwisler, J. Biol. Chem. 252:187-193 (1977).

8.   B. Bizzini, in: "Receptors and Recognition," P. Cuatrecasas, ed., Vol. 1B, pp. 175-217, John Wiley Inc., New York (1977).

9.   J.W. Griffin, D.L. Price, W.K. Engel, and D.B. Drachman, J. Neuropathol. Exp. Neurol. 36:214-227 (1977).

10. D.L. Price, J.W. Griffin, A. Young, K. Peck, and A. Stocks, Science 188:945-947 (1975).

11. K. Stoeckel, M. Schwab, and H. Thoenen, Brain Res. 132:273-285 (1977).

12. E. Yavin, Z. Yavin, W.H. Habig, M.C. Hardegree, and L.D. Kohn, J. Biol. Chem. 256:7014-7022 (1981).

13. P. Lazarovici and E. Yavin, Biochim. Biophys. Acta 812:523-531 (1985).

14. P. Lazarovici and E. Yavin, Biochim. Biophys. Acta 812:532-542 (1985).

15. P. Lazarovici, J.L. Tayot, and E. Yavin, Toxicon 22:401-413 (1984).

16. P. Lazarovici and E. Yavin, Biochemistry, in press.

17. E. Yavin and A. Nathan, Eur. J. Biochem. 154:403-407 (1986).

18. A. Nathan and E. Yavin, Submitted for publication.

19. J. Holmgren, H. Elwing, P. Fredman, and L. Svennerholm, Eur. J. Biochem. 106:371-379 (1980).

20. L.A. Greene and A.S. Tischler, Adv. Cell Neurobiol. 3:373-414 (1982).

21. W. Seifert, in: "Gangliosides in Neurological and Neuromuscular Function, Development and Repair," M.M. Rapport and A. Gorio, eds., pp. 99-117, Raven Press, New York (1981).

22. R.K. Margolis, S.R.J. Salton, and R.U. Margolis, J. Biol. Chem. 258:4110-4117 (1983).

23. P. Lazarovici and E. Yavin, in: "7th Int. Conference on Tetanus," G. Nistico, P. Mastroeni, and M. Pitzurra, eds., pp. 29-48, Gangemi Pub. Co., Rome (1985).

24. A.W. Dalziel, G. Lipka, B.Z. Chowdhry, J.M. Sturtevant and D.E. Schafer, Molec. Cell Biochem. 63:83-91 (1984).

25. G. Schwarzmann, W. Mraz, J. Sattler, R. Schindler, and H. Wiegandt, Hoppe-Seyler's Z. Physiol. Chem. 359:1277-1286 (1978).

26. W. Mraz, G. Schwartzmann, J. Sattler, T. Momoi, B. Seemann and H. Wiegandt, Hoppe-Seyler's Z. Physiol. Chem. 361:177-185 (1980).

27. H. Wiegandt, in: "Advances in Cytopharmacology," B. Ceccareli and R. Clementi, eds., Vol. 3, pp. 17-25, Raven Press, New York (1979).

28. A.L. Stone and E.H. Kolodny, Chem. Phys. Lipids 6:274-279 (1971).

29. B. Cestaro, J. Barenholz, and S. Gatt, Biochemistry 19:615-619 (1980).

30. C.W.M. Grant, in: "Membrane Fluidity in Biology," R.C. Aloia, ed., pp. 131-150, Academic Press, New York (1983).

INTERACTION OF CHOLERA TOXIN WITH ITS RECEPTOR THE MONOSIALOGANGLIOSIDE

GM$_1$: A FLUORESCENCE STUDY

Marc De Wolf, Géry Bastiaens, Albert Lagrou, Guido Van Dessel,
Herwig Hilderson and Wilfried Dierick

RUCA-Laboratory for Human Biochemistry and UIA-Laboratory
for Pathological Biochemistry, University of Antwerp
Groenenborgerlaan 171
B2020 Antwerp, Belgium

INTRODUCTION

Cholera toxin (CT) is an enterotoxin secreted by *Vibrio cholerae* producing its pathological effects by increasing the c-AMP level in intestinal epithelial cells[1,2]. It is an oligomeric protein (Mr $\sim$ 84,000) composed of two structural and functional distinct subunits CT A and CT B (Mr $\sim$ 29,000 and 55,000 respectively). CT B contains five identical polypeptide chains (Mr = 11,600), most likely arranged in a ring-like pentameric configuration and CT A consists of two non-identical polypeptide chains A$_1$ or α-chain (Mr = 23,000) and A$_2$ or γ-chain (Mr = 5,500) linked by a single disulfide bridge (for reviews see refs. 1, 3-5). CT A is synthesized as a single polypeptide chain which is "nicked" between the two cysteïne residues by an extracellular bacterial protease. During this proteolysis two serine residues are removed at the COOH terminus of A$_1$[6]. The first event in the action of CT on cells is the rapid, irreversible binding to receptors on the cell surface. It is generally accepted that the receptor for the toxin is the monosialoganglioside GM$_1$ (for reviews see refs. 3,4,7).

CT B is responsible for the binding of the toxin to GM$_1$ whereas CT A stimulates adenylate cyclase activity by catalyzing the ADP ribosylation of the G$_{s\alpha}$ subunit of the GTP binding stimulatory regulatory protein[8-12]. According to Cassel and Selinger[11] ADP-ribosylation of G$_{s\alpha}$ enhances the GDP-GTP exchange (the turn on reaction) and inhibits the hormone stimulated GTPase activity (the turn off reaction). As a result CT stabilizes the adenylate cyclase activity in its activated state. In contrast to CT, pertussis toxin catalyzes the ADP-ribosylation of the α_i subunit of the inhibitory GTP binding regulatory protein which blocks the hormonal inhibition of adenylate cyclase[13,14]. The modern version of this model takes into account the multimeric nature of the GTP binding proteins (N$_s$,N$_i$)[15]. Both N$_s$ and N$_i$ are heterotrimers each containing an α,β and γ subunit. The β and γ subunits of both N$_s$ and N$_i$ are identical. The α_s subunit possesses a GTP-binding site and is ADP-ribosylated by CT. Receptor mediated GTP-GDP exchange occurring on α_s in the presence of Mg^{2+} is associated with a dissociation of $\beta\gamma$ from GTP-bound α_s which directly activates adenylate cyclase[15,16]. ADP-ribosylation of α_s by CT promotes the dissociation of α_s from $\beta\gamma$. The ADP-ribose α_s-GTP-complex directly converts adenylate cyclase to its activated form[17].

Although the receptor on the cell surface has been identified, several structural and functional aspects of the mechanism of CT action remain to be elucidated. More specifically the precise structures within CT B and CT A that are responsible for the binding and for the catalytic activity remain to be established as well as the mechanism of insertion and/or translocation in the membrane bilayer.

Cholera toxin translocation

The substrate of the catalytic part of CT, the α_s subunit of the stimulatory GTP binding regulatory protein of adenylate cyclase is located on the cytoplasmic side of the plasma membrane with most of it extended into the cytoplasm[18],[19]. Therefore, there must be some mechanism by which the active part of the toxin gains access to its substrate. The characteristic lag period 10 to 90 minutes before onset of activation of adenylate cyclase has been associated with this penetration mechanism[4],[22].

Several mechanisms of entry have been proposed such as the creation of a hydrophilic channel following binding of CT B to GM_1. This channel may be formed by the CT B itself[20] or by intrinsic membrane proteins acting as translocators of the A_1 polypeptide chain[21]. An alternative explanation is that multivalent binding of CT B to several receptors would induce a perturbation in both toxin and membrane structure[22],[23]. This would result in the dissociation of CT into CT A and CT B and subsequent penetration of the A component into the cell membrane.

Is there a conformational change in CT structure following receptor binding ?

Initial evidence in favor of a GM_1 induced conformational change came from the observation that complexation of CT with GM_1 causes a 12 nm "blue shift" in the fluorescence emission spectrum of CT[24]. This perturbation of the fluorescence emission spectrum can be explained in any of two ways : (i) Complex formation between GM_1 and CT induces a *major* conformational change in the toxin structure which is essential for the penetration of the catalytic part of CT A into or through the cell membrane. (ii) Complexation between GM_1 and CT induces a *subtle* local conformational change involving aromatic amino acid side chains responsible for the fluorescence properties of the toxin.

Since the corrected fluorescence emission spectra of CT, CT A and CT B are centered around 335, 330 and 340 nm respectively and since nitrophenyl-sulphenylation of the indole residues, almost completely eliminates the fluorescence it is concluded that the fluorescence of the toxin and its subunits is mainly due to Trp residues. Therefore, the GM_1 induced "blue shift" has to be ascribed to a decrease in polarity of the microenvironment of one or more indole side chains in the toxin molecule. A similar perturbation of the fluorescence emission spectrum is observed upon addition of GM_1 or oligo-GM_1 to CT B. Since each β polypeptide chain of CT B has only one Trp residue located at position 88[25], complexation with GM_1 or oligo-GM_1 must at least perturb the microenvironment of that residue.

In order to determine whether this change in microenvironment of Trp 88 is associated with a more extended conformational change the effects of GM_1 or oligo-GM_1 on the fluorescence of different fluorescent labeled derivatives (dansyl, fluoresceine) of CT and CT B were investigated[26]. Addition of ligand did, however, neither affect quantum yields nor emission spectra of these derivatives. These data therefore argue against the occurrence of a gross conformational change in CT structure. On the other hand chemical modification, fluorescence quenching and energy transfer studies[26-28] revealed that Trp 88 is an important determinant in the association of CT with the carbohydrate moiety of GM_1.

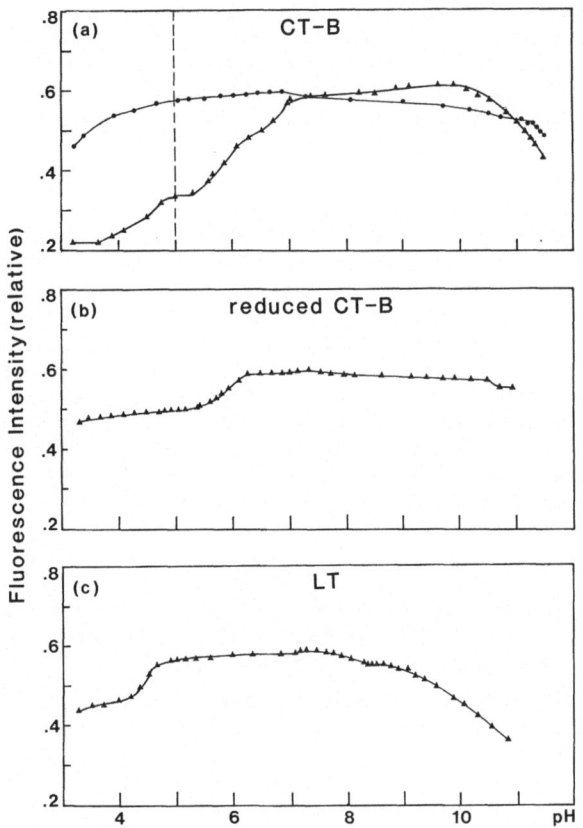

Fig. 1. Fluorimetric titration profiles of :
a) CT B. It is believed that in the pH 7 to
5 range the protonated imidazole groups of
His 13 and His 94 are responsible for quenching.
Below pH 5 the protonated carboxyl group of
Glu 11 is an additional quencher whereas above
pH 9 quenching is probably the result of an
energy transfer from the excited indole moiety
of Trp 88 to the ionized phenol group of Tyr 12.
Complexation of CT B with oligo-GM_1 ($\bullet$) blocks
the quenching pointing to a localisation of
these quenching amino acid side chains in or
near the binding site.
b) Reduced and alkylated CT B. Reductive cleavage of the single disulfide bridge of CT B apparently enlarges the distances between the
lone Trp 88 and the quenching amino acid side
chains except His 94, which eliminates their
quenching potency.
c) The heat labile enterotoxin of *Escherichia
coli* (LT). In LT B which is structurally and
functionally very similar to CT B an arginine
residue is substituted for His 13 and an asparagine residue for His 94, consequently no
quenching is expected in the pH 7 to 5 region.

Structural features inferred from fluorimetric titration experiments

Further details concerning the binding site on the toxin and possible GM_1 induced conformational changes were obtained by performing fluorimetric titration experiments[29]. The fluorescence intensity of CT is highly pH dependent, in the pH range 3 to 7 a marked increase in fluorescence intensity is observed reaching a maximum at neutral pH, corresponding to a quantum yield of 7.6 %.

The pH intensity profiles of CT A and CT B reveal that CT B is mainly responsible for the pH sensitivity of the fluorescence of the holotoxin. The fluorimetric titration curve of CT B (Fig. 1a) shows that the quenching in the pH 3 to 7 range is due to interaction between Trp 88 and two types of ionisable quenching amino acid side chains. In the pH range 5 to 7 it is assumed that fluorescence quenching is mainly due to the interaction between an indole group and a protonated imidazole group (imidazolium group)[30]. The further decrease in fluorescence intensity in the pH 5 to 3 range can be ascribed to deionization of an interacting carboxyl containing side chain[31]. The decrease in fluorescence intensity above pH 9 can be explained by an energy transfer from the excited indole moiety of Trp 88 to ionized tyrosines[32]. In the presence of GM_1 or oligo-GM_1 the quenching of fluorescence intensity of CT and CT B as a function of pH is almost completely abolished (Fig. 1a). Since the fluorimetric titration profile is actually a reflection of the mutual orientations of the quenching amino acid side chains and the indole group involved, a perturbation of this profile may be an indication of conformational changes in the protein. This conformational change could for instance prevent the charge transfer interaction between the indole group of Trp 88 and the imidazolium group.

Another possibility that should be taken into consideration is that complexation with GM_1 or oligo-GM_1 causes a shielding of the quenching amino acid side chains from the interaction with H_3O^+. Evidence for the latter notion came from the observation that addition of GM_1 or oligo-GM_1 to CT B at pH 4 did not reverse the quenching (although binding appears normal as evidenced by the characteristic "blue shift" in the fluorescence emission spectrum). This indicated that in the CT B-GM_1 complex protonated amino acid residues are still able to quench to the same extent the Trp fluorescence. In addition subsequent reversing the pH to 7 of the protonated GM_1 complex did not abolish the quenching, further indicating the lack of exchange of protons with the medium.

In contrast to CT B the fluorimetric titration of CT A reveals a featureless profile indicating that the Trp fluorescence (CT A contains 3 Trp residues) is not significantly altered by any ionisable group displaying a pK_a in the range 4 to 9. Activation of A (by treatment with dithiothreitol) does not affect the lack of pH sensitivity in this pH region. Above pH 9 however, a significant drop in fluorescence intensity of activated CT A is observed. This indicates that activation induces some conformational change in the α-chain which could be essential for expression of catalytic (mono-ADP ribosyltransferase) activity. The free sulfhydryl group (generated in α upon reduction with dithiothreitol) itself is, however, not essential for activity[33].

Amino acid side chains involved in the fluorescence quenching of Trp 88

From the rather high quenching observed in the pH range 7 to 5, one must assume that a rather strong interaction occurs between the imidazolium and indole amino acid side chains. His 94 is a potential candidate for this interaction since it is separated only by one amino acid residue from an "inverse turn" linking two α-helix segments at the C-terminal of the β-polypeptide chain whereas Trp 88 is located immediately adjacent at the opposite site of this inverse turn[34] (Fig. 2). His 13 could also be in close proxi-

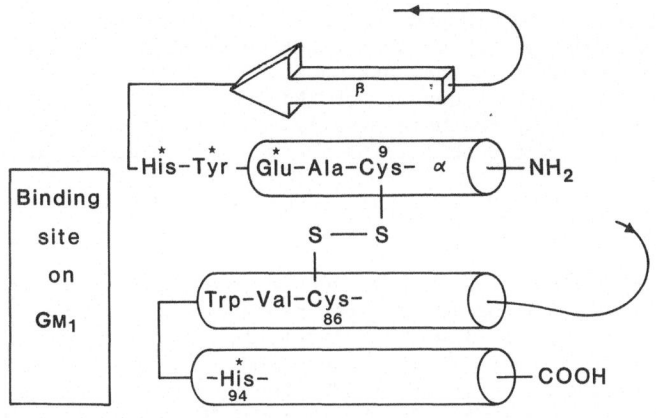

Fig. 2. Topography of amino acid residues involved
in the pH induced intrinsic quenching of the fluorescence of the lone Trp 88 residue. CT B secunda-
ry structure is as proposed by Duffy and Lai[34] using
the predictive method of Chou and Fasman[35]. ☆ Indicates amino acid residues responsible for fluorescence quenching.

mity of Trp 88 because of the single intrachain disulfide bridge between
positions 9 and 86 joining the C- and N-terminal of the β-polypeptide chain.
The titration profile of CT B reveals that in the pH 5 to 7 range the fluorescence intensity-pH relationship is not sigmoidal. This could point to
the involvement of more than one imidazolium group in the quenching process.
This view is substantiated by the observation that opening of the single disulfide bridge strongly perturbs the titration profile (Fig. 1b). In this
case the fluorescence intensity is only pH sensitive in the pH 5.75-6.5
range. Furthermore, a reduction of the extent of quenching is observed.
These findings can be rationalized by assuming that the cleavage of the disulfide bridge enlarges the distances between Trp 88 and His 13 whereas
His 94 is still able to quench the Trp fluorescence. Reductive cleavage of
the disulfide bridge also abolishes the quenching in the acid (pH 3.5 to 5)
and alkaline (pH > 9) regions. A possible explanation for this phenomenon
is that Glu 11 and Tyr 12 are responsible for the quenching in these pH regions and that cleavage of the disulfide bridge also enlarges the distances
between these residues and Trp 88. Additional evidence for the involvement
of both His 13 and His 94 in the Trp 88 fluorescence quenching comes from
fluorimetric titrations of the heat labile enterotoxin of *Escherichia coli*
(LT). LT resembles CT not only functionally but structurally and immunologically as well[36-38]. The nucleotide sequence of the LT B cistron has been
determined by Dallas and Falkow[39]. Translation of this sequence revealed
that LT B and CT B show significant amino acid sequence homology. 80 of the
LT B residues are identical to those of CT B (representing 79 % homology).
The non-homologous amino acids are scattered throughout the primary sequence. Based on circular dichroism studies the overall folding of the polypeptide chains of LT and CT is quite similar[40]. In LT B an arginine residue is
substituted for His 13 and an asparagine residue is substituted for His 94.
Therefore, if our prediction concerning the His residues involved in the
quenching process is correct, no quenching in the pH range 5 to 7 should
occur which is, as indicated in Fig. 1c, indeed the case.

Another point of interest deserves our attention. The particular sequence (Cys-Ala-Glu-Tyr-His) comprises the region (Cys-Ala-Glu-Tyr-) that
has been shown[41,42] to be significantly homologous to the highly preserved
region (Cys-Ala-Gly-Tyr-Cys-) of the α subunit of glycoprotein hormones

such as TSH, LH, hCG and FSH and serine proteases. Since complexation with GM$_1$ prevents the protonation or deprotonation of these amino acid side chains one must assume that they are located in or near the binding site for GM$_1$. This is further supported by the observation that after reductive cleavage of the single intrachain disulfide bridge and alkylation of the free sulfhydryl groups the ability of the toxin to bind GM$_1$ is abolished.

Is there a pH induced conformational change exposing or forming hydrophobic domains on the CT structure ?

The foregoing experiments revealed some new features about the binding site of CT, however, so far these experiments did not provide any evidence for the occurrence of a pH induced conformational change in the CT structure. In another approach to study the occurrence of pH induced conformational change we studied the effect of pH on acrylamide quenching and detergent binding of CT, CT A and CT B. In the pH range 7 to 5 a significant and time dependent increase in fluorescence quenching of CT occurred.

A similar effect of pH on acrylamide quenching of CT B was observed. This indicates that upon lowering the pH the Trp 88 residue becomes more exposed. In order to investigate whether this pH induced exposure of Trp 88 is paralleled by the formation or exposure of more extended hydrophobic domains on the CT structure the binding of the nonionic detergent Bry 96 was measured.

Determination of the ratio of intrinsic fluorescence with Br-Bry 96 (F) relative to that with unbrominated Bry 96 (F$_o$), both in the presence of 0.2M NaCl revealed that at neutral pH only CT A shows some affinity for this mild detergent[43]. In the intact toxin the hydrophobic domain on CT A is apparently shielded from interaction with Br-Bry 96 and receptor binding does not induce the exposure of hydrophobic domains on the CT structure. Since activation of CT A by treatment with dithiothreitol did not affect detergent binding it is suggested that the hydrophobic domain on CT A is located at the inter subunit interface region between the α-polypeptide and one of the β-polypeptide chains. Dissociation of the α-polypeptide chain from the γ-polypeptide does not contribute to the formation of hydrophobic domains but may be important for the expression of catalytic activity. The presence of hydrophobic domains on CT A was also reflected by the ability of phospholipid dispersions to interact with CT A as evidenced by their ability to quench the intrinsic fluorescence of CT A and to modulate the mono-ADP ribosyl transferase activity (activated CT A)[44].

The affinity of CT A for Bry 96 is salt dependent and appears to be low since no quenching detected detergent binding was observed upon reducing the Br-Bry 96 concentration down to 0.005 %. Previous studies using charge-shift electrophoresis[45] were unable to reveal any detergent binding to CT and its subunits. The inherent low ionic strength in the electrophoretic separations and the low affinity probably account for this discrepancy. On the other hand, studies using photoreactive glycolipid and phospholipid compounds incorporated into model membranes demonstrated that only CT A was labelled when the photo-activated label was buried in the phospholipid bilayer[46-47].

Detergent binding of CT A is not pH dependent whereas upon lowering the pH detergent binding to CT and CT B are significantly enhanced. In the pH range 6.5 to 4.2 a gradual and time dependent increase in detergent binding to CT and CT B was observed. In the narrow pH range 4.2 to 4 a marked enhancement of Br-Bry 96 quenching occurred.

Since maximal detergent binding was observed below pH 5 one must assume that the hydrophobicity of CT or CT B is increased independently of further exposure of Trp 88. The pH induced increase of Trp 88 and hydrophobic domains exposure is dependent on the ionic strength, fully reversible and completed in 10 min. (25°C) with a half time of 1.2 min. and 2.0 min. These

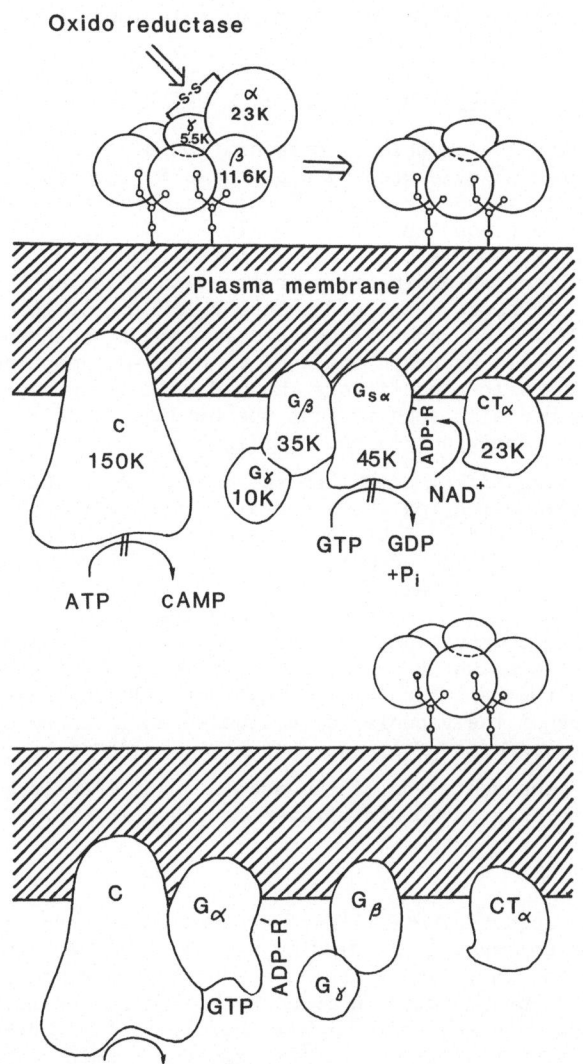

Fig. 3. Schematic representation of the mechanism of translocation of the α subunit of CT through the lipid bilayer and its interaction with the GTP binding stimulatory regulatory protein of adenylate cyclase (N_s). (top) After multivalent binding of the toxin to GM_1 and splitting of the intrachain disulfide bridge between the α and γ subunits by an oxidoreductase, the α subunit dissociates from the stable $β_5γ$ $(GM_1)_5$ complex and its weak hydrophobic nature permits it to interact with the lipid bilayer. After further translocation through the lipid bilayer by a still unknown mechanism the catalytic part of the   subunit gains access to its substrate. (bottom) Following internalization, the α subunit of CT catalyzes the ADP-ribosylation of $α_s$ of N_s which promotes the dissociation of $α_s$ from βγ. The ADP ribose $α_s$-GTP complex directly converts the catalyst of adenylate cyclase to its activated form.

similar kinetics further point to the causal relationship between both phenomena.

Receptor binding stabilizes quaternary structure of CT

Upon decreasing the pH below 4 dissociation of CT B in its constituent monomers is observed as evidenced by a decrease in fluorescence anisotropy of dansylated CT B. At pH 3.2 and 25°C the dissociation is first order with time displaying a rate constant of 0.150 min^{-1}. Complex formation of CT B with GM_1 or oligo-GM_1 completely prevents the dissociation of CT B. This could be the result of shielding from protonation of amino acid residues committed to subunit association or/and cross linking of the β-polypeptide chains by oligo-GM_1. The terminal galactose residue could for instance interact with Trp 88 (via hydrogen bonding) on one β-polypeptide chain and the sialic acid residue with a positively charged amino acid residue (such as Arg 35[34]. This would also suggest that the binding sites on CT B are clefts formed by adjacent β-polypeptide chains (Fig. 3).

This stabilizing effect of GM_1 on the quaternary structure of CT B might offer an explanation for the observation that although a rapid internalization of CT occurs[48] ^{125}I-CT remains persistently (several days) bound to human fibroblasts and is degraded slowly[49]. Since during this time the toxin is probably recycled back to the cell surface the stabilizing effect of GM_1 might protect against dissociation and loss of binding at the acidic pH of intracellular compartments such as endosomes and lysosomes. It is therefore not excluded that this persistent binding of CT enables the CT A to penetrate into the cell both at the level of cell surface as well as the level of membranes of the vacuolar apparatus which is in accordance with the partial protective effects of lysosomotropic amines[50-51].

A schematic representation of the mechanism of translocation of the α subunit of CT through the lipid bilayer is given in Fig. 3.

CONCLUSION

Fluorimetric quenching and titration experiments have been shown to be useful in studies concerning the binding site on CT B for its receptor the monosialoganglioside GM_1. Apart from the lone Trp 88 residue several amino acid residues have been located in or near this binding site. Some of these amino acids reside in a sequence which is homologous to a highly preserved region of the β subunit of glycoprotein hormones. Whether this structural similarity which might be the result of a divergent or convergent mechanism of evolution has a functional counterpart remains to be established.

Our fluorescence studies also support the view that the mere binding of CT to its receptor is not sufficient to induce a conformational rearrangement in the toxin structure allowing the α subunit to interact with the lipid bilayer. Presumably some processing of the toxin is a prerequisite for penetration of the active part into or through the lipid bilayer of the cell membrane. Cleavage of the intrachain disulfide bridge joining the α and γ subunits by a thiol:protein disulfide oxidoreductase might provide such a function. Binding of CT B to GM_1 would serve to bring the α subunit in close proximity of the lipid bilayer. Following dissociation of the α subunit from the stable $β_5γ$ $(GM_1)_5$ complex its weak hydrophobic nature might be sufficient to allow initial interaction with the lipid bilayer. The mechanism of further translocation through the lipid bilayer remains to be elucidated.

ACKNOWLEDGMENTS

These studies were supported in part by the Public Health Service Grant N° 1RO1 AM 32136 from the National Institute of Arthritis, Diabetes, and Digestive and Kidney Diseases, US Public Health Service and by the Belgian FGWO Grant 3.002.83.

REFERENCES

1. R.A. Finkelstein, Cholera, CRC Crit. Rev. Microbiol. 2:553-623 (1973).
2. M. Field, Cholera toxin, adenylate cyclase, and the process of active secretion in the small intestine : The pathogenesis of diarrhea in cholera, in : "Physiology of Membrane Disorders", T.E. Andreoli and J.F. Hoffman, eds., Plenum Publishing Corporation, New York, Chapter 41:877-898 (1978).
3. J. Holmgren, Cholera toxin and the cell membrane, in : "Bacterial Toxins and Cell Membranes", J. Jeljaszewick and T. Wadström, eds., Academic Press, New York, pp. 333-366 (1978).
4. V. Bennett, and P. Cuatrecasas, Cholera toxin : Membrane gangliosides and activation of adenylate cyclase, in : "The Specificity and Action of Animal, Bacterial and Plant Toxins", P. Cuatrecasas and N.F. Greaves, eds., Chapman and Hall, London, pp. 3-66 (1977).
5. C.-Y. Lai, The chemistry and biology of cholera toxin, CRC Crit. Rev. Biochem. 9:171-206 (1980).
6. L.K. Duffy, J.W. Peterson, and A. Kurosky, Isolation and characterization of a precursor form of the 'A' subunit of cholera toxin' FEBS Letters 126:157-166 (1981).
7. S. van Heijningen, Similarities in the action of different toxins, in : "Molecular Action of Toxins and Viruses", C. Cohen and S. van Heijningen, eds., Elsevier Biochemical Press, Amsterdam, 6:169-190 (1982).
8. J. Moss, and M. Vaughan, Activation of adenylate cyclase by choleragen, Annu. Rev. Biochem. 48:581-600 (1979).
9. G.L. Johnson, Cholera toxin action and the regulation of hormone-sensitive adenylate cyclase, in : "Molecular Actions of Toxins and Viruses", P. Cohen and S. van Heijningen, eds., Elsevier, Amsterdam, pp. 33-49 (1982).
10. D. Cassel, and T. Pfeuffer, Mechanism of cholera toxin action : Covalent modification of the guanylnucleotide-binding protein of the adenylate cyclase system, Proc. Natl. Acad. Sci. USA 75:2669-2673 (1978).
11. D. Cassel, and Z. Selinger, Mechanism of adenylate cyclase activation by cholera toxin : Inhibition of GTP hydrolysis at the regulatory site, Proc. Natl. Acad. Sci. USA 74:3307-3311 (1977).
12. J.K. Northup, P.C. Sternweis, M.D. Smigel, L.S. Schleifer, E.M. Ross, and A.G. Gilman, Purification of the regulatory component of adenylate cyclase, Proc. Natl. Acad. Sci. USA 77:6516-6520 (1980).
13. T. Katada, and M. Ui, ADP ribosylation of the specific membrane protein of C_6 cells by islet activaint protein associated with modification of adenylate cyclase activity, J. Biol. Chem. 257:7210-7216 (1982).
14. L. Olansky, G.A. Myers, S.L. Pohl, and E.L. Hewlett, Promotion of lipolysis in rat adipocytes by pertussis toxin : Reversal of endogenous inhibition, Proc. Natl. Acad. Sci. USA 80:6547-6551 (1983).
15. A.G. Gilman, G Proteins and dual control of adenylate cyclase, Cell 36: 577-579 (1984).
16. J.K. Northup, Overview of the guanine nucleotide regulatory protein systems, N_s and N_i, which regulate adenylate cyclase activity in plasma membranes, in : Molecular Mechanisms of Transmembrane Signalling, P. Cohen and M.D. Houslay, eds., Elsevier, Amsterdam, pp. 91-116 (1985).
17. R.A. Kahn, and A.G. Gilman, ADP-ribosylation of G_s promotes the dissociation of its α and β subunits, J. Biol. Chem. 259:6235-6240 (1984).
18. Z. Farfel, H.R. Kaslow, and H.R. Bourne, A regulatory component of adenylate cyclase is located on the inner surface of human erythrocyte membranes, Biochem. Biophys. Res. Commun. 90:1237-1241 (1979).
19. H.R. Kaslow, G.L. Johnson, V.M. Brothers, and H.R. Bourne, A regulatory component of adenylate cyclase from human erythrocyte membranes, J. Biol. Chem. 255:3736-3741 (1980).

20. D.M. Gill, The arrangement of subunits in cholera toxin, _Biochemistry_ 15:1242-1248 (1976).

21. D.M. Gill, Seven toxic peptides that cross cell membranes, _in_ : "Bacterial Toxins and Cell Membranes", J. Jeljaswicz and T. Wadström, eds., Academic Press, London, pp. 291-332 (1978).

22. P.H. Fishman, Role of membrane gangliosides in the binding and action of bacterial toxins, _J. Membrane Biol._ 69:85-97 (1982).

23. L.D. Kohn, E. Consiglio, M. De Wolf, E.F. Grollman, F.D. Ledley, G. Lee, and N.P. Morris, Thyrotropin receptors and gangliosides, _Adv. Exp. Med. Biol._ 125:487-503 (1980).

24. B.R. Mullin, S.M. Aloj, P.H. Fishman, G. Lee, L.D. Kohn, and R.O. Brady, Cholera toxin interactions with thyrotropin receptors on thyroid plasma membranes, _Proc. Natl. Acad. Sci. USA_ 73:1679-1683 (1976).

25. C.Y. Lai, Determination of the primary structure of cholera toxin B subunit, _J. Biol. Chem._ 252:7249-7256 (1977).

26. M.J.S. De Wolf, M. Fridkin, and L.D. Kohn, Tryptophan residues of cholera toxin and its A and B protomers : Intrinsic fluorescence and solute quenching upon interacting with the ganglioside GM_1, oligo-GM_1 or dansylated oligo-GM_1, _J. Biol. Chem._ 256:5489-5496 (1981a).

27. M.J.S. De Wolf, M. Fridkin, M. Epstein, and L.D. Kohn, Structure-function studies of cholera toxin and its A and B protomers. Modification of tryptophan residues, _J. Biol. Chem._ 256:5481-5488 (1981b).

28. M. De Wolf, G. Van Dessel, H.J. Hilderson, A. Lagrou, and W. Dierick, Acrylamide quenching of cholera toxin and its A and B protomers using steady state and dynamic fluorometry, _Arch. internat. Physiol. Biochim._ 91:B11-B13 (1983).

29. M. De Wolf, G. Van Dessel, A. Lagrou, H.J. Hilderson, and W. Dierick, Structural features of the binding site of cholera toxin inferred from fluorescence measurements, _Biochim. Biophys. Acta_, 832:165-174 (1985).

30. M. Shinitzky, and R. Goldman, Fluorometric detection of histidine-tryptophan complexes in peptides and proteins, _Eur. J. Biochem._ 3:139-144 (1967).

31. A. White, Effect of pH on fluorescence of tyrosine, tryptophan and related compounds, _Biochem. J._ 71:217-220 (1959).

32. I. Steinberg, Long-range nonradioactive transfer of electronic excitation energy in proteins and polypeptides, _Annu. Rev. Biochem._ 40:83-114 (1971).

33. J.J. Mekalanos, R.J. Collier, and W.R. Romig, Enzyme activity of cholera toxin. II. Relationships to proteolytic processing, disulfide bond reduction, and subunit composition, _J. Biol. Chem._ 254:5855-5861 (1979).

34. L.K. Duffy, and C.Y. Lai, Involvement of arginine residues in binding site of cholera toxin subunit B, _Biochem. Biophys. Res. Commun._ 91:1005-1010 (1979).

35. P.Y. Chou, and G.D. Fasman, Prediction of protein conformation, _Biochemistry_ 13:222-244 (1974).

36. J.D. Clements, and R.A. Finkelstein, Immunological cross-reactivity between a heat-labile enterotoxin(s) of _Escherichia coli_ and subunits of _Vibrio cholerae_ enterotoxin, _Infect. Immun._ 21:1036-1039 (1978).

37. S.L. Kunkel, and D.C. Robertson, Purification and chemical characterization of the heat-labile enterotoxin produced by enterotoxigenic _Escherichia coli_, _Infect. Immun._ 25:586-596 (1979).

38. W.S. Dallas, and S. Falkow, The molecular nature of heat-labile enterotoxin (LT) of _Escherichia coli_, _Nature (London)_ 277:406-407 (1979).

39. W.S. Dallas, and S. Falkow, Amino acid sequence homology between cholera toxin and _Escherichia coli_ heat-labile toxin, _Nature_ 288:499-501 (1980).

40. J. Moss, J.C. Osborne, Jr., P.H. Fishman, S. Nakaya, and D.C. Robertson, _Escherichia coli_ heat-labile enterotoxin ganglioside specificity and ADP-ribosyltransferase activity, _J. Biol. Chem._ 256:12861-12865 (1981).

41. F.D. Ledley, B.R. Mullin, G. Lee, S.M. Aloj, P.H. Fishman, L.T. Hunt, M.O. Dayhoff, and L.D. Kohn, Sequence similarity between cholera toxin and glycoprotein hormones : Implications for structure activity relationship and mechanism of action, Biochem. Biophys. Res. Commun. 69: 852-859 (1976).

42. A. Kurosky, D.E. Markel, J.W. Peterson, and W.M. Fitch, Primary structure of cholera toxin β-chain : A glycoprotein hormone analog ?, Science 195:299-301 (1977).

43. M. De Wolf, G. Van Dessel, A. Lagrou, H.J. Hilderson, and W. Dierick, Hydrophobic fluorescence quenching of the subunits of cholera toxin : Possible relevance to lipid bilayer penetration, Arch. internat. Physiol. Biochim. 93:B136 (1985).

44. M. De Wolf, L.D. Kohn, H. Depauw, G. Van Dessel, H.J. Hilderson, A. Lagrou, and W. Dierick, Interaction of cholera toxin with thyroidal plasma membranes : Role of phospholipids, Arch. internat. Physiol. Biochim. 90:B110-B112 (1982).

45. W.H.J. Ward, P. Britton, and S. van Heijningen, The hydrophobicities of cholera toxin, tetanus toxin, and their components, Biochem. J. 199: 457-460 (1981).

46. M. Tomasi, and C. Montecucco, Lipid insertion of cholera toxin after binding to GM_1-containing liposomes, J. Biol. Chem. 256:11177-11181 (1981).

47. B.J. Wisnieski, and J.S. Bramhall, Photolabelling of cholera toxin subunits during membrane penetration, Nature 289:319-321 (1981).

48. K.C. Joseph, S.U. Kim, A. Stieber, and N.K. Gonatas, Endocytosis of cholera toxin into neuronal GERL, Proc. Natl. Acad. Sci. USA 75:2815-2819 (1978).

49. P.P. Chang, P.H. Fishman, N. Ohtomo, and J. Moss, Degradation of choleragen bound to cultured human fibroblasts and mouse neuroblastoma cells, J. Biol. Chem. 258:426-430 (1983).

50. M.D. Houslay, and K.R.F. Elliott, Is the receptor-mediated endocytosis of cholera toxin a pre-requisite for its activation of adenylate cyclase in intact rat hepatocytes, FEBS Letters 128:289-292 (1981).

51. J. Hagman, and P.H. Fishman, Inhibitors of protein synthesis block action of cholera toxin, Biochem. Biophys. Res. Commun. 98:677-684 (1981).

INCORPORATION OF GABA/BENZODIAZEPINE RECEPTORS INTO NATURAL

BRAIN LIPID LIPOSOMES: BIOCHEMICAL CHARACTERIZATION

David R. Bristow and Ian L. Martin

MRC Molecular Neurobiology Unit
University of Cambridge Medical School
Hills Road
Cambridge, U.K.

INTRODUCTION

γ-aminobutyric acid (GABA) is one of the major inhibitory neurotransmitters in mammalian CNS. Its action is mediated through two pharmacologically distinct receptor subtypes designed $GABA_A$ and $GABA_B$. Interaction of GABA with the former opens anion channels, located in cell membranes, which allow the passage of chloride ions down their electrochemical gradient; this generally leads to membrane hyperpolarisation. This receptor is therefore amenable to functional analysis using biochemical and electrophysiological techniques. Such analysis should elucidate the mechanism of ligand-induced regulation of the anion channel at the molecular level, identify the minimum subunit requirements for function and provide further physiochemical information on the $GABA_A$ receptor complex.

The biochemical approach to address these questions requires, as a necessary first stage, the solubilisation of the receptor from its membrane location followed by purification to separate the $GABA_A$ complex from other membrane components (reviewed by Stephenson and Barnard, 1986). Finally, the isolated $GABA_A$ receptor complex can be reconstituted into a lipid bilayer structure (such as a liposome) to allow the measurement of channel function using, for example, radioactive ion flux or patch clamp techniques.

161

However, the GABA$_A$ - chloride ionophore complex has an added complication (or fascination) in that it also exhibits allosteric regulatory sites for a number of psychoactive drugs. These include the 1,4-benzodiazepines (BDZ, reviewed by Haefely, 1984), the barbiturates (Macdonald and Barker, 1979), the pyrazolopyridines (eg cartazolate, Beer et al., 1978), a class of putative anxiolytics, and several convulsant compounds such as picrotoxinin (Ticku et al., 1978) and tert-butylbicyclophosphorothionate (TBPS, Squires et al., 1983). Several of these sites appear to express a molecular interaction with each other in that the binding of an agonist at one recognition site induces an allosteric regulation of binding at another site on the receptor complex. An example of such a neurochemical interaction is the facilitation of agonist BDZ binding, reflecting an increase in receptor affinity, when examined in the presence of GABA (Martin and Candy, 1978; Tallman et al., 1978), certain barbiturates (eg pentobarbital, Leeb-Lundberg et al., 1980), the pyrazolopyridines (Beer et al., 1978) or some anions (Martin and Candy, 1978, Costa et al., 1979). The significance of this molecular phenomena is enigmatic, though electrophysiologically (review by Simmonds, 1984), they appear to indicate a functional linkage within the GABA$_A$ receptor complex. It seems reasonable therefore to use these responses as an indication of receptor integrity during the physical trauma of solubilisation and purification.

Solubilisation of the GABA/BDZ receptor has been attempted using a variety of detergents (reviewed by Fischer and Olsen, 1986) with the most successful, based on the retention of several neurochemical interactions in solution, using 3-[(3-cholamidopropyl)dimethylammonio] propane sulphonate (CHAPS) in the presence of soybean phospholipids (Sigel and Barnard, 1984). However, even under these conditions the responses are anomalous and labile (Sigel and Barnard, 1984; Hammond and Martin, 1986) suggesting these conditions are non-ideal.

During our attempts to further protect and stabilise the GABA/BDZ receptor and its associated modulatory sites during CHAPS solubilisation we included a natural brain lipid extract and cholesteryl hemisuccinate (CHS) (Bristow and Martin, 1986). As a consequence of this procedure, we obtained a liposome fraction containing GABA/BDZ receptors which exhibited several modulatory sites. The biochemical characterisation of these liposome incorporated GABA/BDZ receptors from rat cerebellum are described here.

METHODS

The preparation of rat cerebellar synaptosomal membranes (CbSPM) has been previously described (Hammond and Martin, 1986). The extraction of brain lipids briefly involves the vigorous homogenisation of rat cerebella in (2 vol (w/v) of chloroform: methanol, 1:2, v/v) followed by standing on ice for 30 min. To this homogenate was slowly added 2.6 vol (w/v) of chloroform: 2.7 M KCL (3:1, v/v) and the mixture left on ice for a further 30 min. The mixture was then centrifuged (9,700 x g, 10 min) and the lower chloroform layer removed and stored on ice under nitrogen. The upper layer and pellet was re-extracted by homogenisation with 2.6 vol (w/v) of chloroform, centrifuged as above and the chloroform layer removed and pooled with the first extract. The chloroform was then evaporated under a stream of nitrogen and the lipids dried to constant weight under vacuum. The lipids were then dissolved in chloroform to give 10 mg lipids/ml and stored under nitrogen at 4°C until use.

Solubilisation Procedure
The required volume of lipid solution was evaporated under nitrogen and then dissolved in 2.4% (w/v) CHAPS and 2 mM CHS (in 5 mM Tris-citrate buffer (pH 7.1) containing EDTA (1 mM), phenylmethylsulphonyfluoride (100 μM), soybean trypsin inhibitor (50 mg/L) and bacitracin (50 mg/L); henceforth called Tris + inhibitors buffer) to give a 9 mg/ml cerebellar lipid solution. The solubilisation procedure briefly involved diluting this solution 1:1 with CbSPM at 20°C and incubating for 15 min followed by 5 min on ice. The mixture was then centrifuged at 100,000 x g for 1 h which produced a supernatant containing soluble receptors, a lipid layer (liposome fraction) and an insoluble pellet. The lipid layer was removed and

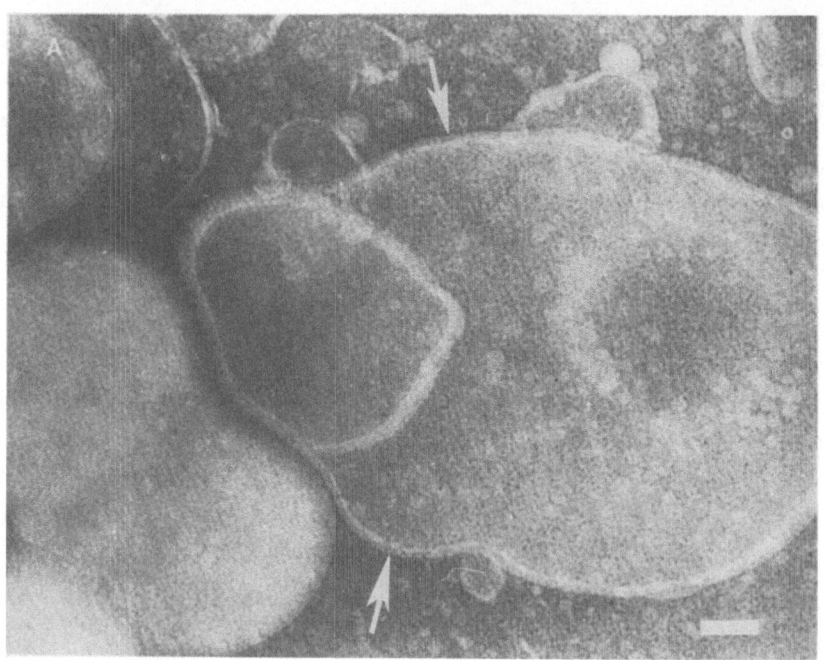

Figure 1A. Electron micrograph showing the liposome fraction at
143,000 x magnification (scale bar = 65 nm). The bilayer structure (arrowed) is shown convincingly at this magnification.

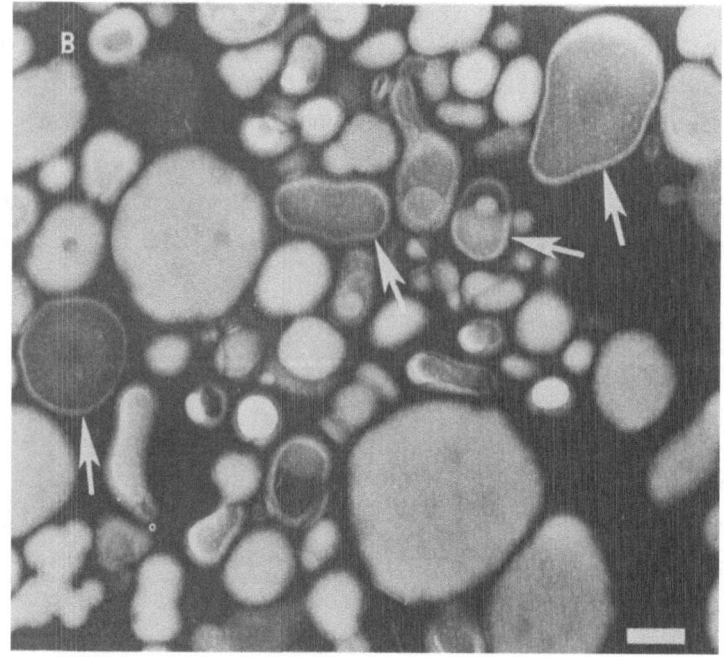

Figure 1B. Electron micrograph of the liposome fraction viewed
at 45,000 x magnification (scale bar = 200 nm). The range of
liposomes formed by this procedure is 100-700 nm. In several
instances, it appears as though the ammonium molybdate has gained
access to the inside of the structures, possibly as a consequence
of drying, and shows the unilamella structure (arrowed).

diluted in Tris + inhibitors buffer and used immediately in the
[^{3}H]flunitrazepam ([^{3}H]FNZ) binding assay or stored at 4°C and
assayed after 48 h. The [^{3}H]FNZ binding assay has been
previously described (Bristow and Martin, 1986). Briefly, the
facilitation of [^{3}H]FNZ (0.5 nM) binding was examined in the
presence of GABA (10^{-4}M + 0.15 M NaCl), cartazolate (10^{-6}M +
0.15 M NaCl) or pentobarbital (10^{-3}M + 0.15 M NaCl). The FNZ
binding parameters (K_D, B_{max}) were obtained from [^{3}H]FNZ
equilibrium saturation analysis (0.25-60 nM).

Electron Microscopy

Samples of the liposome fraction for electron microscopy
were placed on formvar-coated copper grids and left to settle
(45 s, 20°C). The copper grids were then dipped in 2% (w/v)
ammonium molybdate solution and left briefly to stand (45 s,
20°C), followed by a rapid wash in deionised water, and left
finally to dry at room temperature prior to viewing.

RESULTS

Electron microscopic analysis of the lipid layer, which
had been stained with 2% (w/v) ammonium molybdate, revealed the
structures to be unilamellar liposomes (Fig. 1A). The diameter
of the liposomes was calculated to range from 100-700 nm (Fig.
1B). The liposome fraction could be separated into protein
containing liposomes (proteoliposomes) and protein free
liposomes by sucrose gradient centrifugation (5-20%, 100,000 x
g, 2 h). The heavier proteoliposome fraction was found to
contain all the [^{3}H]muscimol binding sites and the majority of
the [^{3}H]FNZ binding sites; both fractions contained liposome
structures. Trypsin digestion of the liposome fraction
(2 mg/ml, 30 min, 20°C) reduced specific [^{3}H]FNZ binding by 64%
± 3 (n=4). A similar treatment of a CHAPS soluble GABA/BDZ
receptor preparation (according to Bristow and Martin, 1986)
resulted in a higher inhibition of [^{3}H]FNZ binding (83% ± 6, n
= 4).

Equilibrium [^{3}H]FNZ saturation analysis of the liposome
fraction gave mono-phasic Hofstee plots (Fig. 2), suggesting a

TABLE 1. [³H]FNZ BINDING PARAMETERS IN CEREBELLAR SYNAPTIC MEMBRANES AND LIPOSOME FRACTION IMMEDIATELY AND 48 HOURS (4°C) AFTER PREPARATION

	FNZ BINDING PARAMETERS	
	K_D (nM)	B_{max} (pmol/mg protein)
CEREBELLAR MEMBRANES	2.3 ± 0.2 (6)	0.87 ± 0.05 (4)
LIPOSOME FRACTION (t = 0)	3.6 ± 0.4 (6)	0.24 ± 0.07 (5)
LIPOSOME FRACTION (t = 48 h, 4°C)	2.9 ± 0.4 (4)	0.15 ± 0.05 (3)

[³H]FNZ saturation analysis (0.25 - 60 nM) was performed on cerebellar synaptic membranes and liposome fraction (immediately after preparation and after 48 h storage at 4°C). The parameters were obtained from a computer assisted Hoftsee analysis of the data (according to Zivin and Waud, 1982) and are the means ± SEM of the numbers of determinations given in parentheses. The plots obtained were all mono-phasic suggesting the existence of only a single binding site in each case.

single binding site population with a dissociation constant (K_D) of 3.6 ± 0.4 nM. This value is very similar to the K_D obtained in CbSPM preparations (Table 1). The affinity of the receptor for FNZ does not significantly change after 48 h storage at 4°C, however, the number of sites (B_{max}) is reduced over this time period by about 40% (Table 1).

The modulation of [³H]FNZ (0.5 nM) binding by GABA, cartazolate and pentobarbital in CbSPM's and in the liposome fractions is illustrated in Fig. 3. The facilitatory responses observed in the liposomes when immediately examined or after storage (48 h, 4°C) are not significantly different (Wilcoxon Rank Sum Test) from those seen in CbSPM's.

DISCUSSION

There are several reports of successful soluble GABA/BDZ receptor reconstitution into lipid vesicles (Schoch et al., 1984; Sigel et al., 1985; Hammond and Martin, in preparation).

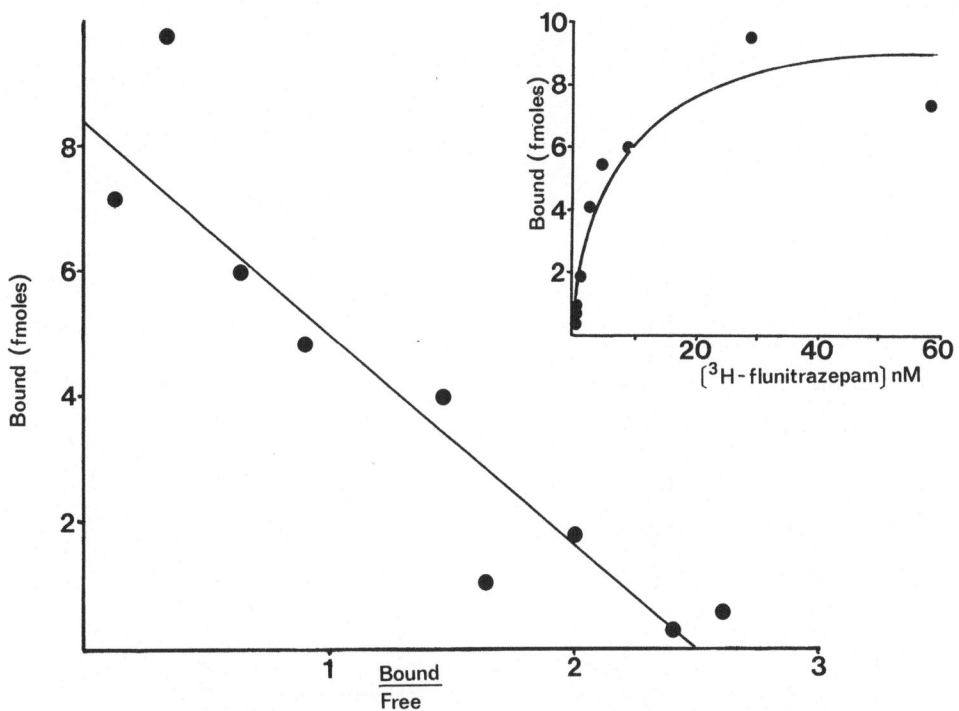

Figure 2. Representitive Hofstee plot of [³H]FNZ saturation data (0.25-60 nM) obtained from the liposome fraction immediately after preparation. On the y-axis, bound represents specific FNZ binding (fmoles), and on the x-axis, bound/free represents the ratio of specific bound to free FNZ. The line is fitted by computer analysis of the data according to Zivin and Waud (1982). The insert shows the untransformed saturation data from the same experiment.

These liposome preparations all appear to incorporate both the GABA and the BDZ binding sites. However, possibly due to the time required for receptor incorporation (usually 36-48 h), several of the accessory binding sites appear to decay during the reconstitution procedure. Interestingly, to date, there have been no reports of a functional GABA-activated anion channel in these preparations. We report here a rapid method to incorporate GABA/BDZ receptors into unilamellar liposomes. This preparation has retained the BDZ binding site with an affinity for [³H]FNZ very similar to that observed in membranes. Several modulatory responses are retained both quantitatively and qualitatively immediately after preparation and for 48 h at 4°C. The preservation of these neurochemical responses in the liposomes is identical to the protection

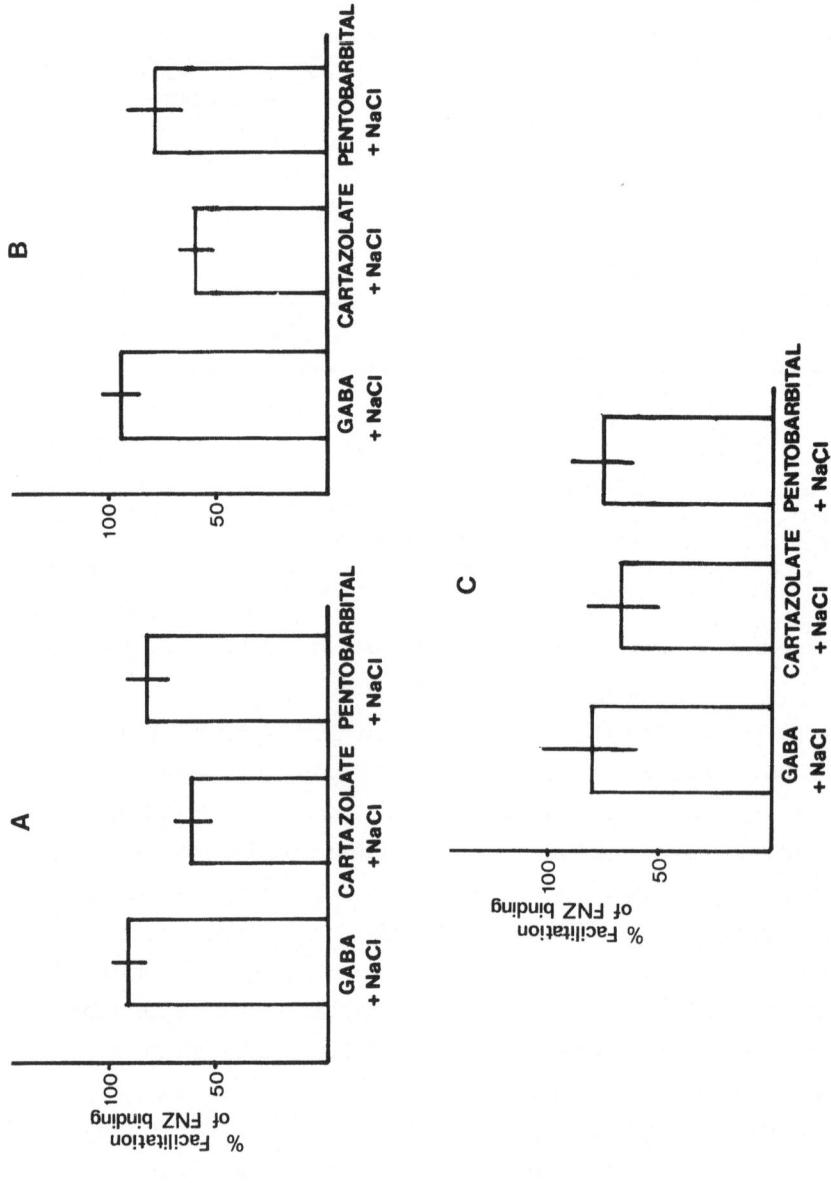

Figure 3. Histogram representations of the % facilitation of [³H]FNZ (0.5 nM) binding induced by GABA, cartazolate and pentobarbital, all in the presence of 0.15 M NaCl. The histograms show the responses in A) CbSPMs, B) Liposome fraction immediately after preparation and C) Liposome fraction after storage (48 h, 4°C). The bars represent the means ± SEM of 6-8 separate experiments. The responses in A) are not significantly different (Wilcoxon test) from those in B) and C) nor are B) and C) significantly different from each other.

elicited by soluble natural brain lipids observed in a CHAPS soluble GABA/BDZ receptor preparation (Bristow and Martin, 1986). Several pieces of evidence reported here indicate that the receptor complex is incorporated into the liposome lamellae. GABA is lipid insoluble, the finding that it elicits a response suggests that the GABA receptor is not simply trapped within the vesicles and must be assessible from the outside of the vesicle. The trypsin degradation data which shows an apparent lower degradation of sites in liposomes than in a soluble receptor preparations also supports receptor incorporation into lamallae, possibly bidirectionally, with the majority of the sites pointing outwards, or otherwise in some way that protects, to some degree, against proteolytic action.

In conclusion, this liposome incorporated GABA/BDZ receptor, which neurochemically appears in a functional state, may prove a useful preparation to examine the significance of these neurochemical interactions with regard to anion channel functionality. It may therefore yield information on the pharmacological prerequisites for $GABA_A$ receptor channel function, or indeed, be a useful preparation to examine other neurotransmitter receptors.

REFERENCES

Beer, B., Klepner, C.A., Lippa, A.S. and Squires, R., 1978, Enhancement of [^{3}H]diazepam binding by SQ 65396; a novel antianxiety agent, Pharmacol. Biochem. Behav., 9:849.

Bristow, D.R. and Martin, I.L., 1986, Solubilisation of the γ-aminobutyric acid/benzodiazepine receptor: Protection and stabilisation of modulatory characteristics using natural brain lipid extracts, Soc. Neurosci. Abstr., 12:664.

Costa, T., Rodbard, D. and Pert, C.B., 1979, Is the benzodiazepine receptor coupled to a chloride anion channel?, Nature, 277:315.

Fischer, J.B. and Olsen, R.W., 1986, Biochemical aspects of GABA/benzodiazepine receptor function, in: "Benzodiazepine /GABA Receptors and Chloride Channels: Structural and Functional Properties", R.W. Olsen and J.C. Venter, eds., Alan R. Liss Inc. New York.

Haefely, W., 1984, Actions and interactions of benzodiazepine
    agonists and antagonists at GABAergic synapses, in:
    "Actions and Interactions of GABA and Benzodiazepines",
    N.G. Bowery, ed., Raven Press, New York.

Hammond, J.R. and Martin, I.L., 1986, Solubilisation of the
    benzodiazepine/GABA receptor complex: comparison of the
    detergents octylglucoside and CHAPS, J. Neurochem.,
    47:1161.

Leeb-Lundberg, F., Snowman, A. and Olsen, R.W., 1980,
    Barbiturate receptor sites are coupled to benzodiazepine
    receptors, Proc. Natl. Acad. Sci. U.S.A., 77:7468.

Macdonald, R.L. and Barker, J.L., 1979, Enhancement of
    GABA-mediated postsynaptic inhibition in cultured
    mammalian spinal neurons: A common mode of anticonvulsant
    action, Brain Res., 167:323.

Martin, I.L. and Candy, J.M., 1978, Facilitation of
    benzodiazepine binding by sodium chloride and GABA,
    Neuropharmacology, 17:993.

Schoch, P., Haring, P., Takacs, B. Stahli, C. and Möhler,
    H., 1984, A GABA/benzodiazepine receptor complex from
    bovine brain: purification, reconstitution and
    immunological characterisation, J. Recept. Res., 4:189.

Sigel, E. and Barnard, E.A., 1984, A γ-aminobutyric
    acid/benzodiazepine receptor complex from bovine cerebral
    cortex: Improved purification with preservation of
    regulatory sites and interactions, J. Biol. Chem.,
    11:7219.

Sigel, E., Mamalaki, C. and Barnard, E.A., 1985, Reconstitution
    of the isolated GABA/benzodiazepine receptor complex into
    phospholipid vesicles, Neurosci. Lett., 61:165.

Simmonds, M.A., 1984, Physiological and pharmacological
    characterisation of the actions of GABA, in: "Actions and
    Interactions of GABA and Benzodiazepines", N.G. Bowery,
    ed., Raven Press, New York.

Squires, R.F., Casida, J.E., Richardson, M. and Saederup, E.,
    1983, [^{35}S] t-butybicyclophosphorothionate binds with high
    affinity to brain specific sites coupled to γ-aminobutyric
    acid-A and anion recognition sites, Mol. Pharmacol.,
    23:236.

Stephenson, F.A. and Barnard, E.A., 1986, Purification and characterisation of the brain GABA/benzodiazepine receptor, in:"Benzodiazepine /GABA receptors and chloride channels: Structural and Functional Properties", R.W. Olsen and J.C. Venter, eds., Alan. R. Liss Inc., New York.

Tallman, J.F., Thomas, J.W. and Gallager, D.W., 1978, GABAergic modulation of benzodiazepine binding site sensitivity, Nature, 274:383.

Ticku, M.K., Ban, M. and Olsen, R.W., 1978, Binding of [^{3}H]-α-dihydropicrotoxinin, a γ-aminobutyric acid synaptic antagonist, to rat brain membranes, Mol. Pharmacol., 14:391.

Zivin, J.A. and Waud, D.R., 1982, How to analyse binding, enzyme and uptake data: The simplest case, a single phase, Life Sci., 30:1407.

INTERACTION OF LAMININ WITH CELL MEMBRANES

G. Risse[a], J. Dieckhoff[b], H.G. Mannherz[b] and K. von der Mark[a]

[a]Max-Planck-Institut für Biochemie, Abteilung für
Bindegewebsforschung, Martinsried/München, BRD
[b]Institut für Anatomie und Zellbiologie, Universität
Marburg, BRD

INTRODUCTION

Laminin ,a major component of basement membranes, is a large $(M_r=10^6)$ glycoprotein with diverse biological functions. By binding to various matrix components including type IV collagen, heparan sulfate proteoglycan and nidogen it plays a key role in the organization of the basal lamina. Furthermore laminin is involved in a variety of interactions of the basement membrane with the adjacent epithelial and mesenchymal cells. In studies on isolated cells and organoids in vitro laminin has been shown to promote cell migration and adhesion, to effect cell shape and polarity, to induce outgrowth of cell processes and neurites and to stimulate proliferation and differentiation (for recent review see 1). Laminin seems to influence predominantly epithelial cell behaviour, but also effects on mesenchymal cells have been described, for example on muscle cell differentiation. We could show that laminin if present as substrate on culture dishes selects for myogenic versus fibroblastic cells and stimulates myodifferentiation[2]. These observations initiated the search for laminin-binding components on the muscle cell surface. Affinity chromatography of detergent solubilised muscle cell membranes resulted in the isolation of a 68kd protein, which upon reconstitution into liposomes binds specifically to laminin[3]. Polyclonal rabbit antibodies raised against the purified protein inhibit attachment of myoblasts to laminin substrates and reduce the binding of radiolabeled laminin to myoblasts (U.Kühl, unpublished results). Using the same isolation procedure a similar protein could be isolated from cell membranes of the EHS basement membrane tumor. In analysing matrix receptors we are aiming towards elucidation of the mechanisms of signal transduction which eventually will lead to the observed biological effects. Recently, a laminin-binding membrane protein, the 5'-nucleotidase from smooth muscle cell membranes, has been characterised. The AMPase activity of this enzyme is stimulated several-fold by laminin but inhibited by fibronectin, another component of the extracellular matrix[4]. This may allow the mechanism of laminin-cell interactions to be examined. The 5'-nucleotidase catalyses the hydrolysis of AMP to phosphate and adenosine which may act as a local hormone or "second messenger". Structural comparison of this enzyme and the muscle-derived laminin receptor revealed a certain degree of homology but

---

Abbreviation: EHS, Engelbrecht-Holm-Swarm

also demonstrated that the two proteins are not identical. In this report we would like to present data on the interaction of myoblasts with laminin and on the biochemical characterisation of laminin-binding cell surface proteins.

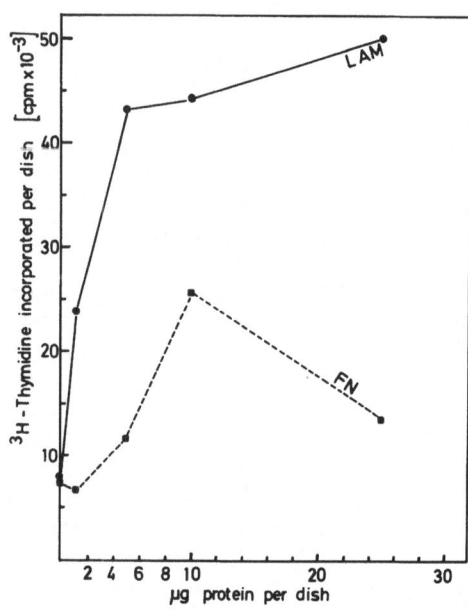

Fig.1: Effect of laminin and fibronectin on ^{3}H -thymidine incorporation of MM14 myoblasts. 6 cm^2 Culture dishes were coated with increasing amounts of laminin or fibronectin, saturated with bovine serum albumin and inoculated with 10^4 MM14 mouse skeletal myoblasts. After 4 hrs 5 Ci ^{3}H-thymidine was added to each dish, and after 20 hrs TCA-insoluble thymidine was measured using the filter technique. Concomitant with the higher incorporation of ^{3}H-thymidine cells also proliferate more rapidly on laminin (not shown).

RESULTS AND DISCUSSION

Role of Laminin in Myoblast Differentiation

The biological effects of laminin on cell behaviour were investigated in cultures of skeletal myoblast. During the complex process of myoblast differentiation in vitro laminin was shown to have numerous functions. Thus it not only serves as substrate for the adhesion of myoblasts[2] but also stimulates division and proliferation of these cells (Fig.1). Furthermore laminin induces dramatic morphological changes. Skeletal myoblasts plated onto laminin substrates elongate and assume a spindle shape typical for differentiated, fusion-capable myoblasts while they remain round or more fibroblastic on fibronectin[2]. As a consequence of the enhanced proliferation the fusion rate on laminin is considerably higher than on fibronectin. Owing to the high affinity of myoblasts to laminin, myoblasts and muscle fibroblasts can be separated by selective adhesion to laminin or fibronectin substrates (Fig.2).

174

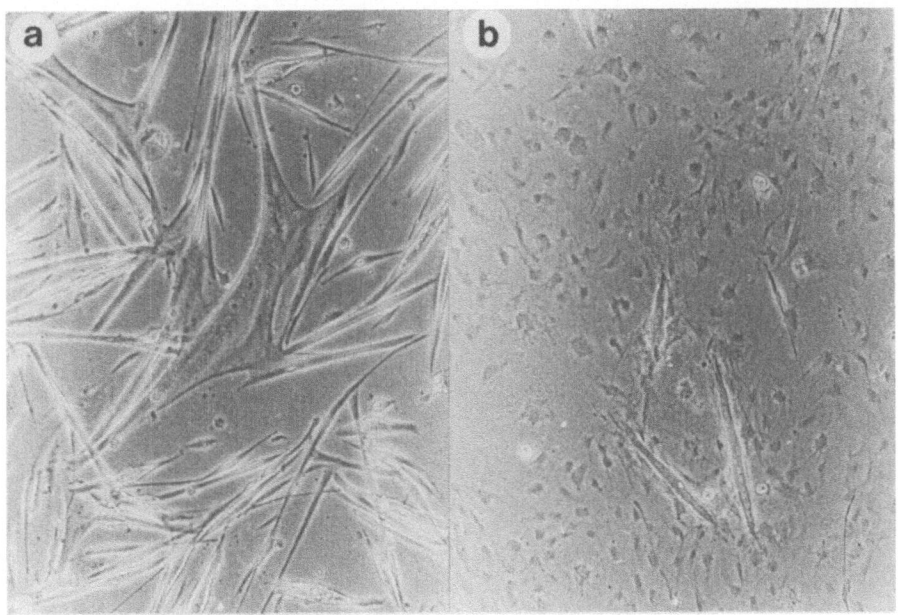

Fig.2: Selection of myogenic cells from mouse thigh muscle cells on laminin/type IV collagen. Primary 1d mass cultures of embryonic mouse thigh muscle were dissociated and allowed to adhere after a brief recovery period to type IV collagen/laminin (A) or fibronectin/type I collagen (B) for 15 min in serum-free medium. The non-adherent cells were decanted and the adherent cells were grown for another 12 days in DMEM containing horse serum and embryo extract. On laminin/type IV collagen only myogenic cells developed which showed a high fusion rate after 12 days, while on fibronectin/ type I collagen fibrogenic cells dominated. Cells were photographed after 12d (X80).

Characterisation of Laminin-binding Proteins

The effects of laminin on cell adhesion, cell proliferation and cell morphology suggested the existence of specific recognition sites for laminin on the muscle cell surface. First information about the nature of such a laminin-binding protein was obtained by blotting muscle cell membrane proteins after separation by SDS-electrophoresis with 125-I-labeled laminin revealing two prominent bands corresponding to proteins of M_r = 92 and 68kd (Fig.3a).
This technique also allowed us to get further information about the binding sites within the laminin molecule. Thus we demonstrated that the E8 fragment of laminin, located in the long arm of the molecule (Fig.4), is apparently also involved in the binding to the 68kd protein (Fig.3b-d). Recently we confirmed the cell binding activity of E8 using cell attachment experiments (S.Goodman, in prep.). Terranova et.al.[5] have located a major cell binding site in the E1 fragment of laminin. Our results as shown here indicate the existence of at least one further binding site in the E8 region of laminin.
The 68kd component could be purified by affinity chromatography of detergent-solubilised muscle cell membranes as described[3]. In brief, muscle cell membranes were obtained from mouse thigh muscle after thorough homogenization and purified by several steps of centrifugation, including a sucrose density gradient, according to Cates and Holland[8]. The laminin-

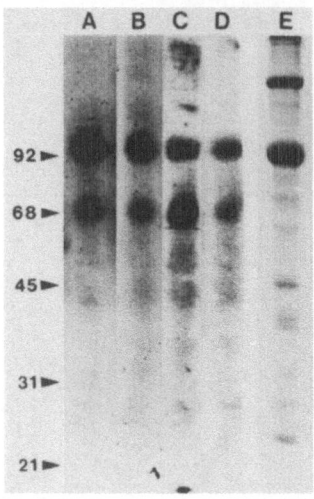

Fig.3: Binding of 125-I-labeled laminin and laminin fragments to plasma membranes of mouse muscle. Plasma membranes were collected on a sucrose gradient and separated by SDS-polyacrylamide gel electrophoresis. After transfer to nitrocellulose paper by electroblotting, strips were incubated with radioiodinated laminin (a,b), E8 (c), or E1 (d), resp., washed and autoradiographed (for location of the E8 and E1 fragments within the laminin molecule see Fig.4). In (a) the incubation was in absence, in (b) in presence of Ca^{2+}. In this type of experiment the binding of laminin to the 68kd receptor appears not to be Ca^{2+}- dependent. The E8 fragment (c) binds stronger to the 68kd protein than the E1 fragment. The binding to the 92kd molecule (= -Actinin) might be unspecific and due to the very high concentration of this protein in the membrane protein preparation (see (e): Coomassie blue staining of total plasma membranes after electrophoresis).

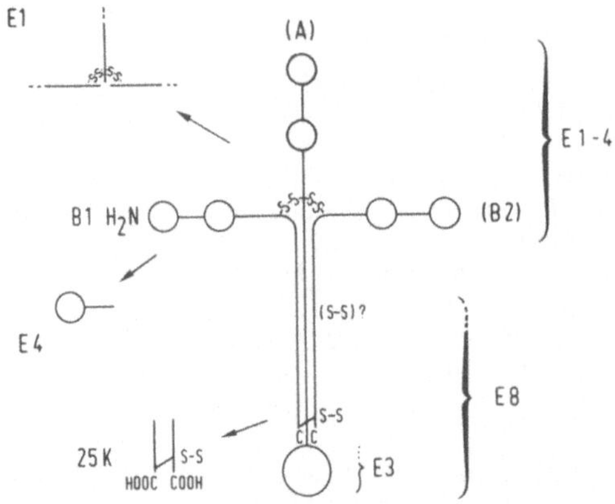

Fig.4: Model of the laminin molecule. This model is based on rotary shadow images and amino acid sequences. The presumptive localization of its major proteolytic fragments as obtained by cleavage with elastase is indicated. (by courtesy from Dr.R.Deutzmann).

binding protein was isolated from the membrane fraction by affinity chromatography on laminin, covalently linked to sepharose 4B, first in 0.1% NP40, then in 0.1% deoxycholate. After absorption with fibronectin sepharose and rechromatography on laminin sepharose the 68kd protein migrated in the reduced form as a single band on SDS-gel electrophoresis.
Using the same isolation procedure a protein of the same molecular weight (Fig.5) was isolated from the EHS-tumor.

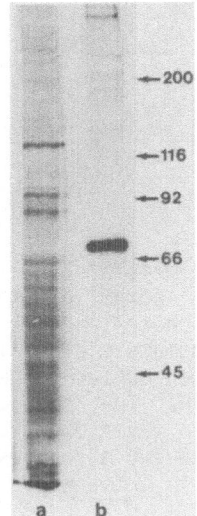

Fig.5: SDS-polyacrylamide gel of the laminin-binding protein from EHS tumor after chromatography on laminin sepharose (B). (A): Total EHS membrane proteins as starting material collected from the 17/40% interface of a discontinuous sucrose gradient. Molecular weight markers in kilodalton. 10% acrylamide gel, silver staining.

The 68kd laminin receptor shares with serum albumin not only the size; under conditions of low stringency albumin also binds to laminin sepharose. For this reason, laminin receptor preparations are sometimes contaminated with albumin. Both proteins are, however, clearly distinct molecules by their amino acid composition and two-dimensional tryptic peptide maps (Fig.6).

5'-Nucleotidase and its Relationship to the Laminin Receptor

The general problem in discussing the role of matrix receptors such as the laminin-binding protein desribed here is the complete lack of knowledge of their biological function. It is unlikely that such receptors are restricted in their function to the binding property only. The reported effects of laminin on cell shape or proliferation would require the onset of transmembrane signals involved in the intracellular events of cytodifferentiation. An interesting possibility to explain the mechanism of laminin-cell interactions arose from the observation that the AMPase activity of the 5'-nucleotidase from smooth muscle is enhanced by laminin but inhibited by fibronectin[4]. 5'-Nucleotidase (EC 3.1.3.5.) is a membrane-bound ectoenzyme present in many cells such as smooth muscle[8] liver, glia and tumor cells; it is thought to be a transmembrane protein[8], catalysing the hydrolysis of AMP to free adenosine and inorganic phosphate (for review see 9). 5'-Nucleotidase now binds laminin and fibronectin, either when

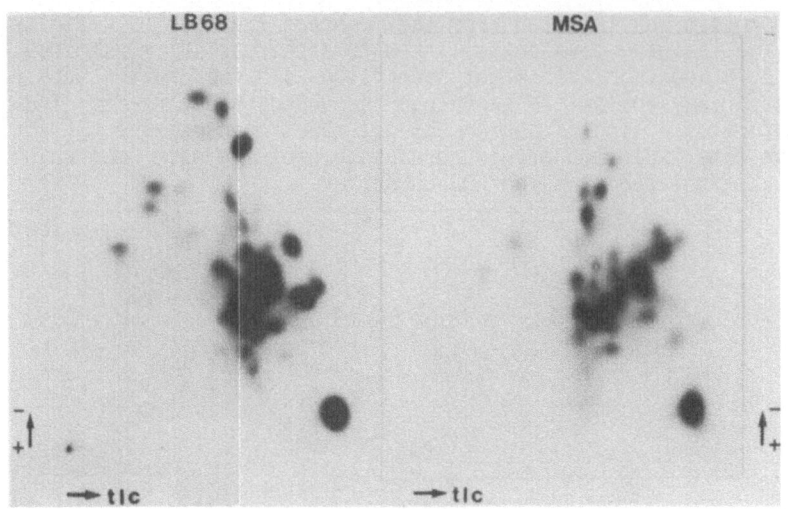

Fig.6: Two-dimensional peptide maps of the laminin-binding protein from EHS tumor (LB 68) and mouse serum albumin (MSA). The proteins were purified by SDS-electrophoresis on a 10% polyacrylamide slab gel and stained briefly with Coomassie blue. The 68kd bands were cut out, labeled with 125-I-Bolton-Hunter reagent and digested with a mixture of trypsin/chymotrypsin. The resulting peptides were separated twodimensionally by thin layer electrophoresis on silicagel sheets and chromatography[7]. The two molecules are clearly not identical.

solubilised in detergent or when integrated into liposomes[4]. The enzyme activity is increased (2-4fold) in the presence of laminin, and, in contrast, decreased (by 80%) in the presence of fibronectin. Collagens type I, III and IV do not influence the enzyme activity[4]. Identical amino acid composition and the observation that the M_r 79kd 5'-nucleotidase is

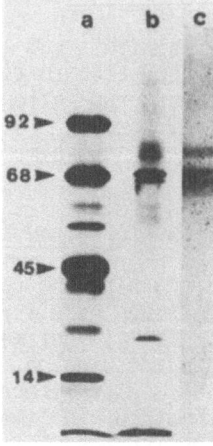

Fig.7: Partial conversion of the 76k-form of 5'-nucleotidase to a 68k-derivative. After several weeks of storage, the majority of the 5'-nucleotidase is converted to the 68k-derivative (c) which shows a similar band pattern as compared to a preparation of laminin receptor from EHS-tumor (b). A specific conversion of the 76kd nucleotidase to a 68kd form can be seen after digestion with endoglycosidase F (not shown). a = MW standard

converted to a 68kd molecule upon storage (Fig.7) and after treatment with endoglycosidase F suggested that both proteins may be related or identical; however, two-dimensional peptide mapping of the 5'-nucleotidase from chick gizzard and the chick muscle laminin receptor demonstrated only a certain degree of homology but not the identity of these two proteins (Fig.8). The reported activation of 5'-nucleotidase by laminin may open new possibilities to examine the mechanism of laminin-cell interactions. The adenosine produced by the enzyme may be taken up by the cell and subsequently affect intracellular phophorylation and differentiation processes.

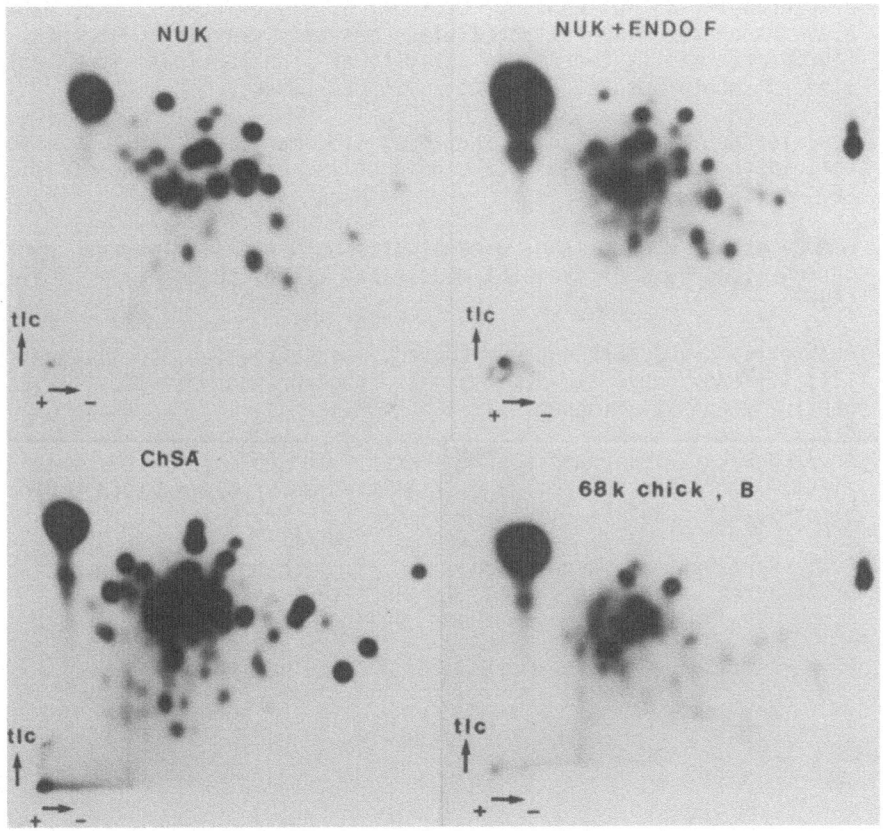

Fig.8: Comparison of 5'nucleotidase (from chicken gizzard) and laminin receptor (from embryonal chick muscle) by two-dimensional fingerprinting. The proteins were purified by SDS-electrophoresis on a 10% polyacrylamide gel and stained briefly with Coomasie blue. The 68kD bands of endoglycosidase F digested nucleotidase, of the laminin receptor from chick muscle and of chick serum albumin, resp., and the 76kD band of native nucleotidase were cut out, iodinated with Bolton-Hunter reagent and digested with a mixture of trypsin/chymotrypsin. The resulting peptides were analysed by 2D-mapping.
NUK = native nucleotidase (MW = 76kD)
NUK + ENDO F = endoglycosidase F digested nucleotidase (MW = 68kD)
68kD chick, B = laminin receptor from chick muscle
ChSA = chick serum albumin

References

(1) K.v.d. Mark and U. Kühl, Laminin and its receptor, Biochim. biophys. Acta 823:147 (1985).

(2) U. Kühl, M. öcalan, R. Timpl and K.v.d. Mark, Role of laminin and fibronectin in selecting myogenic versus fibrogenic cells from skeletal muscle cells in vitro, Develop. Biol., in the press.

(3) H. Lesot, U. Kühl and K.v.d. Mark, Isolation of a laminin-binding protein from muscle cell membranes, Eur. Mol. Biol. Organ. J. 2:861 (1983).

(4) J. Dieckhoff, J. Mollenhauer, U. Kühl, B. Niggenmeyer, K.v.d. Mark and H.G. Mannherz, The extracellular matrix proteins laminin and fibronectin modify the AMPase activity of 5'nucleotidase from chicken gizzard smooth muscle, FEBS Lett. 195:824 (1986).

(5) V.P. Terranova, C.N. Rao, T. Kalebic, I.M. Margulies and L.A. Liotta, Laminin receptor on breast carcinoma cells, Proc. Natl. Acad. Sci. USA 80:444 (1983).

(6) G.A.Cates and P.C.Holland, Biosynthesis of plasma-membrane proteins during myogenesis of skeletal muscle in vitro, Biochem. J. 174:873 (1978).

(7) P.T.Kelly, K.v.d.Mark and G.W.Conrad, Identification of collagen types I,II,III and V by two-dimensional fingerprints of 125- I-labeled peptides, Analyt. Biochem. 112:105 (1981).

(8) A. Zachowsky, H.W. Evans and A. Paraf, Immunological evidence that plasma-membrane 5'nucleotidase is a transmembrane protein, Biochim. biophys. Acta 664:121 (1981).

(9) G.W. Kreutzberg, M. Reddington and H. Zimmermann, "Cellular Biology of Ectoenzymes", Springer Verlag, Berlin (1986).

SOLUBILIZATION OF THE FUSICOCCIN RECEPTOR AND A PROTEIN KINASE FROM HIGHLY

PURIFIED PLASMA MEMBRANE FROM OAT ROOTS

A.H. De Boer, T.L. Lomax, R.P. Sandstrom and R.E. Cleland

Department of Botany, University of Washington
Seattle, WA 98195

ABSTRACT

Fusicoccin (FC), a diterpine glucoside phytotoxin, that stimulates electrogenic H^+ extrusion in higher plants binds specifically to a plant plasma membrane (PM) protein. A highly purified preparation of plasma membrane vesicles has been prepared from oat roots and used to study the FC-binding protein (receptor). In these right-side out vesicles the FC-binding site is protected from trypsin degradation, but disruption of the vesicle integrity with Triton X-100 or solubilization of the receptor renders it susceptible to trypsin degradation. The receptor has been successfully solubilized with octyl-glucoside (90 % recovery) and can be assayed effectively by binding to glass-fiber filters coated with 1 % polyethylinimine.

A Mg^{2+}-dependent Ca^{2+}-activated protein-kinase was solubilized from the PM vesicles through sonication. This shows that the kinase is a peripheral membrane protein. The soluble kinase has a sharp pH optimum at pH 7.

The enzyme that is thought to be activated by FC, the plasma membrane ATPase, was present in the PM vesicles with a very high specific activity (4 - 5 µmol/mg protein. min). We show evidence that the detergent activition of this enzyme is not simply due to breaking the vesicle integrity. A conformational change or removal of a regulatory subunit may cause stimulation of the H^+-ATPase and this may be of importance in our study of the mode of action of fusicoccin.

INTRODUCTION

Plant growth substances have been known for a long time and numerous reports describe their effects on plant growth and performance. However, virtually nothing is known about the molecular mechanism of primary action of plant hormones.

A tool in unravelling this molecular mechanism may be a phytotoxin produced by the fungus Fusicoccum amygdali Del., called fusicoccin (FC). Although FC is not a natural plant hormone it has attracted much attention because it mimics the effects of the naturally occurring plant growth hormone auxin in many respects[1].

A primary effect of FC is hyperpolarization of the membrane electrical potential and increased proton extrusion. It is assumed that this is brought about through activation of a H^+-ATPase located in the plasma membrane (PM). The PM contains a protein with an external domain[2] that specifically binds FC with high affinity ($K_d = 10^{-10} - 10^{-9}$ mol.l^{-1})[3]. This protein may act as a FC-receptor and play a role in the signal transduction across the PM. Although it has not been shown yet that the FC-binding protein possesses receptor function we will call it "receptor" for convenience.

With its location in the PM this receptor could belong to the class of cell surface receptors that have been studied in great detail in mammalian cells, e.g. the epidermal growth factor (EGF) receptor[4]. One of the immediate consequences of growth factor/receptor interaction is a rapid stimulation of intracellular receptor kinase activity that results in autophosphorylation of the receptor at several tyrosine residues as well as phosphorylation of other membrane proteins[4]. Recently it has been shown that also FC stimulates protein phosphorylation in plant cells[5]. A secondary effect of EGF/receptor interaction is an increase of cytoplasmic pH and subsequent stimulation of (Na,K)-ATPase activity. FC stimulates H^+-ATPase activity although the reports on the effect of FC on the cytoplasmic pH are contradictory[6,7].

A number of properties of the FC-receptor are known so far: it is a glycoprotein with a M.W. of around 80 kDa. The ligand binding is sensitive to mercurials[8], to acid phophatase and α-mannosidase[9], to trypsin treatment[10] and to a yet unidentified, endogenous ligand[11]. A phenomenon of particular interest is the sensitivity of the binding to phosphatase, since this is an indication that the receptor is a phosphorylated protein and that it must be in the phosphorylated state for maximum binding. The soluble auxin receptor, isolated from tobacco callus, also seems to loose binding capacity when dephosphorylated (K.R. Libbenga, pers. comm.). Interestingly, the opposite is true for mammalian PM-receptors; e.g. receptor "desensitization" follows receptor phosphorylation at threonine residues[12,13].

Despite the fact that the FC-binding protein seems to be located exclusively in the PM, no study has made use of highly purified PM as the starting material to characterize the FC-receptor. The reason for this is not obvious since a method for purifying plant PM is available[14]. Our choice to start with highly purified PM has proven to be advantageous as shown in the Results section.

The goal of our project is the purification and reconstitution of the FC-receptor, along with the H^+-ATPase, into liposomes in order to study the direct interaction between these two proteins. At the same time we are investigating whether there is a PM-bound protein-kinase that has the FC-receptor as target molecule, since phosphorylation/dephosphorylation may be a mechanism for the regulation of receptor sensitivity.

MATERIAL AND METHODS

PM-extraction Five day old oat roots, grown in 1 mM CaSO$_4$ at 25 $^\circ$C were ground in a Waring blender in 2 volumes of buffer: 250 mM sucrose, 10 mM Tris/HCl (pH 7.5) and 1 mM EDTA. The grindate was filtered through 4 layers of cheesecloth and centrifuged 20 min at 10,000 g. The resulting supernatant was centrifuged for 30 min at 100,000 g to obtain the microsomal pellet. The resuspended pellet was layered on a dextran/polyethylene glycol 2-phase system and after 3 consecutive separations[14], purified plasma membrane was obtained. The yield was around 20 µg protein per g fresh weight.

ATPase assay The assay was carried out in 0.5 ml assay buffer with 2

5 mM Na-ATP and 30 mM Mes/Tris (pH 6.5) with or without 0.0125 % Triton X-100. The reaction proceeded at 37 ^{O}C for 20 min and was stopped with 1 ml, 2 % H_2SO_4, 5 % SDS and 0.5 % ammonium-molybdate. [P_i] was measured spectrophotometrically (A_{660}) after adding 50 μl of 10 % ascorbic acid.

Electron Spin Resonance The ESR measurements were done as descibed by Lomax and Mehlhorn[15].

^{3}H-FC binding assay. Membrane-bound: Each assay tube (total volume 0.5 ml) contained: 4 μg protein, buffer A (250 mM sucrose, 25 mM Mes/Tris (pH 7), 1 mM $MgCl_2$, 0.5 mM EDTA and 0.5 mg/ml DTT) and 10^{-9} M ^{3}H-FC (plus 10^{-6} M FC to measure non-specific binding). Samples were incubated for 1 h at 30 ^{O}C and were then pipetted into 2 ml of "wash-buffer" (250 mM sucrose, 10 mM Mes/Tris, pH 6.5) present on GSTF filters and filtered by vacuum. Filters were washed twice with 5 ml of wash-buffer.
Soluble: Binding conditions were the same as for the membrane bound receptor except that buffer A contained 25 mM KF. After the 1 h incubation period the sample was mixed with 5 ml glycine/KOH buffer (pH 9.5) and filtered over GF/B filters (soaked for 4 h in 1 % polyethylinimine (PEI)) without vacuum. The filters were washed once with 5 ml of the glycine buffer and the filters were counted.

Acetone solubilization Twenty volumes of (-20 ^{O}C) acetone was added to a 20 μl PM sample (1 mg/ml) while stirring. The precipitate was washed once with an equal amount of acetone and dried under N_2-gas. To the precipitate 0.5 ml of buffer A plus 150 mM KCl and 0.1 % Triton was added and the pellet was resuspended through 1 min sonication at 100 W. The sample was centrifuged at 100,000 g for 30 min and the supernatant was used to assay for the soluble receptor.

Phosphorylation. Membrane-bound: The standard reaction mixture (0.2 ml for 2 replicates) contained: 50 μg protein, 1 mM [γ-^{32}P]-ATP in buffer B (250 mM sucrose, 5 mM $MgCl_2$, 1 mM EGTA, 1.055 mM $CaCl_2$ (to give 0.1 mM free Ca^{2+}) and 50 mM Mes/Tris, pH 7.0). The reaction was started with addition of ATP, allowed to proceed for 20 min at 25 ^{O}C and stopped by addition of 2 ml ice-cold 10 % TCA + 20 mM Na-pyrophosphate. This was filtered through HAWP filters and the filters were washed twice with the same TCA solution; filters were dried and counted.
Soluble: The kinase was solubilized in 0.2 ml of buffer B minus $CaCl_2$ with 1 min sonication on ice (protein concentration: 2.5 mg/ml). After centrifugation at 100,000 g the supernatant was collected. The reaction mixture contained: 20 μl supernatant, 0.1 mM [γ-^{32}P]-ATP, 1 mg/ml Histone IIIS and buffer B. The reaction was started with ATP and proceeded for 10 min at 30 ^{O}C. The rest was similar to the membrane bound kinase assay.

Trypsin digestion Samples (0.15 mg protein/ml) were incubated at 25 ^{O}C in Tris/HCl (100 mM, pH 8.0) buffer with 0.1 mg trypsin/ml (protein : trypsin = 1.6 : 1). The digestion was stopped with trypsin inhibitor (trypsin : trypsin inhibitor = 1 : 2).

Protein- assay: proteins were measured according to Bradford[16].

Chemicals Fusicoccin was a kind gift from professor Ballio from Rome. Radioactive dihydrofusicoccin (^{3}H-FC, spec. act. 35.2 Ci/mmol), was prepared by Amersham U.K., [γ-^{32}P]-ATP (spec. act. 3000 Ci/mol), was obtained from NEN Research Products U.S.A.. Histone IIIS (calf thymus), Trypsin (type IX from porcine pancreas) and Trypsin inhibitor (type IS from soybean) were obtained from Sigma.

RESULTS

Characterization of PM-vesicles

Plasma-membrane vesicles were isolated from oat roots using aequous phase partitioning. This results in a PM fraction which is virtually free of other contaminating membranes. The PM is characterized by a very high specific activity of vanadate sensitive, Mg^{2+}-dependent and K^+-stimulated ATPase (4 - 5 µmol P_i/mg. min.). The ATPase has a sharp pH-optimum at pH 6.5 and is highly specific for ATP.

The presence of low concentrations of Triton X-100 or lysolecithin is required to obtain a fully activated ATPase (Fig. 1A). This phenomenon of detergent activated ATPase has been used to argue that aequous phase partitioning results in sealed vesicles with a right-side out orientation[14], assuming that detergents are necessary to make the hydrolytic site of the ATPase accessible to ATP. We questioned this explanation of the so-called "latency" as it is known, for example, that detergents activate the mitochondrial F_o-F_1 ATPase through removal of an inhibitory subunit[17].

In order to understand the detergent activation we first checked the degree of sealedness of the vesicles using the technique of Electron Spin Resonance (ESR) spectroscopy[15]. The vesicles behaved as ideal osmometers and their iso-osmolar volume (6.4 µl/mg protein) agreed well with the packed vesicle volume. This indicates that nearly all the vesicles were sealed to molecules the size of the quencher ($TMA_2MnEDTA$, M.W. about 500). The vesicles were able to maintain a pH-gradient although there was a slow decay over time (0.025 pH units/min). The maximum measured pH-gradient was about half that of the imposed pH-gradient. This indicates that a sizable portion of the vesicles was sealed to protons as well.

When the sealed volume was reduced through titration with Triton X-100 (Fig. 1B), vesicles were fully permeable at lower Triton concentrations than those necessary for activation of the ATPase. Thus, the detergent activation of the ATPase is not just due to making the hydrolytic site of the ATPase accessible to ATP but also to some other, as yet unknown, process. This does not invalidate the aforementioned conclusion that the main body of the

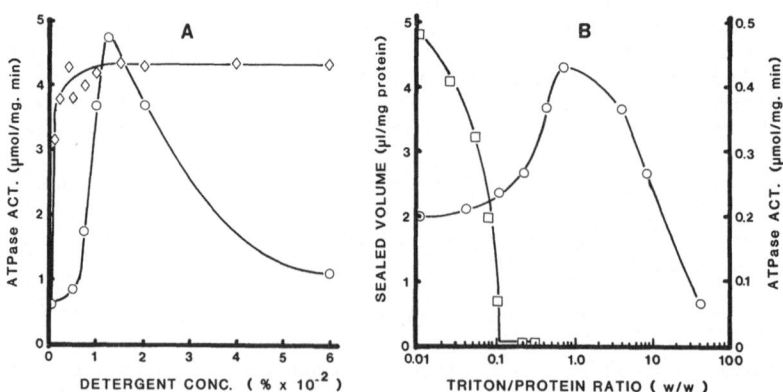

Fig. 1. A: Effect of Triton X-100 (circles) and lysolecithin (diamonds) on the ATPase activity; [protein] = 4 µg/ml. B: Sealed vesicle volume (squares) and ATPase activity (circles) as affected by Triton X-100; [protein] = 2.12 mg/ml. The high protein concentration needed for the ESR measurements resulted in a rapid depletion of ATP and consequently in an underestimation of the Triton activated ATPase activity.

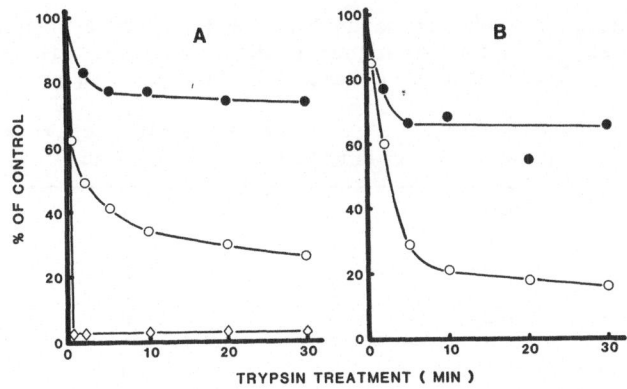

Fig. 2. Trypsination of the FC-receptor (A) and the ATPase (B). Closed circles = native vesicles, open circles = + 0.0125 % Triton X-100, diamonds = soluble FC-receptor.

vesicles is sealed and right-side out but shows that this conclusion was based upon a false assumption.

Trypsin reduced the ATPase activity in native vesicles by about 35 % and by 85 % in the presence of low levels of Triton X-100 (Fig. 2B). Because the trypsin sensitive, catalytic site faces the cell interior, we can conclude that 65 % of the vesicles have a right-side out orientation and are trypsin sealed. Interestingly, the trypsin sensitivity of FC-binding showed the same pattern as that of the ATPase (Fig. 2A). When solubilized, the receptor was very sensitive to trypsin: in 30 sec all FC-binding was abolished.

The membrane bound FC-receptor

Due to the purity of our PM we were able to assay for membrane-bound ^{3}H-FC by filtration through GSTF-filters (Ø 0.22 µm); clogging of the filter was no problem with 4 µg of protein per 0.5 ml assay tube. A comparison of the traditional centrifugation assay[10,11] with the filtration method showed that recovery of specifically bound FC was more than double with the filtration method (non-specific binding was less than 10 % in both cases).

FC-binding is a time dependent process: at 30 °C ([^{3}H-FC] = 10^{-9} M) binding of FC to the PM-bound protein was saturated in 1 h and binding remained stable for 4 h (the time range tested). The FC-binding was rather pH insensitive with a binding maximum at pH 7. Sonication of the vesicles in the presence of free Ca^{2+} for up to 60 sec had no significant effect on binding. This indicates that the FC-binding protein is an integral membrane protein.

Detergent solubilization of the FC-receptor

For further identification and purification the receptor has to be solubilized from the membrane. The first problem encountered in the solubilization procedure is the assay method for the soluble receptor: how to separate free ^{3}H-FC from receptor bound ^{3}H-FC ? We compared 3 methods: precipitation with polyethylene glycol (PEG) 6000, filtration on a Biogel P-6 column and filtration through glass-fiber filters coated with polyethylinimine (PEI). The latter method worked very well when the samples were

Table 1. Binding of ^{3}H-FC to solubilized and membrane-bound receptor after treatment with 1 % detergent (Detergent : protein = 125 : 1). Control = binding to the native vesicles: 12.2 pmol/mg protein.

Detergent	% detergent in assay	% of initial activity		
		solubilized	in pellet	inactivated
CHAPS	0.2	13	21	66
Na-cholate	0.4	15	28	57
Digitonin	0.1	18	35	47
Lubrol PX	0.2	26	17	57
Octyl-glucoside	0.2	91	9	0
Triton X-100	0.1	27	21	52
Zwittergent 3-14	0.1	3	3	94
Acetone (+Triton)	0.1	82	16	2

pipetted into high pH buffer (pH 9.5 - 10.5) prior to filtration. At pH 8, on the other hand, recovery on the glass-fiber filters was only 30 %. Non-specific binding was very low (< 2%) and pH independent. Precipitation with PEG 6000 was not pursued because recovery was very low. Filtration over Biogel P-6 gave a good separation of free and bound ligand, but recovery of the receptor was less than 50 % of that bound by the glass-fiber filters. It is important to note that the filters did recover 100 % of the ^{3}H-FC/receptor complex collected in the column fractions.

Of the 7 detergents that we tested only octyl-glucoside (OG) solubilized 90 % of the receptor without any loss in total binding activity (Table 1). The amount of detergent present during the binding assay is important; for example there is a rapid drop in activity at OG concentrations exceeding 0.2 %. Also effective is acetone solubilization in the presence of 0.1 % Triton X-100. It should be noted that solubilization of the acetone powder is difficult in the absence of Triton.

To retain full activity of the soluble receptor it is essential to sonicate the sample during solubilization with OG. The release of the receptor from the membranes is not improved by sonication, rather sonication seems to help maintain the soluble receptor in an active form.

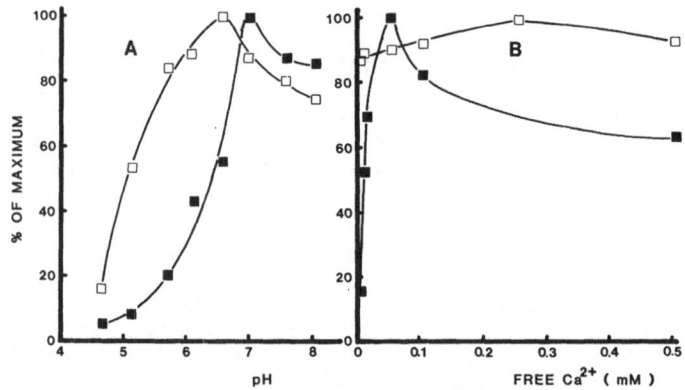

Fig. 3 Protein-kinase activity as a function of pH (A) and free Ca^{2+} (B) Open symbols = membrane bound kinase, closed symbols = soluble kinase.

Plasma-membrane bound protein-kinase activity

Many membrane receptors are phosphorylated proteins[4], hence our interest in the presence of a protein-kinase bound to the PM. To our knowledge only one paper describes the properties of a plant protein-kinase identified as PM-bound[18].

Protein kinase activity was demonstrated in our PM vesicles by the time-dependent, temperature-sensitive incorporation of $^{32}P_i$ from $[\gamma-^{32}P]$-ATP into the membrane proteins. Since the kinase was presumed to be located on the inside of the membranes, the supply of ATP to the kinase might limit its reaction rate. As expected, the activity could be increased either by increasing the ATP concentration, or by limiting the competition for ATP by the ATPase by inhibition of the ATPase with 0.1 mM vanadate. Latency of the kinase could not be established because Triton X-100 inactivated the enzyme, even at low levels. The conditions that we chose for further experiments were 1 mM ATP with an incorporation time of 20 min.

Effect of pH and ions

The kinase is strongly pH dependent with a pH-optimum between 6.5 and 7 (Fig. 3A). The presence of NaCl or KCl up to 100 mM has no significant effect on kinase activity. The kinase is highly Mg^{2+} dependent: the optimum Mg^{2+} concentration of 4 mM brings about a 9-fold increase in activity. In its membrane bound form there was only a small Ca^{2+} dependency of the kinase (Fig. 3B). It cannot be excluded that even in the presence of 1 mM EGTA the membrane binds enough Ca^{2+} to activate the kinase; viz. in soluble form the kinase appears to be strongly Ca^{2+} activated, as shown later.

The presence of native phosphatase might reduce the apparent $^{32}P_i$-incorporation. However, the inclusion of 10 mM F^- (F^- is a potent phosphatase inhibitor) rather reduced (15 %) than increased the amount of $^{32}P_i$-incorporated. This indicates that at pH 7 phosphatase activity is not important.

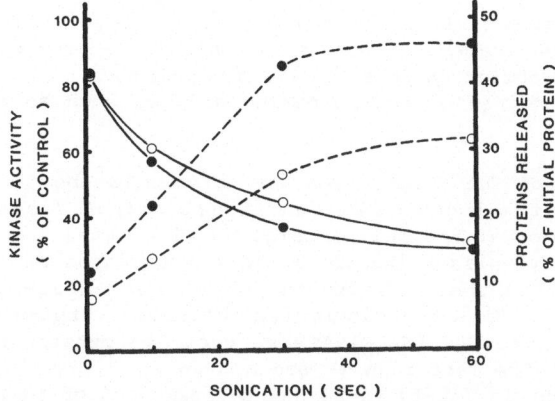

Fig. 4. Effect of sonication on membrane bound kinase activity (solid lines) and release of proteins from the membrane (dotted lines). Open circles = sonication in the presence of 0.1 mM free Ca^{2+} (= buffer B, pH 6.6), closed circles = sonication in the absence of Ca^{2+} (= buffer B − Ca^{2+}, pH 8.0). After sonication the samples were centrifuged at 100,000 g and the pellet was resuspended in buffer B (pH 7) for the kinase assay. The supernatant was used to measure the loss in proteins. Control is the sample before sonication and centrifugation.

Solubilization of the kinase

In contrast to the ATPase and the FC-receptor the kinase seems to be a peripheral protein . Sixty seconds of sonication reduced membrane bound kinase activity with 70 % (Fig. 4). The attachment of the kinase to the membrane is not stronger in the presence of Ca^{2+} although this is certainly true for other membrane-bound proteins. Even centrifugation is enough to remove 20 % of the kinase and 10 % of membrane proteins in general.

The kinase released from the membrane through sonication can be recovered in the soluble fraction and assayed with Histone III-S as substrate. After solubilization the kinase shows strong Ca^{2+} activation and its activity is strongly reduced in the acidic pH range just as when the kinase was still membrane bound (Fig. 3).

DISCUSSION

A relevant question is, "why do most plant cells have a FC-receptor protein in their PM if it is likely that these cells will never see FC ?" Some natural function for this protein is likely, and evidence is accumulating that there is an endogenous ligand that can compete with FC[11]. The identity and role of this ligand are still to be determined.

Our study shows that it is advantageous to start with highly purified plasma-membrane because:
- The localization of the enzymes studied is known,
- The specific activity of the enzymes is very high, viz.:
  * Mg^{2+}-dependent, K^+-stimulated ATPase: 4 - 5 µmol/min. mg protein. Higher values have only been shown after solubilization and further purification of the ATPase.
  * FC-binding: 13 pmol/mg protein. This equals 0.10 % of the total PM protein assuming a M.W. of 80,000 for the receptor[10]. Pesci et al.[8] reported 4.3 pmol FC bound per mg protein in a membrane fraction obtained through sucrose density centrifugation at 160,000 g from maize coleoptiles.
  * Protein-kinase activity: 5 - 10 nmol $^{32}P_i$-incorporated/mg protein. min. Hetherington and Trewavas[19] report 2.3 nmol $^{32}P_i$-incorporated/mg. min in a microsomal fraction enriched in PM from pea buds.
- There is a low degree of contaminating enzymes such as phosphatases and proteases.

Trypsination of functional membrane enzymes has been used as a technique to determine the location of functional enzyme domains: i.e. at the cytoplasmic side or at the cell exterior[20]. The catalytic site of the PM-ATPase faces the cytoplasm. Therefore, in a population of 100 % right-side out sealed vesicles there should be no loss in ATPase activity due to trypsin. We used porcine trypsin instead of bovine trypsin, as the latter produced inactivation of (Na,K)-ATPase from the outside in right-side out sealed vesicles[20]. The fact that we see a 30 - 35 % loss in ATPase activity in 30 min (Fig. 2B) is likely to be due to leaky and/or inside-out vesicles. The observed non-latent ATPase activity, usually 15 % of the maximal, Triton stimulated activity, does not necessarily reflect the same percentage of leaky or inside-out vesicles. If the Triton activation of the ATPase is not just due to opening sealed right-side out vesicles (as suggested by Fig. 1B) then the 15 % non-latent ATPase must be associated with more than 15 % of the vesicle population.

Binding of FC occurs at the outside of the plasma membrane[2]. Nevertheless, we only observed a rapid loss in FC-binding activity after the vesicles had been made leaky with Triton X-100 or after solubilization. This indicates that in the native state the FC-binding site is protected on the outside from trypsin digestion, probably due to the presence of a carbohydrate moiety. Does the receptor have a cytoplasmic domain that is essential for the activity of the binding site located at the cell exterior ? The receptor sensitivity to trypsination after the vesicles have been opened with detergent and after receptor solubilization suggests that this is the case. Another possibility is that the detergent unmasks a trypsin sensitive site at the outside of the PM. We favour the former explanation because a sample that showed only 50 % latency of the ATPase (as opposed to 85 % in normal samples) lost 50 % of both FC-binding capacity and ATPase activity. Moreover, Stout and Cleland[10] observed a strong reduction in FC-binding to PM enriched vesicles after trypsin digestion without detergent. These vesicles were obtained through sucrose density centrifugation; a procedure known to yield mainly leaky vesicles.

In the light of the trypsin experiments it is interesting to note that cell surface receptors in mammalian cells can regulate ligand binding to the extracellular part through a conformational change of the inner protein part. After binding of the ligand to the EGF-receptor, the activated protein kinase C phosphorylates the receptor at a threonine residue located close to the cytoplasmic membrane surface. This causes a large decrease in the affinity of the receptor for EGF without changing the total number of binding sites[12,21]. The same mechanism of "desensitization" of the receptor response has been suggested for the α_1-adrenergic receptor[13].

The successful solubilization of the FC-receptor with OG opens the way to further purification. OG has a preference for forming mixed micelles with lipids instead of forming pure micelles, which may explain the effect of sonication. OG is characterized by a high critical micelle concentration (CMC; 0.73 % = 25 mM) and forms small micelles (M.W. = 8000). Thanks to its high CMC-value OG can be easily removed by dilution and dialysis; this is important for further purification of the receptor.

The interesting study of Blowers et al.[18] shows that the PM contains an autophosphorylating protein-kinase that is Ca^{2+}/calmodulin dependent with a M.W. of around 18 kDa. We have shown the Ca^{2+}-dependency of the soluble kinase originating from the PM of oat roots but further purification is necessary to study its calmodulin dependency because the presence of endogenous calmodulin makes addition of exogenous calmodulin ineffective in activation[19]. Preliminary evidence indicates that our kinase may also be a protein in the 15 - 18 kDa range. A protein-kinase of such small size is unknown in mammalian cells as far as we know. It is either unique to plant PM or it is a loosely attached catalytic subunit of a larger molecule.

The next step in our research will be the purification of the solubilized FC-receptor and protein-kinase. The interesting question is whether the kinase can use the receptor as a target molecule and whether phosphorylation/dephosphorylation events affect the receptor affinity for the ligand. Unravelling the amino acid sequence of these proteins is not merely of academic interest but may also elucidate important functional domains in the protein. It is becoming clear now that domains with an essential biochemical role are highly conserved through evolution.

Acknowledgements. We thank Dr. R.J. Mehlhorn for his assistance with the ESR spectroscopy measurements. This work was supported by NSF Grant DMB 8502021.

REFERENCES

1. E. Marre, Fusicoccin: a tool in plant physiology, Ann. Rev. Plant Physiol. 30: 273 (1979).

2. P. Aducci, R. Federico and A. Ballio, Interaction of a high molecular weight derivate of fusicoccin with plant membranes, Phytoph. Medit. 19: 187 (1980).

3. P. Aducci, M. Coletta and M. Marre ; An improved Scatchard analysis of fusicoccin-binding to maize coleoptile membranes, Plant. Sci. Lett. 33: 187 (1984).

4. T. Hunter and J.A. Cooper, Protein-tyrosine kinases, Ann. Rev. Biochem. 54: 897 (1985).

5. L. Tognoli and R. Colombo, Protein phosphorylation in intact cultured sycamore (Acer pseudoplatanus) cells and its response to fusicoccin, Biochem. J. 235: 45 (1986).

6. A. Bertl and H. Felle, Cytoplasmic pH of root hair cells of Sinapis alba recorded by a pH-sensitive micro-electrode. Does fusicoccin stimulate the proton pump by cytoplasmic acidification ?, J. Exp. Bot. 36: 1142 (1985).

7. R.J. Reid, L.D. Field and M.G. Pitman, Effects of external pH, fusicoccin and butyrate on the cytoplasmic pH in barley root tips measured by ^{31}P-nuclear magnetic resonance spectroscopy, Planta 166: 341 (1985).

8. P. Pesci, S.M. Cocucci and G. Randazzo, Characterization of fusicoccin binding to receptor sites on cell membranes of maize coleoptile tissues, Plant Cell Env. 2: 205 (1979).

9. P. Aducci, A. Ballio, L. Fiorucci and E. Simonetti, Inactivation of solubilized fusicoccin-binding sites by endogenous plant hydrolases, Planta 160: 422 (1984).

10. R.G. Stout and R.E. Cleland, Partial characterization of fusicoccin binding to receptor sites on oat root membranes. Plant Physiol. 66: 353 (1980).

11. P. Aducci, G. Crosetti, R. Federico and A. Ballio, Fusicoccin receptors. Evidence for endogenous ligand, Planta 148: 208 (1980).

12. K.D. Brown, P. Dicker and E. Rozengurt, Inhibition of epidermal growth factor binding to surface receptors by tumor promotors, Bioch. Biophys. Res. Comm. 86: 1037 (1979).

13. L.M.F.Leeb-Lundberg, S. Cotecchia, J.W. Lomasney, J.F. DeBernardis, R.J. Lefkowitz and M.G. Caron, Phorbol esters promote α_1-adrenergic receptor phosphorylation and receptor uncoupling from inositol phospholipid metabolism, Proc. Natl. Acad. Sci. 82: 5651 (1985).

14. C.H. Larsson, Plasma membranes, in: "New Series Vol. 1: Cell Components, p 85", H.F. Linskens and J.F. Jackson, eds., Springer-Verlag, Berlin.

15. T.L. Lomax and R.J. Mehlhorn, Determination of osmotic volumes and pH gradients of plant membrane and lipid vesicles using ESR spectroscopy, Biochim. Biophys. Acta. 821: 106 (1985).

16. M.M. Bradford, A rapid and sensitive method for the quantitation of microgram quantities of protein utilizing the principle of protein dye-binding, Anal. Biochem. 72: 248 (1976).

17. H.R. Lötscher, C. De Jong and R.A. Capaldi, Interconversion of high and low adenosine tri-phosphatase activity forms of Escherichia coli F_1 by the lauryldimethylamine oxide, Biochem. 23: 4140 (1984).

18. D.P. Blowers, A. Hetherington and A. Trewavas, Isolation of plasma-membrane-bound calcium/calmodulin-regulated protein kinase from pea using Western blotting, Planta 166: 208 (1985).

19. A.M. Hetherington and A. Trewavas, Activation of a pea membrane protein kinase by calcium ions, Planta 161: 409 (1984).

20. B. Forbusch III, Characterization of right-side out membrane vesicles rich in (Na,K)-ATPase and isolated from dog kidney outer medulla, J. Biol. Chem. 257: 12768 (1982).

21. C.S. King and J.A. Cooper, Effects of protein kinase C activation after epidermal growth factor binding on epidermal growth factor receptor phosphorylation, J. Biol. Chem. 261: 10073 (1986).

BACTERIORHODOPSIN: MOLECULAR BIOLOGY OF THE LIGHT ACTIVATED PROTON AND DIVALENT CATION RECEPTOR IN THE MEMBRANES OF HALOBACTERIA

L. Packer

Membrane Bioenergetics Group
University of California
Berkeley, California

BACKGROUND

The study of membrane receptors, their specificity and their relation to membrane transport and signalling ranks among the most important of the current unsolved problems in biological research. One membrane transport system that has received unprecedented attention during the last dozen years as a model system for such investigations has been bacteriorhodopsin, a retinal-containing protein which exists in specialized regions of the envelope membranes of halobacteria called "purple membrane." Certain mutants of halobacteria also produce similar "white membranes" containing the apoprotein only. In both instances, there is evidence that the proteins are organized in a hexagonal arrangement in tightly packed arrays.[1,2]

Bacteriorhodopsin, and the apoprotein after reconstitution by the addition of all-trans retinal, are capable of initiating a photochemical reaction cycle that accepts protons at one surface of the membrane and results in the release of protons at the other side of the membrane in a process that is strictly dependent on the energy derived from photon absorption.[3] In addition to binding protons, bacteriorhodopsin binds divalent cations such as calcium and magnesium, which has a profound effect on the binding of protons and the photocycle.[4] When the bivalent cations are removed, "blue membranes" are formed. Blue membranes cannot use their photochemical reactions to continuously release and take up protons.[5]

Because the specialized regions of the halobacteria membrane contain only this protein intermixed with 14-16 lipids per protein molecule, and is remarkably stable, the system is well suited for answering fundamental questions concerning the function of membrane receptors. The chapter will include a description of the system and a discussion of the current status of knowledge on this membrane receptor.

INTRODUCTION TO MOLECULAR BASIS OF PROTON TRANSPORT

The molecular basis of transport still represents one of the major unsolved problems in membrane biology. This chapter will outline some criteria that researchers would want to know about for any membrane transport system and describe mainly from the work we have done how bacteriorhodopsin fulfills these criteria. Also to be described is how laser light scattering methods can be used to characterize purple and white membrane systems and reconstituted liposome systems containing these membranes and bacteriorhodopsin.

The accompanying chart lists (Fig. 1) some of the aspects of a membrane proton pump one would like to know: 1) what are the energy sources, 2) what are the working groups of the reaction cycle, 3) how is the direction of the flux down the conduction pathway determined, 4) which group carries the protons or the charge, 5) what are the conduction pathways to the reaction cycle or the charge, 6) what are the conformational constraints, 7) is mobility of the working groups necessary for function, 8) are proton channels present, and if so, how are they regulated and are they required to overcome restrictions due to limited mobility of the working groups, 9) does the binding of any other cations or ligands regulate the activity of the membrane proton pump? These are some of the questions one could in principle ask about for every membrane transporting system.

THE BACTERIORHODOPSIN PHOTOCYCLE

Soon after its discovery, bacteriorhodopsin was recognized to undergo a photoreaction cycle when techniques that have been available for the study of visual photoreceptors were applied to bacteriorhodopsin (Fig. 2). It was found that all-trans retinal present in bacteriorhodopsin undergoes an isomerization to the 13 cis configuration in 50 micro seconds during which the formation of the M_{412} species occurred.[3] Also during this process, the Schiff base nitrogen, which is the site of attachment of the retinal chromorphore to the protein, becomes deprotonated. Then the system relaxes and reprotonates back to the original chromophore in a process requiring about 5 milliseconds.

The purple membrane is easily recognized in and isolated from halobacteria. Halobacteria burst when suspended in water and release their purple membranes which can readily be collected and purified by sucrose gradient centrifugation. Fig. 3 shows a characteristic appearance of a purple membrane in freeze fracture electron microscopy.

What is the Energy Source?

What are the Working Groups of the Reaction Cycle?

How is the Direction of Flux Down the Conduction Pathway Determined?

Which Groups Carry H (or H^+)?

What are the Conduction Pathways for the Reaction Cycle and Charge?

What are the Conformational Constraints (Mobility Domains) of the Working Groups?

Are Proton Channels Required to Overcome Restrictions Due to Limited Mobility of Working Groups?

Fig. 1. Molecular Aspects of a Membrane Proton Pump.

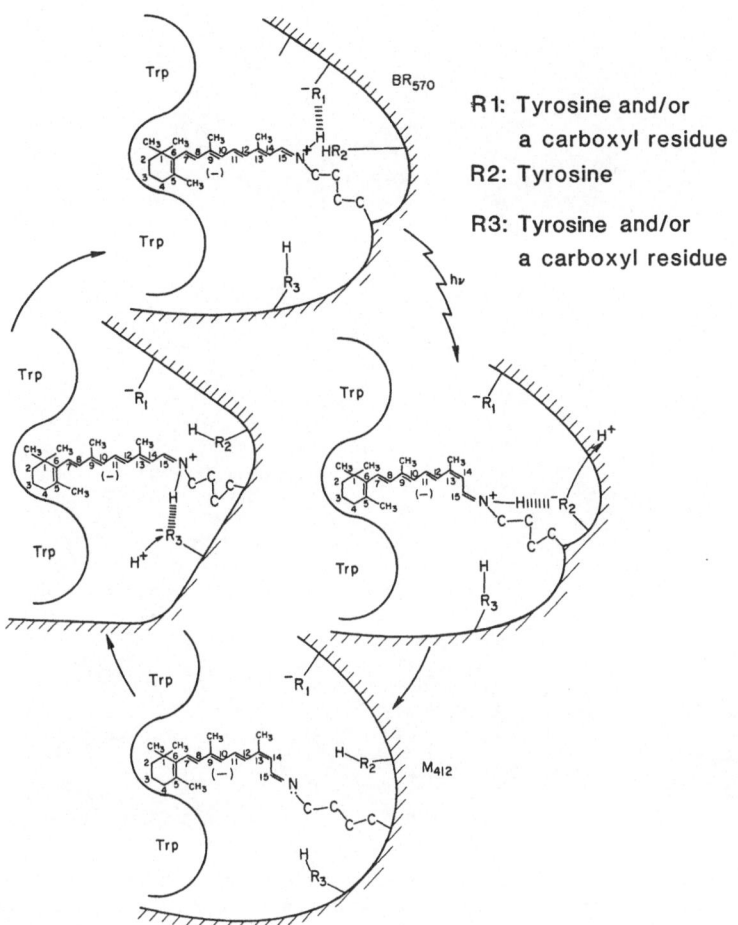

R1: Tyrosine and/or
    a carboxyl residue
R2: Tyrosine

R3: Tyrosine and/or
    a carboxyl residue

Fig. 2. Scheme Showing Possible Involvement of Various Residues in
Bacteriorhodopsin and Its Chromophore in the Photoreaction Cycle and Proton
Pumping Process.

The scheme shows photoisomerization of the side chain of retinal (13-cis form
and deprotonation of the Schiff base nitrogen atom during formation of M
causing a change in association of the proton on the Schiff base withy nearby
negatively charged residues in bacteriorhodopsin. This allows the proton to
travel to othyer residues and be released. After reprotonation oecurs in a
process requiring a longer time. Other residues in the protein receiving
protons from the inner cytoplasmic surface of the membrane result in the
reprotonation of the Schiff base nitrogen and restoration of the original
light-adapted form of the retinal chromophore in its all-trans configuration.

SIZE AND CHARGE OF PURPLE MEMBRANES

We have applied laser light scattering techniques to study the size and
charge distribution of purple membranes. The Laser light scattering
techniques (shown in Fig. 4) were used in two modes.[6] At 90 degrees, one can
derive a profile of the size distribution of the membrane particles (QELS).
This method is particularly useful for large particles like vesicles or
cells. At low angles of light scattering, if one applies an electric field,
the doppler shift of the frequency can be exploited to determine the mobility
of particles and, thus, to directly measure the electrophoretic mobility
which is related to the surface charge (LDV).

193

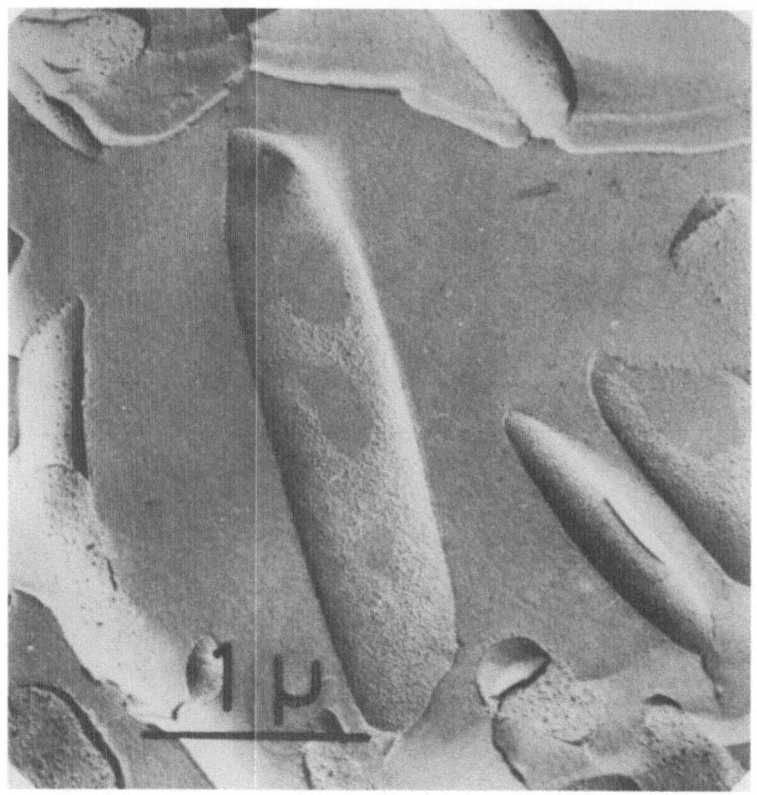

Fig. 3. Freeze fracture Electron Micrograph showing presence of patches in the surface of the cytoplasmic membrane where bacteriorhodopsin molecules are concentrated in purple, as in this case, or in white membrane preparations.

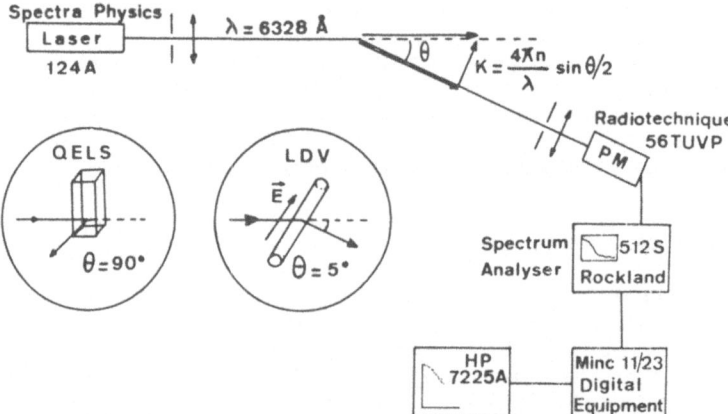

Fig. 4. Scheme illustrating how laser light scattering studies may be used to obtain size distributions and electrophoretic mobilities of membrane particles and vesicles.

The typical spectrum obtained at 90° by QELS of isolated purple membrane can be used to determine the size distribution profile shown in Fig. 5. The results show that Purple membranes are larger than was reported previoulsy by electro-microscopy. Furthermore, trypsin-treated preparations became so large that accurate size measurements by QELS could not be made. On the other hand, white membrane preparations exhibit an average diameter of 80 nanometers. This is more in the range of what is reported in the literature for the size of the purple membranes as seen by electron microscopy. We examined the membrane preparations, Le-Fort Tran et al.,[7] with the electron microscope and found that the native purple membrane shows some degree of aggregation (Fig. 6). Moreover, if the preparations are trypsin-treated even larger aggregates are formed. The white membrane, on the other hand, does not show this aggregation. When measuring the proton release and uptake by membrane preparations the extent of aggregation is important to the determination of accurate stoicheometries, since only the protons released to the bulk medium are detected under usual conditions.

Proton release by bacteriorhodopsin or by the opsin reconstituted with retinal can be measured by using laser flash photolysis techniques in conjunction with pH indicator dyes. Results of Robinson et al.[8] showed that in native bacteriorhodopsin about 1.5 protons are released per M412 species formed in the photocycle. If the preparations were trypsin treated, this number was significantly smaller. This would be expected if the aggregation state prevented the free exchange of protons between the membranes and the medium. With white membranes, the ratio is as high as 4. The white membranes yield higher proton release ratios even after trypsin treatment most likely because the membrane are dispersed and not aggregated. White membranes have been reported to have a different complement of lipids which may prevent the aggregation behavior seen in wild type preparations, Lam et al.[9]

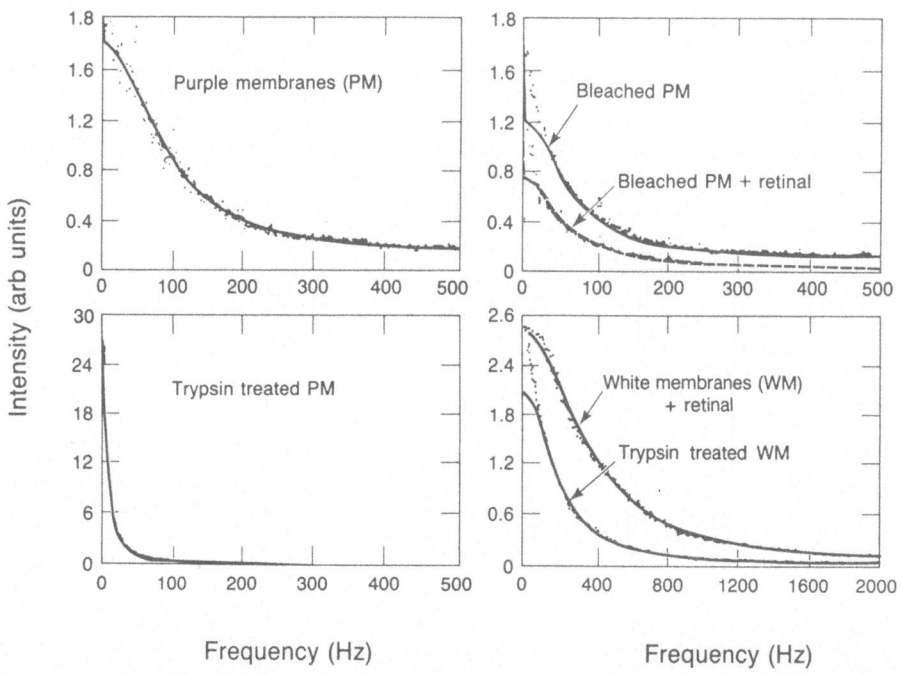

Fig. 5. Laser light scattering traces in the quasi-elastic light scattering mode (QELS) demonstrating the size distribution of various purple and white membrane preparations.

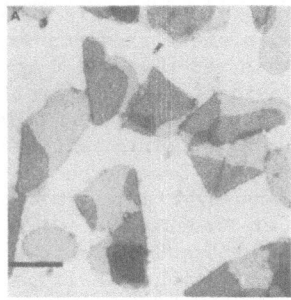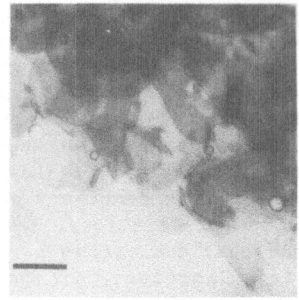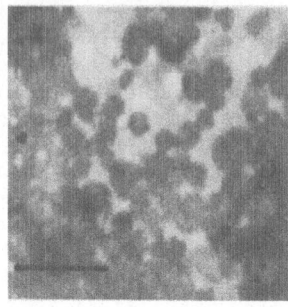

Fig. 6. Negatively stained Electron Micrographs of isolated purple and white membranes (right photo), in their native (left photo), and trypsin-treated (center photo) configuration.

MEMBRANE SURFACE CHARGE

If a proton gradient is established by bacteriorhodopsin, it is likely that participating groups on the purple membrane surface should alter the surface charge density during activity. Deprotonation of these groups results in a negative charge contribution on the membrane surface, which can be followed conveniently with spin probes. Carmelli et al.[10] used amphipathic, positively charged spin probes which partition between the medium and lipid domain of the membrane. A charge in partitioning as a result of surface charge alteration can be followed. A characteristic partitioning spectrum exhibiting a composite of the sharp spectrum of probe in the isotropic environment and the broad based spectrum of lipid bound probe is shown in Fig. 7. The decreases in line height of the aqueous signal upon illumination can be followed to give kinetic and quanitative determination of changes in surface charge.

One can observe this kind of membrane charge change more directly with laser light scattering techniques because purple membranes show a linear relationship between the LDV spectrum with the applied electric field (Fig. 8) due to an electrophoretic mobility response of the membranes to the voltage. With this method light-induced increases in the negative surface

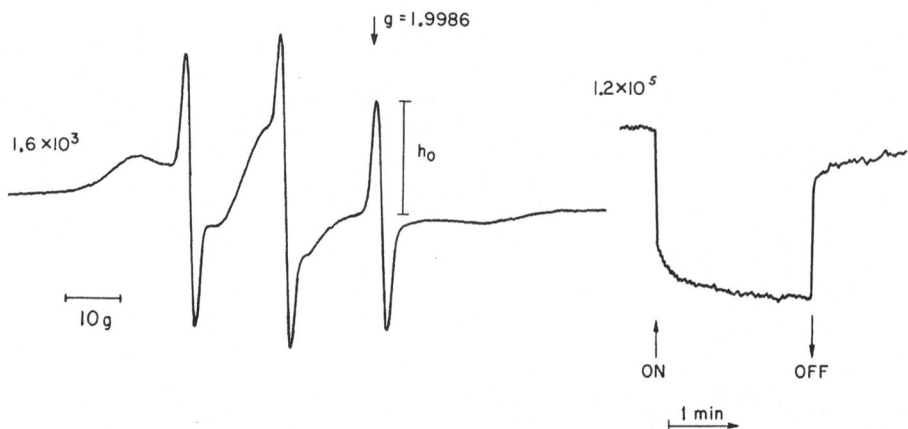

Fig. 7. Light induced changes in partitioning of the amphipathic cat-12 probe occurring after illumination or return to darkness of suspensions of purple membrane measured by laser flash EPR studies. After Carmelli et al.[10]

196

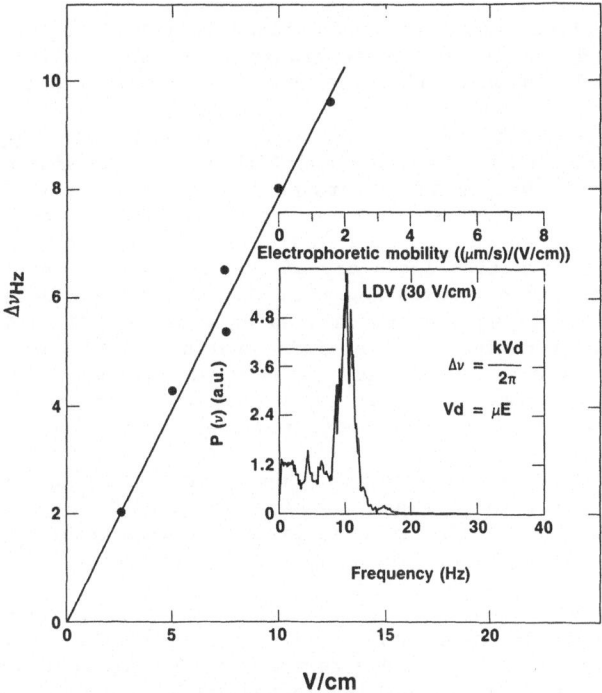

Fig. 8. Mobility of purple membrane preparations as a function of applied electric field measured by laser light scattering methods in the Laser Doppler Velicometry Mode (LDV). After Arrio et al.[6]

charge occur. This effect is most pronounced under conditions where photocycling is slow and high steady rate leads of the M_{412} species accumulate.[11]

The electrophoretic mobility technique is useful to characterize bacteriorhodopsin reconstituted liposomes. We have shown[12] that LDV spectra of a suspension of purple membranes incorporated into liposomes is very similar to that obtained using purple membranes except that the scale and sample are different. In the dark, a single peak is observed. In the light, one peak which does not shift is observed, and additionally, three peaks are observed that shift. Two of these peaks shift to the right and one peak falls to the left of the original peak seen in the dark. These new peaks disappear when the sample is returned to dark. Thus this method can be used to observe different populations of liposomes streaming in the electric field with different mobilities. The spectra may reflect heterogeneity in the orientation of the reconstituted liposomes, and perhaps lack of reconstitution for the sample which did not change its mobility in the light, and also varying extents of incorporation of the purple membrane into liposomes during their preparation.

AMINO ACID SIDE CHAIN GROUPS POSSIBLY INVOLVED IN H^+ TRANSLOCATION

Among the groups that would be suspected to be functionally important will be those groups which can be reversibly protonated or deprotonated such as tyrosine, amino acids within carboxylic acide sidegroups, lysine and arginine; these are the main ones we have considered. Extensive modification of lysine groups resulted in no change in activity;[13,14] modification of the tyrosine groups, however, show profound changes in activity.[15,16] The primary structure and location of these groups in bacteriorhodopsin is shown in Fig. 9 where the location of tyrosine residues is highlighted.

Investigations to elucidate the rate of photocycling often employ laser flash photolysis techniques. Chemical modification of the side chains of tyrosine by iodination changes the pH properties of tyrosine residues.[14] The rate of deprotonation as measured by M_{412} increases, and the rate of reprotonation slows (shown in Fig. 10), indicating tyrosine is probably involved at both the formation and decay of the M_{412} species of the photocycle. There is other evidence which suggests the involvement of tyrosine residues in the photocycling and proton pumping processes.[16] Fig. 11 shows a schematic view of how tyrosine is reversibly protonated. After the bacteriorhodopsin first absorbs a photon and forms the K species, and then all the other subsequent steps thermally decay in the dark, and before the end there is a change in at least one more tyrosine residue.

The role of carboxyl groups have also been investigated by chemical modification.[17] Fig. 12, for example, shows a hydrophobic type of reagent which reacts with buried protein carboxyl groups. An exogeneus spin label can be attached to the carboxyl groups by these techniques.[18,19]

Trypsin treatment of labelled bacteriorhopodsin results in the release of a protein fragment containing spin labels (Fig. 13). This fragment corresponds to the cytoplasmic carboxy terminus tail. Removal of the labelled fragment results in a decrease in the aqueous componant of the ESR signal in the membrance fraction. Spectra of supernatent containing the labelled tail detached from the rest of the molecule exhibit a high degree of label mobility. Herz et al.[19] have determined the distance of the remaining buried label is about 16-20 A° in the interior and that its modification inhibits the activity of the proton pump.

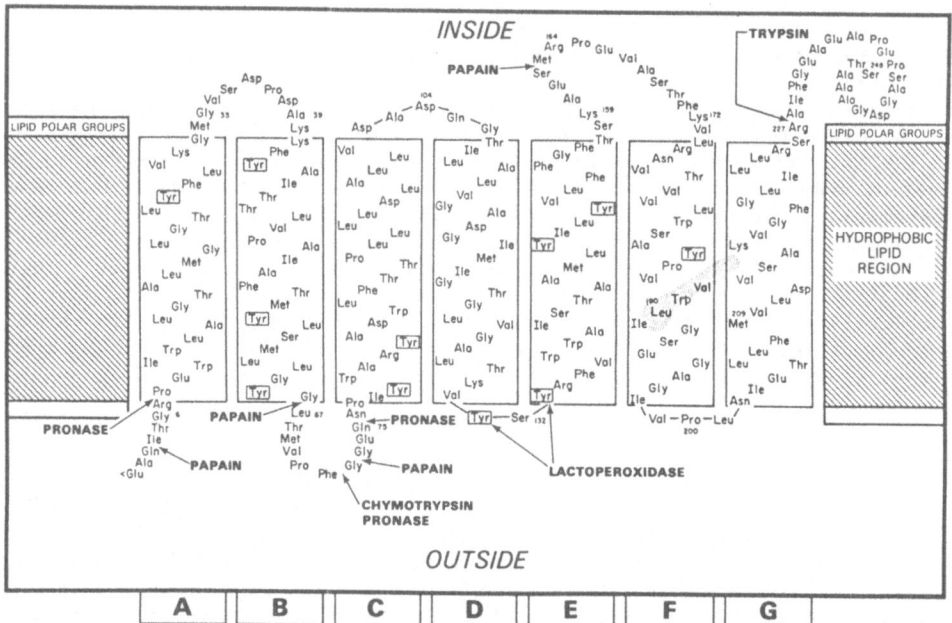

Fig. 9. Structure of bacteriorhodopsin highlighting the distribution of tyrosine residues in the primary structure of the bacteriorhodopsin molecule.

PATHWAY OF LIGHT-DRIVEN PROTON TRANSLOCATION
ACROSS THE PURPLE MEMBRANE

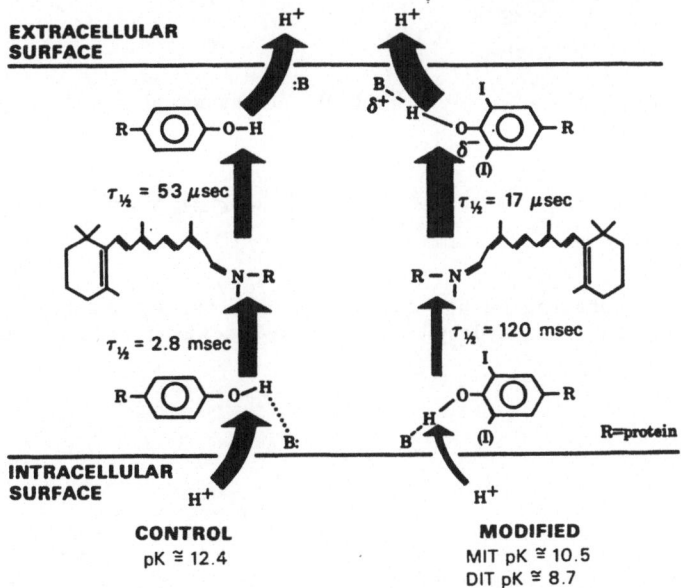

Fig. 10. Diagram showing the effect of iodination on acceleration of the deprotonation phase of the photocycle of bacteriorhodopsin and inhibition of the M_{412} decay reprotonation phase of the photocycle. <u>Scherrer</u> et al.[15]

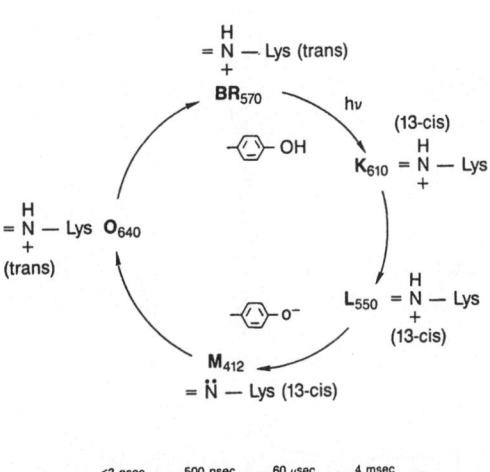

Fig. 11. Role of tyrosine protonation in the photocycle of bacteriorhodopsin. After Lam et al.[16]

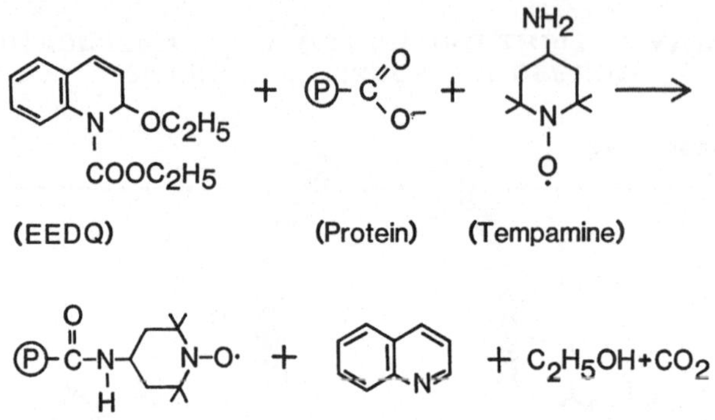

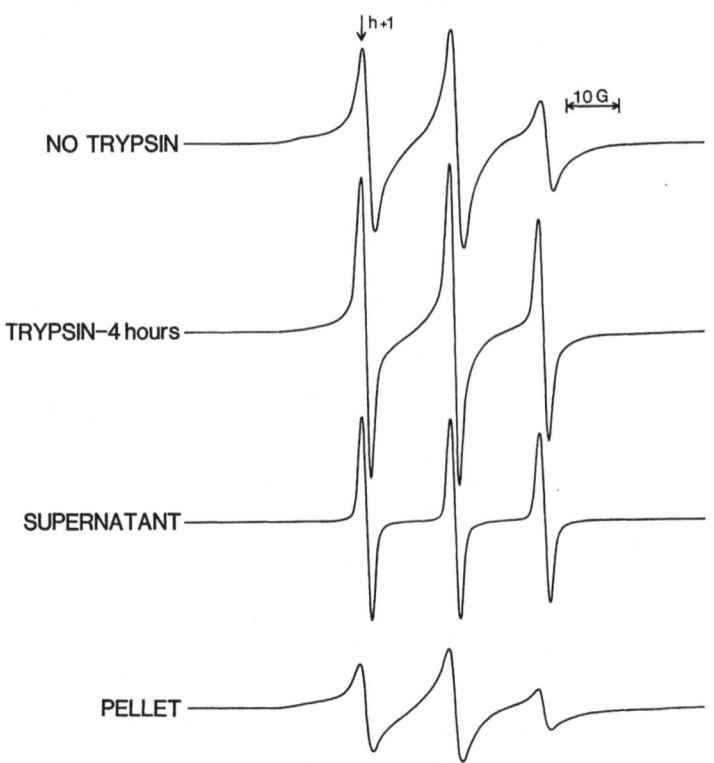

Fig. 12. Scheme showing method for spin labelling of carboxyl residues of bacteriorhodopsin using hydrophobic carboxyl activating reagents. After Herz et al.[19]

Fig. 13. Distribution of spin labelled carboxyl residues in trypsin-treated purple membranes. After Herz et al.[19]

Some photocycle activity measurements are shown in Fig. 14. This is a flash photolysis experiment showing the decay of the M_{412} species. The control decay rate is the most rapid. After treatment of the preparation with ethylacetimdate, a reagent which reacts with almost all of the epsilon amino groups of lysine, there is not much change in the M_{412} decay rate. However, after treatment with a carboxyl activating reagent, without an external nucleophile present, there is a huge decrease in the rate of the M_{412} decay. If, however, during the modification reaction the carboxyl activating system is swamped with externally added nucleophile (GME, glycine methyl ester), it is later observed the photocycle is still inhibited, but not nearly as much as if the nucleophile was a residue inside the protein. Use of a nucleophile inside the protein causes a crosslinkage to occur between an amino group of lysine and a nearby carboxyl group. The results, then suggest that a short cross link is occurring which greatly slows M_{412} decay phase of the photoreaction cycle: the implication is that mobility of such groups in the protein is probably restricted. To prove it, purple membranes were reacted with ethyl acetimidate; this placed on the exposed lysine amino groups a bulky substituent. Then purple membranes were reacted further with the carboxyl activating agent, and then found that the M_{412} decay rate was the same as if we had added the nucleophile externally. This provided further proof that our interpretation of the results of the experiment was correct, namely that crosslinking accounted for the larger inhibition of activity.

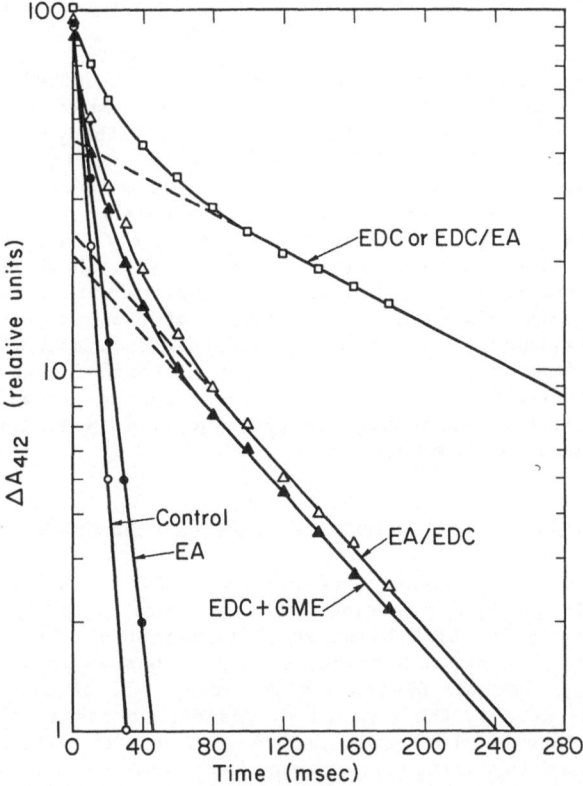

Fig. 14. Laser flash photolysis studies of native and chemically modified bacteriorhodopsin in purple membranes after treatment with carboxyl activating reagents in the absence and presence of an externally added nucleophile. The effects of lysine modification are also shown. After Herz et al.[19]

PRESENCE OF AN H$^+$ CHANNEL

We were also interested in trying to find out whether there was a channel through which protons could pass in bacteriorhodopsin. To test this idea, liposomes reconstituted with bacteriorhodopsin were prepared, loading the vesicles inside during the preparation procedure with potassium. Then we measured the pH changes that occurred after valinomycin was added. Valinomycin together with potassium provides a diffusion potential, the driving force to see if protons could move through across the liposome by passage through the lipid belayer or through bacteriorhodopsin.[20] Results are summarized in Table I. We first cross-linked the purple membrane preparation with glutaralderhyde before liposomes were formed. This was done so we could crosslink the bacteriorhodopsin either in the light-adapted or dark-adapted state. It was subsequently found that the light-adapted state was much more readily able to cause a movement of protons when we added valinomycin, whereas it is less active in this regard in the dark-adapted samples which showed in these cases, almost half the activity. This result seemed consistent with the fact that it is known that the dark adapted bacteriorhodopsin is a 50-50 mixture of cis and trans form of retinal, and it is the all-trans form only which pumps protons.

Other controls (not shown in Table I) showed that after valinyocin addition, neither oligomyoin or DCCD, inhibitors of Proton ATPase in membranes, showed any effect on H$^+$ exchange activity. However, when all-trans retinal is added, it slows the passive proton exchange (Table I), suggesting that retinal may block this channel through which the protons seem to be passing.

This has been further tested in other ways as described by Dencher et al.[21], (cf. also chapter by Pethgas et al. in this volume). These investigators used bacteriorhodopsin liposomes loaded with pyranine dyes. pH jump experiments were performed by mixing the liposomes together with a solution at a different pH. With bleached preparations the pH decay rate is much faster then when native bacteriorhodopsin is studied. Thus a more direct method for establishing a pH gradient also shows that protons can pass go through bacteriorhodopsin in the absence of retinal. Both methods show that stepwise reconstituion of bleached purple membranes with all-trans retinal progressively inhibit the rate and the extent of the proton exchange in the dark caused by valinomycin addition to the bleached liposome preparations (cf. Table I). Furthermore, if one treats the bleached preparation with iodine (cf. Table I), which would modify the tyrosine groups, H$^+$ exchange activity is inhibited. This suggests tyrosine resides in the "channel" may be involved in passive H$^+$ movements.

BIVALENT CATIONS AND BACTERIORHODOPSIN STRUCTURE AND FUNCTION

It has long been known that bacteriorhodopsin in the purple membrane is capable of forming a "blue membrane".[4,22,23] There are now at least seven different methods known which bring about formation of the blue membrane, such as treatment with acid, heating, extensive washes in the presence of chelating agents, electric fields and so forth. All of these methods lead to the same result: namely, the removal of residual amounts of bivalent cations bound to the surface of the purple membrane. These bivalent cations are thought to be bound to negatively charged residues such as the carboxyl residues of glutamate and aspartate. These residues are mainly located on the inner cytoplasmic surface of the purple membrane. Blue membranes do not carry out proton pumping activity. The formation of purple membranes from blue membranes can be accomplished by adding back to the preparations a variety of cations. Monovalent ions like sodium can restore the purple color but at very high concentrations, whereas rather low concentrations, 2 to 4 moles of cation per bacteriorhodopsin is required for bivalent cations. Trivalent cations such as lanthanum work at even lower concentrations. These findings are interesting since they suggest that a regulatory affect on the activity of bacteriorhodopsin as a proton pump may be brought about by factors that affect the reversible binding and modulation of the content of bivalent cations on surface groups. It is not yet clear whether the importance of the binding of bivalent cations is to establish the proper conformation of the retinal chromorphore in its active site, or whether a more specific effect on proton translocation is brought about by this binding. Studies are underway at the present time in a number of laboratories to clarify more precisely the effects of bivalent cation binding on the photocycle in the blue membrane and on the activity of the purple membrane in proton translocation.

Table I. Summary of factors affecting the rate and extent of passive proton transport through bleached bacteriorhodopsin liposomes. Effects of chemical modification, inhibitors, and retinal reconstitution of bleached bacteriorhodopsin are demonstrated. After Konishi and Packer ().

THE PROTON CHANNEL IN BACTERIORHODOPSIN

	DARK-VALINOMYCIN INDUCED K^+-LOADED LIPOSOMES - H^+ UPTAKE		PROTEOLIPOSOMES ATTACHED TO LIPID IMPREGNATED MILLIPORE FILTER MEMBRANES
	RATE (ng/mg Protein/min)	EXTENT (ng/mg Protein)	PHOTOPOTENTIAL mV
LIPOSOMES ONLY, CONTROL	6.3	3.4	-
NATIVE BACTERIORHODOPSIN	8.2	4.8	90
BLEACHED BACTERIORHODOPSIN	353.0	76.9	20
BLEACHED BACTERIORHODOPSIN IODINATED	26.3	30.4	92
BLEACHED BACTERIORHODOPSIN + ALL-TRANS RETINAL	70.4	32.3	-
LIGHT INDUCED			
NATIVE BACTERIORHODOPSIN,	715.0	120.0	
98 PERCENT BLEACHED BACTERIORHODOPSIN	0.0	0.0	

BLEACHING: PURPLE MEMBRANES IN 0.3M HYDROXYLAMINE SOLUTION pH 7.0, EXPOSED TO LIGHT FOR 5 HOURS AT 10°

IODINATION: 1M IODINE SOLUTION ADDED IN 0.2M Na-BORATE BUFFER, pH 8.5 TO AN IODINE:BACTERIORHODOPSIN MOLAR RATIO = 20:1

ASSAY: 500mM K^+-LOADED LIPOSOMES AT 20°: ADDITIONS WHERE INDICATED, ALL-TRANS RETINAL 0.3 μM, VALINOMYCIN 0.66 μM

SUMMARY OF THE MOLECULAR ASPECTS OF THE ACTIVITY OF BACTERIORHODOPSIN AS A H^+ PUMP

It is clear that in the case of bacteriorhodopsin, we have some definitive information on the criteria required for membrane proton pump. To review these briefly: 1) First, the energy source very clearly is photon absorbtion. 2) The functional group is all-trans retinal. 3) The photo-reaction cycle is also fairly well known. 4) There is evidence that carboxyl residues and tyrosine residues are involved in the proton translocation, and also that the Schiff-base proton of the lysine residue where retinal is attached to the protein is involved. 5) What about the conduction pathways for the functional groups and charge? This is the area that is largely unknown. We don't know yet the linkage between the photocycle and the various functional groups that have been identified. They appear to protonate and deprotonate in the time frame of the photocycle, but it is not known which one is connected to which, and we probably don't know all the groups involved. 6) Concerning the question of whether conformational constraints or mobility is required, two kinds of cross-linking studies suggest that mobility of the groups is required. Short intramolecular cross-links appear to be very inhibitory to function. 7) Evidence exists for passive proton channels in bleached bacteriorhodopsin and these channels may aid in the activity of the functional groups. Possibly a proton channel works in concert with these functional groups. It can be conceived that both mechanisms are probably involved. 8) Concerning other ligands, there is evidence that bivalent cations affect charge at the membrane surface and this affects the activity of bacteriorhodopsin as a proton pump.

REFERENCES

1. Henderson, R., and Unwin, P.N.T., Three Dimensional Model of Purple Membrane Obtained by Electron Microscopy, Nature, 257 (1975).

2. Mukohata, Y., Sugiyama Y. Kaji, Y., Usukura, J., and Yamada, E., The White Membrane of Crystalline Bacterhiodopsin in Halobacterium Halobium Strain R.M.W. and its Conversion into Purple Membrane by Exogenous Retinal, Photochem. Photobio. 33 (1981).

3. Nagle, S.F., Parodi, L.A., and Lozier, R.H., Procedure for Testing Kinetic Models of the Photocycle in Bacteriorhodopsin, Biophys. J. 38 (1982).

4. Chang, C.H., Chen J.G., Govindjee R., and Ebrey T., Cation Binding by Bacteriorhodopsin, Proceedings of the Nat. Acad. Sciences., 82 (1984).

5. Tsuji, K. and Rosenheck, K., Cation Binding by Bacteriorhodopsin, FEBS Lett, 98 (1974).

6. Arrio, B., Johannin, G., Volfin, P., Lefort-Tran, M., Packer, L., Robinson, A.E., and Hrabeta, E., Aggregation of Purple Membrane Fragments Following Cleavage of the C-Terminal Tail of Bactriorhodopsin by Proteolysis, Arch. Biochem. Biophys., 246:1 (1986).

7. Lefort-Tran, M., Pouphile, M., Arrio, B., Johannin, G., Volfin, P., and Packer, L., Stacking of Purple Membranes in vitro, in: Ion Interactions of Biological Systems, New York, Plenum Press, (1985).

8. Robinson, A.E., Hrabeta, E., and Packer, L., Measurement of Proton/M_{412} Ratios in Suspensions of Purple and White Membranes from Halobacterium halobium, in: "Ion Interactions of Biological Systems," Plenum Press, New York (1985).

9. Lam, E., Fry, I., Packer, L., and Mukohata, Y., Comparison of the 0650 Photo-intermediate and Acid-induced Species in Membrane Patches from Halogacterium Halobium S, and R, MW Strains, FEBS Lett. 146:1 (1982).

10. Carmeli, C., Quintanilha, A.T., and Packer, L., Surface Charge Changes in Purple Membranes and the Photoreaction Cycle of Bacteriorhodopsin, Proc. Natl. Acad. Sci. USA 77:4707-4711 (1980).

11. Packer, L., Arrio, B., Johannin, G. and Volfin, P., Surface Charge of Purple Membranes Measured by Electrophoretic Light Scattering, Biochem. Biophys. Res. Commun. 122:252-258 (1984).

12. Packer, L. and Arrio, B., Characterization of purple membranes containing bacteriorhodopsin in reconstitute lipid systems, in: "Proceedings from conference on °New Technological Applications of Phospholipid Bilayers, Thin Films, and Vesicles','Plenum Press, New York (1986).

13. Konishi, T., Tristram, S. and Packer, L., The Effect of Cross-Linking on Photocycling Activity of Bacteriorhodopsin, Photochem. Photobiol. 29:353-358 (1979).

14. Lam, E., and Packer, L., Effect of Fluorescamine Modification of Purple Membranes on Exciton Coupling and Light-To-Dark Adaptation, Biochem. Biophy. Res. Comm. 101:464-471 (1981).

15. Scherrer, P., Packer, L. and Seltzer, S., Effect of Iodination of the Purple Membrane on the Photocycle of Bacteriorhodopsin, Arch. Biochem. Biophys. 212:589-601 (1981).

16. Lam, E., Seltzer, S., Katsura, T. and Packer, L., Light-Dependent Nitration of Bacteriorhodopsin, Arch. Biochem. Biophys. 227 :321-328 (1983).

17. Packer, L., Tristram, S., Herz, J.M., Russell, C. and Borders, C.L., Chemical Modification of Purple Membranes: Role of Arginine and Carboxylic Acid Residues in Bacteriorhodopsin, FEBS Lett. 108:243-248 (1979).

18. Herz, J.M. and Packer, L., Structural Involvement of Carboxyl Residues in the Photocycle of Bacteriorhodopsin, FEBS Lett. 131:158-164 (1981).

19. Herz, J.M., Mehlhorn, R.J., and Packer, L., Topographic Studies of Spin-Labeled Bacteriorhodopsin: Evidence for Buried Carboxyl Residues and Immobilization of the COOH-Terminal Tail, J. Biol. Chem. 258:9899-9907 (1983).

20. Konishi, T. and Packer, L., A Proton Channel in Bacteriorhodopsin, FEBS Lett. 89 (1978).

21. Dencher, N.A., Burghaus, P.A., and Grzesiek, S., Active and Passive Proton Translocation Across Bacteriorhodopsin, VNU Press (1986).

22. Chang, C.H., Jonas, R., Melchiore, S., Govindjee, R., and Ebrey, T.G., Mechanism and Role of Divalent Cation Binding of Bacteriorhodopsin Biophys. J. Vol. 49 (1986).

23. Ariki, M. and Lanyi, J.K., Characterization of Metal Ion-binding Sites in Bacteriorhodopsin, J. Bio. Chem. 261:18 (1986).

INFLUENCE OF THE CHROMOPHORE RETINAL AND THE STATE OF AGGREGATION ON THE PROTON/HYDROXYL ION FLUX ACROSS BACTERIORHODOPSIN

Petra A. Burghaus and Norbert A. Dencher

Biophysics Group, Department of Physics
Freie Universität Berlin
Arnimallee 14, D-1000 Berlin 33, FRG

ABSTRACT

The influence of the chromophore retinal in bacteriorhodopsin on the passive proton/hydroxide ion flux through the protein was examined. Bacteriorhodopsin was reconstituted into unilamellar lipid vesicles. Transmembrane pH-gradients were quickly established across the vesicular membrane and the induced kinetics of the fluorescence changes of the entrapped dye pyranine were compared for vesicles with incorporated native, chromophore-free, and regenerated protein. The H^+/OH^- diffusion across the apoprotein was considerably faster than through the protein containing covalently bound retinal. A linear dependence between the ΔpH decay time and the degree of regeneration was observed.

INTRODUCTION

Transmembrane proteins which actively or passively translocate protons/hydroxide ions are important components of biological membranes. Although little is known about structure and molecular mechanism, it is generally assumed that these proteins contain a specific transmembrane pathway conducting the ions across the protein moiety and an active center or gate controlling the ion flux. In the present study we have examined the properties of the H^+/OH^- pathway in the light-energized H^+-pump bacteriorhodopsin (BR). BR, the only protein of the purple membrane of Halobacterium halobium is aggregated as trimers in a two-dimensional hexagonal lattice. Its single polypeptide chain of 248 amino acids traverse the membrane via seven α-helical segments. BR's chromophore retinal is bound to lysine 216 forming a protonated Schiff's base and has two known functions. It is the antenna for the absorption of photons and its isomerization from all-trans to a cis isomer is part of the active transport process[1,2,3].

To find out whether the chromophore retinal also participates in the H^+ pathway through BR was the primary aim of the present study. We have approached this question by comparing the rate of H^+/OH^- ion diffusion through the native and the chromophore-free protein. Our results clearly demonstrate that retinal hinders the passive H^+/OH^- translocation through BR.

SAMPLES AND EXPERIMENTAL PROCEDURES

Purple membranes from <u>Halobacterium</u> <u>halobium</u> S_9 or ET 1001 were isolated and purified essentially as previously described[4,5]. Dimyristoylphosphatidylcholine (DMPC), dimyristoylphosphatidylglycerol (DMPG) and phosphatidylserine (from bovine brain, PS) showed a single spot in thin-layer chromatography. BR was reconstituted in large unilamellar (100 - 500 nm) vesicles by solubilizing the purple membranes with the nonionic detergent Triton X-100 (0.15 - 0.5 % w/w). A mixture of 95 % DMPC and 5 % negatively charged lipid (DMPG or PS) was dissolved in chloroform and spread homogeneously as a thin film on the glass wall of a round bottom flask by evaporation of the solvent ($< 10^{-3}$ Torr, > 12 h). The lipid was then resuspended in an appropriate volume of the detergent solution containing monomeric BR to give the desired lipid to protein ratio (ranging from 30 to 150 m/m). Removal of the detergent was performed in the dark by prolonged dialysis at 5°C in the presence of Bio-Beads SM-2. The vesicles were purified by density gradient centrifugation[6,7].

The chromophore-free apoprotein bacterioopsin (BO) was prepared by illumination of a BR-lipid vesicle suspension in the presence of 0.2 M hydroxylamine. The chromophore was completely converted to retinaloxime and excessive hydroxylamine was removed by dialysis, however, most of the retinaloxime remained attached to the membrane[5,8]. BR was stepwise regenerated by incubation of the bleached vesicles with an appropriate amount of ethanolic all-<u>trans</u> retinal solution. (The ethanol concentration in the sample never

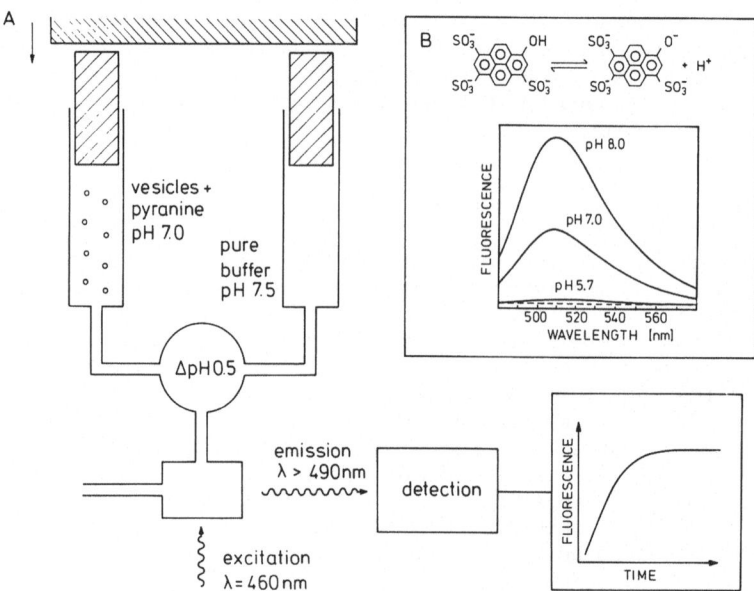

Fig. 1. (A) Schematic representation of the experimental setup and conditions. Stopped-flow unit and fluorescence spectrometer used for the generation of transmembrane pH-gradients and subsequent measurement of changes of the fluorescence emission intensity of the pH sensitive dye pyranine entrapped in the internal aqueous bulk phase of proteoliposomes. (B) Structural formula of pyranine and pH-dependence of its fluorescence emission spectrum; excitation wavelength of 450 nm.

exceeded 1 %.) The extent of regeneration was determined spectroscopically. In all experiments discussed retinal was predominantly in the all-<u>trans</u> configuration.

Transmembrane pH-gradients (ΔpH = 0.4 - 1.0) were generated by rapidly mixing (< 5 ms) equal amounts of a buffered (1 mM phosphate buffer + 50 mM K_2SO_4) vesicle suspension and a buffer solution of different pH. Mixing was performed in a stopped-flow unit as schematically shown in Fig. 1A. The kinetics of the H^+/OH^- diffusion across the vesicular membrane in response to the imposed gradient was monitored using the pH-sensitive fluorescent dye pyranine (8-hydroxy-1,3,6-pyrenetrisulfonate).

The negatively charged dye was entrapped in the vesicle interior by short sonication of preformed vesicles in the presence of the dye. Part of the BR-vesicle suspension was bleached after pyranine entrapment and for native and bleached samples the external free pyranine was subsequently removed by dialysis. Upon excitation of pyranine at 460 nm the intensity of fluorescence emission at wavelength above 490 nm is directly proportional to the concentration of the deprotonated molecule (see Fig. 1B) and hence reflects the pH of the vesicle's internal aqueous bulk phase. Further details of these procedures are described elsewhere[9].

H^+/OH^- DIFFUSION ACROSS NATIVE AND CHROMOPHORE-FREE BACTERIORHODOPSIN

Upon establishment of a transmembrane pH-gradient across the reconstituted vesicles the protons/hydroxide ions permeate across both the lipid bilayer and the protein. In order to examine any influence of the chromophore retinal on the ion pathway through the protein moiety, vesicles with incorporated native BR, BO, and BR regenerated from BO and retinal are compared. The samples with native BR serves as a standard, then the chromophore was bleached and thereafter stepwise regenerated by addition of appropriate amounts of retinal. In all preparations tested, the H^+/OH^- flux through BO is considerably faster than the one through native BR and through regenerated BR. The results of a typical experiment are illustrated in Fig. 2. Vesicles with a molar lipid to protein ratio of 95 were subjected to a pH-jump of 0.45 units. Measurements were carried out at 8°C. At this temperature all BR molecules are aggregated (see below). Both the regenerated (> 20 %) and the native samples require a fit of the kinetic fluorescence traces with two exponential rate constants. For native samples the faster one has an amplitude of about 10 % of the overall signal. The data of bleached samples are evaluated with one exponential. Upon removal of the chromophore, an about 9-fold decrease in the overall lifetime of the fluorescence signal occurs. That indicates a drastic acceleration of the H^+/OH^- diffusion through the protein: Stepwise regeneration of the chromophore leads to a corresponding linear increase of the predominant lifetime. The short lifetime of small amplitude remains essentially unchanged upon regeneration. - These results show that covalently bound retinal induces a hindrance of the H^+/OH^- flux across BR. They are in qualitative but not quantitative agreement with a previous report by Konishi and Packer[10].

Also after extrapolation to 100 % regeneration there still remains a significant difference in the ΔpH decay rate between native BR and maximal regenerated BR (80 - 90 %). The retinaloxime remaining attached to the membrane in the regenerated samples or the nonregenerated (10 - 20 %) bacterio-opsin molecules that must be more permeable for H^+/OH^- ions than regenerable BO molecules might be the reason for this observation. The added retinal solution itself does not affect the permeability of the lipid bilayer, as was proven by control experiments.

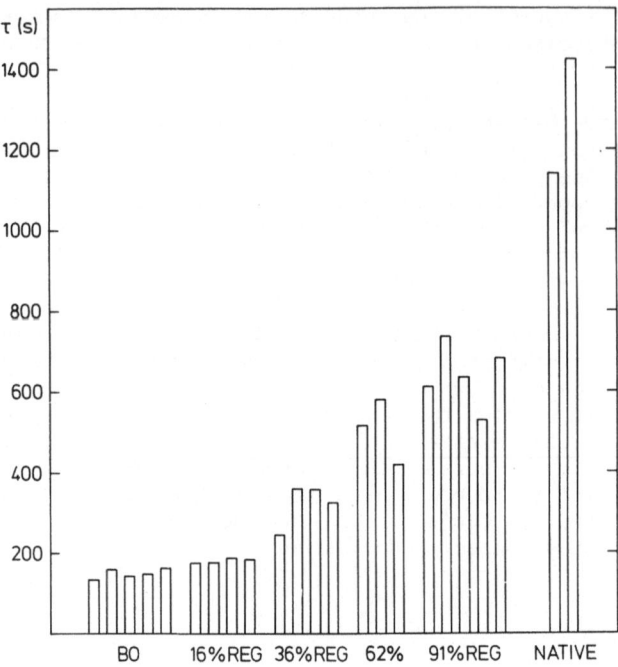

Fig. 2. Effect of covalently bound retinal on the H^+/OH^- diffusion through BR. The decay time (i.e., the predominant
component comprising 60 - 100 % of the amplitude of the
overall process) of the pH-gradient established across
reconstituted vesicles is plotted for preparations containing native BR (native), the chromophore-free protein
moiety BO, and to increasing extent regenerated BR.
$\Delta pH = 0.4$; 8°C; 1 mM phosphate buffer pH 7.0, 50 mM
K_2SO_4, and 1.5 mM pyranine in the vesicle interior.

Removal of the chromophore might perturb the protein lattice of the
purple membrane. Therefore, it could be argued that the accelerated H^+/OH^-
diffusion rate of chromophore-free samples may be due to the induced lattice
perturbation and is not an intrinsic property of the protein itself. To examine the influence of the lattice structure on the H^+/OH^- flux kinetics,
measurements were performed on monomeric protein molecules that undergo fast
translational and rotational diffusion within the vesicle bilayer. Reconstituted BR-lipid vesicles with a lipid to protein ratio of 110 show a temperature dependent transition of the state of aggregation of BR in the lipid
bilayer. At temperatures below the gel to liquid-crystalline phase transition of DMPC (23°C) the BR molecules are aggregated as trimers whereas they
are monomeric at higher temperatures. Measurements with these vesicles at
30°C confirm the difference between chromophore-free and regenerated BR. The
kinetic parameters obtained for BO-samples are τ_1 = 3.0 s with an amplitude
of 65 % and τ_2 = 18 s, A_2 = 35 %. After regeneration of 80 % of the BR molecules the predominant lifetime τ_1 increases by a factor of 2.4 and τ_2
1.3-fold. Obviously, the difference between chromophore-free and regenerated
samples is smaller as compared to aggregated BR but this may be due to the
strongly increased permeability of the lipid phase at higher temperatures.
These measurements show that the observed hindrance of the H^+/OH^- diffusion
across BR in the presence of covalently bound retinal is independent of the
state of aggregation of the BR molecules and consequently an intrinsic property of the protein itself.

CONCLUSION

The results presented in this report show that retinal either is located in the proton/hydroxide ion path across the protein moiety of BR connecting both surfaces of the membrane or indirectly controls this path by inducing conformational changes in the protein upon binding. In the presence of the chromophore, passive H^+/OH^- diffusion through BR is hindered.

ACKNOWLEDGEMENT

This investigation was supported by the Deutsche Forschungsgemeinschaft (SFB 312/B4, Heisenberg Grant De 300/1).

REFERENCES

1. W. Stoeckenius and R. A. Bogomolni, Bacteriorhodopsin and Related Pigments of Halobacteria, Ann. Rev. Biochem. 52:587 (1982).

2. N. A. Dencher, The Five Retinal-Protein Pigments of Halobacteria: Bacteriorhodopsin, Halorhodopsin, P 565, P 370, and Slow-cycling Rhodopsin, Photochem. Photobiol. 38:753 (1983).

3. J. K. Lanyi, Bacteriorhodopsin and Related Light-energy Converters, in: "New Comprehensive Biochemistry", L. Ernester, ed., Elsevier/North Holland, Amsterdam pp. 315 (1984).

4. D. Oesterhelt and W. Stoeckenius, Isolation of the Cell Membrane of Halobacterium halobium and its Fractionation into Red and Purple Membrane, Methods Enzymol 31:667 (1974)

5. P.-J. Bauer, N. A. Dencher and M. P. Heyn, Evidence for Chromophore-Chromophore Interactions in the Purple Membrane from Reconstitution Experiments of the Chromophore-free Membrane, Biophys. Struct. Mech. 2:79 (1976).

6. N. A. Dencher and M. P. Heyn, Preparation and Properties of Monomeric Bacteriorhodopsin, Methods Enzymol. 88:5 (1982).

7. M. P. Heyn and N. A. Dencher, Reconstitution of Monomeric Bacteriorhodopsin into Phospholipid Vesicles, Methods Enzymol. 88:31 (1982).

8. D. Oesterhelt, M. Meentzen and L. Schuhmann, Reversible Dissociation of the Purple Complex in Bacteriorhodopsin and Identification of 13-cis and all-trans-Retinal as its Chromophores, Eur. J. Biochem. 40:453 (1973).

9. N. A. Dencher, P. A. Burghaus and S. Grzesiek, Determination of the Net Proton-Hydroxide Ion Permeability across Vesicular Lipid Bilayers and Membrane Proteins by Optical Probes, Methods Enzymol. 127:746 (1986).

10. T. Konishi and L. Packer, A Proton Channel in Bacteriorhodopsin, FEBS Lett. 89:333 (1978).

COMPLETE TRACKING OF PROTON FLOW IN THYLAKOIDS

THE UNIT CONDUCTANCE OF CF$_0$ IS GREATER THAN 10 fS

Gerald Schoenknecht, Holger Lill,
Siegfried Engelbrecht and Wolfgang Junge

Biophysik, Fachbereich Biologie/Chemie
Universitaet Osnabrueck, Postfach 4469
D-4500 Osnabrueck, Germany (FRG)

ABSTRACT

We investigated the proton conductance of the channel portion of chloroplast ATP synthase (CF$_0$) in thylakoids which were depleted from the soluble portion of the ATP synthase (CF$_1$) by EDTA treatment. Proton pumps were stimulated by short flashes of light. Proton flux through CF$_0$ was measured spectrophotometrically in three different ways: as proton efflux from the lumen (via neutral red), charge flow across the membrane (via electrochromism) and proton influx into the medium (via phenol red). Hence we completely tracked the protons on their way from the lumen through CF$_0$ into the medium.

A first treatment with EDTA removed up to 12% of total CF$_1$ without increasing the proton permeability of the membranes. A 2nd EDTA treatment removed further 20% of CF$_1$ and increased the proton permeability of membranes by 3 orders of magnitude. In control thylakoids a pH transient decayed with a relaxation time of 20-30 s. After the 2nd EDTA wash the decay time was 85 ms if the electric potential difference was shunted (e.g. by added gramicidin) and 7 ms if it was present. We calculated a lower limit of the protonic unit conductance of CF$_0$ under the assumption that all exposed CF$_0$ were proton conducting. It was 10 fS, corresponding to the passage of 6200 protons/s per CF$_0$ (at 100 mV electric driving force) and by orders of magnitude higher than so far reported for any F$_0$ channel.

INTRODUCTION

Proton-translocating ATP synthases of bacteria, chloroplasts and mitochondria are composed of two parts: the membrane embedded F$_0$, acting as proton channel, and the extrinsic F$_1$, containing the active site(s) for ATP synthesis or hydrolysis. Only F$_0$ plus F$_1$ can act as a proton translocating ATP synthase, while F$_1$ alone carries only ATPase activity [1,2]. F$_1$ is composed of 5 subunits: α, β, γ, δ and ε. 3 subunits (a, b and c) are found for bacterial F$_0$, while 4 subunits (IV, I and III plus II) were recently reported for CF$_0$ [3,4]. The mechanism of proton conduction through F$_0$ is still poorly understood despite an impressive body of genetical and biochemical information [5-7]. Published values of the proton conductance of F$_0$ (mainly bacterial F$_0$ reconstituted in liposomes) are very low, some percent of a femto-Siemens (fS) [6,8]. At 100 mV driving force this corresponds to less than 100 protons/s and F_0. Compared to that, turnover numbers of more than 900 protons/s and F_0 are required to support observed photophosphorylation rates as high as 1100 µmol ATP/mg chl per h (see Tab. 1) (calculated with 1 $CF_0-CF_1/1000$ chl [9,10] and a stoichiometry of 3 protons/ATP [11,12]) (chl: chlorophyll).

We investigated the proton conductance of exposed CF_0 in CF_1-depleted thylakoids. By flash excitation a voltage in the order of 50 mV [review 13] and a pH difference of 0.06 units [14,15] was generated across the thylakoid membrane. This caused a transient proton current from the lumen into the medium, which was competely tracked by spectrophotometric techniques: (i) we measured the proton efflux from the lumen via neutral red. Under selective buffering of the suspending medium (by BSA) small flash-induced transients of surface pH at the lumenal side of thylakoids are specifically and quantitative indicated by this membrane soluble dye [16,14,17]. Recent criticism to this measuring technique [18] was answered and shown to be unwarranted [19]. (ii) We measured charge flow across the membrane via field indicating absorption changes of intrinsic pigments at 522 nm (electrochromism) [20,review 13] and (iii) the proton appearence in the medium via phenol red, a hydrophilic pH-indicating dye [21] specific for the external suspending medium [32].

MATERIAL AND METHODS

10-14 days old pea seedlings were homogenized in 200 ml grinding medium (400 mM sorbitol, 10 mM tricine/NaOH, pH 7.8). The homogenate was filtered through a nylon mesh (20 µm) and centrifuged for 7 min at 1200 g. The pellet was incubated with EDTA (1 mM), tricine/NaOH (3 mM), pH 7.8 at a chlorophyll concentration of 0.3 - 0.6 mM for 10 min on ice and subsequently centrifuged for 10 min at 20000 g ("1st EDTA wash"). For other samples the same incubation procedure was repeated ("2nd EDTA wash"). Final dilution and preparation of stacked thylakoids ("control") was as in [22]. The degree of extraction of CF_1 was determined by immunoelectrophoresis, both for the supernatant and for the pellet, as in [23]. It was checked via ATPase activity in the supernatant [24] (assuming a specific activity of 30 µmol P_i/mg chl per min as measured for purified pea CF_1 [23]). Standard procedures were used for determinations of protein concentration [25] and chlorophyll concentration [26]. ATP synthase activity with PMS was measured by the luciferin-luciferase system from LKB [27]. Rates of O_2 evolution under continuous saturating illumination were determined by a Clark electrode.

Flash spectrophotometric experiments were carried out in a setup as described [28,29]. 15 ml sample in a cuvette of 2 cm pathlength contained 20 µM chlorophyll, 10 mM NaCl and 200 µM hexacyanoferrate (III) as electron acceptor, pH 7.3. Further additions are indicated in the legends. Samples were exited by saturating flashes of light at 5 s intervals and 20 or 40 transients were averaged. The electric potential difference across the membrane was measured via electrochromic absorption changes at 522 nm [20]. pH-transients in the suspending medium were determined by the absorption changes of phenol red (15 µM) at 559 nm [21]. pH-transients at the lumenal surface of the thylakoid membranes were measured via neutral red (15 µM plus 2.6 mg/ml BSA) at 548 nm [16,14,17]. In both cases the pH transients were obtained by subtraction of the signal recorded in the absence of the dye from that one recorded in the presence of the dye.

RESULTS AND DISCUSSION

Table 1.CF$_1$ contents, coupling ratio of the rate of O$_2$ evolution and rates of cyclic photophosphorylation as function of EDTA treatment

Preparation	CF$_1$ content (% of total)		O$_2$ coupling ratio	ATP synthesis rate (μmol ATP per mg chl per h)
	Immunoelectrophoresis Pellet Supernatant	Via ATPase supernatant		
Control	100 0	0	3.5	1110
1st EDTA wash	92 8	9	3.6	970
2nd EDTA wash	70 22	23	1.0	227

O$_2$ evolution was measured with 30 μM chlorophyll and 2.2 mM hexacyanoferrate (III) without and with 0.5 μM nigericin. ATP synthesis rates were determined with PMS (50 μM) as cofactor.

1st EDTA Wash Resulted in up to 12% Extraction of CF$_1$ Without Generating Proton Leaks

5-12% of total CF$_1$ were extracted during the 1st EDTA wash, while the O$_2$ coupling ratio [ratio of O$_2$ evolution with and without nigericin (0.5 μM)] was unaffected and the ATP synthesis rate was only slightly diminished (Tab. 1). Both indicated that the membrane remained proton-tight. This was confirmed by spectrophotometric measurements: a long lasting acidification of the lumen (Fig.1,d), only little acceleration of the electric field decay (Fig.1,e) and a long lasting alkalinisation of the medium (Fig.1,f). Incubation with DCCD (N,N´-Dicyclohexylcarbodiimide), which is known to block proton conduction through F$_0$ [30,31] had no influence (not shown). The diminuation of the extent of spectrophotometric signals after EDTA treatments was caused by desactivation of electron transport chains (ETC), as evident from a proportional reduction of the rate of uncoupled O$_2$ evolution. The acceleration of the rise of pH transient in the medium (Fig.1,f) is related to unstacking of thylakoids by EDTA as established in [22,32].

Hexacyanoferrate (III) (200 μM) was used as specific electron acceptor for photosystem I. It did not compete with plastoquinone for photosystem II (see Fig.1 in [22]) and, due to its low pK, did not bind protons upon reduction. So, flash excitation of both photosystems caused the uptake of only 1 proton per ETC from the medium (at photosystem II) and the release of 2 protons per ETC into the lumen (see Fig.2) [21,22,review 13]. The net result was an acidification by 1 proton per ETC. Therefore, with control thylakoids we observed an acidification of the lumen (Fig.1,a) and an alkalinization of the medium (Fig.1,c) which relaxed in 20-30 s reflecting the passage of protons from the lumen into the medium to produce the expected net acidification (not visible in the time domain of Fig.1). Thylakoids which had undergone one EDTA wash showed the alkalinization of the medium (Fig.1,f) with a relaxation time of approx. 14 s (measured at a longer time scale, not shown), proving that the membrane was almost as proton tight as in the control.

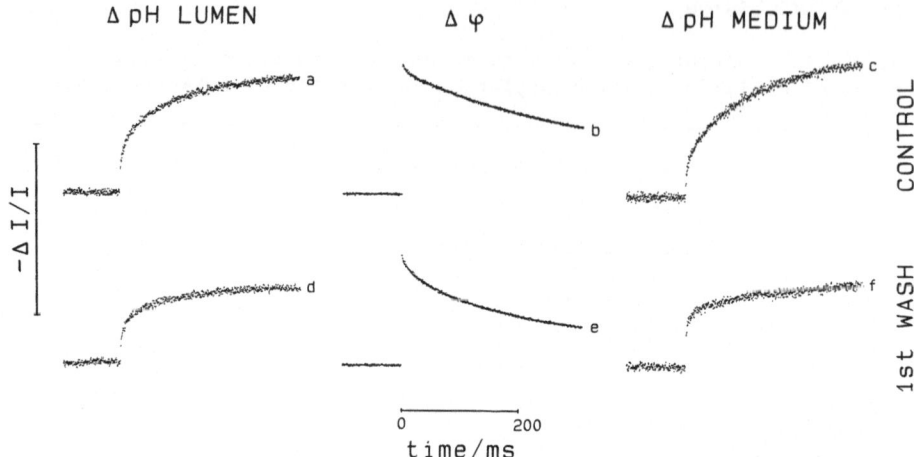

Fig. 1. Time course of the absorption change of neutral red indicating transients of the pH in the thylakoid lumen (ΔpH LUMEN). Time course of the field indicating electrochromic absorption change at 522 nm (Δφ) and time course of the absorption change of phenol red indicating transients of the pH in the suspending medium (ΔpH MEDIUM). Signals of control thylakoids (top) are compared with those of thylakoids once washed with EDTA (bottom). Positively directed signals indicate decreased absorption (see y-axis) which represents acidification of the lumen, generation of a transmembrane voltage and alkalinization of the medium respectively. The bar indicates a ΔI/I of 1.8×10^{-3} for ΔpH LUMEN (548 nm), a ΔI/I of 6×10^{-3} for Δφ (522 nm) and ΔI/I of 2.4×10^{-3} for ΔpH MEDIUM (559 nm).

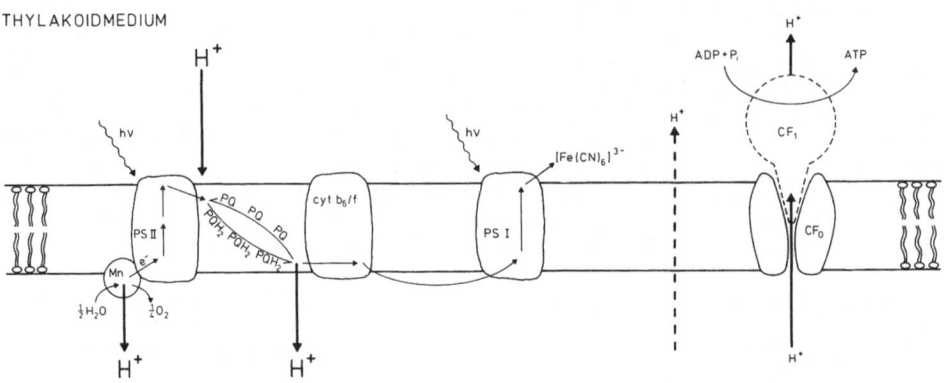

Fig. 2. Schematic view of electron and proton flow in thylakoids. Electron flow from the wateroxidising enzyme (Mn) and photosystem II (PS II) proceeds via the quinone pool (PQ/PQH_2) and the cytochrom b_6/f complex (cyt b_6/f) to photosystem I (PS I) and finally to hexacyanoferrate. 2 protons per ETC are released into the lumen, while only 1 proton is taken up from the suspending medium. The 2 protons per ETC leave the lumen by leak conductances across the membrane or through CF_0.

2nd EDTA Wash Led to a Further 20% of CF$_1$ Extraction and Caused Proton Conductance through CF$_0$

After the 2nd EDTA wash the O$_2$ evolution was accellerated and insensitiv to added uncouplers and the ATP synthesis rate dropped to 20% of controls (Tab. 1). Likewise the flash spectrophotometric measurements showed a drastically increased proton leakage from the lumen into the medium: nearly no acidification of the lumen (Fig.3,h), the decay time of the electric potential difference dropped from more than 100 ms to 7 ms (Fig.3,k), and instead of an alkalinization of the medium (lasting for 14 s) a rapid acidification was observed (Fig.3,m), accelerated by 3 orders of magnitude as compared to the control. Incubation with DCCD (25 μM, 10 min) blocked the high proton conductance: the O$_2$ evolution was recoupled and flash spectrophotometric signals became comparable to those of once EDTA-washed thylakoids (Fig.3,g,i,l). The proportion of the alkalinization of the medium (+DCCD) and of the net acidification (-DCCD) was approx. 1/1 (Fig.3,l/m) as expected with hexacyanoferrate (III) as electron acceptor. Further established blocking agents of F$_0$ channels [33], namely tributyltin$^+$, triphenyltin$^+$ (5 μM) [34] and venturicidin (50 μg/mg protein) [35] all acted similarly to DCCD, corroborating the view that the high proton permeability after 2 EDTA washes was caused by proton translocation through CF$_0$.

The rationale for adding gramicidin (Fig.3,n-s) was to eliminate the electric portion (F$\Delta\phi$) of the electrochemical driving force ($\Delta\tilde{\mu}(H^+)$) for protons and to leave over the chemical portion (-2.3RTΔpH).

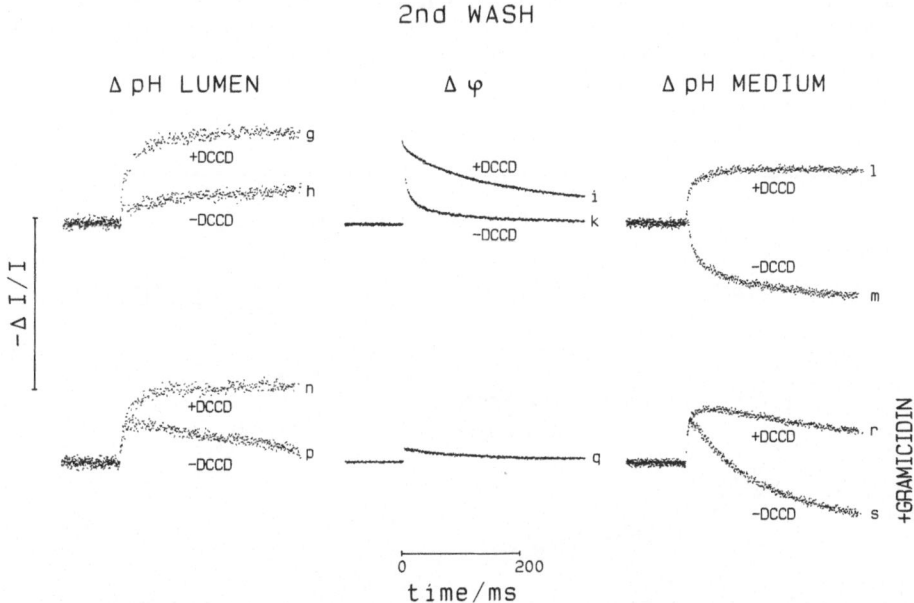

Fig. 3. Time course of ΔpH LUMEN, Δφ and ΔpH MEDIUM (as in Fig.2) for twice EDTA-washed thylakoids in the absence (top) and in the presence of gramicidin (3 nM) (bottom). The pairs of traces result from a measurement without DCCD (lower trace in each pair) and another measurement after incubation with DCCD (25 μM, 10 min) (upper one). The time course of Δφ in the presence of gramicidin (trace q) is not affected by incubation with DCCD. The bar indicates a ΔI/I of 0.9 ×10^{-3} for ΔpH LUMEN (548 nm) and the same as in Fig. 2 for the rest.

Gramicidin (3 nM) shunted the electric potential difference in less than 1 ms (Fig.3,q) regardless of wether thylakoids were EDTA-washed or not, while it only negligibly accelerated the relaxation of a pH difference across the membrane (Fig.3,r). This elimination of electric force by gramicidin distinctly decelerated the proton flow through CF_0, as obvious from a brief acidification of the lumen (Fig.3,p) and alkalinisation of the medium (Fig.3,s) preceding the net acidification.

The Kinetics of Proton Flow through CF_0

The time course of proton flow through CF_0 is reflected by the difference between the two pH transients within each pair of pH transients in Fig.3. So, by subtracting within each pair the pH transient measured in the presence of DCCD from that measured without DCCD (h-g,m-l,p-n,s-r) we got the time course of proton flow through CF_0. Scaled up to the same amplitude this differences are shown in Fig.4. It is obvious that the time course was the same, wether viewed from the lumen or viewed from the medium, both, in the presence as well as in the absence of gramicidin (Fig.4).

Kinetic analysis for exponentials of the signals in Fig.4 showed that the proton flux in the absence of gramicidin followed a biphasic decay, with a fast phase at 7.5 ms and a slow one at 60 ms their extents being approximately equal. These relaxation times found a corollary in the following: (i) 7 ms was the decay time of the electric potential difference of thylakoids washed twice with EDTA (Fig.3,k). (ii) 85 ms was the relaxation time of the then monoexponential proton flow in the presence of gramicidin (Fig.4, right). Hence the fast phase (at 7.5 ms) was attributable to electric discharge of the membrane capacitance with protons as charge carriers (electric driving force), while the slow phase (at 60 ms) reflected the relaxation of the concentration difference of protons between lumen and medium (chemical driving force). In contrast to our previous work [40], here, the electric relaxation was measured via proton flow, which allowed to calculate the protonic unit conductance of CF_0.

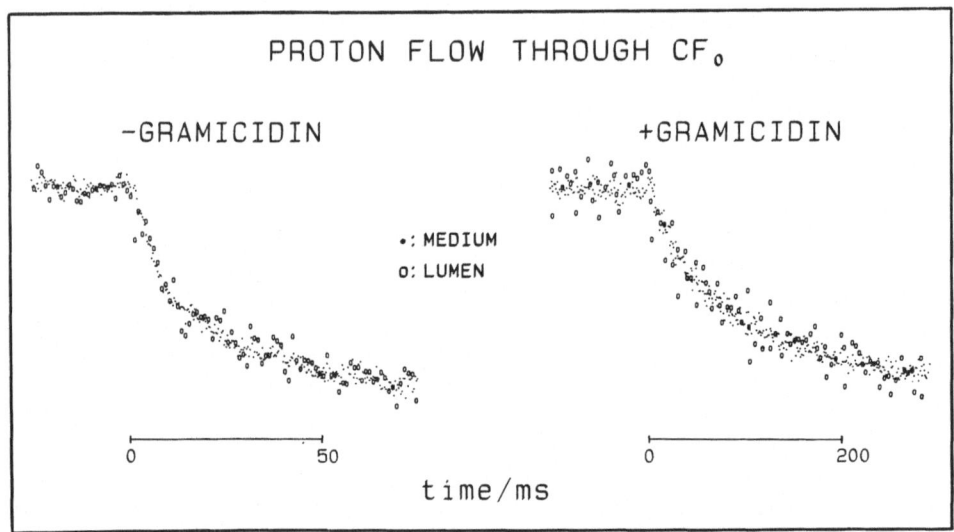

Fig. 4. Time course of proton flow through CF_0 as viewed from the lumen (o) and as viewed from the medium (•) in the absence (left) and in the presence of gramicidin (right). The time axis in the left was spread by 4 !

The Protonic Unit Conductance of CF_0 is Greater than 10 fS

The specific conductance of the thylakoid membrane for protons, $\hat{\sigma}$ (in S/cm^2), followed from the capacitor equation

$$\hat{\sigma} = \hat{c}/\tau,$$

with $\hat{c}$ denoting the specific capacitance (1 $\mu F/cm^2$, as usually assumed for biological membranes) and τ the relaxation time of the electric potential difference (7 ms). The data implied a specific conductance of 140 $\mu S/cm^2$. This was related to the surface density of exposed CF_0. We took accepted figures for the frequency of CF_0-CF_1 in terms of CF_1/chl, approx. 1/1000 [9,10] and for the membrane area per chlorophyll 2.2 nm^2 [36], and we assumed that the proton conductance was attributable to all CF_0 which were exposed by CF_1 extraction, approx. 30%. Then the density of exposed CF_0 in the membrane was 13.6×10^9 CF_0/cm^2. As a consequence the unit conductance was 10 fS. (This was a time- and ensemble averaged unit conductance in contrast to "open channel-" and "true single channel conductance".) Almost certainly this underestimated the unit conductance since the tightness of the membrane after the 1st EDTA wash had already demonstrated that not all exposed CF_0 were necessarily proton-conducting.

CONCLUSIONS

Proton flow across thylakoid membranes as mediated by exposed CF_0, the channel portion of the ATP synthase, was studied by three different spectrophotometric techniques. The correspondence of the kinetics obtained by these measurements showed that the proton flow was completely tracked. A lower limit for the unit conductance of CF_0 for protons was obtained under the assumption that all exposed CF_0 were conducting. This value was 10 fS corresponding to the passage of 6200 protons/s at 100 mV electrical driving force. Even this lower limit exceeded so far published values for the conductance and the turnover numbers of F_0 channels by orders of magnitude. It was very satisfying that it also exceeded the highest turnover numbers which were reported for the intact synthase, e.g. for CF_0-CF_1 (900 protons/s). Further studies showed that only a few percent of exposed CF_0 were proton conducting, thus the true unit conductance may be as high as 170 fS [38]. This observation together with the complete lack of increased proton conductance after the 1st EDTA wash indicated that the majority of exposed CF_0 may have been blocked, possibly by subunit δ of CF_1, which remained bound on CF_0 acting as a plug to the channel as supported elsewhere [23,39].

ACKNOWLEDGEMENTS: We thank K. Schuermann and H. Kenneweg for technical assistance and for the photographs. This work was financially supported by the Deutsche Forschungsgemeinschaft (Sonderforschungsbereich 171, Projekt B3). Parts of this work were previously published in [37].

REFERENCES

1. McCarty,R.E. & Moroney,J.V. (1984) in: The Enzymes of Biological Membranes Vol. IV (ed. Martonosi,A.N.) pp. 383-413, Plenum Press, New York
2. Vignais,P.V. & Satre,M. (1984) Molec. and Cell. Biochem. 60, 33-70
3. Hennig,J. & Herrmann,R.G. (1986) Mol. Gen. Genet. 203, 117-128
4. Cozens,A.L.,Walker,J.E.,Phillips,A.L.,Huttly,A.K. & Gray,J.C. (1986) EMBO J. 5, 217-222

5. Sebald,W. & Hoppe,J. (1981) Curr. Top. Bioenerg. 12, 1-63
6. Fillingame,R.H. (1981) Curr. Top. Bioenerg. 11, 35-106
7. Hoppe,J. & Sebald,W. (1984) Biochim. Biophys. Acta 768, 1-27
8. Sone,N., Hamamoto,T. & Kagawa,Y. (1981)
     J. Biol. Chem. 256, 2873-2877
9. Berzborn,R.J., Mueller,D., Roos,P. & Andersson,B. (1981)
     in: Photosynthesis III. (ed. Akoyunoglou,G.) pp. 107-120
     Balaban Int. Sci. Serv., Philadelphia, Pa.
10. Strotmann,H., Hesse,H. & Edelmann,K. (1973)
     Biochim. Biophys. Acta 314, 202-210
11. Junge,W., Rumberg,B. & Schroeder,H. (1970)
     Eur. J. Biochem. 14, 575-581
12. Davenport,J.W. & McCarty,R.E. (1981) J. Biol. Chem. 256, 8947-8954
13. Junge,W. & Jackson,J.B. (1982) in: Photosynthesis, Vol. 1
     (ed. Govindjee) pp. 589-646, Academic, New York
14. Junge,W., Auslaender,W., McGeer,A.,J. & Runge,T. (1979)
     Biochim. Biophys. Acta 546, 121-141
15. Wille,B. & Lavergne,J. (1982) Photobiochem. Photobiophys. 4, 131-144
16. Auslaender,W. & Junge,W. (1975) FEBS Lett. 59, 310-315
17. Hong,Y.Q. & Junge,W. (1983) Biochim. Biophys. Acta 722, 197-208
18. DeWolf,F.A., Groen,B.H., VanHoute,L.P.A., Peters,F.A.L.J., Krab,K. &
     Kraayenhof,R. (1985) Biochim. Biophys. Acta 809, 204-214
19. Junge,W., Schoenknecht,G. & Foerster,V. (1986)
     Biochim. Biophys. Acta -in press-
20. Junge,W. & Witt,H.T. (1968) Z. Naturf. 23b, 244-254
21. Junge,W. & Auslaender,W. (1973) Biochim. Biophys. Acta 333, 59-70
22. Polle,A. & Junge,W. (1986) Biochim. Biophys. Acta 848, 257-264
23. Engelbrecht,S., Lill,H. & Junge,W. (1986)
     Eur. J. Biochem. -in press-
24. Sakurai,H., Shinohara,K., Hisabori,T. & Shinohara,K. (1981)
     J. Biochem. 90, 95-102
25. Sedmak,J.J. & Grossberg, S.E. (1977) Anal. Biochem. 79, 544-552
26. Arnon,D.J. (1949) Plant Physiol. 24, 1-15
27. Schmidt,G. & Graeber,P. (1985) Biochim. Biophys. Acta 808, 46-51
28. Junge,W. (1976) in: Chemistry and Biochemistry of Plant Pigments,
     Vol. 2 (ed. Goodwin) pp. 233-333, Academic, New York
29. Foerster,V., Hong,Y.Q. & Junge,W. (1981)
     Biochim. Biophys. Acta 638, 141-152
     Biochim. Biophys. Acta 807, 238-244
36. Thomas,J.B., Minnaert,K. & Elbers,P.D. (1956)
     Acta Bot. Neerl. 5, 314-321
37. Schoenknecht,G., Junge,W., Lill,H. & Engelbrecht,S. (1986)
     FEBS Lett. 203, 289-294
38. Lill,H., Engelbrecht,S., Schoenknecht,G. & Junge,W. (1986)
     Eur. J. Biochem. -in press-
39. Junge,W., Hong,Y.Q., Quian,L.P., & Viale,A. (1984)
     Proc. Natl. Acad. Sci. USA 81, 3078-3082
40. Schmid,R., Shavit,N. & Junge,W. (1976)
     Biochim. Biophys. Acta 430, 145-153

ESSENTIAL ROLE OF ARGININE RESIDUES IN THE INTERACTION OF F_0 WITH F_1

IN ESCHERICHIA COLI ATP SYNTHASE

Karlheinz Altendorf, Karl Steffens, Erwin Schneider
and Roland Schmid

Mikrobiologie, Fachbereich Biologie/Chemie
Universität Osnabrück, 4500 Osnabrück (FRG)

INTRODUCTION

In a wide variety of organisms the ATP synthase (F_1F_0; EC 3.6.1.34) plays a key role in energy metabolism. This enzyme has been described for mitochondria, chloroplasts and bacteria (for reviews, see refs. 1-3). Among these the ATP synthase complex of Escherichia coli has been most extensively characterized by both biochemical and genetic studies. As in all other ATP synthase systems, the E. coli complex is composed of two entities: the F_1 sector is a peripheral protein and catalyzes the synthesis of ATP, whereas the F_0 part is embedded in the cytoplasmic membrane and serves as a proton translocator. The ATP synthase works in a reversible manner, which means that it is also capable of building up an electrochemical proton gradient driven by ATP hydrolysis.

The properties of the F_1 part, which is well characterized have been reviewed recently (4,5). Less data are available for the F_0 part (6,7). After purification and incorporation into phospholipid vesicles, the F_0 complex catalyzes passive proton uptake (8-11). Reconstitution with F_1 enables the system to carry out an ATP-driven proton translocation, which is sensitive towards DCCD[1] (9,10). F_0 is composed of three different subunits designated as a (M_r = 30,276), b (M_r = 17,265) and c (M_r = 8,288) (12-14). The subunits have been proposed to occur in the unusual stoichiometry of $a_1b_2c_{10\pm1}$ (15). In agreement with this proposal recent cross-linking studies have indicated the presence of ab, ab_2 and b_2 oligomers (16,17). The primary structure of all subunits has been deduced from the DNA sequence of the genes in the unc operon (12-14). Subunit c is very hydrophobic and can be extracted from whole cells with chloroform/methanol. It is probably embedded in the membrane in a "hairpin"-like structure. The binding site (an aspartic acid residue) for the specific inhibitor DCCD is located within a stretch of hydrophobic amino acids (for review, see ref. 18). The topology of subunits a and b in the membrane is also not well established. Both from computer models (19-21) and from biochemical studies (17, 22, 23) subunit b appears to be an amphipathic molecule composed of a short (30 amino acids) hydrophobic stretch at the N-terminal end, anchored in the membrane, and a large hydrophilic part extending into the water phase. It might play a role in binding of F_1, since it is

---

[1]DCCD, N,N'-dicyclohexylcarbodiimide

protected against trypsin treatment by bound F_1 (17,22). Also cross-linking studies suggest that subunit b is in close proximity of subunit β of F_1 (16). Subunit a is very hydrophobic and may span the membrane several times. Modification with [^{35}S] diazoniumbenzenesulfonate and cross-linking experiments suggest that this subunit possesses hydrophilic domains and is in spatial proximity of subunits b and β of F_1 (16).

In this communication we have investigated the contribution of arginine residues in F_0 to proton translocation and F_1 binding. To this end F_1-stripped membrane vesicles or purified F_0, incorporated into liposomes, were treated with arginine-specific reagents (24,25) and the effects of this treatment on F_1 binding and proton translocation were analyzed. The results support our recent finding (26) that arginine residues in F_0 are essential for the interaction with F_1, but do not play a role in proton translocation. To locate the relevant arginine residue(s) we have investigated the pattern of [^{14}C] phenylglyoxal incorporation after radioactive labelling of F_0 liposomes. Since the data suggest that all F_0 subunits may contribute to the observed effects, arginine residues in subunit b are indicated to play a predominant role in F_1 binding.

MATERIALS AND METHODS

Bacterial growth

E. coli K-12 was grown on a minimal medium with 0.4 % glucose as described in (27). E. coli strain KY 7485, which carries the complete unc operon on a heat-inducible lysogen λ-phage (28,29), overproduces the ATP synthase several times. This strain was grown on a glucose containing minimal medium (30), supplemented with thiamine (0.5 µg/ml), arginine (84 µg/ml) and guanine (45 µg/ml) as described in (31).

Preparations

Everted membrane vesicles of E. coli were prepared as in (32). F_1-depleted membranes were obtained as follows: Everted membranes were incubated for 1 h at room temperature in 1 mM Tris-HCl, pH 8.0, 0.5 mM EDTA, 2.5 mM 2-mercaptoethanol, 10 % (v/v) glycerol (32) and centrifuged for 1 h at 180,000 x g. Subsequently, membranes were resuspended in the same buffer and stirred overnight at 4 °C. The resulting membrane preparation exhibited less than 5 % of the original ATPase activity. F_1F_0 (33) and F_1 (34) were prepared as described. F_0 was isolated as in (10) or directly from F_1-depleted membranes of E. coli KY 7485 (11).

Modification of membranes with phenylglyoxal or 2,3-butanedione

F_1-stripped membranes were resuspended at 2 mg/ml of 50 mM Mops[2]-NaOH, pH 7.9, 10 mM MgCl$_2$ and 10 % (v/v) glycerol. Aliquots (1 ml) were incubated with the indicated concentrations of phenylglyoxal or 2,3-butanedione, respectively (see figure legends). The reaction was stopped by the addition of 5 ml 50 mM Mops-NaOH, pH 7.0, 10 mM MgCl$_2$, 50 mM arginine-HCl and 10 % (v/v) glycerol. The membranes were collected for 45 min at 180,000 x g. The pellets were washed once in 50 mM Mops-NaOH, pH 7.0, 10 mM MgCl$_2$, 10 % (v/v) glycerol and finally suspended to 0.5 mg/ml in the same buffer. For all experiments 1 M stock solutions of dicarbonylreagents were freshly prepared in the appropriate incubation buffer containing 50 % (v/v) ethanol.

---

[2]Mops, 3-(N-morpholino)propanesulfonic acid

222

Modification of F_0 liposomes

F_0 liposomes were prepared as described (10). Modification with 2,3-butanedione was carried out as follows: 200 µl F_0 liposomes (80 µg F_0) were diluted by 600 µl 20 mM Tricine[3]-NaOH, pH 8.0, 46.6 mM borate, 3.33 mM $MgSO_4$ together with different concentrations of 2,3-butanedione. The mixtures were incubated for 60 min at 25 °C. The reaction was stopped by the addition of 4 ml 15 mM Tricine-NaOH, pH 8.0, 35 mM borate, 2.5 mM $MgSO_4$, 40 mM arginine-HCl and subsequent centrifugation for 15 min at 180,000 x g. The liposomes were washed in 1 ml 15 mM Tricine-NaOH, pH 8.0, 35 mM borate, 2.5 mM $MgSO_4$ and resuspended in 200 µl of the same buffer. 150 µl were used to determine passive proton uptake. The remaining material was reconstituted with F_1 and assayed for ATPase activity.

The modification of F_0 liposomes with phenylglyoxal was performed in the absence of borate as in (26). All data for proton uptake and F_1 binding were corrected by the values obtained with identically treated liposomes lacking F_0.

Reconstitution procedures

F_1-stripped membranes were reconstituted as follows: 5 units (1 unit = 1 µmol P_i x min^{-1}) of F_1 ATPase were added to 1 mg of membranes (0.5 mg/ml) and the mixture was incubated for 15 min at 37 °C. Subsequently, 5 ml of ice-cold buffer was added, and membranes were collected by centrifugation. The ATPase activity of the supernatant was determined to estimate the amount of unbound F_1. The pellets were washed once, resuspended in 0.5 ml of the same buffer and finally assayed for ATPase activity. Reconstitution of F_0 liposomes with F_1 was carried out as described in (26). Liposomes, which had been pretreated with borate-containing buffers, were reconstituted with F_1 in a medium containing 15 mM Tricine-NaOH, pH 8.0, 35 mM borate and 2.5 mM $MgSO_4$ under the same conditions as in (26).

Labelling of F_0 liposomes with [14C] phenylglyoxal

Liposomes were prepared from acetone/ether purified soybean phospholipids (35) and F_0 was incorporated as in (10). Integration of F_1F_0 into liposomes was done in the same way using the lipid/protein ration 80:1. An aliquot of each liposome preparation was examined for proton uptake or ATP-driven proton translocation, respectively. The modification with [14C] phenylglyoxal was carried out as indicated in the figure legends. To stop the reaction, each assay was diluted with 5 ml 50 mM Mops-NaOH, pH 7.0, 2 mM $MgSO_4$, 50 mM arginine-HCl and centrifuged for 15 min at 180,000 x g. After washing the liposomes once with 5 ml 50 mM Mops-NaOH, pH 7.0, 2 mM $MgSO_4$ they were resuspended in 100 µl of the same buffer. Prior to analysis on SDS[4] polyacrylamide gel electrophoresis the lipids were extracted from the proteoliposomes by the method of Huang et al. (36). The liposomes were precipitated by addition of 1 ml methanol and centrifuged at 12,000 x g for 7 min. The precipitates were dried under a stream of nitrogen, suspended in 1 ml toluene/ether (2:1) and sonicated for 15 s in an ultrasonic bath. After centrifugation for 15 min at 12,000 x g the pellets (protein) were dried under a stream of nitrogen and stored at -20 °C until use.

---

[3]Tricine, N-[2-hydroxy-1,1-bis(hydroxymethyl)ethyl]-glycine
[4]SDS, sodium dodecyl sulfate

SDS polyacrylamide gel electrophoresis

SDS gel electrophoresis was carried out according to Weber and Osborn (37) with the following modifications: a slab chamber (9x12 cm) of 1 mm thickness was used. The running gel solutions were made of 9 ml 128 mM phosphate buffer, containing 2 % (w/v) SDS, adjusted to pH 6.9 with NaOH, 3.3 ml 3 % (w/v) polyacrylamide, 2.2 ml 45 % (w/v) acrylamide / 1.2 % (w/v) N,N'-methylenebisacrylamide, 5.5 ml H_2O, 10 μl N,N,N',N'-tetramethyl-ethylenediamine and 9 ml 128 mM phosphate buffer, containing 2 % (w/v) SDS, adjusted to pH 6.9 with NaOH, 3.3 ml 3 % (w/v) polyacrylamide, 6.6 ml 45 % (w/v) acrylamide / 1.2 % (w/v) N,N'-methylenebisacrylamide, 1.1 ml H_2O, 10 μl N,N,N',N'-tetramethylethylenediamine. After a 5 min vacuum treatment 80 μl of 10 % (w/v) freshly-prepared ammonium persulfate was added to each solution. A 5-15 % acrylamide gradient was formed by an ice-cooled two-funnel gradient mixer containing 9 ml of each solution. The polymerized running gel was overlaid by a stacking gel formed from a solution containing 9 ml 128 mM sodium phosphate buffer, pH 6.9 with 2 % (w/v) SDS, 2.2 ml 45 % (w/v) acrylamide / 1.2 % (w/v) N,N'-methylenebisacrylamide, 8.8 ml H_2O, 10 μl N,N,N',N'-tetramethylethylenediamine and 80 μl 10 % (w/v) ammonium persulfate. The electrode solution was composed of 46 mM sodium phosphate buffer, pH 6.9 and 0.1 % (w/v) SDS. The gel was run for 18 h at room temperature.

Assays

Protein concentration and ATPase activity were carried out as described (26). Passive proton uptake measurements were carried out as described in (10). ATP-driven proton translocation was detected by ACMA[5] fluorescence quenching. F_0 liposomes (10 μg F_0, reconstituted with F_1) or membrane vesicles (200 μg) were suspended in 20 mM Tricine-NaOH, pH 8.0, 10 mM $MgCl_2$, 300 mM KCl to a final volume of 2 ml and assayed as described in (10).

Chemicals

Phenylglyoxal and phospholipids were purchased from Sigma (München, FRG). 2,3-butanedione was supplied by Serva (Heidelberg, FRG), [^{14}C] phenylglyoxal by Amersham Buchler (Braunschweig, FRG). ACMA was a generous gift from Dr. Overath (Tübingen, FRG). All other chemicals were of analytical grade. Escherichia coli strain KY 7485 was a generous gift of Drs. Fillingame and Kanazawa.

RESULTS

Effects of modification of F_0 by arginyl reagents on the rebinding of F_1

Everted membranes stripped of F_1 lost capacity of rebind F_1 when incubated with increasing concentrations of phenylglyoxal. This is demonstrated by a decrease in membrane-associated ATPase activity (Fig. 1). Since the ATPase activity was recovered in the supernatant, this decrease could not be due to inhibition of bound F_1 by the modification of F_0 or by residual trace amounts of the reagent. As it turned out from experiments shown in Fig. 2, the inhibition of the reconstitution of membrane-bound DCCD sensitive ATPase activity as well as ATP-driven proton translocation was time dependent with higher rates at increased phenylglyoxal concentrations. In another set of experiments, arginine residues of the F_0 sector in F_1-stripped membranes were modified by 2,3-butanedione. As shown in Fig. 3, this reagent was effective only in the presence of borate, which presumably

---

[5]ACMA, 9-amino-6-chloro-2-methoxyacridine

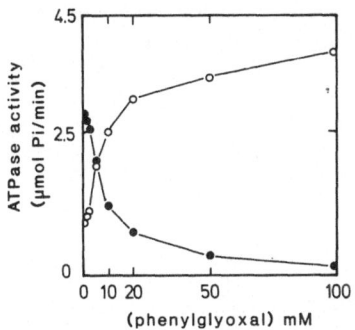

Fig. 1. Effect of phenylglyoxal treatment on the binding of
F$_1$ ATPase to everted membrane vesicles.
Everted membranes depleted of F$_1$ were modified with
different concentrations of phenylglyoxal for 45 min
at 37 °C and reconstituted with F$_1$ as described in
Materials and Methods. After reconstitution the distribution of ATPase activity between membranes (1 mg, ●)
and supernatant (o) was determined.

stabilizes the reaction product of guanidinium groups with 2,3-butane-
dione (25). In the presence of 35 mM borate the inhibition of F$_1$ binding
as well as reconstitution of ATP-driven proton translocation by 2,3-bu-
tanedione was comparable to that observed with phenylglyoxal. By contrast,
borate at 35 mM decreased the inhibitory potential of phenylglyoxal

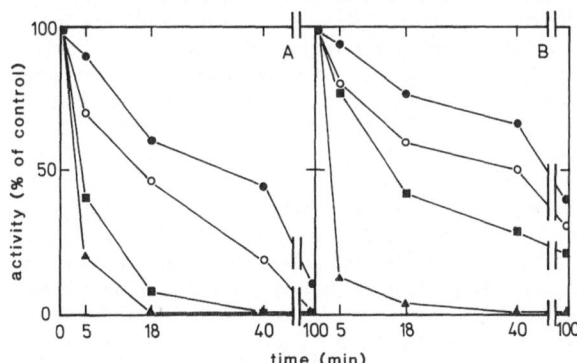

Fig. 2. Time course of inhibition of F$_1$ binding at different
phenylglyoxal concentrations
A, ATP-driven proton translocation (control value: 44 %
fluorescence quenching x 200 µg^{-1}). B, DCCD-sensitive
membrane-bound ATPase activity (control value: 1,1 µmol P$_i$
x min^{-1} x mg^{-1}). Phenylglyoxal concentrations in the
modification assays were 5 mM (●), 15 mM (o), 40 mM (■)
and 80 mM (▲). Stripped membranes were modified with
different concentrations of phenylglyoxal at 37 °C for
the times indicated and reconstituted with F$_1$ as described in Materials and Methods. Incubation of reconstituted membranes with DCCD was done as follows: membranes (20 µg) were incubated for 30 min at 37 °C in
0.5 ml 50 mM Tris-HCl, pH 8.0, 20 mM MgCl$_2$, 80 µM
DCCD and then assayed for ATPase activity. ATP-driven
proton translocation was measured by ACMA fluorescence
quenching as described in Materials and Methods.

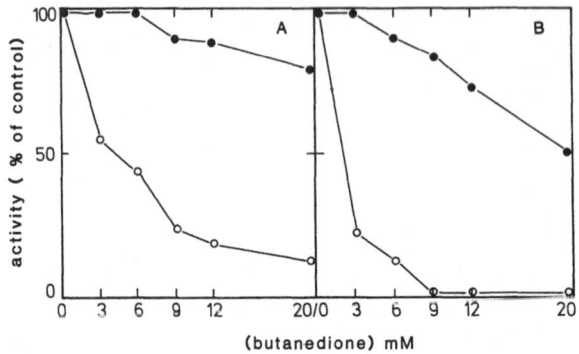

Fig. 3. Chemical modification of F_1-stripped membranes
with 2,3-butanedione in the presence or absence
of borate.

A, binding of DCCD-sensitive ATPase activity. Control
values were 4.67 μmol P_i x min^{-1} x mg^{-1} in the presence (o) or 4.55 μmol P_i x min^{-1} x mg^{-1} in the
absence (●) of borate. B, ATP-driven proton translocation. Control values were 51.7 % fluorescence
quenching x 200 $μg^{-1}$ in the presence (o) or 43.3 %
fluorescence quenching x 200 $μg^{-1}$ in the absence (●)
of borate.

Membrane vesicles depleted of F_1 were treated with the
indicated concentrations of 2,3-butanedione for 60 min
at 25 °C and reconstituted with F_1 as described in
Materials and Methods. When the treatment was carried
out in the presence of borate, the following experimental modifications were introduced: membranes were
incubated with the reagent in 15 mM Tricine-KOH,
pH 8.0, 35 mM borate, 10 mM $MgCl_2$, 50 mM KCl. The
reaction was stopped with 15 mM Mops-KOH, pH 7.0,
35 mM borate, 100 mM $MgCl_2$, 50 mM KCl, 50 mM arginine-
HCl. The washing steps and the reconstitution with
F_1 were done as in Materials and Methods using as
buffer 15 mM Mops-KOH, pH 7.0, 35 mM borate, 10 mM
$MgCl_2$, 50 mM KCl. Fluorescence quenching measurements
were carried out as in Materials and Methods, but
using a buffer containing 50 mM Mops-KOH, pH 7.4,
35 mM borate, 50 mM KCl, 10 mM $MgCl_2$, when vesicles
were modified in the presence of borate.

by 40 % (data not shown). Surprisingly, another arginine reagent, cyclo-
hexanedione, had no effect whatsoever of F_1 binding. We cannot explain
this, but similar differences in the reactiveness of dicarbonyl reagents
have also been observed previously (25).

Since modification of F_1-stripped membranes by phenylglyoxal prevented the binding of F_1 to F_0, we tested whether bound F_1 could protect
the relevant arginine residues against attack by the reagent. This indeed
seemed to be the case to a significant degree (Fig. 4). Thus, the presence
of F_1 diminished the inhibitory effect of modification by 20 mM phenylglyoxal on a subsequent reconstitution by about 40 %, when assayed either
as DCCD-sensitive ATPase activity (Fig. 4 A) or as ATP-driven H^+ translocation (Fig. 4 B).

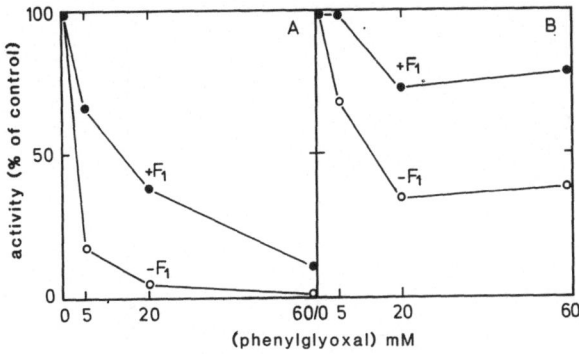

Fig. 4. Influence of bound F_1 on the modification
by phenylglyoxal
A, ATP-driven proton translocation. Control values
were 30.2 % fluorescence quenching x 200 μg^{-1} for
"+F_1" (●) and 31 % fluorescence quenching x 200 μg^{-1}
for "-F_1"(o). B, binding of DCCD-sensitive ATPase
activity. Control values were 1.02 μmol P_i x min^{-1}
x mg^{-1} for "+F_1" (●) and 0.95 μmol P_i x min^{-1} x mg^{-1} for
"-F_1" (o).
Everted membranes (5 mg) were incubated twice for
45 min at O °C in 5 ml 1 mM Tris-HCl, pH 8.0, 10 mM
EDTA, 10 % (v/v) glycerol to remove F_1 ATPase ("-F_1").
After centrifugation for 45 min at 180,000 x g the
membranes were suspended in 1 ml 50 mM Mops-NaOH,
pH 7.9, 10 mM $MgCl_2$, 10 % (v/v) glycerol and treated
for 45 min at 37 °C with the indicated concentrations
of phenylglyoxal. The reaction was stopped by the
addition of 5 ml 50 mM Mops-NaOH, pH 7.0, 10 mM $MgCl_2$,
50 mM arginine-HCl, 10 % (v/v) glycerol and subsequent
centrifugation. In membranes designated as "+F_1" the
phenylglyoxal incubation preceded the removal of F_1.
After these procedures both membrane preparations were
devoid of ATPase activity. Reconstitution with F_1 and
determination of DCCD-sensitive ATPase activity (B)
and ATP-driven proton translocation (A) was done as
described in Materials and Methods.

Effects of modification of F_0 by arginyl reagents on passive H^+ translocation

To investigate whether the modification of the F_0 sector by dicarbonyl
reagents caused an inhibition of proton translocation, we measured passive
proton uptake into liposomes containing the purified F_0 complex (10). The
results obtained by incubation of F_0 liposomes with increasing concentrations of phenylglyoxal are shown in Fig. 5 A. Proton uptake was nearly
unaffected whereas the binding of F_1 was strongly inhibited by the reagent,
in agreement with the results obtained with everted membranes. As reported
recently (26), this is further supported by the fact, that proton uptake of
F_0 liposomes, pretreated with 20 mM phenylglyoxal was still fully sensitive
towards DCCD but F_1 could only block the activity by 30 %.

In a similar experiment, F_0 liposomes were treated with 2,3-butanedione
in the presence of borate. As shown in Fig. 5 B, like phenylglyoxal 2,3-
butanedione caused a strong inhibition of F_1 binding, the proton uptake
activity, however, remained unaffected or was even slightly increased.

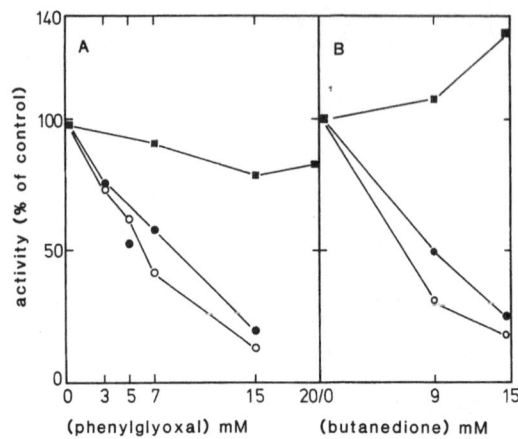

Fig. 5. Chemical modification of F_0 liposomes with phenylglyoxal
and 2,3-butanedione
F_0 liposomes were treated with the indicated concentrations
of phenylglyoxal (A) or 2,3-butanedione (B) and subsequently reconstituted with F_1 as described in (26).
Determination of liposome-bound ATPase activity (●,
control values: 5.7 μmol P_i x min^{-1} x mg F_0^{-1} for A;
16.6 μmol P_i x min^{-1} x mg F_0^{-1} for B), ATP-driven fluorescence quenching (o, control values: 57 % quenching
x min^{-1} x 10 μg F_0^{-1} for A; 47 % quenching x min^{-1}
x 10 μg F_0^{-1} for B) and proton uptake (■ , control values:
4.9 μmol H^+ x min^{-1} x mg F_0^{-1} for A; 4.15 μmol H^+ x
min^{-1} x mg F_0^{-1} for B) were done as described in
Materials and Methods.

The modified F_0 liposomes were still sensitive to DCCD with regard to
proton uptake, but the blockage of the channel by F_1 was greatly reduced
(data not shown).

Labelling of F_0 with [^{14}C] phenylglyoxal

To identify the subunit(s) that might be involved in binding of F_1,
F_0 containing liposomes were incubated with [^{14}C] phenylglyoxal up to
60 min. After removing excess of radiolabel and lipid, the F_0 subunits
were separated by SDS polyacrylamide gel electrophoresis. The separation
system employed is a modification of the procedure of Weber and Osborn
(37), which was especially suited for our purposes, because at pH 6.9
the reaction product of arginine residues with phenylglyoxal was sufficiently stable for detection by autoradiography. This gel system is linear
for proteins with molecular weights between 92- and 17-kDa (data not shown).
It is interesting to note that in this gel system subunit a runs with its
real molecular weight of 30,000 (1), in contrast to the system according
to Laemmli (38), where it appears at a molecular weight of 24,000 (1).
In Fig. 6 B, the autoradiograph of F_0 subunits treated for different times
with [^{14}C] phenylglyoxal (20 mM) is shown. Subunit b was most heavily
labelled. In this case the reaction seemed to be complete after 15 min,
whereas in subunit c the incorporation of radioactivity continued until
60 min. This observation is also supported by scintillation counting of
gel slices (Fig. 7). Assuming a quantitative extraction of all F_0 subunits
from liposomes and taking into account that the number of arginine residues
for F_0 in b and c is equal (based on a stoichiometry of $a_1b_2c_{10\pm1}$, ref. 15)
the results indicate that subunit c was less accessible to the reagent

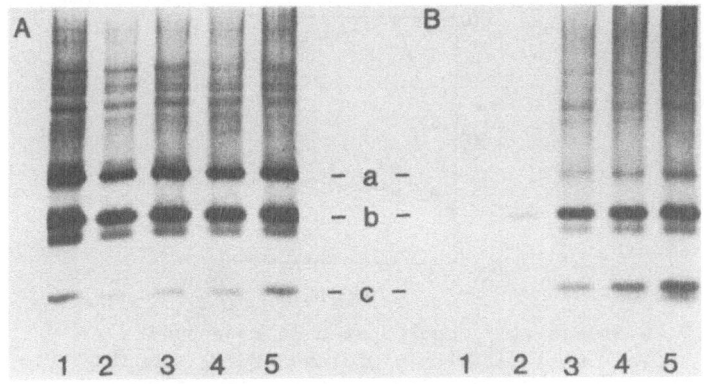

Fig. 6. Gel electrophoresis of radiolabelled F_0
125 μl F_0 liposomes (50 μg F_0) were mixed with 125 μl
100 mM Mops-NaOH, pH 7.9, 4 mM $MgSO_4$ and incubated for
the times indicated at 37 °C with 20 mM [^{14}C] phenylglyoxal (1,6 μCi x μmol^{-1}). The assays were further
treated as described in Materials and Methods. The
extracted protein was incubated for 15 min at room
temperature in 20 μl probe solution containing 80 mM
sodiumphosphate buffer, pH 6.9, 3 % (w/v) SDS, 20 %
(w/v) sucrose, 0.005 % (w/v) bromphenol blue, 140 mM
2-mercaptoethanol and then subjected to gel electrophoresis. After running the gel was fixed for 1 h
with 10 % (w/v) trichloroacetic acid, stained for
1.5 h in 50 % (v/v) methanol, 5 % (v/v) acetic acid,
0.25 % (w/v) coomassie blue and destained overnight
in 10 % (v/v) isopropanol, 10 % (v/v) acetic acid
(Fig. 6 A). After drying the gel was exposed for
three weeks at -80 °C to an X-ray film (Kodak, x-omat
S) for autoradiography (Fig. 6 B). The lanes are
showing F_0 incubated with the radiolabel for 0 min
(1), 5 min (2), 15 min (3),30 min (4),and 60 min (5).

than subunit b. Under the same considerations the labelling pattern of
subunit a corresponds to its arginine content of 4 residues per copy.
A protein of molecular weight of 14,000, which is of unknown identity and
sometimes copurified with F_0 (10) was also labelled. Another major band,
which emerged under subunit c (Fig. 6 B) was probably due to unremoved
phospholipid, which was contaminated with [^{14}C] phenylglyoxal (39). This
band was not stained with coomassie blue (Fig. 6 A).

   To check whether bound F_1 could change the labelling pattern in F_0,
liposomes containing F_0 as well as F_1F_0 were treated with [^{14}C] phenylglyoxal for 45 min, causing 80 % inhibition of F_1 binding. In both cases
roughly the same amount of F_0 protein was present. As shown in Fig. 8 B,
the incorporation of radioactivity in subunit b was clearly reduced in
the presence of F_1. Similar results were obtained, when F_0 as well as
F_1F_0 were treated with [^{14}C] phenylglyoxal for only 15 min. After SDS
gel electrophoresis the protein bands corresponding to subunits a, b
and c were cut out of the gel, solubilized and counted. Compared to F_0,
about half of the amount of radioactivity was found in subunit b, when
the F_1 part was present (138 cpm vs. 235 cpm). It should be mentioned
however, that under both labelling conditions used, no significant difference in the labelling pattern of subunits a and c could be observed.

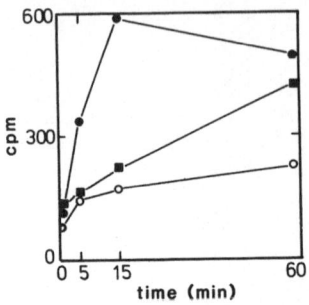

Fig. 7. Radioactivity incorporated in F_0 subunits
The dried gel with radiolabelled F_0 subunits (Fig. 6)
was soaked for 1 day in water/ethanol/glycerol
(36.5 : 12.5 : 1). The bands of the F_0 subunits were
cut out and dissolved overnight at 55 °C with 0.5 ml
30 % (v/v) H_2O_2. Subsequently, the samples were diluted
to 10 ml with "Quick Szint 212" (Zinsser, FRG) and
counted in a Packard ß-scintillation counter.
● , subunit $\underline{b}$; o , subunit $\underline{a}$; ■ , subunit $\underline{c}$.

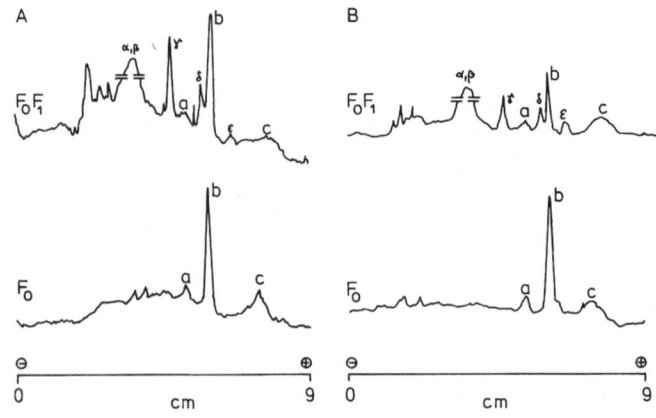

Fig. 8. Effect of bound F_1 on the modification of F_0 with
[^{14}C] phenylglyoxal
A, densitogram of protein stained gel. B, densitogram
of the corresponding autoradiograph. 150 μl of F_0
liposomes (60 μg F_0) or 540 μl F_1F_0 liposomes (218 μg
F_1F_0) were mixed with 37.5 μl or 135 μl 250 mM Mops-
NaOH, pH 7.9, 25 mM $MgSO_4$, respectively, and incubated
for 45 min at 37 °C with 20 mM [^{14}C] phenylglyoxal
(1.6 μCi x $μmol^{-1}$). All further treatments were done
as described in Materials and Methods. For electrophoresis the protein, devoid of lipids was suspended
in 20 μl denaturation buffer. 5 μl and 15 μl of each
assay were applicated to gel electrophoresis carried
out as in Fig. 6. The bands with subunit $\underline{b}$ were cut out
from the 15 μl containing lanes and subjected to
scintillation counting as in Fig. 7. The remaining
part of the gel was exposed to an X-ray film for
6 weeks at -80 °C.

230

DISCUSSION

Chemical modification of amino acid residues is a powerful tool to study structure and function of proteins. In this context the reaction of dicarbonyl reagents with guanidinium groups is often exploited to achieve an alteration of arginine residues under mild experimental conditions. This is documented for phenylglyoxal and 2,3-butanedione (24,25).

Sone et al. (40) first demonstrated an inhibition of F_1 binding and proton translocation after chemical modification of the reconstituted F_0 part from the thermophilic bacterium PS3 by phenylglyoxal. Loo et al. (41) investigated the role of arginine residues of the immobilized subunit c of F_0 from E. coli with solid-phase radioimmunoassay. They concluded that modified arginine residues of the polar loop of subunit c interfere with the binding of F_1 to the DCCD binding protein. More convincing evidence that the polar loop of subunit c in F_0 might be in contact with F_1 stems from genetic studies (42).

In our study we have used F_1-stripped membranes as well as F_0 incorporated into liposomes to investigate the role of arginine residues in the membrane sector of the ATP synthase. The modification of F_1-stripped everted membranes with phenylglyoxal inhibited the binding of F_1 to F_0 as well as the reconstitution of ATP-driven proton translocation. Similar results have also been reported by Bragg (43). It was always observed that ATP-driven proton translocation was slightly more sensitive towards the reagent than the reconstitution of DCCD-sensitive ATPase activity (Fig. 2). This observation may indicate that the inhibition of F_1 binding is due to an "all or none" effect, since only vesicles, saturated with F_1 are registrated by the ACMA fluorescence quenching method. On the other hand, a disturbation of the membrane integrity due to the ambiguous solubility of the reagent, cannot be excluded. Phenylglyoxal is described to be very hydrophilic when analyzed by distribution in a hexane/water system, but it was also shown to be soluble in octanol and be retained by phospholipid vesicles (39).

Since arginine residues of F_0 are interacting with F_1, some degree of protection by bound F_1 was expected. The observation, that membranes were partially sensitive towards the reagent even in the presence of bound F_1 (Fig. 4) suggests that also (an) arginine residue(s) that is/are not shielded by the catalytic complex is/are important in the binding process. If this is true, the modification of this/these residue(s) probably interfered with the F_1 binding by a conformational change of the F_0 complex. These findings correspond to the observation that the labelling pattern of F_0 is influenced by bound F_1, when F_0 liposomes are treated with [^{14}C] phenylglyoxal (Fig. 8 B). The decrease of bound radioactivity in subunit b in the presence of F_1 indicates that arginine residues of this subunit are in close contact to the catalytic complex. The residual modification of F_0 subunits in the presence of F_1 is in agreement with the observation, that bound F_1 only partially protected F_0 against the inhibitory effects of phenylglyoxal (Fig. 4).

The reaction of 2,3-butanedione with F_1-stripped membranes had the same consequences on F_1 binding as shown for phenylglyoxal. The dependence on borate is best interpreted by forming a stabilizing complex, which consists of one molecule each of borate and 2,3-butanedione per arginine residue (25).

Phospholipid vesicles reconstituted with F_0 were used to investigate the behaviour of the membrane sector to dicarbonyl reagents in an artificial, fully defined environment. The highly uniform orientation of F_0 in the liposomes (10) simultaneously allowed to monitor the changes of F_1

binding capacity and proton translocation properties. The inhibition of F_1 binding was slightly more sensitive towards arginine modification in F_0 liposomes compared to membrane vesicles (Fig. 1,5), but occurred in the same concentration range. Taken together the data from both systems indicate that dicarbonyl reagents under our experimental conditions only impaired F_1 binding and did not affect proton translocation through F_0. As we have mentioned before (26), the discrepancy between our results and those of Sone et al. (40) regarding the proton translocation through F_0 may be due to the different experimental conditions.

An evaluation of the data from the labelling experiment of F_0 liposomes with $[^{14}C]$ phenylglyoxal (Fig. 6,7) suggests, that the inhibition of F_1 binding to F_0 is a complicated process, in which all subunits of the membrane sector might be involved. The time course of incorporation of the radiolabel into subunits a and b (Fig. 7) shows that the reaction with these proteins in nearly complete after 15 min. Thereby, the higher amount of bound radioactivity in subunit b reflects its higher content of arginine residues, which are presumably all exposed to the water phase (19-21). It is interesting to note that the inhibition of F_1 binding to stripped membranes reached about 50 % after 15 min at a concentration of 40 mM phenylglyoxal (Fig. 2). Therefore, it is an important finding that subunit c continued to incorporate radioactivity until 60 min reaction time, what corresponds again to 80 % inhibition. The reaction pattern might account for the further inhibition of the F_1 binding after the completion of the reaction with subunits a and b. This conclusion corresponds well to the results of Loo et al. (41) mentioned above. On the other hand, the role of arginine residues of subunit c in F_1 binding is difficult to understand. It might be possible that a modification of these residues induces a conformation in the F_0 part, which is incompatible with F_1 binding. In this context it is of special interest that the contribution of altered arginine residues in subunit c to the inhibition of F_1 binding had only minor effect on proton translocation through F_0.

The total number of arginine residues, as calculated from the amino acid sequences of the subunits and their suggested stoichiometry in F_0 (15) is 20 residues per subunits b and c, respectively, and 4 residues per a (12-14). In subunits b and c, these residues are located in hydrophilic regions which are assumed to face the F_1 side of the membrane (18,19). From predictions of the secondary structure of subunit a, three of the four arginine residues are located on one side of the membrane (21). However, the orientation of subunit a in the membrane is still unknown. Provided that all arginine residues of the F_0 subunits have been labelled, the total amount of radioactivity in subunit a should be about five times less than that in subunit b or c. However, from the data shown in Fig. 7 one can clearly see that this is not the case. Therefore, we conclude that under conditions at which 80 % inhibition of F_1 binding occurred, not all arginine residues in subunits b and c have been modified. Moreover, the data also indicate that the arginine residues in subunits a and b appear to be accessible more easily to phenylglyoxal and might therefore be responsible for the initial phase of inhibition.

ACKNOWLEDGEMENTS

We thank Drs. Fillingame and Kanazawa for providing us with samples of E. coli strain KY 7485. We thank Brigitte Herkenhoff for excellent technical assistance, Mrs. Johanna Petzold for typing the manuscript and Dr. Tilly Bakker-Grunwald for a critical revision. This work was supported by the Deutsche Forschungsgemeinschaft, the Niedersächsische Ministerium für Wissenschaft und Kunst and the Fonds der Chemischen Industrie.

REFERENCES

1. A. E. Senior and J. G. Wise, The proton-ATPase of bacteria and mitochondria, $\underline{J}$. $\underline{Membrane}$ $\underline{Biol}$. 73:105 (1983).

2. N. Nelson, Proton-ATPase of chloroplasts, $\underline{Curr}$. $\underline{Top}$. $\underline{Bioenerg}$. 11:1 (1981).

3. M. Futai and H. Kanazawa, Structure and function of proton-translocating adenosine triphosphatase (F_0F_1): Biochemical and molecular biological approaches, $\underline{Microbiol}$. $\underline{Rev}$. 47:285 (1983).

4. S. D. Dunn and L. A. Heppel, Properties and functions of the subunits of the $\underline{Escherichia}$ $\underline{coli}$ coupling factor ATPase, $\underline{Arch}$. $\underline{Biochem}$. $\underline{Biophys}$. 210:421 (1981).

5. P. V. Vignais and M. Satre, Recent developments on structural and functional aspects of the F_1 sector of H^+-linked ATPases, $\underline{Mol}$. $\underline{Cell}$. $\underline{Biochem}$. 60:33 (1984).

6. E. Schneider and K. Altendorf, The proton-translocating portion (F_0) of the $\underline{E}$. $\underline{coli}$ ATP synthase, $\underline{Trends}$ $\underline{Biochem}$. $\underline{Sci}$. 9:51 (1984).

7. J. Hoppe and W. Sebald, The proton conducting F_0-part of bacterial ATP synthases, $\underline{Biochim}$. $\underline{Biophys}$. $\underline{Acta}$ 768:1 (1984).

8. R. S. Negrin, D. L. Foster, and R. H. Fillingame, Energy-transducing H^+-ATPase of $\underline{Escherichia}$ $\underline{coli}$. Reconstitution of proton translocation activity of the intrinsic membrane sector, $\underline{J}$. $\underline{Biol}$. $\underline{Chem}$. 255:5643 (1980).

9. P. Friedl and H. U. Schairer, The isolated F_0 of $\underline{Escherichia}$ $\underline{coli}$ ATP-synthase is reconstitutively active in H^+-conduction and ATP-dependent energy-transduction, $\underline{FEBS}$ $\underline{Lett}$. 128:261 (1981).

10. E. Schneider and K. Altendorf, ATP synthetase (F_1F_0) of $\underline{Escherichia}$ $\underline{coli}$ K-12. High-yield preparation of functional F_0 by hydrophobic affinity chromatography, $\underline{Eur}$. $\underline{J}$. $\underline{Biochem}$. 126:149 (1982).

11. E. Schneider and K. Altendorf, Subunit b of the membrane moiety (F_0) of ATP synthase (F_1F_0) from $\underline{Escherichia}$ $\underline{coli}$ is indispensable for H^+ translocation and binding of the water-soluble F_1 moiety, $\underline{Proc}$. $\underline{Natl}$. $\underline{Acad}$. $\underline{Sci}$. USA, 81:7279 (1984).

12. N. J. Gay and J. E. Walker, The atp operon: nucleotide sequence of the promoter and the membrane proteins and the δ subunit of $\underline{Escherichia}$ $\underline{coli}$ ATP-synthase, $\underline{Nucleic}$ $\underline{Acids}$ $\underline{Res}$. 9:3919 (1981).

13. H. Kanazawa, K. Mabuchi, T. Kayano, T. Noumi, T. Sekiya, and M. Futai, Nucleotide sequence of the genes for F_0 components of the proton-translocating ATPase from $\underline{Escherichia}$ $\underline{coli}$: prediction of the primary structure of F_0 subunits, $\underline{Biochem}$. $\underline{Biophys}$. $\underline{Res}$. $\underline{Commun}$. 103:613 (1981).

14. J. Nielsen, F. G. Hansen, J. Hoppe, P. Friedl, and K. v. Meyenburg, The nucleotide sequence of the atp genes coding for the F_0 subunits a, b, c and the F_1 subunit δ of the membrane bound ATP synthase of $\underline{Escherichia}$ $\underline{coli}$, $\underline{Mol}$. $\underline{Gen}$. $\underline{Genet}$. 184:33 (1981).

15. D. L. Foster and R. H. Fillingame, Stoichiometry of subunits in the H^+-ATPase complex of $\underline{Escherichia}$ $\underline{coli}$, $\underline{J}$. $\underline{Biol}$. $\underline{Chem}$. 257:2009 (1982).

16. J. P. Aris and R. D. Simoni, Cross-linking and labeling of the $\underline{Escherichia}$ $\underline{coli}$ F_1F_0-ATP synthase reveal a compact hydrophilic portion of F_0 close to an F_1 catalytic subunit, $\underline{J}$. $\underline{Biol}$. $\underline{Chem}$. 258:14599 (1983).

17. J. Hermolin, J. Gallant, and R. H. Fillingame, Topology, organization, and function of the psi subunit in the F_0 sector of the H^+-ATPase of $\underline{Escherichia}$ $\underline{coli}$, $\underline{J}$. Biol. Chem. 258:14550 (1983).

18. W. Sebald and J. Hoppe, On the structure and genetics of the proteolipid subunit of the ATP synthase complex, Curr. Top. Bioenerg. 12:1 (1981).

19. J. E. Walker, M. Saraste, and N. J. Gay, E. coli F_1-ATPase interacts with a membrane protein component of a proton channel, Nature (Lond.) 298:867 (1982).

20. H. U. Schairer, J. Hoppe, W. Sebald, and P. Friedl, Topological and functional aspects of the proton conductor, F_0, of the Escherichia coli ATP-synthase, Biosci. Rep. 2:631 (1982).

21. A. E. Senior, Secondary and tertiary structure of membrane proteins involved in proton translocation, Biochim. Biophys. Acta 726:81 (1983).

22. D. S. Perlin, D. N. Cox, and A. E. Senior, Integration of F_1 and the membrane sector of the proton-ATPase of Escherichia coli. Role of subunit "b" (uncF protein), J. Biol. Chem. 258:9783 (1983).

23. J. Hoppe, P. Friedl, H. U. Schairer, W. Sebald, K. v. Meyenburg, and B. B. Jørgensen, The topology of the proton translocating F_0 component of the ATP synthase from E. coli K12: studies with proteases, EMBO J. 2:105 (1983).

24. K. Takahashi, The reaction of phenylglyoxal with arginine residues in proteins, J. Biol. Chem. 243:6171 (1968).

25. J. F. Riordan, Functional arginyl residues in carboxypeptidase A. Modification with butanedione, Biochemistry 12:3915 (1973).

26. K. Steffens, E. Schneider, B. Herkenhoff, R. Schmid, and K. Altendorf, Chemical modification of the F_0 part of the ATP synthase (F_1F_0) from Escherichia coli. Effects on proton conduction and F_1 binding, Eur. J. Biochem. 138:617 (1984).

27. B. D. Davis and E. S. Mingioli, Mutants of Escherichia coli requiring methionine or vitamin B_{12}, J. Bacteriol. 60:17 (1950).

28. T. Miki, S. Hiraga, T. Nagata, and T. Yura, Bacteriophage λ carrying the Escherichia coli chromosomal region of the replication origin, Proc. Natl. Acad. Sci. USA 75:5099 (1978).

29. D. L. Foster, M. E. Mosher, M. Futai, and R. H. Fillingame, Subunits of the H^+-ATPase of Escherichia coli. Overproduction of an eight-subunit F_1F_0-ATPase following induction of a λ-transducing phage carrying the unc operon, J. Biol. Chem. 255:12037 (1980).

30. S. Tanaka, S. A. Lerner, and E. C. C. Lin, Replacement of a phosphoenolpyruvate-dependent phosphotransferase by a nicotinamide adenine dinucleotide-linked dehydrogenase for the utilization of mannitol, J. Bacteriol. 93:642 (1967).

31. M. Senda, H. Kanazawa, T. Tsuchiya, and M. Futai, Conformational change of the α subunit of Escherichia coli F_1 ATPase: ATP changes the trypsin sensitivity of the subunit, Arch. Biochem. Biophys. 220:398 (1983).

32. G. Vogel and R. Steinhart, ATPase of Escherichia coli: Purification, dissociation, and reconstitution of the active complex from the isolated subunits, Biochemistry 15:208 (1976).

33. P. Friedl, C. Friedl, and H. U. Schairer, ATP synthetase of Escherichia coli K12: Purification of the enzyme and reconstitution of energy-transducing activities, Eur. J. Biochem. 100:175 (1979).

34. M. Futai, P. C. Sternweis, and L. A. Heppel, Purification and properties of reconstitutively active and inactive adenosine triphosphatase from Escherichia coli, Proc. Natl. Acad. Sci. USA 71:2725 (1974).

35. N. Sone, M. Yoshida, H. Hirata, and Y. Kagawa, Reconstitution of vesicles capable of energy transformation from phospholipids and adenosine triphosphatase of a thermophilic bacterium, J. Biochem. (Tokyo) 81:519 (1977).

36. K. S. Huang, H. Bayley, and H. G. Khorana, Delipidation of bacteriorhodopsin and reconstitution with exogenous phospholipid, Proc. Natl. Acad. Sci. USA 77:323 (1980).

37. K. Weber and M. Osborn, The reliability of molecular weight determinations by dodecyl sulfate-polyacrylamide gel electrophoresis, J. Biol. Chem. 244:4406 (1969).

38. U. K. Laemmli, Cleavage of structural proteins during the assembly of the head of bacteriophage T4, Nature (Lond.) 227:680 (1970).

39. N. Latruffe, M. S. El Kebbaj, C. Moussard, and Y. Gaudemer, Permeability of inner mitochondrial membrane to arginine reagents, FEBS Lett. 144:273 (1982).

40. N. Sone, K. Ikeba, and Y. Kagawa, Inhibition of proton conduction by chemical modification of the membrane moiety of proton translocating ATPase, FEBS Lett. 97:61 (1979).

41. T. W. Loo, H. Stan-Lotter, D. Mackenzie, R. S. Molday, and P. D. Bragg, Interaction of Escherichia coli F_1-ATPase with dicyclohexylcarbodiimide-binding polypeptide, Biochim. Biophys. Acta 733:274 (1983).

42. M. E. Mosher, L. K. White, J. Hermolin, and R. H. Fillingame, H^+-ATPase of Escherichia coli. An uncE mutation impairing coupling between F_1 and F_0 but not F_0-mediated H^+ translocation, J. Biol. Chem. 260:4807 (1985).

43. P. D. Bragg, The ATPase complex of Escherichia coli, Can. J. Biochem. Cell Biol. 62:1190 (1984).

SOLUTE TRANSPORT ACROSS BACTERIAL MEMBRANES

W.N. Konings, A.J.M. Driessen, M.G.L. Elferink and
B. Poolman

Department of Microbiology
University of Groningen
9751 NN Haren, The Netherlands

INTRODUCTION

The cell-envelope of bacteria is composed of a cytoplasmic membrane which is surrounded by a cellwall, murien- or peptidoglycan layer, and outside the cellwall in Gram-negative bacteria by an outer membrane (lipopolysaccharide layer). Both the outer membrane and the cellwall are freely permeable for small solutes and do not form an osmotic barrier of the cell. This function is exclusively fulfilled by the cytoplasmic membrane.

Solutes can cross this cytoplasmic membrane by three physical mechanisms (1): "passive diffusion" in which solute movement down its concentration gradient occurs without specific interaction with membrane components; "facilitated diffusion" which involves a specific interaction with membrane components and in which solute translocation also occurs down its concentration gradient; and "active transport", a process in which the solute interacts with a specific carrier molecule in the membrane and translocation of solute occurs at the expense of metabolic energy against its electrochemical gradient. Active transport comprises the transport processes in which electrochemical energy is the driving force for solute translocation and also the transport processes which utilize chemicalm redox or light energy.

These active transport processes can be classified in the following groups (Fig. 1):

1. Primary transport systems: transport by enzyme system(s) which convert chemical or light energy into electrochemical energy. These transport systems comprise the electrogenic proton pumps (the electron transfer systems and the Ca^{2+}, Mg^{2+}-stimulated ATPase) and the phosphate-bond driven solute transport system. The proton pumps usually translocate protons from the cytoplasm to the external medium. These processes lead to the generation of an electrochemical gradient of protons ($\Delta\tilde{\mu}_{H^+}$). This $\Delta\tilde{\mu}_{H^+}$ exerts a force on the protons, the proton motive force ($\Delta\tilde{\mu}_{H^+}/F$, Δp) which is composed of an electrical potential ($\Delta\psi$) and a chemical gradient of protons across the membrane ($Z\Delta pH$ expressed in mV which equals $\frac{2.3\ RT}{F}$ (pH_{in} - pH_{out})). In equation:

$$\Delta\tilde{\mu}_{H^+}/F = \Delta p = \Delta\psi - Z\Delta pH \qquad (mV)$$

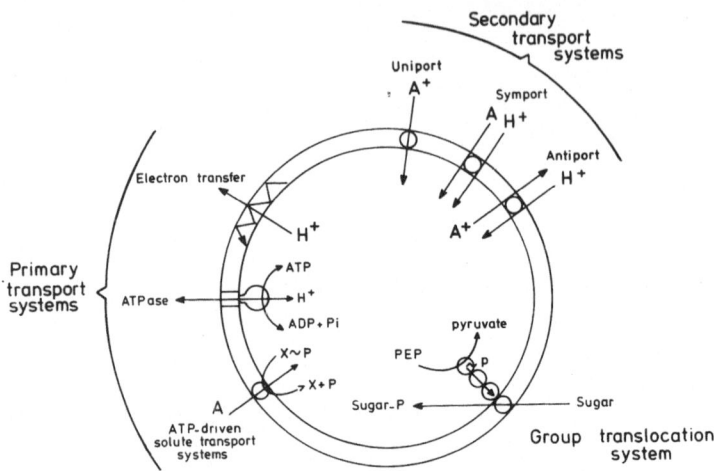

Fig. 1. Schematic presentation of the different solute transport
systems found in bacteria. The ATP-driven transport systems do
not necessarily use ATP but possibly other phosphate-bond
energy functions as the energy source.

2. Secondary transport systems: solute transport by these systems
is driven by the electrochemical gradients of the translocated solutes
(see ref. 1 for a detailed analysis of the driving forces). In addition
to the chemical gradient(s) of the translocated solute(s) ($\Delta\bar{\mu}_A/F$ and
$-Z\Delta pH$) the electrical potential will be a driving force if net-charge
is translocated. The following general equation holds for the driving
force of secondary transport processes:

$$\text{driving force: } \Delta\bar{\mu}_A/F + (n + m)\, \Delta\psi - n\, Z\Delta pH \qquad (mV)$$

in which A is the translocated solute with charge m;

$$\Delta\bar{\mu}_A/F = Z \log (A)_{in}/(A)_{out}$$

and n is the number of translocated protons.

3. Group translocation: the solute is substrate for a specific
enzyme system in the membrane and the enzymatic reaction results in a
chemical modification of the solute and release of the product at the
other side of the membrane. The major (if not only) group-translocation
systems found in bacteria are the PEP-dependent sugar transport systems
(PTS).

In Gram-negative bacteria the binding-protein dependent transport
systems are found. These binding proteins perform their functions in
the periplasmic space between the outer membrane and the cytoplasmic
membrane. The driving force for solute transport by these systems appears to be ATP or another form of phosphate-bond energy. These systems
belong therefore to the primary transport systems. Since this periplasmic space is absent in Gram-positive bacteria binding-protein
dependent transport systems are are not found in these organisms.

MODELSYSTEMS FOR SOLUTE TRANSPORT STUDIES

Studies on solute transport performed with intact cells have supplied valuable information about the biochemical nature, specificity

and kinetic properties of solute transport systems. However, it became evident that these studies hardly allow unambiquous conclusions about the mechanism of energy coupling to solute transport. Cells often contain endogenous energy reserves and even in the absence of externally added energy source(s), rapid acculumation of solutes can be observed. Treatment of the cells in order to deplete these endogenous energy reserves can result in cell preparations in which the addition of energy sources stimulates solute transport. But even then it remains difficult to obtain conclusive information about the exact nature of the direct driving force for solute transport. Another problem associated with studies on solute transport in intact cells is the metabolism of the solutes under study. Studies have been performed with non-metabolizable solutes or in mutants that are blocked in the main metabolic pathways of the solute, but the main problems outlined above make interpretations in terms of the mechanism of energy-coupling difficult. Furthermore, measurements of the proton motive force in intact cells are severely complicated due to non-specific concentration-dependent binding of $\Delta\psi$ probes, like tetraphenylphosphonium (Ph_4P^+) to different cell elements (2). This binding can lead to an overestimation of the true forces and makes studies on the quantitative relationships between the proton motive force and solute transport in intact cells difficult if not impossible.

In order to obtain detailed information about the mechanism of energy-coupling to solute transport a more defined system is required. This has been developed by the isolation of closed cytoplasmic membrane vesicles from intact cells, in which the functional and structural properties of the cytoplasmic membrane have been retained. Kaback (3) devised a method for the isolation of cytoplasmic membrane vesicles from the Gram-negative bacterium Escherichia coli. The isolation procedure comprises two essential steps: (i) the conversion of the bacterial cell into an osmotically sensitive form, and (ii) the controlled lysis of this sphaeroplast in the presence of nucleases and a chelating agent. The first step is usually accomplished by incubating the cells in a hypertonic medium with lysozyme which hydrolyses the cellwall of the cell envelope. This cellwall allows the cell to resist a considerable tugor pressure and contributes to the shape of the cell. In order to enable lysozyme to perform its action on this layer in Gram-negative organisms, the outer membrane has to be removed at least partially. Since Mg^{2+} plays an important role in maintaining the structural integrity of the outer membrane, destabilisation of the outer membrane can be achieved by removal of divalent cations by chelating agents like EDTA. In contrast to Gram-negative bacteria, the cellwall of Gram-positive bacteria is directly accessible to lysozyme. In a hypertonic medium, cells are converted into protoplasts. In the second step of the isolation procedure, the protoplasts are lysed by dilution in a hypotonic medium in the presence of deoxyribonuclease (DNase) and ribonuclease (RNase), in order to hydrolyze the liberated DNA and RNA (Fig. 2). The membrane vesicles obtained are washed extensively and collected by differential centrifugation. Some problems arise, however, for organisms that excrete proteolytic enzymes, such as Bacillus subtilis. Due to the action of these enzymes, the transport activity of the membrane vesicles was often only a fraction of the activity of intact cells. A rapid isolation procedure was devised in which the cells are treated with lysozyme in a hypotonic medium. In this way the protoplast formation step can be circumvented and higly active membrane vesicles are obtained (4).

Membrane vesicles obtained by osmotic lysis have the same orientation as the cytoplasmic membrane of the intact micro-organism (4). They are almost completely devoid of cytoplasmic constituents and less than

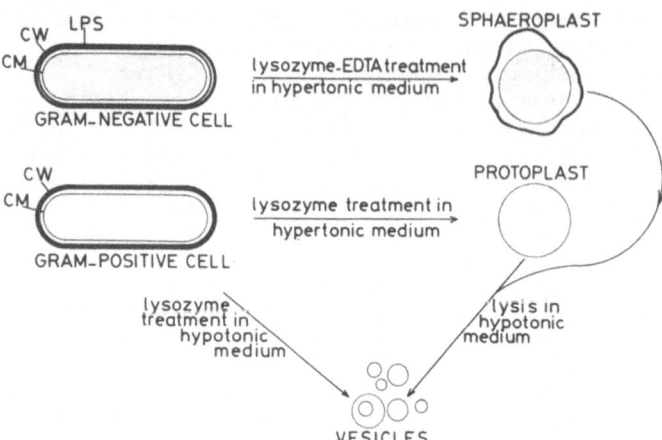

Fig. 2. Scheme of the isolation procedure of membrane vesicles from Gram-positive bacteria and Gram-negative bacteria. CW, cellwall (peptidoglycan layer); CM, cytoplasmic membrane; OM, outer membrane.

5% of the nucleic acids and less than 1-2% of cytoplasmic enzymes are found in these vesicles. Several membrane-associated enzymes and integrated membrane functions, such as phospholipid synthesis, electron transfer, ATP hydrolysis and synthesis and active transport are, however, maintained in these membrane vesicles (3,5,6).

Membrane vesicles are excellent modelsystems for studies of energy transducing processes, if a Δp of the desired magnitude and polarity can be generated. Major Δp-generating mechanisms in Gram-positive bacteria are (i) the oxidation of electrondonors via electron transfer chains (1), or (ii) ATP hydrolysis via the membrane-bound ATPase (7). Transport studies with membrane vesicles derived from different bacteria revealed that solutes are mainly transported via secondary transportsystems (1,3,5,7).

Many bacterial membranes, such as those of fermentative bacteria contain only the proton translocating ATPase as primary proton pump. Since these membrane vesicles have the same orientation as the cytoplasmic membrane of intact cells, the catalytic site of the ATPase is located at the inner surface of the membrane. ATP is membrane-impermeable and the addition of ATP to the incubation mixture cannot result in the generation of a Δp. Several procedures are available to generate artificially a Δp or one of its components. An electrical potential (inside negative) can be generated by a valinomycin-mediated outwardly-directed potassium movement (8) or by the addition of membrane-permeable anions, such as thiocyanate, chlorate and tetraphenylboron (9). A pH gradient (inside alkaline) can efficiently be generated by the imposition of an outwardly directed acetate diffusion gradient. This method makes use of the high membrane permeability of the protonated form of acetate, acetic acid (10). Efflux of acetic acid results in an alkalization of the intravesicular space.

Although a Δψ or ΔpH of considerable magnitude can be generated with these procedures, their transient character severely limits the application of these procedures for studies on solute transport. To overcome these experimental difficulties a procedure has been developed for the incorporation of a powerful Δp-generating system into bacterial membranes. This has been achieved by fusion of membrane vesicles with liposomes in which these Δp-generating systems have been incorporated (Fig. 3) (11-13). Such a Δp-generating system should be: (i) obtainable in large quantities; (ii) easy to isolate and purify; (iii) highly active with a high turnover rate; (iv) simple to reconstitute into liposomes, and (v) capable of generating a Δp of considerable magnitude and defined polarity. The light-induced proton pump Bacteriorhodopsin and the redox-linked proton pump cytochrome c oxidase meet these requirements. Bacteriorhodopsin can be isolated from cytoplasmic membranes of Halobacterium halobium in which it is found in patches (purple membranes) organized in a two-dimensional crystalline lattice (14). Reconstitution is accomplished by co-sonication of these purple membranes and phospholipids. This procedure results in the formation of Bacteriorhodopsin proteoliposomes in which an everted Δp, Δψ, inside positive, and ΔpH, inside acid, is generated upon illumination (11).

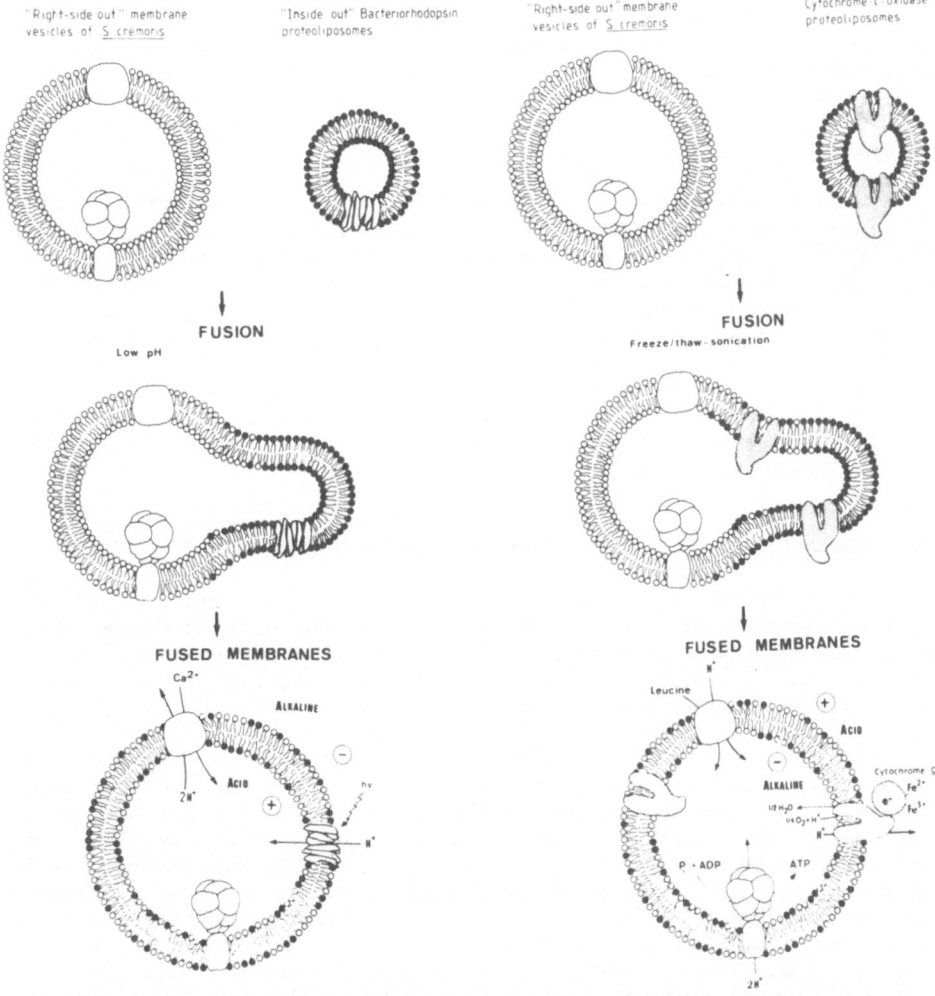

Fig. 3. Scheme of the insertion of proton motive force generating systems into bacterial membrane vesicles by membrane fusion. A. Fusion between bacteriorhodopsin proteoliposomes and bacterial membranes. B. Fusion between cytochrome c oxidase proteoliposomes and bacterial membranes.

Cytochrome $\underline{c}$ oxidase can be isolated in large quantities from beef-heart mitochondria. It can be purified by successive ammonium sulphate precipitation (15). Reinsertion into liposomes is accomplished by detergent dialysis (16). Although cytochrome $\underline{c}$ oxidase is slightly asymmetrically reconstituted into the liposomes, the use of the membrane impermeable electron donor cytochrome $\underline{c}$ always assures the generation of a Δp, $\Delta\psi$, inside negative, and $\overline{\Delta pH}$, inside alkaline. Only those cytochrome c oxidase molecules with their cytochrome c binding site located at the outer surface are reduced by cytochrome $\underline{c}$. Both Δp-generating systems have been incorporated into membrane vesicles of the fermentative lactic acid bacterium Streptococcus cremoris by membrane fusion (11-13).

Fusion between bacterial membranes and proteoliposomes can be induced by a freeze/thaw step (12) or by low pH incubation (11). Following membrane fusion, the membranes are sonicated for a short period of time to decrease the stability and to decrease the passive ion-permeability of the membranes. Fusion of cytochrome $\underline{c}$ oxidase proteoliposomes with membrane vesicles of S. cremoris by freeze/thaw-sonication results in hybrid membrane vesicles in which a high Δp, $\Delta\psi$, inside negative, and ΔpH, inside alkaline, is generated by the addition of the electrondonor system ascorbate/N,N,N',N'-tetramethyl-p-phenylane-diamine (TMPD)/cytochrome $\underline{c}$. This Δp can drive secondary transport of several amino acids (12,13).

Low pH induced fusion of Bacteriorhodopsin proteoliposomes with S. cremoris membrane vesicles results in hybrid membranes in which a Δp, $\Delta\psi$, inside positive, and ΔpH, inside acid, is generated upon illumination. This polarity of the Δp makes this system only suitable for studies of extrusion systems. In these membranes, upon illumination the uptake of calcium via a Ca^{2+}/H^+ exchange mechanism was observed (11).

Since the incorporated systems can generate a constant Δp for a long period of time, studies on the relation between the Δp and the steady state accumulation of amino acid solutes by secondary transport systems can be performed.

THE REGULATION OF SECONDARY SOLUTE TRANSPORT BY ELECTRON TRANSFER

The mechanism of energy coupling to secondary transport of several solutes has been studied extensively in bacteria such as Escherichia coli, Rhodopseudomonas sphaeroides and Streptococcus cremoris. These studies have demonstrated that rather complex interactions can exist between the different energy transducing systems in bacteria.

For instance, in the phototrophic bacterium Rhodopseudomonas sphaeroides evidence has been presented that the existence of a proton motive force alone is not sufficient for solute uptake and that also turnover of the cyclic electron transfer chain is necessary (17). The initial rate of uptake of the amino acid alanine has been measured at a constant light intensity (= constant rate of cyclic electron transfer) but varying magnitudes of the proton motive force. Under these conditions the rate of transport increased exponentially with the proton motive force. However, when the proton motive force was kept constant and the rate of electron transfer was varied, the rate of transport increased linearly with the light intensity (Fig. 4). At low light intensities there was no uptake of alanine, even when the proton motive force was high. These results demonstrate that the electron transfer chain functions not only as a generator of a proton motive force but also influences directly the activity of the carrier.

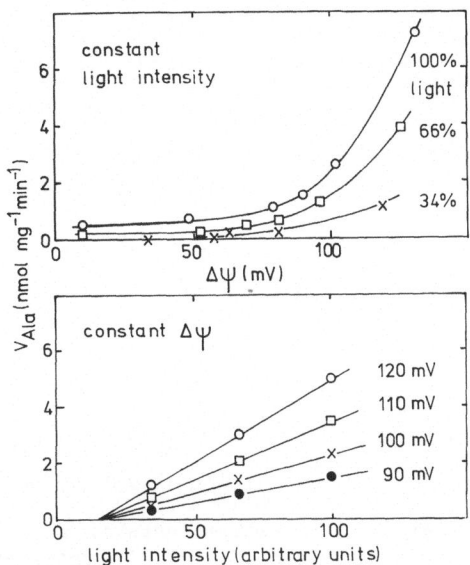

Fig. 4. The relation between V_{ala} and $\Delta\psi$ at constant light intensity and between V_{ala} and light intensity at constant $\Delta\psi$ in cells of Rps. sphaeroides.
Upper pannel: the alanine uptake rate (V_{ala}) as a function of $\Delta\psi$ at constant light intensity. Lower pannel: V_{ala} as a function of light intensity at constant $\Delta\psi$ values. Data from the upper pannel were replotted. Taken from Elferink et al. (1983a), with permission.

Such an interaction between the electron transfer system and solute transport carriers is not specific for Rps. sphaeroides. In a strain of Rps. sphaeroides in which the E. coli transport protein for lactose (the M-protein) was incorporated via genetic transformation a similar relation between the rate of cyclic electron transfer and lactose transport was observed (18). Kinetic analysis of the changes in the initial rate of both alanine and lactose uptake indicated that the regulation is due to a light dependent change of the activity of the carrier molecules in the membrane (18,19).

A very attractive experimental system for investigating the role of electron transfer and the proton motive force in solute transport was found in membrane vesicles of E. coli which contain PQQ-dependent glucose dehydrogenase. Glucose dehydrogenase is found in E. coli as the apoenzyme which can be converted into the active holoenzyme by the addition of its prosthetic group pyrollo-quinolinequinone (PQQ) (20). This system is ideal for transport studies, because the enzyme is coupled to the respiratory chain and its activity can be increased by adding increasing amounts of PQQ to E. coli membrane vesicles. Therefore, the rate of electron flow and the redox state of the components of the electron transfer chain can be varied by adding varying amounts of PQQ.

In Fig. 5 the result of such an experiment is shown. In the absence of a ΔpH (due to the action of nigericin) the rate of lactose uptake decreased at constant $\Delta\psi$ value, with the respiration rate, while $\Delta\mu_{lactose}$ remained constant. At lower respiration rates (lower PQQ concentrations) both the $\Delta\psi$ and $\Delta\mu_{lactose}$ decreased. This decrease of the uptake rate was steeper since also the driving force ($\Delta\psi$) decreased at lower rates of electron transfer (20).

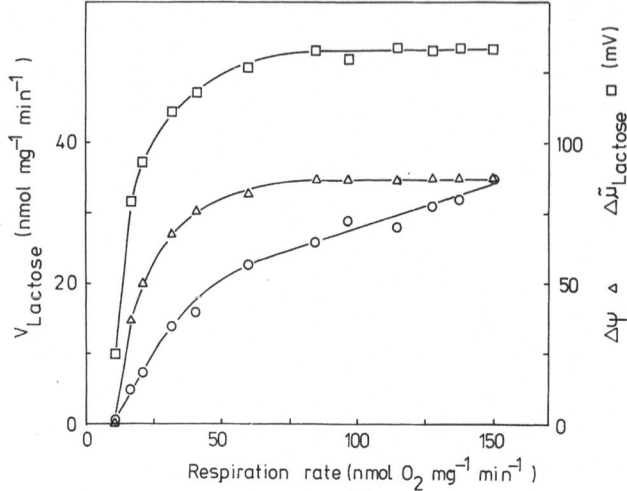

Fig. 5. The initial rate of lactose uptake, steady state accumulation level and $\Delta\psi$ as a function of the respiration rate in E. coli membrane vesicles energized by PQQ-dependent glucose oxidation. The membrane vesicles were pretreated with 1 µM nigericin to collaps the ΔpH component of the Δp.
Δ, electrical potential; o, initial rate of lactose uptake; $\square$, steady state level of lactose accumulation. From J.M. van Dijl, M.G.L. Elferink and K.J. Hellingwerf, unpublished results.

REGULATION OF SECONDARY SOLUTE TRANSPORT BY DITHIOL-DISULFIDE INTER-CHANGE

A large number of reports has appeared on the involvement of sulf-hydryl groups in the function of membrane proteins such as carriers, energy transducing enzymes and receptor proteins. Dithiol-disulfide interconversions have been reported to play an essential role in many solute transport and energy transducing systems in bacteria, mitochon-dria and chloroplasts. Moreover, many reports have demonstrated that the sensitivity of various transport systems to SH reagents is changed either by addition of substrates or by energization of the membrane or by both (for review see 21). Such sulfhydryl reagent sensitivity has also been reported for the phosphoenolpyruvate dependent glucose trans-port system (22) and the lactose transport system of Escherichia coli (23). These two transport systems operate by completely different mechanisms. The lactose transport system is a secondary transport system which translocates lactose in symport with protons. The PEP-dependent glucose transport system (glucose PTS) couples the transport of glucose to the hydrolysis of PEP via a series of phosphoryl-group transfer reactions (24). Evidence has been presented for the involve-ment of redox sensitive dithiol groups in the membrane bound enzyme II of the glucose PTS and that the redox state of these groups control the activity of the enzyme (25).

The activity of the lactose carrier has also been shown to be subject to redox control of dithiol-disulfide groups (26) as is indi-cated by the following experiments. The accumulation of L-proline and lactose in right-side out membrane vesicles of E. coli can be energized by ascorbate-phenazine methosulfate, D-lactate or an artificially imposed chlorate or potassium diffusion potential. This uptake could be

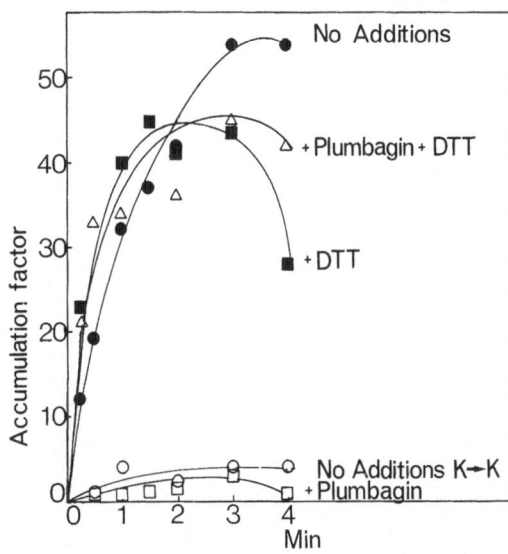

Fig. 6. Effects of 0.5 mM plumbagin and 10 mM dithiothreitol on uptake of L-proline by membrane vesicles of E. coli energized by a valinomycin induced potassium diffusion potential. Experiments were performed at room temperature. No additions: with a potassium gradient; plumbagin + DTT: with a potassium gradient, incubated with 0.5 mM plumbagin and then with 10 mM dithiothreitol; DTT, with a potassium gradient, incubated with 10 mM dithiothreitol; No additions K ⟶ K: no potassium gradient; + plumbagin: with a potassium gradient, incubated with 0.5 mM plumbagin. From W.N. Konings and G.T. Robillard (1982) with permission.

inhibited to the uptake levels observed under non-energized conditions by lipophilic oxidants like plumbagin (0.5 mM), phenazine methosulfate (PMS) (5 mM), menadione (0.6 mM) and 1,2-naphthoquinone (0.3 mM) (26). Strong inhibitions of uptake were also observed with polar oxidants like ferricyanide (10 mM) provided that the membrane vesicles were pre-incubated for several minutes with these oxidants (26,27). The inhibitions exerted by these oxidants could not be explained by a decreased flow of electrons to oxygen and/or a decreased proton motive force. In all cases the addition of dithiothreitol (DTT) to the membrane vesicle suspension restored fully the transport activity (Fig. 6). These data demonstrate that the activity of the carriers for L-proline and lactose of E. coli can be regulated by altering the redox potential in the membrane in a similar way as was demonstrated for the glucose-PTS. Kinetic studies of L-proline transport energized by a chlorate diffusion potential in the presence of different concentrations of plumbagin demonstrate that oxidation of the carriers leads to a decrease of the V_{max} and thus to inactive carriers (Fig. 7). A similar conclusion has been drawn from studies on the binding of galactosides to the lac carrier of E. coli in the presence of plumbagin (28).

The redox process involves the conversion of dithiols to disulfide or vice versa in the carrier proteins. In the proline carrier two sets of dithiol/disulfides appear to play a role: one set located at the outer surface, the other at the inner surface of the cytoplasmic membrane (27). This conclusion is based on the following observations. In right-side out membrane vesicles electron transfer in the respiratory chain leads to the generation of a proton motive force, inside negative

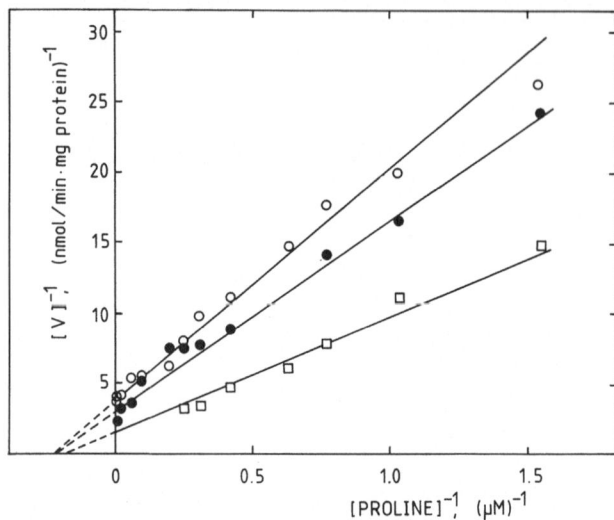

Fig. 7. The effect of plumbagin on the kinetics of proline transport in membrane vesicles of <u>E. coli</u> ML 308-225. Membrane vesicles were incubated in the presence of 50 μM (●), 100 μM (o) plumbagin or without further additions (▢) for 30 min on ice prior to the uptake assay. Proline transport was driven by a chlorate-diffusion potential at pH 8.0 as described in ref. 27.

and alkaline and to the conversion at the outer surface of a disulfide to a dithiol as is shown by the increased inhibition of proline transport by the membrane impermeable dithiol reagent thorin. The inhibition exerted by thorin was completely reversed by dithiolthreitol. A similar but irreversible inhibition was observed with the membrane impermeable SH-reagent glutathione hexane maleimide. Pretreatment of the membrane vesicles with ferricyanide or with thorin protected against glutathione hexane maleimide inhibition since the transport activity of L-proline was fully restored after DTT treatment. Both SH-reagents therefore appear to react with the same SH-groups of a dithiol located at the outer surface.

Evidence for the involvement of dithiols located at the inner surface of the cytoplasmic membrane was obtained from studies with inside-out membrane vesicles. These inside-out membrane vesicles accumulate L-proline if a chlorate diffusion potential (inside negative) is imposed. Ferricyanide reversibly inhibits this proline uptake, which appears to be the result of oxidation of SH-groups since ferricyanide also protects against the irreversible inhibition by glutathione hexane maleimide. A similar protection against glutathione hexane maleimide inactivation in these inside-out membrane vesicles could be achieved by imposition of a proton motive force by electron flow, inside positive and acid, indicating that a dithiol is converted to a disulfide upon energization. These results strongly indicate that two redox sensitive dithiol groups play a role at least in the carrier of proline.

This redox control of the transport proteins combined with the control exerted by electron transfer suggest the existence of a redox interaction of the transport proteins with components of the electron transfer system. Such an interaction could possibly operate as is schematically shown in Fig. 8. According to this scheme electrons can be transferred from an electron transfer intermediate to a redox centre located at the outer surface thereby reducing this redox centre. This

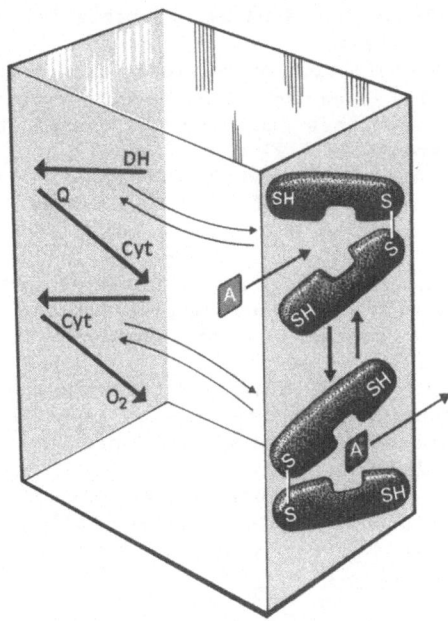

Fig. 8. Hypothetical scheme of the redox interaction between the respiratory chain and a carrier protein as described in text.

reduction causes an opening of the carrier and an exposure of the solute binding site to the outer surface. When solute and proton(s) are bound electron transfer from the outer redox center to the inner redox center occurs which leads to an oxidation of the center at the outer surface and a reduction of the center at the inner surface. This results in a closing of the carrier at the outer surface and an exposure of the binding site to the inner surface. When solute and proton(s) are released in the cytoplasm electrons are transferred from the inner redox center to intermediate(s) of the electron transfer system with redox potential(s) higher than the first intermediate. In solute transport systems which are not regulated by electron transfer such redox transitions require intramolecular electron transfer between the redox centers only (see also 29). The redox sensitivity of the carrier proteins suggests that in intact cells a redox control coupled to glutathion or thioredoxine is very likely to occur.

REFERENCES

1. W.N. Konings, and P.A.M. Michels, Electron transfer-driven solute translocation across bacterial membranes, in: Diversity of Bacterial Respiratory Systems", C.J. Knowles, ed., CRC Press Inc., Boca Raton (1980).
2. L.S. Lolkema, K.J. Hellingwerf, and W.N. Konings, The effect of "probe-binding" on the quantitative determination of the proton motive force in bacteria, Biochim.Biophys.Acta 681:85 (1982).
3. H.R. Kaback, Membrane vesicles, electrochemical ion-gradients, and active transport, Curr.Top.Membr.Trans. 16:393 (1982).
4. W.N. Konings, A. Bisschop, M. Veenhuis, and C.A. Vermeulen, New procedure for the isolation of membrane vesicles of Bacillus subtilis and an electron microscopy study of their ultrastucture, J.Bacteriol. 116:1456 (1973).

5. H.R. Kaback, Transport studies in bacterial membrane vesicles, Science, 186:882 (1974).

6. W.N. Konings, Energization of solute transport in membrane vesicles from anaerobically grown bacteria, Meth.Enzymol. 56:378 (1979).

7. F.M. Harold, Membranes and energy transduction in bacteria, Curr.Top.Bioenerg. 6:83 (1977).

8. H. Hirata, K. Altendorf, and F.M. Harold, Role of an electrical potential in the coupling of metabolic energy to active transport by membrane vesicles of Escherichia coli, Proc.Natl.Acad.Sci. USA 70:1804 (1973).

9. B. Poolman, W.N. Konings, and G.T. Robillard, The location of redox-sensitive groups in the carrier protein of proline at the outer and inner surface of the membrane in Escherichia coli, Eur.J.Biochem. 135:41 (1983).

10. J.R. Lancaster, and P.C. Hinkle, Studies on the galactoside transporter in inverted membrane vesicles of Escherichia coli. I. Symmetrical facilitated diffusion and proton coupled transport. J.Biol.Chem. 252:7657 (1977).

11. A.J.M. Driessen, K.J. Hellingwerf, and W.N. Konings, Light-induced generation of a proton motive force and Ca^{2+}-transport in membrane vesicles of Streptococcus cremoris fused with bacteriorhodopsin proteoliposomes, Biochim.Biophys.Acta 808:1 (1985).

12. A.J.M. Driessen, W. de Vrij, and W.N. Konings, Incorporation of beef heart cytochrome c oxidase as a proton motive force generating mechanism in bacterial membrane vesicles, Proc.Natl.Acad.Sci. USA 82:7555 (1985).

13. A.J.M. Driessen, W. de Vrij, and W.N. Konings, Functional incorporation of beef-heart cytochrome c oxidase into membrane vesicles of Streptococcus cremoris, Eur.J.Biochem. 154:617 (1986).

14. W. Stoeckenius, The purple membrane of salt-loving bacteria, Sci.Am. 243:38 (1976).

15. C.A. Yu, L. Yu, and T.E. King, Studies on cytochrome c oxidase. J.Biol.Chem. 250:1383 (1975).

16. P.C. Hinkle, J.J. Kim, and E. Racker, Ion transport and respiratory control in vesicles formed from cytochrome c oxidase and phospholipids, J.Biol.Chem. 247:1338 (1972).

17. M.G.L. Elferink, I. Friedberg, K.J. Hellingwerf, and W.N. Konings, The role of the proton motive force and electron flow in light-driven solute transport in Rhodopseudomonas sphaeroides, Eur.J.Biochem. 129:583 (1983).

18. M.G.L. Elferink, K.J. Hellingwerf, F.E. Nano, S. Kaplan, and W.N. Konings, The lactose carrier of Escherichia coli functionally incorporated in Rhodopseudomonas sphaeroides obeys the regulatory conditions of the phototrophic bacterium, FEBS Lett. 164:198 (1983).

19. M.G.L. Elferink, The interaction between electron transfer, proton motive force and solute transport in bacteria, Ph.D. Thesis. University of Groningen (1985).

20. B.J. van Schie, K.J. Hellingwerf, J.P. van Dijken, M.G.L. Elferink, J.M. van Dijl, J.G. Kuenen, and W.N. Konings, Energy transduction by electron transfer via a pyrollo-quinoline quinone-dependent glucose dehydrogenase in Escherichia coli, Pseudomonas aeruginosa, and Acinetobacter calcoaceticus (var. lwoffi), J.Bacteriol. 163:493 (1985).

21. A. Fonyo, SH-group reagents as tools in the study of mitochondrial anion transport, J.Bioenerg.Biomembr. 10:171 (1978).

22. R. Haguenauer-Tsapis, and A. Kepes. Unmasking of an essential thiol during function of the membrane band enzyme II of the phosphoenolpyruvate glucose phosphotransferase system of Escherichia coli, Biochim.Biophys.Acta 465:118 (1977).

23. D.E. Cohn, G.J. Kaczorowski, and H.R. Kaback, Effect of the proton electrochemical gradient on maleimide inactivation of active transport in Escherichia coli membrane vesicles, Biochemistry 20:3308 (1981).

24. P.W. Postma, and S. Roseman. The bacterial phosphoenol pyruvate: sugar phosphotransferase system, Biochim.Biophys.Acta 457:213 (1976).

25. G.T. Robillard, and W.N. Konings. Physical mechanism for regulation of phosphoenol pyruvate-dependent glucose transport activity in Escherichia coli, Biochemistry 20:5025 (1981).

26. W.N. Konings, and G.T. Robillard. The physical mechanism for regulation of protein solute transport in Escherichia coli, Proc.Natl. Acad.Sci. USA 79:5480 (1982).

27. B. Poolman, W.N. Konings, and G.T. Robillard. The location of redox-sensitive groups in the carrier protein of proline at the outer and inner surface of the membrane in Escherichia coli, Eur.J.Biochem. 135:41 (1983).

28. J.M. Neuhaus, and J.K. Wright. Chemical modification of the lactose carrier of Escherichia coli by plumbagin, phenylarsine oxide or diethylpyrocarbonate affects the binding of galactoside, Eur.J.Biochem. 127:597 (1983).

29. G.T. Robillard, and W.N. Konings, The role of dithiol-disulphide interchange in solute transport and energy transduction processes, Eur.J.Biochem. 127:597 (1982).

REGULATION OF FREE INTRACELLULAR Ca^{2+} IN RESTING AND CON A-STIMULATED MOUSE SPLENOCYTES

K.V.S. Prasad[+], M.S. Peppler[*] and J.G. Kaplan[+]

+Department of Biochemistry
*Department of Medical Microbiology
University of Alberta
Edmonton, Alberta, Canada, T6G 2H7

ABSTRACT

Free intracellular Ca^{2+} concentration $f[Ca^{2+}]_i$ was measured using quin 2-loaded (Tsien et al., 1982a,b) mouse splenocytes. Resting cells had a $f[Ca^{2+}]_i$ of 145.4 ± 48.9 nM. Addition of (5 µg/ml) concanavalin A (Con A), a polyclonal mitogen, to splenocyte culture doubled the $f[Ca^{2+}]_i$ in about 5 min. Deprivation of external Ca^{2+} completely abolished this effect. In cells suspended in Na^+-free buffer in absence of mitogen, a steady increase in $f[Ca^{2+}]_i$ was observed suggesting the presence of a Na^+/Ca^{2+} antiport system at the plasma membrane: the $f[Ca^{2+}]_i$ doubled in Na^+-free buffer in about 10 min. Addition of Con A to these cells resulted in a further two-fold increase in $f[Ca^{2+}]_i$. Addition of monensin (20 µM), a Na^+ ionophore, also increased the basal $f[Ca^{2+}]_i$ without affecting the Con A-stimulated increase. However ouabain (0.5 mM) had no effect on the basal Ca^{2+} levels during the short period in which it was tested. These results indicate that under physiological conditions, Na^+/Ca^{2+} antiport contributes to Ca^{2+} efflux and Con A stimulates Ca^{2+} influx thereby raising the $f[Ca^{2+}]_i$. Cells treated with pertussis toxin (50-450 ng/ml) did not show the usual increase in $f[Ca^{2+}]_i$ upon addition of Con A, suggesting the involvement of G_i or a similar GTP-binding protein, in Ca^{2+} mobilization.

INTRODUCTION

Lymphocytes stimulated by mitogenic lectins like Con A are a useful model for the investigation of the signals leading to proliferation (Chaly et al., 1986). Increased Na^+, K^+-ATPase-mediated K^+ influx (Quastel and Kaplan, 1970), Na^+ influx (Segel et al., 1978; Prasad et al., 1986) and Ca^{2+} influx (Whitney and Sutherland, 1972) are some of the important early events that take place soon after the addition of mitogenic lectins to lymphocytes. Bard et al., (1978) have shown that mitogenic activation of DNA synthesis in lymphocytes at 48 h requires the presence of extracellular Ca^{2+} for the first 20 h period. Recent advances in the measurement of intracellular free calcium using fluorescent calcium chelators like quin 2 (Tsien et al., 1982) have permitted attempts to elucidate the role of Ca^{2+} as a second messenger. Fertilization of sea urchin eggs (Whitaker and Steinhart , 1982), stimulation of lymphocytes by mitogenic lectins (Tsien et al., 1982a,b; Hesketh el al., 1983) or by antiimmunoglobulin antibody (Tax et al., 1983) caused an increase in $f[Ca^{2+}]_i$ within minutes. Attempts are being made to evaluate the role of Ca^{2+} in the overall signal transduction pathway leading to cell growth and proliferation (Hesketh et al., 1983). Cells are usually exposed to an external Ca^{2+} concentration of 0.5 to 2.0 mM and the $f[Ca^{2+}]_i$ is around 100 nM; in spite of the enormous gradient across the cell membrane, healthy cells regulate their $f[Ca^{2+}]_i$ within narrow limits. It was thus of interest to evaluate some of the components involved in the regulation of $f[Ca^{2+}]_i$ both in resting and Con A-stimulated mouse splenocytes.

MARERIALS AND METHODS

Thymide (methyl-^{3}H) was obtained from New England Nuclear. Ouabain, monensin, Triton X100, Con A, EGTA were purchased from Sigma. Tissue culture media and reagents were obtained from GIBCO and trypan blue from MCB Reagents Ltd. Quin 2 and quin 2/AM were purchased from Calbiochem and Chelex 100 from Biorad. Pertussis toxin was isolated from Bordetella pertussis according to previously published methods (Sekura et al., 1983), and biological activity tested by goose erythrocyte agglutination and clumping of Chinese hamster ovary cells (Hewlett et al., 1983).

Preparation of Mouse Splenocytes

Balb/c, male mice, 8-12 weeks old, were killed by cervical dislocation and their spleens were disrupted through a wire mesh into a few drops of fetal calf serum (FCS). The cells were then transferred to a sterile tube along with the washings (for which RPMI 1640 was used); they were then spun at 500 x g for 5 min. The pellet was initially suspended in a few drops of FCS and then diluted to 12 ml with ice cold 0.83% NH_4Cl and kept at room temperature for not more than 5 min. This treatment ensured the lysis of red blood cells. After carefully layering 1 ml of fetal calf serum (FCS) at the bottom of the tube, the cells were again spun at 500 x g for 5 min. and washed twice with RPMI 1640 to remove traces of NH_4Cl. Cells were cultured in RPMI 1640 medium with FCS (10%), glutamine (2mM), penicillin (40 IU/ml) and streptomycin (40 g/ml) at a density of 2×10^6 cells per ml. Viability was >95% as assessed by trypan blue exclusion. Experiments were carried out later the same day, after allowing the cells to sit at 37 C in 5% CO_2 for about 2 h. H^3-TdR uptake was measured at 48 h after stimulation with Con A, as described (Prasad et al., 1986)

Measurement of $f[Ca^{2+}]_i$ was exactly according to the procedure of Tsien et al., 1982a,b. Na^+-free buffer contained an equimolar amount of choline chloride in place of NaCl and Na_2HPO_4 was replaced by K_2HPO_4, maintaining the overall $[K^+]$ at 5 mM. Ca^{2+}-free buffer was prepared by omitting $CaCl_2$. Glass distilled water treated with Chelex 100 was used for the preparation of all buffers. In experiments involving the

use of pertussis toxin, cells (2×10^6 ml) were treated with
the toxin for 2 h at 37 C and 5% CO_2, centrifuged at 500 x g
to remove the free toxin and then loaded with quin 2/AM.

RESULTS

   Fig. 1 despicts a typical fluorescence recording of
Ca^{2+}-related changes in quin 2-loaded mouse splenocytes.
Addition of $CaCl_2$ (1mM) to quin 2-loaded cells suspended in
Ca^{2+}-free buffer increased the fluorescence to a higher plateau. Further increase was obtained by the addition of Con A.
A level of 145.4 ± 48.9 nM $f[Ca^{2+}]_i$ in resting cells was
elevated to 267.7 ± 54.5 nM by Con A in about 5 min. However,
this increase was completely abolished in cells suspended in
Ca^{2+}-free buffer, indicating that Con A-mediated increase is
mainly through the mobilization of external Ca^{2+}.
   In the absence of external Na^+, a slow increase in
$f[Ca^{2+}]_i$ was observed (Fig. 2); the basal level of $f[Ca^{2+}]_i$
doubled in about 10 min (Table 1). This suggests the presence
of a Na^+/Ca^{2+} antiport system in the plasma membrane. Monensin, a Na^+ ionophore, and ouabain, an inhibitor of Na^+, K^+-
ATPase, are both known to increase the intracellular Na^+ in
lymphocytes (Prasad et al., 1986). The addition of these
compounds to resting cells suspended in the control buffer

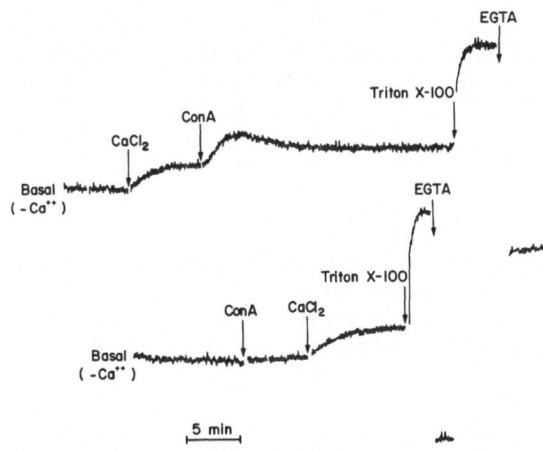

Fig. 1. Effect of Con A on $f[Ca^{2+}]_i$ in mouse
        splenocytes.

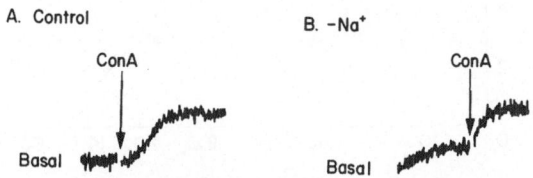

Fig. 2. Effect of $[Na^+]_e$ on $f[Ca^{2+}]_i$ in resting
        and Con A-stimulated mouse splenocytes

should considerably decrease or abolish the Na^+ gradient across the membrane and this should in turn decrease the activity of Na^+/Ca^{2+} antiport. As expected, addition of monensin (20 μM) to resting cells caused a significant increase in $f[Ca^{2+}]_i$ (Table 1). However ouabain (0.5 mM) did not show any effect during the short time (10 min) of exposure. Decrease in the external concentration of Na^+ was accompanied by an increase in the rate of change in the $f[Ca^{2+}]_i$, which reached a maximum of 9.0 nM/min at zero external Na^+ and a minimum (no change) at 146 mM Na^+ (Fig.3). These data (Table 1) clearly demonstrate the presence of Na^+/Ca^{2+} antiport which contributes to the efflux of Ca^{2+}. The absence of external Na^+ or the presence of monensin and ouabain, had no effect on the Con A-stimulated increase in $f[Ca^{2+}]$ (Table 1 and Fig. 2).

Table 1. Effects of various treatments on $f[Ca^{2+}]_i$ mouse splenocytes

	Control	$-Na^+$ ext.
	(nM)	
Resting	145.4 ± 48.9	312.2 ± 68.9[+]
Con A (5 μg/ml)	267.7 ± 54.5[*]	642.8 ± 164.4[+]
Ouabain (0.5 mM)	150.4 ± 11.8	278.2 ± 78.1
Ouabain + Con A	387.3 ± 45.6[*]	632.0 ± 149.9
Monensin (20 μM)	237.6 ± 66.4[*]	430.2 ± 80.8
Monensin + Con A	385.7 ± 84.7[*]	812.7 ± 130.5

Each value represents mean ± S.D. drawn from at least four different experiments. Paired 't' test was used for statistical analysis.
[+]These increases are statistically significant at P < 0.05 as compared to their resting cell values.
[*]These increases as compared to controls by 't' test are statistically significant at P < 0.05.

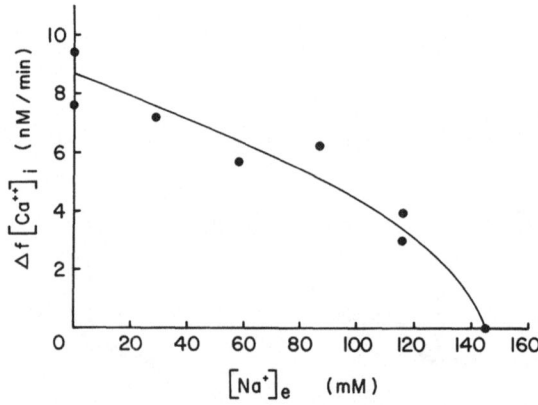

Fig. 3. Effect of $[Na^+]_e$ on the rate of increase in $f[Ca^{2+}]_i$ in mouse splenocytes.

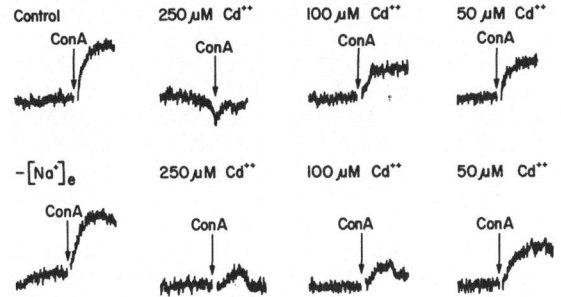

Fig. 4. Effect of Cd^{2+} on $f[Ca^{2+}]_i$ in mouse splenocytes suspended in control and Na^+-free buffer.

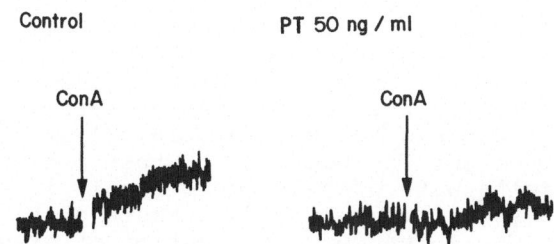

Fig. 5. Effect of pertussis toxin (PT) treatment on $f[Ca^{2+}]_i$ in Con A-stimulated mouse splenocytes.

Addition of $CdCl_2$ (250 uM) to resting mouse splenocytes suspended in control or Na^+-free buffer prevented the Con A-stimulated increase in $f[Ca^{2+}]_i$ (Fig. 4). The slow rise in $f[Ca^{2+}]_i$ observed in cells suspended in Na^+-free buffer was decreased, indicating competition between Cd^{2+} and Ca^{2+}. Resting lymphocytes treated with pertussis toxin (50 nM) for 2 h failed to respond to Con A in terms of Ca^{2+} mobilization (Fig. 5) and a substantial decrease in ^{3}H-TdR incorporation at 48 h was noted (data not shown).

DISCUSSION

The basal $f[Ca^{2+}]_i$ levels as well as the 1.8-fold increase induced by Con A are comparable to those reported in lymphoid cells by others (Tsien et al., 1982a,b: Hesketh et al., 1985). The latter, working on mouse thymocytes, reported a slight, transient, increase in $f[Ca^{2+}]_i$ upon addition of Con A to mouse thymocytes suspended in virtually Ca^{2+}-free buffer (<1 μM Ca^{2+}) obtained by the addition of EGTA (1mM). However, in the absence of external Ca^{2+}, we observed no change in the level of intracellular Ca^{2+} upon addition of Con A. In some of the experiments, the measurements were done within 2 min of transfer of cells from control to Ca^{2+}-free buffer. Thus, in the case of mouse splenocytes, it appears that the intracellular Ca^{2+} stores are not accessible for mobilization by Con A and most of the increase by Con A can be attributed to mobilization of extracellular Ca^{2+}.

Apart from an active Ca^{2+}pumping system (Averdunk and Gunther, 1980; Lichtman et al., 1981; Sarkadi et al., 1982), a Na^+/Ca^{2+} exchange activity has also been demonstrated in rabbit thymocyte plasma membrane vesicles (Ueda, 1983). A number of experimental observations have shown the presence of an active Na^+/Ca^{2+} antiport system in the plasma membrane of whole mouse splenocytes. For example, Ca^{2+} influx was sensitive to external Na^+ concentration. Furthermore, the fact that $f[Ca^{2+}]_i$ rose when external $[Na^+]$ was decreased, indicates that Na^+/Ca^{2+} antiport contributes to the steady-state maintenance of $f[Ca^{2+}]_i$ under normal conditions. We observed that monensin (20 uM) caused much greater increase in $f[Ca^{2+}]_i$ in cells suspended in Na^+-free buffer compared to those in control buffer. However, while monensin mainly translocates Na^+ (which depends on the Na^+ gradient across the plasma membrane), it is quite possible that in the absence of external Na^+ it directly facilitates the entry of Ca^{2+}, thereby increasing the $f[Ca^{2+}]_i$. Ouabain, a cardiac glycoside, is known to increase intracellular Na^+ through its inhibition of Na^+,K^+-ATPase (Prasad et al., 1986). This should result in a decrease in Na^+/Ca^{2+} exchange due to decrease in the Na^+ gradient across the cell membrane and thereby cause a rise in $f[Ca^{2+}]_i$. However, for the short duration during which mouse splenocytes were exposed to ouabain (0.5 mM), no change in $f[Ca^{2+}]_i$ was observed. The possibility that prolonged incubation of the cells with ouabain would result in an increase in $f[Ca^{2+}]_i$ has yet to be verified. Hesketh et al., (1985) speculated that Con A-induced mobilization of external Ca^{2+} could be due to an inhibitory action of the mitogen on Ca^{2+} efflux. However, this explanation is not valid in our system. We have shown that, in mouse splenocytes, Con A caused as great an increase in $f[Ca^{2+}]_i$ in cells suspended in Na^+-free buffer, where the Na^+/Ca^{2+} exchange that is the major contributor to Ca^{2+} efflux is non-functional, as in controls in which this system was active. Thus, a putative inhibition of Ca^{2+} efflux by Con A could scarcely have caused such an increase in $f[Ca^{2+}]_i$ under conditions where Ca^{2+} efflux was already highly inhibited by inactivity of the Na^+/Ca^{2+} antiport.

Cd^{2+} was observed to quench the fluorescence of the quin 2-Ca^{2+} complex in vitro. However, the fluorescence level in quin 2 loaded intact cells exposed to Cd^{2+} was not altered. Our data indicate that Cd^{2+} had a much greater effect on Con A -stimulated increase in $f[Ca^{2+}]_i$ than on the $f[Ca^{2+}]_i$ of resting cells. This could then mean that the plasma membrane Ca^{2+} channels affected by Con A are highly accessible to competition by Cd^{2+} ions.

Guanine nucleotide binding proteins couple hormonal and neurotransmitter receptors to the stimulation and inhibition of adenylate cyclase; pertussis toxin monoADP ribosylates the subunit of G_i, thereby inhibiting its function (for references see reviews by Gilman, 1984; Spiegel et al., 1985).In neutrophils, pertussis toxin inhibited f-Met-Leu-Phe-mediated responses, such as increased PIP breakdown (Verghese et al., 1985), rise in intracellular Ca^{2+} (Verghese et al., 1985; Goldman et al., 1984; Molski et al., 1984), superoxide production (Verghese et al., 1985), lysosomal secretion (Okajima and Ui, 1984; Bokoch and Gilman, 1984) and chemotaxis (Verghese et al.,1985). Recently Gierschik et al. (1986)

have demonstrated the presence of an immunochemically distinct G protein, (compared to previously identified substrates like transducin, G_i and G_o) which is monoADP ribosylated by pertussis toxin. Both f-Met-Leu-Phe and Con A can induce an increase in $f[Ca^{2+}]_i$ in neutrophils; however in that system, pertussis toxin treatment prevented only the f-Met-Leu-Phe-induced increase, not the Con A-stimulated increase in $f[Ca^{2+}]_i$ (Verghese et al.,   1985). These observations are in contrast to our findings in mouse splenocytes: in the latter, the Con A-stimulated increase in $f[Ca^{2+}]_i$ due mainly to mobilization of extracellular Ca^{2+}, was completely inhibited by pretreatment with pertussis toxin. A plausible hypothesis that could account for the difference between the two cell systems is that mobilization of Ca^{2+} in neutrophils by Con A may utilize a mechanism not involving the G protein.

The following data suggest that G proteins may be involved in the generation of proliferative signals in the lymphocyte system: 1) the increased $f[Ca^{2+}]_i$ in Con A-stimulated lymphocytes, 2) the requirement for extracellular Ca^{2+} during the first 20 h of stimulation in order for the cells to enter S phase (Bard et al.,1978), 3) the effect of pertussis toxin treatment: treated cells neither showed the usual Con A-induced increase in $[Ca^{2+}]$ nor the ^{3}H-TdR incorporation that occured at 48 h in the control cells stimulated with mitogen. It will be of interest to determine the effects of pertussis toxin treatment on other components of the signal pathway such as increased Na^+ influx and transcription of mRNA for IL 2 receptors.

ACKNOWLEDGEMENTS

This research has been supported by operating grants from Nateral Sciences and Engineering Research Council of Canada and by equipment grants from Alberta Heritage Foundation for Medical Research (AHFMR); K.V.S. Prasad is a post-doctoral fellow of AHFMR. We thank Ms. Susan Edwards for her technical help and Ms. Susan Smith for her expert typing.

REFERENCES

Averdunk, R., and Gunther, Th., 1980, Effect of Concanavalin A on Ca^{2+} binding, Ca^{2+} uptake and the Ca^{2+} ATPase of lymphocyte plasma membranes, Biochem. Biophys. Res. Commun., 97: 1146.

Bard, E., Colwill, R., L'Anglais, R.L., and Kaplan, J.G., 1978, Response of human lymphocytes to mitogen: at what stage is there a requirement for Ca^{2+}? Can. J. Biochem., 56:900.

Bokoch, G.M., and Gilman, A.G., 1984, Inhibition of receptor - mediated release of arachidonic acid by pertussis toxin, Cell, 39:301.

Chaly, N., Brown, D.L., and Kaplan, J.G., 1986, The lymphocyte at rest and at work, Bioessays, 4:272.

Gierschik, P., Falloon, J., Milligan, G., Pines, M., Gallin, J.I., and Spiegel, A., 1986, Immunochemical evidence for a novel pertussis toxin substrate in human neutrophils, J. Biol. Chem., 261:8058.

Gilman, A., 1984, G proteins and dual control of adenylate cyclase, Cell, 36:577.

Goldman, D.W., Gifford, L.A., Bourne, H.R., and Goetzel, E.J., 1984, Pertussis toxin inhibits the active human neutrophils (Ns) by chemotactic factors (CFs), J. Cell Biol., 99:278a.

Hesketh, T.R., Smith, G.A., Moore, J.P., Taylor, M.V., and Metcalfe, J.C., 1983, Free cytoplasmic calcium concentration and the mitogenic stimulation of lymphocytes , J. Biol. Chem., 258:4876.

Hesketh, T.R., Moore, J.P., Morris, J.D.H., Taylor, M.V., Rogers, J., Smith, G.A., and Metcalfe, J.C., 1985, A common sequence of calcium and pH signals in the mitogenic stimulation of eukaryotic cells, Nature, 317: 481.

Hewlett, E.L., Sauer, K.T., Myers, G.A., Cowell, J.L., and Guerrant, R.L., 1983, Induction of a novel morphological response in Chinese hamster ovary cells by pertussis toxin,Infect. Immun., 40:1198.

Lichtman, A.H., Segel, G.B., and Lichtman, M.A., 1981, Calcium transport and calcium-ATPase activity in human lymphocyte plasma membrane vesicles, J. Biol. Chem., 256:6148.

Molski, T.F.P., Naccache, P.H., Marsh, M.L., Kermode, J., Becker, E.L., and Sha'afi, R.I., 1984, Pertussis toxin inhibits the rise in the intracellular concentration of free calcium that is induced by the chemotactic factors in rabbit neutrophils: possible role of G proteins in calcium mobilization, Biochem. Biophys. Res. Commun., 124:644.

Okajama, F., and Ui, M., 1984, ADP-ribosylation of the specific membrane protein by islet-activating protein, pertussis toxin, associated with inhibition of a chemotactic peptide induced arachidonate release in neutrophils, J. Biol. Chem., 259:13863.

Prasad, K.V.S., Severini, A., and Kaplan, J.G., 1987, Sodium ion influx: an early component of the mitogenic signal, Arch. Biochem. Biophys., In Press.

Quastel, M.R., and Kaplan, J.G., 1970, Early stimulation of K$^+$ uptake in lymphocytes treated with PHA, Exp. Cell. Res., 63:230.

Sarkadi, B., Enyedi, A., Szasz, I., and Gardos, G., 1982, Active calcium transport and calcium-dependent membrane phosphorylation in human peripheral blood lymphocytes, Cell Calcium, 3:163.

Segel, G.B., Simon, W., and Lichtman, M.A., 1979, Regulation of sodium and potassium transport in phytohemagglutinin-stimulated human blood lymphocytes, J. Clin. Invest., 64:834.

Sekura, R.B., Fish, F., Manclark, C.R., Meade, B., and Zhan, Y.L., 1983, Pertussis toxin, affinity purification of a new ADP-ribosyltransferase, J. Biol. Chem., 258: 14647.

Spiegel, A.M., Gierschik, P., Levine, M.A., and Downs, R.W. Jr., 1985, Clinical implications of guanine nucleotide-binding proteins as receptor-effector couplers, N. Engl. J. Med., 312:26.

Tax, W.J.M., Willems,H.W., Reekers, P.P.M., Capel, P.J.A., and Koene, R.A.P., 1983, Polymorphism in mitogenic effect of IgGl monoclonal antibodies against T3 antigen on human T-cells, Nature, 304:445.

Tsien, R.Y. Pozzan, T., and Rink, T.J., 1982a, T-cells mitogens cause early changes in cytoplasmic free Ca^{2+} and membrane potential in lymphocytes, Nature, 295:68.

Tsien, R.Y., Pozzan, T., and Rink, T.J., 1982b, Calcium homeostatis in intact lymphocytes: cytoplasmic free calcium monitored with a new intracellularly trapped fluorescent indicator, J. Cell Biol., 94:325.

Ueda, T., 1983, Na^{+}-Ca^{2+} exchange activity in rabbit lymphocyte plasma membranes, Biochim. Biophys. Acta, 734:342.

Verghese, M., Smith, C.D., and Snyderman, R., 1985, Potential role for a guanine nucleotide regulatory protein in chemo-attractant receptor mediated polyphosphoinositide metabolism. Ca^{2+} mobilization and cellular responses by leukocytes, Biochem. Biophys. Res. Commun.,127:450.

Whitaker, M.J., and Steinhard, R.A., 1982, Ionic regulation of egg activation, Q. Rev. Biophys., 15:593.

Whitney, R.B., and Sutherland, R.M., 1972, Enchanced uptake of calcium by transforming lymphocytes, Cell Immunol. 5:137.

POSSIBLE INVOLVEMENT OF PHOSPHORYLATION/DEPHOSPHORYLATION

IN THE REGULATION OF EPITHELIAL SODIUM CHANNELS

David S. Lester, Carol Asher and Haim Garty

Department of Membrane Research
Weizmann Institute of Science
Rehovot, 76100, Israel

INTRODUCTION

Luminal Na^+ entry in tight epithelia is mediated by amiloride-inhibitable Na^+ channels (1). These channels are modulated by several factors, including antidiuretic hormones, such as vasopressin. In the toad bladder and related epithelia it was demonstrated that vasopressin binds to a basolateral membrane receptor and, consequently, activates adenylate cyclase resulting in a rise in intracellular cyclic AMP (cAMP) (2). This rise brings about a two-fold increase in the number of conducting channels (3). The natriferic action of the hormone can also be mimicked by exogenous cAMP (2,4). The increase in Na^+ transport was shown to be preceded by an activation of the cytosolic Type II cAMP-dependent protein kinase (cAMPPK) (5), change in cell Ca^{2+} (6,7), and dephosphorylation of a 50-55 KDa protein (8). This phosphoprotein was tentatively identified as the regulatory subunit of the Type II cAMPPK (9).

The mechanism(s) by which the increased level of cell cAMP and accompanying events regulate the channel density remain unclear. Molecular processes involved in the hormonal regulation of apical Na^+ transport can be explored in isolated membrane vesicles in which the channel-mediated flux is assessed under simplified, controlled conditions. Using such techniques it has been demonstrated that the channel conductance is coupled to the cytoplasmic Ca^{2+} activity (10,11). The channel-mediated flux was high if the membranes were isolated from cells preincubated in Ca^{2+}-free EGTA buffer, and low if the cells were exposed to 10^{-5}M free Ca^{2+} (10,11).

In the present study we demonstrate that: (a) preincubation of whole bladders with 8-bromocAMP (8BrcAMP) increases the channel-mediated flux in vesicles derived from these cells, (b) this activation is not the result of a direct action of the cAMPPK on the channel, (c) Ca^{2+} decreases the intensity of a 53-55 KDa phosphoprotein, presumably by activating a phosphatase. However, dephosphorylation of this protein by cytosolic phosphatases partially purified from toad bladder epithelial cells was not observed.

MATERIALS AND METHODS

Materials

Valinomycin, phorbol myristate acetate, 8BrcAMP, p-nitrophenolphosphate (substrate 104), ATP (vanadate free), leupeptin, PMSF and soybean trypsin inhibitor were obtained from Sigma Chemical Co., Dowex 50Wx8 (50-100 mesh Tris form) from Fluka, and ammonium vanadate from Merck. Amiloride HCl was a gift from Merck Sharp and Dohme Co.. ^{22}NaCl (200μCi/ml, carrier free) and [γ-^{32}P]-ATP (3,000 Ci/mmol) were purchased from Amersham Radiochemicals. All other chemicals were of analytical grade.

Animals

Toads (Bufo marinus, Mexican origin) were obtained from Lemberger, WI, and kept in tanks.

Vesicle Preparation

Vesicles were prepared as described by Garty and Asher (10). Briefly, urinary bladders were excised from double-pithed toads that had their ventricles perfused with 500 ml NaCl-Ringer solution. The bladders were rinsed and the epithelial cells were scraped free using a microscope slide and suspended in an homogenizing medium containing (in mM) 90 KCl, 5 Tris (pH 7.8), 45 sucrose, 10 MgCl$_2$ and 10 EGTA or 0.01 CaCl$_2$. Cells were washed twice, then resuspended in the appropriate buffer and incubated at 0°C or 25°C for 45 min. They were then pelleted, resuspended in Ca^{2+}-free buffer and homogenized for 10 sec by a Polytron homogenizer (Kinematica, Luzern, Switzerland). Nuclei and unbroken cells were spun down at 1,000xg for 5 min. The vesicles were sedimented by centrifuging for 1 h at 27,000xg. The microsomal pellets were resuspended in minimal volumes for the transport assay. The supernatants were collected and used for the characterization and isolation of cytosolic phosphatase activity.

Transport Assay

^{22}Na$^+$ uptake by toad bladder microsomes was measured at 25°C as described previously (12). Microsomes preloaded with KCl were eluted from a Dowex column with 175 mM sucrose to remove external K$^+$. Upon addition of valinomycin (3μM), a K$^+$ diffusion potential was established, and the vesicles were immediately added to reaction vials containing ^{22}Na$^+$ plus or minus amiloride (1.5μM). The time course of uptake was measured over 3.5 min. The reaction was stopped by pipetting 150μl of assay mixture onto a chilled Dowex column and washing the vesicles with ice-cold 175 mM sucrose into counting vials. The Na$^+$ channel activity (pmol ^{22}Na·mg protein^{-1}·min^{-1}) was defined as the difference between total and amiloride-insensitive fluxes. Protein content was assayed in the eluted vesicles by the Bradford protein assay (13).

Protein Phosphorylation

Cells were suspended and preincubated as described above. Instead of homogenization, cells were suspended in 20mM Tris/HCl pH 7.8 to induce lysis. At the same time leupeptin (2.5μg/ml), PMSF (0.1mM), and soybean trypsin inhibitor (25μg/ml) were added to inhibit proteolytic activity. 40mM Tris (pH 7.8) with 20mM EGTA or 20μM free Ca and 2mM vanadate was immediately added, as well as, 20μM [γ-^{32}P]-ATP (250μCi/ml). The mixture was left for 10 min at 25°C and then diluted 5-fold with ice cold buffer containing 6mM unlabelled ATP. The samples were centrifuged at 27,000xg for 1 hr. The pellet and supernatant were separated and washed twice with trichloroacetic acid (final concentration 10%) followed by a 1ml acetone wash. Samples were suspended in 100μl SDS-sample buffer (with 5% β-mercaptoethanol), left at room temperature for 25 min, boiled for 2.5 min and then stored at -20°C until loaded on gels. The samples were run on 15% Laemmli (14) gels, then dried and exposed to autoradiographic film.

Phosphatase Assay

Vesicles and supernatants were prepared as described above. Vesicles were lysed in 20mM Tris (pH 7.5) buffer. Final assay concentrations were 10mM $MgCl_2$, 10mM EGTA or 100μM free $CaCl_2$ and 10mM para-nitrophenolphosphate (pNPP). Assays were carried out in plastic multiwell titer plates (300μl total volume). The reaction was monitored, kinetically, by repeated reading on a Biotek (Model 310) Elisa Reader at 405nM, with a 600nM reference wavelength.

RESULTS

It is well-documented that cAMP activates the channel-mediated Na^+ transport in toad bladder epithelium (1,3). However, Garty and Asher (11) reported that vesicles derived from scraped cells incubated with 8BrcAMP, do not exhibit increased rates of Na^+ uptake. We found that when whole bladders were incubated with 8BrcAMP, a stable increase in the channel-mediated flux measured in vesicles can be observed. This activation, however, is apparent only if the cells are incubated in EGTA for 30 min at 25°C prior to their homogenization (Figure 1). The above finding indicates that 8BrcAMP activates channels by inducing a sustained modification of membranal protein and that the natriferic response can be studied in membrane vesicles.

The only known function of cAMP is the induction of protein phosphorylation by activating cAMPPK. To determine if 8BrcAMP activation of transport in vesicles was a result of direct phosphorylation of the channel or another membrane component(s), purified Type II cAMPPK from bovine heart and partially purified Type II cAMPPK from toad bladder were trapped in vesicles with 8BrcAMP, ATP, vanadate and protease inhibitors. Neither were found to have any effect on EGTA-activated or Ca^{2+}-inhibited Na^+ flux (results not shown). Under the same conditions, the trapped cAMPPK was shown to be active as histone fragment 2b was phosphorylated. Thus, the cAMPPK does not directly phosphorylate the channel and appears to work via phosphorylation of a cytosolic intermediate.

To identify if there was phosphorylation of a cytosolic intermediate(s) when Na^+ transport was activated, lysates of cells preincubated in Ca^{2+}-free medium were incubated with [$\gamma^{32}P$]-ATP. The membranes and cytosol were separated and run on SDS-gels. The predominant cytosolic phosphoprotein under these conditions has a similar M_r (=55KDa) to the Type II cAMPPK regulatory subunit (Figure 2, column 1). Other bands of 40K, 28K and 16K were found. Since Ca^{2+} was previously shown to inhibit Na^+ flux in vesicles (10,11) we tested the effect of Ca^{2+} in this preparation. to see if it may act via the reverse pathway as EGTA-activation, i.e. dephosphorylation. Figure 2, columns 2,3 and 4 demonstrate that in the presence of Ca^{2+} (10^{-5}M) either in the preincubation or lysis buffer, greatly reduces the intensity of phosphorylation of the 55KDa protein by [$\gamma^{32}P$]-ATP. The labelling of this protein is even totally abolished when Ca^{2+} treated cells are lysed in Ca^{2+} lysis buffer. These results suggest the presence of a

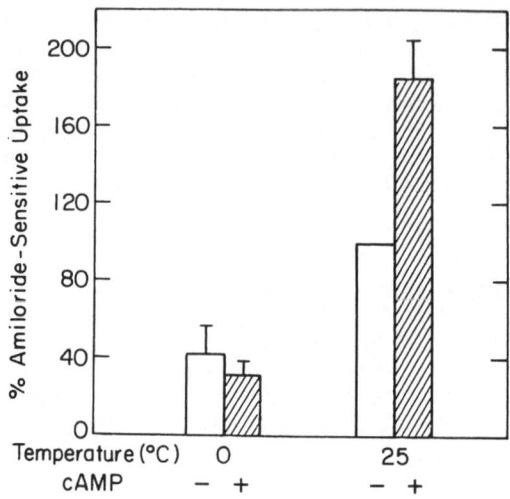

Figure 1. cAMP activation of amiloride-sensitive sodium uptake. Paired
hemibladders were incubated with or without 1 mM 8BrcAMP at 25°C for
15 min. Bladders were rinsed in Ca^{2+}-free medium, and cells scraped, and
divided into two equal portions incubated at 0°C or 25°C for 45 min.
Vesicles were prepared and Na^+ flux assayed as described in "Materials and
Methods". Data are expressed as percent of the amiloride-sensitive flux in
vesicles derived from cells incubated at 25°C in the absence of 8BrcAMP.
Mean ± S.E. of 3 experiments is presented.

Ca^{2+}-activated, vanadate-insensitive phosphatase acting upon the cAMPPK
regulatory subunit. Since leupeptin is present in the lysis medium, disappearance of the band is not due to Ca^{2+}-proteases.

The presence of phosphatases in the membrane and cytosol was examined
using pNPP as a substrate. In the presence of EGTA and vanadate, all
pNPPase activity was inhibited, while if Ca^{2+} was present, between 15-30%
of membrane activity was not inhibited by vanadate (results not shown).
These results indicate that the membrane fraction does indeed contain a
Ca^{2+} actiated, vanadate insensitive phosphatase which might be responsible
for the dephosphorylation observed in Figure 2, columns 2,3 and 4.

Upon solubilization of membranes by Triton X-100 or cholate (1%), the
membrane-associated Ca^{2+}-activated, vanadate-insensitive PNPPase activity
was lost. Fractionation of cytosolic PNPPases, and subsequent characterization demonstrated two pNPPase activities. Analysis of pH optima revealed
that pNPPase 2 is slightly activated by Ca^{2+} at a basic pH (Figure 3).
This pNPPase is also partially inhibited (66%) by 1mM vanadate (results not
shown). However, trapping of this phosphatase in EGTA-activated vesicles
had no effect on Na^+ transport no matter what conditions were used (unpublished observations). Thus, a membrane-associated, Ca^{2+}-activated, vanadate-insensitive phosphatase can be identified but determination of its
role in deactivation of the Na^+ channel has yet to be established.

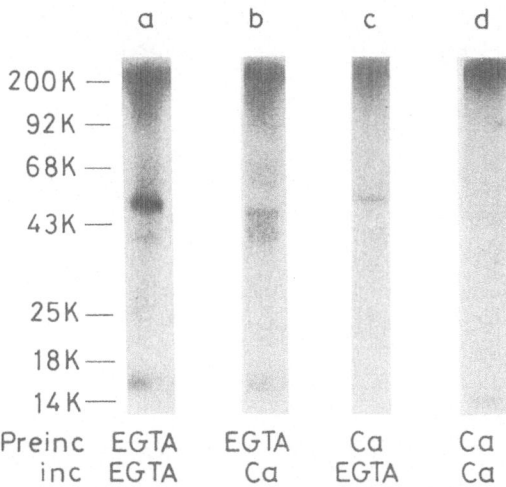

Figure 2. EGTA-dependent phosphorylation of a 55KDa soluble protein. Cells (1/4 bladder) were preincubated in EGTA or Ca^{2+} (10^{-5}M) and then lysed in low osmolarity EGTA or Ca^{2+} buffer containing $[\gamma^{32}P]$-ATP ($250\mu Ci/ml$), protease inhibitors, and vanadate (1 mM). Cytosolic samples were run in 12% SDS-gels, dried and exposed to autoradiographic film.

DISCUSSION

In the present study we have demonstrated that incubating bladders with 8BrcAMP produces a sustained activation of Na^+ channels which can be detected in vesicles derived from these cells. Thus, the well documented natriferic effect of vasopressin and cAMP must involve either the insertion of channels in the apical surface or a stable modification of apical protein(s) or lipid(s). As cAMPPK is the receptor molecule for cAMP we investigated the possibility that cAMPPK activates Na^+ channels by directly phosphorylating the channel protein. However, interacting the cytoplasmic face of the isolated membrane with purified cAMPPK failed to activate transport. The trapped enzyme was active as determined by phosphorylation of histone fragment 2b trapped in the vesicles together with cAMPPK. Including phosphatase, partly purified from the cell cytoplasm, in the vesicles failed to alter the transport rate as well. Thus, it appears that the modulation of channels by cAMP does not involve a direct phosphorylation/dephosphorylation of the channel protein. In searching for mediating phosphoproteins we observed a 55KDa protein that was the predominant phosphoprotein if the cells were pretreated and lysed in EGTA. Phosphorylation was strongly reduced after exposure of the cells and/or the lysate to Ca^{2+}. A similar protein was previously observed by Walton et al. (8) and was tentatively identified as the regulatory subunit of cAMPPK (9). The autophosphorylation of the regulatory subunit is indicative of the activation of the cAMP-dependent enzyme. Thus, if the 55KDa protein is indeed the regulatory subunit of cAMPPK, additional as yet unidentified phosphoproteins must be involved in the natriferic action. It is interesting that Ca^{2+} shown to inhibit channels (10,11) can induce dephosphorylation of this pro-

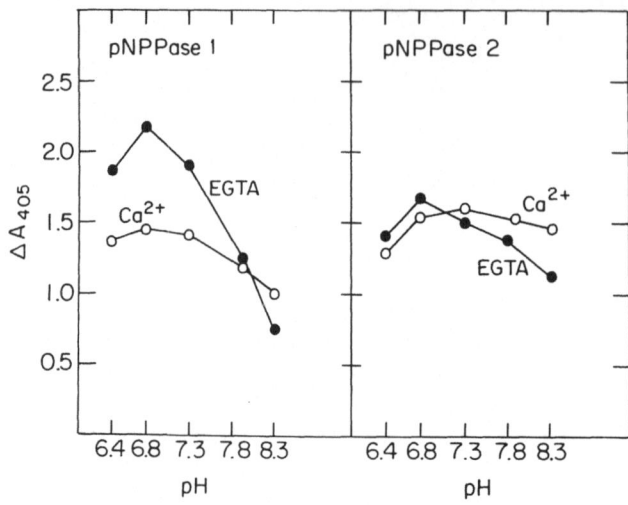

Figure 3. pH profiles of cytosolic pNPP-ases in the presence and absence of Ca^{2+}. Activity was measured by following the release of para-nitrophenol spectrophotometrically at 405 nm. Activity units are ΔA_{405}/mg protein/ min.

tein. The data of Figure 2 predict the presence of a Ca-activated, vanadate-insensitive phosphatase which indeed was identified in the membranal fraction of toad bladder cells. Its characterization and purification proved difficult because of large variability among preparations and its inactivation upon detergent solubilization. The effect of this Ca^{2+}-activated phosphatase combined with the activation of phosphodiesterase by Ca^{2+} (4) may function to terminate the natriferic action of vasopressin.

The major question not answered by the present and previous experiments is how protein phosphorylation is linked to the activation or induction of channels. Several mechanisms, some of which are being currently investigated in our laboratory, may be involved in this process. For example, it has been proposed that cAMPPK may phosphorylate a microfilament crosslinker which could effect trafficking of intracellular vesicles towards the plasma membrane and ultimate insertion (4). However, cytochalasin B, a microfilament disrupting agent (15) had no effect on Na^+ transport (11). In contrast, Ca^{2+}, also capable of microfilament disruption, inhibits Na^+ transport (10,11). Another possibility is that intermediate filaments interact directly (anchoring) with the channel, causing changes in orientation of the Na^+ channel in the membrane. As calmodulin antagonists and colchicine had no effect on Na^+ transport in vesicles (11) it is unlikely that this process is involved in the modulation of transport.

Other unresolved issues are why activation of transport in vesicles can be induced only by incubating whole bladders with 8BrcAMP and why this activation is apparent only after incubating cells scraped from the cAMP activated bladder in EGTA medium at 25°C. The insensitivity of scraped cells to 8BrcAMP may reflect impairment of cellular mechanisms by permeabilization and damage of the cells during scraping. Such potential impairment had been reported before by at least two groups (17,18). The second question is more difficult to answer since the nature of activating channels by incubating cells in EGTA is not yet understood. The data clearly indicate however, that incubating bladders with 8BrcAMP does not activate the same pool of channels influenced by exposing scraped cells to EGTA since the effects of the two treatments were synergistic (Figure 1).

In summary the current data suggest that cAMP activates Na^+ channels by inducing a stable change in apical proteins and/or lipid. This change is not a direct phosphorylation of the channel protein. Ca may regulate this process by activating a membranal enzyme which dephosphorylates the regulatory subunit of cAMPPK. 8BrcAMP and Ca-free solutions apparently activate channels by different mechanisms and the molecular events involved in either process are not yet clear.

ACKNOWLEDGEMENTS

This work was supported by NIH (DK-36328) and BSF (84-00066) grants to H.G.. D.L. is a Stiftung Volkswagenwerk Research Fellow and was supplied funds from the Ministry of Immigration, State of Israel. Thanks to Orna Yaeger for technical assistance and Celina Gross for expert typing.

REFERENCES

1. Sariban-Sohraby, S. and Benos, D.J., Am. J. Physiol. 250:C175–C190 (1986).
2. J. S. Handler and J. Orloff, Ann. Rev. Physiol. 43:611-624 (1981).
3. Lindemann, B., Ann. Rev. Physiol. 46, 497-515 (1984).
4. D. A. Ausiello and J. Orloff, in: "Cyclic Nucleotides," J. W. Kebabian and J. A. Nathanson, eds., Springer-Verlag, New York, pp 271-301 (1982).
5. D. Schlondorff and N. Franki, Biochim. Biophys. Acta 628:1-12 (1980).
6. E. Kelepouris, Z.S. Agus and M. M. Civan, J. Memb. Biol. 88:113-121 (1985).
7. R. M. Burch and P. V. Halushka, Am. J. Physiol. :F939-945 (1984).
8. K. G. Walton, R. J. DeLorenzo, P. F. Curran and P. Greengard, J. Gen. Physiol. 65:153-177 (1975).
9. A. Y. C. Liu, U. Walter and P. Greengard, Eur. J. Biochem. 114:539-548 (1981).
10. H. Garty and C. Asher, J. Biol. Chem. 260:8330-8335 (1985).
11. H. Garty and C. Asher, J. Biol. Chem. 261:7400-7406 (1986).
12. H. Garty, J. Memb. Biol. 82:269-279 (1984).
13. M. M. Bradford, Anal. Biochem. 72:248-254 (1976).
14. U. K. Laemmli, Nature(London) 277:680-682 (1970).
15. M. Schliwa, "The cytoskeleton. An introductory survey," Springer-Verlag, Vienna (1986).
16. S. Pontremolli, E. Melloni and B. L. Horecker, Curr. Topics Cell. Reg. 27:293-303 (1985).
17. R.M. Burch and P.V. Haluska, Am. J. Physiol. 243:F593-F597 (1982).
18. H.J. Rodriguez, D.W. Scholer, M.L. Parkerson and S. Klahr, Am. J. Physiol. 238:F140-F149 (1980).

GLYCOGEN METABOLISM IN SMOOTH MUSCLE

Theodore G. Sotiroudis, Stathis Nikolaropoulos
and Athanasios E. Evangelopoulos

The National Hellenic Research Foundation
Institute of Biological Research
48 Vassileos Constantinou Avenue, Athens 116 35, Greece

INTRODUCTION

In the last few years there has been a surge of interest in smooth muscle biochemistry. There are two major reasons why this tissue is of particular interest. First, although the total high energy phosphagen pool (phosphocreatine + ATP) is substantially lower than that of skeletal muscle, however energy metabolism in smooth muscle is finely tuned to meet contractile energy requirements[1]. Second, it has recently become evident that reversible phosphorylation of myosin, via an enzyme cascade similar to that involved in the activation of glycogen phosphorylase, plays a major role in regulating the interaction of actin and myosin in smooth muscle[2,3].

Glycogen is the most conspicuous source of endogenous carbohydrate metabolism observed in the smooth muscle cells of different organs. Although its function is not fully understood, glycogen has been considered as one of the main available energy stores which takes part in maintaining the vascular tone under stimulated conditions in the absence of glucose[4]. A large number of proteins are known to be involved in glycogen metabolism and its regulation[5]. Glycogenolysis is catalyzed by phosphorylase and debranching enzyme and glycogen synthesis from UDPG by glycogen synthase and branching enzyme[6]. Studies of the control of glycogen metabolism in mammalian skeletal muscle have revealed that a regulatory network of a number of protein kinases and phosphatases and several thermostable regulatory proteins determine the activation state of glycogen phosphorylase and glycogen synthase, the rate-limiting enzymes of this metabolic process[5]. Nevertheless, little is known on the catalytic and structural characteristics of the enzymes involved in the cascade of reactions associated with glycogen synthesis or degradation in smooth muscle and the overall picture of the control of glycogen metabolism in this type of muscle is not well defined.

It is the purpose of this review article to briefly summarize what is known about glycogen metabolism and its regulation in smooth muscle, with particular emphasis on the chemical and regulatory properties of the key enzymes of glycogen metabolism, which include phosphorylase and phosphorylase kinase, glycogen synthase and synthase kinases, and protein phosphatases. Moreover current knowledge of the relationships between smooth muscle contractile activity and glycogen metabolism will be also considered, although the functional (pharmacological and physiological) as well as structural heterogeneity of smooth muscles does not permit facile generalizations between different smooth muscles in the same species or between different species[7]. Finally, recent findings concerning functional compartmentation of glycogen metabolism within smooth muscle cells will be addressed.

GLYCOGEN PHOSPHORYLASE AND PHOSPHORYLASE KINASE

Available information about the properties of glycogen phosphorylases of different tissues suggests that more than two phosphorylase isozymes exist in mammalian tissues. Electrophoretic, immunological and other physicochemical and catalytic studies have shown that five phosphorylase isozymes can be separated and were designated as phosphorylase L, LI, I, II and III[8]. The L and III enzymes were the only forms found in liver and skeletal muscle respectively, while the I enzyme was dominant in brain, lung and smooth muscles. The II and LI enzymes were found to be the hybrid molecules between I and III enzymes and between I and L enzymes respectively[8].

Preliminary immunological and/or kinetic studies on glycogen phosphorylase from bovine tracheal and intestinal smooth muscles have shown that the smooth muscle enzymes differ from their skeletal counterpart, although controversial results regarding their affinity for AMP were reported[9,10]. Victorova and Ramensky[11,12], were first able to isolate an homogeneous smooth muscle phosphorylase preparation from cow myometrium using hydrophobic and affinity chromatographic procedures. Their results have clearly shown that uterine isozyme I presented significant pecularities which distinguish smooth muscle phosphorylase from skeletal muscle and liver isozymes (Table 1).

Table 1. Comparison of Properties of Uterine, Skeletal Muscle and Liver Phosphorylase

	Phosphorylase		
	Uterine	Skeletal Muscle	Liver
Tetramerization[a]	-[12]	+[17]	-[13]
Ka for AMP(b-form)	7.3 μM[12]	50 μM[15]	0.36 mM[16,b]
Na$_2$SO$_4$- Activation (in absence of AMP)	-[12]	+[14]	+[13]
AMP-Activation	+[12]	+[15]	-[13]
Km for Glucose-1-P (a-form)	43 mM[12]	1.6 mM[17]	0.7 mM[18,c]

[a]After phosphorylation with phosphorylase kinase.

[b]($A_{0.5}$) at which v=Vmax/2, determined from Hill plots ; in presence of 0.15 M acetate.

[c]In presence of 5 mM AMP.

Uterine phosphorylase b is a dimer with a Mr of~200,000, but the conversion of phosphorylase b to a was not accompanied by tetramerization, as in the case of liver isozyme[13]. Smooth muscle inactive phosphorylase differs also from liver and skeletal muscle enzymes in that it undergoes additional activation by Na$_2$SO$_4$ only in the presence of AMP[13,14]. Moreover, the Km of phosphorylase a of the myometrium for the substrate glucose-1-P is 27-fold and 61-fold higher than the Km for the skeletal and liver enzymes respectively, while its Ka for the activator AMP is significantly lower (Table 1). It is suggested that the forms cited in the literature as brain phosphorylase, heart isozyme 1, fetal isozyme and uterine smooth muscle phosphorylase should be considered as the same protein species.

Phosphorylase kinase is one of the regulatory enzymes involved in the cascade of reactions associated with glycogenolysis. The enzymes from rabbit[19,20], dogfish[21] and chicken[22,23],skeletal muscle, bovine heart[24] and rat liver[25] have been purified to apparent homogeneity and are known to be all hexadecamers composed of four types of subunits $(\alpha\beta\gamma\delta)_4$, although there appear to exist some distinct differences among the above characterized isozymes.

Since smooth muscle phosphorylase kinase has not been purified or studied to any extend, we had undertaken the purification and characterization of the enzyme from chicken gizzard to facilitate understanding of the hormonal control of glycogen metabolism in smooth muscle. Phosphorylase kinase was purified to near homogeneity by a procedure involving glycerol density gradient ultracentrifugation, 5´AMP-Sepharose 4B and calmodulin-Sepharose 4B affinity chromatography steps[26]. It was interesting as well as puzzling that gizzard phosphorylase kinase showed a completely different subunit pattern to that of rabbit skeletal muscle enzyme (one main protein band of 61 kDa and no protein staining at the position of γ subunit) upon dodecyl sulfate acrylamide gel electrophoresis, although the Mr of chicken gizzard and rabbit skeletal muscle kinases were found to be similar. Gizzard phosphorylase kinase showed a high pH 6.8/8.2 activity ratio of 0.53, it was stimulated by Ca^{2+}, inhibited up to 80% by EGTA and it was activated about 1.9-fold by calmodulin. In a recent report Tsutou et al.[27] using a partially purified uterine smooth muscle phosphorylase kinase preparation have also reached the conclusion that their smooth muscle kinase was activated about 1.5-2.0-fold by exogenous calmodulin and that Ca^{2+}-activation of the enzyme was greatly inhibited by EGTA. Further characterization of gizzard phosphorylase kinase has shown that the Km value for ATP was 0.45 mM, while the $K_{0.5}$ for rabbit muscle phosphorylase b was extremely low, more than 100-fold lower than the Km of activated skeletal muscle phosphorylase kinase for its protein substrate (Table 2).In this respect, it is interesting to report that the gizzard kinase activity was drastically decreased at phosphorylase b concentrations greater than 0.5 mg/ml (at 10 mg/ml of the protein substrate a 5-fold lower activity was observed)[26]. This phenomenon may be due either to the existence of more than one type of binding sites for the protein substrate or that the effector molecules present in the assay mixture (ATP, Mg^{2+}) induce on phosphorylase b a different conformation (at high protein substrate concentrations),which negatively affects b to a conversion. The possibility that the biphasic effect of phosphorylase b on gizzard kinase activity may be due to the presence of an unknown inhibitor can rather be excluded, because phosphorylase b (four times crystallized) chromatographed on a DEAE Sephadex A-50 column (pH 8.2) and gel filtrated through a Sepharose 6B column, was still able to present the inhibitory effect (unpublished results).

A very interesting feature of chicken gizzard phosphorylase kinase is that the enzyme cannot be activated by phosphorylation with cAMP-dependent protein kinase or by autophosphorylation[26], a property also shared by dogfish[21] and chicken[23] skeletal muscle phosphorylase kinase. Preliminary experiments of Ozawa[28,29] with partially purified smooth muscle phosphorylase kinase have shown that the enzyme was not stimulated by cAMP, in accordance with our results, while in contrast, Mohme-Lundholm[10] using an undefined tracheal muscle phosphorylase kinase preparation reported that the kinase was activated by cAMP. Although Mohme-Lundholm´s results are not enough experimentally supported, in fact, it is possible that mammalian smooth muscle phosphorylase kinase can be regulated by a phosphorylation-dephosphorylation process similar to that found in skeletal muscle kinase, and that the inability of chicken gizzard kinase to show such a regularoty pattern may be a common property of ancient vertebrate muscles[21,23].

Table 2. Comparison of Properties of Phosphorylase Kinases Isolated from Various Tissues

Phosphorylase Kinase[a]

	Chicken Gizzard	Rabbit Muscle	Dogfish Muscle	Chicken Muscle	Rat Liver	Bovine Heart
M_r	1.3x10⁶ [26]	1.3x10⁶ [20]	1.2x10⁶ [32]	1.3x10⁶ [22]	1.3x10⁶ [25]	1.3x10⁶ [24]
Subunit Pattern	61 kDa[26] (main protein band)	α,145 kDa;[20] β,130 kDa; γ,45 kDa; δ,17 kDa;	α,138 kDa;[32] β,118 kDa; γ,45 kDa; δ,17 kDa;	α,140 kDa;[22] β,129 kDa; γ,44 kDa; δ,17 kDa;	α,140 kDa;[25] β,116 kDa; γ,45 kDa; δ,17.5 kDa;	α,134 kDa;[24] β,125 kDa; γ,48 kDa; δ,?
Activity Ratio (6.8/8.2)	0.5[26]	0.07[20]	0.3-0.5[21]	0.1-0.4[23]	Optimum pH=[25] 6.8-7.2	0.04-0.2[24]
Km for ATP	0.45 mM[26]	0.4 mM[20]	2 mM[21]	0.2 mM[23]	0.1 mM[33]	0.2 mM[24]
Km for Muscle Phsophorylase	<0.04 mg/ml[26]	4 mg/ml[19,b]	16.5 mg/ml[21]	4 mg/ml[23]	2.5 mg/ml[33]	5.5 mg/ml[35]
Activation by Phosphorylation	-[26]	+[19]	-[21]	-[23]	+[25]	+[24]
Activation by Calmodulin and/or Troponin	+[26]	+[20]	-[21]	+[23]	+[34]	-[35]

[a]Nonactivated phosphorylase kinase; [b]Activated phosphorylase kinase

Concerning other regulatory properties of chicken gizzard phosphorylase kinase, the enzyme activity at pH 6.8 was activated 2-fold by limited proteolysis[26], inhibited by ethanol and was not significantly affected by heparin (unpublished results). In contrast, rabbit muscle phosphorylase kinase activity at pH 6.8 is known to be drastically stimulated by proteolysis[19], organic solvents[30] and heparin[31]. Table 2 presents a comparison of significant properties of phosphorylase kinases isolated from various tissues.

GLYCOGEN SYNTHASE AND GLYCOGEN SYNTHASE KINASES

Glycogen synthase is the rate-limiting step in mammalian glycogenesis and hence is central in the hormonal control of glycogen metabolism. It has been established that glycogen synthase is regulated by a phosphorylation-dephosphorylation mechanism, the phosphorylation being catalyzed by at least three types of protein kinases:(i) cyclic nucleotide-dependent protein kinases, (ii) Ca^{2+}-dependent protein kinases and (iii) Ca^{2+}- and cyclic nucleotide-independent protein kinases[36].

Although glycogen synthase has been purified extensively from a number of eucaryotic sources[36] and at least two different isozymes of glycogen synthase are present in a number of tissues examined by immunoblot analysis[37], most structural and functional knowledge stems from studies of the rabbit skeletal muscle enzyme[36,38], while very little is known on the enzymatic system involved in glycogen synthesis in smooth muscles. Preliminary studies on smooth muscle glycogen synthesis and its control have been performed in uterine and endometrial tissues[39,40], but to our knowledge, glycogen synthase has not been purified or studied to any extend from any type of pure smooth muscle. However, Huang and Robinson[41] have reported the extensive purification of glycogen synthase from human placenta, a tissue which is known to contain large amounts of vascular and extravascular smooth muscle[42]. The human placental glucose-6-P-dependent form of glycogen synthase (D-form) can be converted to the glucose-6-P -independent form (I-form) by incubati the synthase with a copurified synthase phosphatase which shows a stringent requirement for Mn^{2+} [41]. The D to I conversion can be reversed by the addition of cAMP-dependent protein kinase. It is noteworthy that the purified D-form enzyme, which is free of synthase phosphatase activity, shows a unique activation by metal sulfates in the absence of glucose-6-P[41].

A number of cAMP-dependent and independent glycogen synthase kinases has been shown to be present in several tissues[43]. Two isozymic forms of cAMP-dependent protein kinase (types I and II) have been identified in coronary arterial smooth muscle[44], while in a variety of smooth muscles,cAMP-dependent protein kinase has been postulated to play a pivotal role in the β-adrenergic control of contractile activity,and alterations in the extent of phosphorylation of the type II regulatory subunit could affect the response of this tissue to stimulation by β-adrenergic agents[45]. Recently, a multisubstrate Ca^{2+}- and cyclic nucleotide-independent kinase (Mr=47,000) was purified from bovine aortic smooth muscle and has been shown to phosphorylate a number of proteins, which include glycogen synthase, phosphorylase kinase and type II regulatory subunit of cAMP-dependent protein kinase[46]. Phosphorylation of skeletal muscle glycogen synthase by this enzyme was polycation modulable. Glycogen synthase converted to D-form following phosphorylation in either the presence (7 mol ^{32}P/mol synthase) or absence (4 mol ^{32}P/mol synthase) of polylysine. These results suggest that the enzyme may participate in regulati arterial glycogen metabolism. In addition, vascular smooth muscle has been shown to contain protein kinase F_A, the activation factor of ATPMg-dependent protein phosphatase, which has been identified as glycogen synthase kinase 3[47,5].

Smooth muscle phosphorylase kinase may also be considered as a potential Ca^{2+}-dependent glycogen synthase kinase, knowing that skeletal muscle phosphorylase kinase efficiently phosphorylates and inactivates glycogen synthase[36], although the substrate specificity of this smooth muscle kinase towards glycogen synthase has not been examined. For completeness, cGMP-dependent protein kinase and Ca^{2+}- and phospholipid-dependent protein kinase (protein kinase C) must also be cited as two other kinases which are present in smooth muscles[48] and are able to phosphorylate glycogen synthase[36]. Of course, assesment of the physiological significance of glycogen synthase phosphorylation by protein kinase C and cGMP-dependent protein kinase must be made critically, all the more so since it is now evident that glycogen synthase is an extremely promiscuous protein kinase substrate in vitro. Nonetheless, the fact that the smooth muscle of vas deferens together with spleen and brain tissues contain the highest levels of protein kinase C, from a large number of tissues examined[48], strongly suggests a direct action of protein kinase C to phosphorylate and inactivate glycogen synthase in several types of smooth muscle.

PHOSPHOPROTEIN PHOSPHATASES

Recent studies indicate that four enzymes termed protein phosphatases 1, 2A, 2B and 2C account for virtually all the protein phosphatase activity in cellular extracts,acting on phosphorylated proteins involved in the control of glycogen metabolism and other metabolic pathways[49]. Four phosphatases designated SMPI-IV have been identified in the smooth muscle of turkey gizzard[50,51]. Three of these have been purified to apparent homogeneity : SMP-I consists of three subunits (60 kDa, 55 kDa, 38 kDa) in a molar ratio 1:1:1 and it is a protein phosphatase $2A_1$. SMP-II is a Mg^{2+}-dependent monomeric enzyme (Mr=43,000) and could be designated as protein phosphatase 2C. SMP-IV is composed of two subunits (58 kDa and 40 kDa), has a Mr of 150,000 and cannot be classified as either a type 1 or type 2 phosphatase, because although it dephosphorylates the β-subunit of phophorylase kinase, is not inhibited by the heat stable inhibitor-2 and has low activity toward phosphorylase a. The ability of SMP-I and SMP-IV to dephosphorylate phosphorylase a and the observation that SMP-IV dephosphorylates preferentially the β subunit of phosphorylase kinase may suggest a role of these enzymes in glycogen metabolism[50,51]. In addition, it has been suggested that SMP-III and IV are involved in the process of relaxation in vivo[50].

Multiple forms of protein phosphatases have also been identified in mammalian vascular smooth muscle. Werth et al.[52] described an aortic phosphatase consisting of two subunits (67 kDa and 38 kDa)in a molar ratio 1:1, which was active against native myosin and exhibited relatively low activity against phosphorylase a.The properties and subunit structure of this enzyme indicate that it is a protein phosphatase $2A_2$. DiSalvo et al[53] have also identified several protein phosphatases in aortic smooth muscle. One of these was the multisubstrate ATPMg-dependent protein phosphatase, which is apparently the inactive form of the major physiologically relevant phosphatase involved in coordinating glycogen synthesis and breakdown[54] and which did not dephosphorylate myosin light chains (MLC) in either the presence or absence of modulator protein[55]. A second, apparently "latent" phosphorylase phosphatase, which migrated as a protein of Mr of 130,000 during sucrose density gradient centrifugation, has been shown to exist in vascular smooth muscle and its activity against phosphorylase a was markedly stimulated by histone-H_1 and polylysine. This "latent" phosphorylase phosphatase activity has been also identified as a major inhibitor-1 phosphatase in vascular smooth muscle[56], while its substrate protein phosphatase inhibitor-1 was also shown to be present in the same tissue[56]. Since the physical properties of this enzyme are not yet known, comparison

Table 3. Characteristics of Smooth Muscle Phosphatases

Muscle	Subunit Composition[a]	Total Mr	Activators/Inhibitors	Substrates	Type[b]	References
Turkey Gizzard	I. 60 kDa; 55 kDa; 38 kDa (1:1:1)	165,000	-	MLC[c] (100)[d]; Phosphorylase Kinase(α Subunit)(19); Inhibitor 1 (21);Glycogen Synthase(Sites 3)(13); Phosphorylase a (4)	$2A_1$	51,59
	II. 43 kDa (one subunit)	43,000	Mg^{2+}-Dependent	MLC (100); Phosphorylase Kinase (α Subunit)(50); Glycogen Synthase:Sites 3 (4),Site 2(42),Site 1(12); Inhibitor 1 (11); Phosphorylase a (1)	2C	51,60
	III. Unknown	-	-	Myosin>MLC	-	50,58
	IV. 58 kDa; 40 kDa	150,000	Ca^{2+}, Mg^{2+}	MLC (62);Myosin (100); Heavy Meromyosin (92); Phosphorylase a (2); Phosphorylase Kinase (β Subunit) (1)	Cannot be classified	50
Bovine Aorta	Unknown	140,000	ATPMg-Dependent	Phosphorylase a	-	47,55
	Unknown	130,000	"Latent", Histone and Polylysine Activators	Phosphorylase a (100); Inhibitor 1 (9)	-	56,62
	72 kDa; 53 kDa; 35 kDa	63,000	Histone and Poly-lysine Activators	MLC;Phosphorylase a	Type 2 for MLC Type 1 for Phosphorylase a	57
	67 kDa; 38 kDa (1:1)	-	-	MLC(50);Myosin(100) Phosphorylase a(9)	$2A_2$	52

[a]Molar ratio in parentheses ; [b]Classification system proposed by Cohen's group[49] ; [c]Abbreviation:MLC, myosin light chain(s); [d]Values in parentheses represent relative activities (%).

275

with other smooth muscle phosphatases is not possible at this stage. Recently, an apparent defferent form of aortic polycation modulated phosphatase(s) was also identified[57], which migrated as a protein of Mr of 63,000 and showed a subunit structure (72 kDa, 53 kDa, 35 kDa) similar to that found for SMP-I[51]. This aortic phosphatase exhibits relatively low basal activity against phosphorylase a, 5-9-fold lower than that expressed against MLC. However, low concentrations of polylysine (or histone H_1) stimulate phosphorylase phosphatase activity 6-15-fold, while MLC phosphatase activity is virtually abolished[57]. The close parallel of ion exchange chromatographic patterns of MLC phosphatase and phosphorylase phosphatase in aortic smooth muscle suggests that the enzyme(s) may define an important functional link in coordinating contractility and glycogen metabolism in vascular smooth muscle[58]. Some important characteristics of protein phosphatases from different smooth muscles are summarized in Table 3.

THE HORMONAL CONTROL OF SMOOTH MUSCLE GLYCOGEN METABOLISM

COORDINATION WITH CONTRACTILITY

The utilization of glycogen by skeletal and cardiac muscle during contractile activity is well documented, however relatively few studies have been directed towards an elucidation of its role as an energy source and of the mechanisms controlling its synthesis or degradation in smooth muscle. Whereas cardiac and skeletal muscle phosphorylase activation has been shown to be induced by β-adrenergic stimulants, this has not been a consistent observation with all smooth muscles. Epinephrine-induced phosphorylase activation was demonstrated in uterine, tracheal, intestinal but not in arterial smooth muscle[63]. An increase in phosphorylase a activity was also observed following β-adrenergic stimulation of rat portal vein[64] and rabbit stomach[65] smooth muscle. The absence of phosphorylase activation in responce to β-adrenergic stimulation in arterial muscle may indicate that the phosphorylase activating system of this smooth muscle differs significantly from that of other muscle types[63]. It possibly involves a phosphorylase kinase isozyme similar to that we found in chicken gizzard smooth muscle[26], which is unable to be regulated by a protein phosphorylation-dephosphorylation mechanism, in as much as cAMP or dibutyryl cAMP in concentrations up to 100 μM were ineffective in changing the percentage of phosphorylase a activity of the rabbit aorta[63]. In any case, β-adrenergic agonists do not affect either resing tension or MLC phosphorylation[63,66].

It is now well supported that the inward current of the action potential in vertebrate smooth muscle is entirely different from that of nerve or skeletal muscle and is carried by Ca^{2+} [29]. It is released from intracellular storage sites or enters through channels gated by potential or by receptor stimulation[67]. The intracellular concentration of Ca^{2+} regulates the contraction-relaxation cycle in smooth muscle, mainly through the regulation of a kinase catalyzed phosphorylation of the 20 kDa phosphorylatable light chain of myosin, allowing actin-myosin interactions to occur[2,3]. In parallel,it is known that phosphorylase kinase can be activated at the same range of Ca^{2+} concentration,which activated the contractile system[29]. Silver and Stull[66] studying the relationship between MLC phosphorylation, phosphorylase a formation and isometric development during cholinergic (carbachol) stimulation of intact tracheal smooth muscle, they showed that contraction of the muscle is accompanied by increases in phosphorylation of MLC and phosphorylase. Maximal increase in phosphorylation of both substrates preceded the maximal development of isometric tension, while long-term maintenance of isometric tension was accompanied by an immediate decline in MLC phosphate content and a slower decrease of phosphorylase a formation.

Recent studies with coronary arterial smooth muscle[68] suggest that the decline in phosphorylase a formation, after long periods of stimulation, may also occur in vascular smooth muscle, when a distinct contractile agent (KCl) was used. Nevertheless, Silver and Stull[69] have reported for this particular type of stimulation (KCl depolarization) in tracheal smooth muscle, that the extent of phosphorylase a formation was maintained at a maximal value. In addition, it has been demonstrated that the activation of phosphorylase during K[+]-induced contraction of vascular smooth muscle is independent of the cAMP system[68], a result obtained also during the spontaneous contraction of uterine smooth muscle[70]. Thus, the tight temporal coupling found between activation of phosphorylase and development of isometric force in smooth muscles suggests that these events may be functionally coordinated. Such coordination is unlikely to involve activation of cAMP-dependent protein kinases and may be linked to the fact that stimulation of phosphorylase kinase and initiation of contraction are both Ca^{2+}-dependent processes[26,3]. The critical role of Ca^{2+} as a key regulator of phosphorylase activity in smooth muscle is also indicated by the experiments of Pettersson[71], who showed that phosphorylase activation by anoxia or 2,4 dinitrophenol, in bovine mesenteric artery and rabbit colon, is not dependent on cAMP, AMP, ATP or ADP and that mitochondrial Ca^{2+}-release is one of the regulatory factors of the anoxic induced glycogenolysis. In addition, nitroprusside and 8-Br-cGMP, two compounds supposed to be involved in the removal of Ca^{2+} from the cytoplasm of smooth muscle cell, through until now unknown cGMP-dependent mechanisms, have been shown to inhibit contraction and phosphorylase activation induced by α-adrenergic stimulation of KCl[72]. As far as the nature of Ca^{2+}-mediated process of phosphorylase activation in smooth muscle is concerned, Silver and Stull[73] observed that the calmodulin antagonist fluphenazine did not inhibit KCl-mediated phosphorylase a formation in intact tracheal smooth muscle. This finding suggests that calmodulin may be bound to tracheal phosphorylase kinase in the absence of Ca^{2+}, as in the case of rabbit muscle enzyme[20]. Nevertheless, such a functional property might not be a feature of other smooth muscle phosphorylase kinases[26].

ORGANIZATION OF GLYCOGEN METABOLIZING ENZYMES IN SMOOTH MUSCLE

It now seems generally accepted that structural protein of skeletal muscles, inner surfaces of cytoplasmic membrane and the surface of the membranes of subcellular structures may serve as support providing reversible binding of cytoplasmic enzymes[74]. In skeletal muscle, a significant amount of glycogen metabolizing enzymes is associated with glycogen and represent the so called "glycogen particles", which as demonstrated by electron microscopy, are either free or membrane bound[75]. In addition, it is known that the glycolytic enzymes of skeletal muscle form a multienzyme complex, which leads to the compartmentation of glycolytic process[74].

Recently, Lynch and Paul[76,77,1] have shown that exogenous glucose (and not glycogen) is the sole precursor of aerobic lactate production both in unstimulated and KCl-activated porcine carotid arteries, in spite of the substantial glycogenolysis observed in the last case. Moreover, glycogen utilization is coordinated with the increase in the rate of oxidative metabolism, which is associated with the stimulation of mechanic activity[77]. When ouabain was used to inhibit lactate production, the activity of phosphorylase a was elevated. The paradox of activation of phosphorylase in the presence of an inhibition of glycolysis, lead the authors[77] to the hypothesis that glycolysis and glycogenolysis operate in separate compartments in vascular smooth muscle.

Given the lack of basic informations concerning the molecular properties of key enzymes of glycogen metabolism in smooth muscles, their subcellular localization and their mode of interaction with glycogen, it is impossible at the present time to define the nature of the compartment in which glycogenolysis predominates in vascular smooth muscle or to extend the proposed compartmentation hypothesis to other types of smooth muscle.

CONCLUDING REMARKS

A striking feature of recent investigations into the regulation of contractile tone in smooth muscles has been the discovery that phosphorylation-dephosphorylation mechanisms play a major role in the control of actin-myosin interaction. This is a principal reason why this tissue is of particular interest especially for the study of glycogen metabolism, a metabolic process known to be synchronized with contractile energy requirements in various muscle types.

Although relatively insignificant progress has been made in the study of molecular properties of glycogen metabolizing enzymes in smooth muscle, however interesting pecularities and intricate questions concerning the regulation of this metabolic pathway have emerged during the last few years research work. Is smooth muscle phosphorylase kinase structurally and functionally dissimilar to the skeletal muscle isozyme and if so, how may isoforms of phosphorylase kinase exist in the various smooth muscle tissues? Our data[26] indicate that if hormonal control of glycogenolysis exists in chicken gizzard smooth muscle, this control cannot be exercised by a direct interaction between the cAMP-dependent protein kinase and phosphorylase kinase, as has been shown in mammalian skeletal muscle and liver tissues. Is this a unique feature of primitive vertebrate tissues or it is a property shared also by a number of mammalian smooth muscle cells? Is the hormonal control of glycogen synthesis in smooth muscle mediated through a phosphorylation-dephosphorylation mechanism similar to that found in skeletal muscle and if so which is the exact role of the various glycogen synthase kinases and phosphatases?.

Finally, the recent finding[76] that carbohydrate metabolism is compartmentated in vascular smooth muscle rise the question whether this compartmentation includes the formation of functional "glycogen particles" associated with membrane and/or contractile components of the smooth muscle cell. Additional work with purified enzymes and regulatory proteins of smooth muscle(s) glycogen metabolism cascade are needed to resolve the details of this metabolic system and to further understand the coordination of metabolic and contractile activity in smooth muscle.

REFERENCES

1. R.M. Lynch and R.J. Paul, Energy metabolism and transduction in smooth muscle, Experientia 41:970 (1985).
2. R.S. Adelstein, M.D. Pato and M.A. Conti, The role of phosphorylation in regulating contractile proteins, Adv.Cycl.Nucl.Res. 14:361 (1981).
3. K.E. Kamm and J.T. Stull, The function of myosin and myosin light chain kinase phosphorylation in smooth muscle, Ann.Rev.Pharmacol.Toxicol. 25:593 (1985).
4. R.J. Paul, Chemical energetics of vascular smooth muscle, in:"Handbook of Physiology. The Cardiovascular System,"Am.Physiol.Soc., Bethesda, MD (1980).
5. P.Cohen, Protein phosphorylation and the control of glycogen metabolism in skeletal muscle, Phil.Trans.R.Soc.Lond.B302:13 (1983).
6. F. Huijing, Glycogen metabolism and glycogen-storage diseases, Physiol. Rev. 55:609 (1975).
7. R.E. Garfield and A.P. Somlyo, Structure of smooth muscle, in:"Calcium and Contractility," A.K. Grover and E.E. Daniel, eds., The Humana Press, Inc., Clifton, New Jersey (1985).

8. S. Yonezawa and S.H. Hori, Electrophoretic studies on the phosphorylase isozymes, J.Histochem.Cytochem. 23:745 (1975).

9. E. Bueding, N. Kent and J. Fischer, Tissue specificity of glycogen phosphorylase b of intestinal smooth muscle, J.Biol.Chem. 239:2099 (1964).

10. E. Mohme-Lundholm, Smooth muscle phosphorylase and enzymes affecting its activity, Acta Physiol.Scand. 59:74 (1963).

11. L.N. Viktorova and E.V. Pamensky, Glycogen phosphorylase of smooth muscles, isolation and certain properties, Dokl.Acad.Nauk.SSSR 222: 1463 (1975).

12. L.N. Viktorova and E.V. Ramensky, Further molecular and catalytic characterization of uterine phosphorylase b, FEBS Lett. 115:239 (1980).

13. M.M. Appleman, E.G. Krebs and E.H. Fischer, Purification and properties of inactive liver phosphorylase, Biochemistry 5:2101 (1966).

14. H.D. Engers and N.B. Madsen, The effect of anions on the activity of phosphorylase b, Biochem.Biophys.Res.Commun. 33:49 (1968).

15. T.G. Sotiroudis, C.T. Cazianis, N.G. Oikonomakos and A.E. Evangelopoulos, Effect of sodium cholate on the catalytic and structural properties of phosphorylase b, Eur.J.Biochem.131:625 (1983).

16. W. Stalmans and G. Gevers, The catalytic activity of phosphorylase b in the liver, Biochem.J. 200:327 (1981).

17. A.E. Melpidou and N.G. Oikonomakos, Effect of glucose-6-P on the catalytic and structural properties of glycogen phosphorylase a FEBS Lett.154:105 (1983).

18. B. Lederer and W. Stalmans, Human liver glycogen phosphorylase.Kinetic properties and assay in biopsy specimens, Biochem.J. 159:689 (1976).

19. G.M. Carlson, P.J. Bechtel and D.J. Graves, Chemical and regulatory properties of phosphorylase kinase and cyclic AMP-dependent protein kinase, Adv.Enzymol. 50:41 (1979).

20. P. Cohen, Phosphorylase kinase from rabbit skeletal muscle, Meth. Enzymol. 99:243 (1983).

21. S.Pocinwong, H. Blum, D. Malencik and E.H. Fischer, Phosphorylase kinase from dogfish skeletal muscle. Purification and properties, Biochemistry 20:7219 (1981).

22. I.E. Adreeva, G.V. Silonova, N.B. Livanova, T.B. Eronina, V.E. Morozov and B.F. Poglazov, Purification, quaternary structure and certain immunological properties of phosphorylase kinase from chicken skeletal muscles, Biokhimiya 50:1504 (1985).

23. I.E.Andreeva, N.B. Livanova, T.B. Eronina and B.F. Poglazov, Regulatory properties of phosphorylase kinase from chicken skeletal muscles, Biokhimiya 50:1646 (1985).

24. R.H. Cooper, H.S. Sul, E. McCullough and D.A. Walsh, Purification and properties of the cardiac isoenzyme of phosphorylase kinase, J.Biol. Chem. 255:11794 (1980).

25. T.D. Chrisman, J.E. Jordan and J.H. Exton, Purification of rat liver phosphorylase kinase, J.Biol.Chem. 257:10798 (1982).

26. S. Nikolaropoulos and T.G. Sotiroudis, Phosphorylase kinase from chicken gizzard. Partial purification and characterization, Eur.J.Biochem. 151:467 (1985).

27. A. Tsutou, S. Nakamura, A. Negami, K. Mizuta, E. Hashimoto and H. Yamamura, Calcium- and calmodulin-dependent phosphorylase kinase activity in porcine uterine smooth muscle, Biochem.Biophys.Res. Commun. 126:544 (1985).

28. E. Ozawa, Energetics of smooth muscle, J.Jap.Med.Ass. 72:1322 (1974).

29. S. Ebashi, Ca^{2+} in biological systems, Experientia 41:978 (1985).

30. T.J. Singh and J.H. Wang,Stimulation of glycogen phosphorylase kinase from rabbit skeletal muscle by organic solvents, J.Biol.Chem. 254: 8466 (1979).

31. Z. Hessová, M. Varsányi and L.M.G. Heilmeyer, Jr, Dual function of calmodulin (δ) in phosphorylase kinase, Eur.J.Biochem. 146:107(1985).

32. D.A. Malencik and E.H. Fischer, Structure, function and regulation of phosphorylase kinase, in:"Calcium and Cell Function",W.Y. Cheung, ed., Academic Press, New York (1982).

33. J.R. Vandenheede, H. DeWulf and W. Merlevede, Liver phosphorylase b kinase. Cyclic AMP-mediated activation and properties of the partially purified rat liver enzyme, Eur.J.Biochem. 101:51 (1979).

34. S. Nakamura, A. Tsutou, K. Mizuta, A. Negami, T. Nakaza, E. Hashimoto and H. Yamamura, Calcium-calmodulin-dependent activation of porcine liver phosphorylase kinase, FEBS Lett. 159:47 (1983).

35. S.D. Killilea and N.M. Ky, Purification and partial characterization of bovine heart phosphorylase kinase, Arch.Biochem.Biophys. 221: 333 (1983).

36. P.J. Roach, Glycogen synthase and glycogen synthase kinases, Curr.Top. Cell.Regul. 20:45 (1982).

37. H.R. Kaslow, D.D. Lesikar, D. Antwi and A.W.H. Tan, L-type glycogen synthase. Tissue distribution and electorphoretic mobility, J.Biol. Chem. 260:9953 (1985).

38. P. Cohen, The role of protein phosphorylation in neural and hormonal control of cellular activity, Nature 296:613 (1982).

39. A. Rubulis, R.D. Jacobs and E.C. Hughes, Glycogen synthesis in mammalian uterus, Biochim.Biophys.Acta 99:584 (1965).

40. A. Milwidsky and A. Gutman, Glycogen metabolism of normal human myo-metrium and leiomyoma. Possible hormone control, Gynecol.Obstet. Invest. 15:147 (1983).

41. K.-P. Huang and J.C. Robinson, Purification and properties of the glucose-6-phosphate-dependent form of human placental glycogen synthase, Arch.Biochem.Biophys. 175:583 (1976).

42. R.J. Babcoc, Smooth muscle in the human placenta, Am.J.Obst.Gynec. 105:612 (1969).

43. K.K. Schlender and E.M. Reimann, Glycogen synthase kinases distribution in mammalian tissues of forms that are independent of cyclic AMP, J.Biol.Chem. 252:2384 (1977).

44. P.J. Silver, C. Schmidt-Silver and J. DiSalvo, β-adrenergic regulation and cAMP kinase activation in coronary arterial smooth muscle, Am.J.Physiol. 242:H177 (1982).

45. C.W. Scott and M.C. Mumby, Phosphorylation of typeII regulatory subunit of cAMP-dependent protein kinase in intact smooth muscle, J.Biol. Chem. 260:2274 (1985).

46. J. DiSalvo, D. Gifford and A. Kokkinakis, A multisubstrate Ca^{2+} and cyclic nucleotide independent kinase from vascular smooth muscle. Modulation of activity by polycations, Biochem.Biophys.Res.Commun. 136:789 (1986).

47. J. DiSalvo, J.M. Jiang, J.R. Vandenheede and W. Merlevede, The ATPMg-dependent phosphatase is present in mammalian vascular smooth muscle, Biochem.Biophys.Res.Commun. 108:534 (1982).

48. J.F. Kuo, R.G.G. Andersson, B.C. Wise, L. Mackerlova, I. Salomonsson, N.L. Brackett, N. Katoh, M. Shoji and R.W. Wrenn, Calcium-dependent protein kinase:Widespread occurrence in various tissues and phyla of the animal kingdom and comparison of effect of phospholipid, calmodulin and trifluoperazine, Proc.Natl.Acad.Sci. U S A 77:7039 (1980).

49. T.S. Ingebritsen and P. Cohen, Protein phosphatases:Properties and role in cellular regulation, Science 221:331 (1983).

50. M.D. Pato and E. Kerc, Purification and characterization of a smooth muscle myosin phosphatase from turkey gizzards, J.Biol.Chem. 260: 12359 (1985).

51. M.D. Pato, R.S. Adelstein, D. Crouch, B. Safer, T.S. Ingebritsen and P. Cohen, The protein phosphatases involved in cellular regulation. 4. Classification of two homogeneous myosin light chain phosphatases from smooth muscle as protein phosphatase-2A₁ and 2C and a homo-geneous protein phosphatase from reticulocytes active on protein

synthesis initiation factor eIF-2 as protein phosphatase-2A$_2$, Eur.J.Biochem. 132:283 (1983).

52. D.K. Werth, J.R. Haeberle and D.R. Hathaway, Purification of a myosin phosphatase from bovine aortic smooth muscle, J.Biol.Chem. 257: 7306 (1982).

53. J. DiSalvo, D. Gifford and A. Kokkinakis, Modulation of aortic protein phosphatase activity by polylysin, Proc.Soc.Exp.Biol.Med. 177:24 (1984).

54. W. Merlevede, J.R. Vandenheede, J. Goris and S.-D. Yang, Regulation of ATPMg-dependent protein phosphatase, Curr.Top.Cell.Regul. 23:177 (1984).

55. J. DiSalvo, D. Gifford, J.R. Vandenheede and W. Merlevede, Spontaneously active and ATPMg-dependent protein phosphatase activity in vascular smooth muscle, Biochem.Biophys.Res.Commun. 111:912 (1983).

56. E. Waelkens, J. Goris, J. DiSalvo and W. Merlevede, Inhibitor-1 phosphatase activity in vascular smooth muscle, Biochem.Biophys.Res. Commun. 120:397 (1984).

57. J. DiSalvo, D. Gifford and A. Kokkinakis, Properties and function of a bovine aortic polycation modulated protein phosphatase, in:"Advances in Protein Phosphatases, "W.Merlevede and J.DiSalvo,eds., Leuven University Press, Leuven (1985).

58. J. DiSalvo, G. Gifford and M.J. Jiang, Properties and function of phosphatases from vascular smooth muscle, Fed.Proc. 42:67 (1983).

59. M.D. Pato and R.S. Adelstein, Purification and characterization of a multisubunit phosphatase from turkey gizzard smooth muscle. The effect of calmodulin binding to myosin light chain kinase on dephosphorylation. J.Biol.Chem. 258:7047 (1983).

60. M.D. Pato and R.S. Adelstein, Characterization of a Mg^{2+}-dependent phosphatase from turkey gizzard smooth muscle, J.Biol.Chem. 258: 7055 (1983).

61. H. Onishi, J. Umeda, H. Uchiva and S. Watanabe, Purification of gizzard myosin light chain phosphatase, and reversible changes in the ATPase and superprecipitation activities of actomyosin in the presence of purified preparations of myosin light chain phosphatase and kinase, J.Biochem. 91:265 (1982).

62. J. DiSalvo, E. Waelkens, D. Gifford, J. Goris and W. Merlevede, Modulation of latent protein phosphatase activity from vascular smooth muscle by Histone-H$_1$ and polylysine, Biochem.Biophsy.Res. Commun. 117:493 (1983).

63. D.H. Namm, The activation of glycogen phosphorylase in arterial smooth muscle, J. Pharmacol.Exper.Ther. 178:299 (1971).

64. R.J. Paul and P. Hellstrand, Dissociation of phosphorylase a activation and contractile activity in rat portal vein, Acta Physiol.Scand. 121:23 (1984).

65. J. Debowy, Adrenergic regulation of phosphorolysis and hydrolysis of glycogen in smooth muscle of rabbit stomach in situ, Arch.Immunol. Ther.Exper. 25:863 (1977).

66. P.J. Silver and J.T. Stull, Regulation of myosin light chain and phosphorylase phosphorylation in tracheal smooth muscle, J.Biol.Chem. 257:6145 (1982).

67. T.B. Bolton, Calcium exchange in smooth muscle, in:"Control and manipulation of calcium movement," J.R. Parratt, ed., Raven Press, New York (1985).

68. P.E. Galvas, C. Kuettner, R.J. Paul and J. DiSalvo, Temporal relationships between isometric force, phosphorylase and protein kinase activities in vascular smooth muscle, Proc.Soc.Exp.Biol.Med. 178:254 (1985).

69. P.J. Silver and J.T. Stull, Phosphorylation of myosin light chain and phosphorylase in tracheal smooth muscle in response to KCl and carbachol, Mol.Pharmacol. 25:267 (1984).

70. J. Diamond, Phosphorylase, calcium and cyclic AMP in smooth muscle contraction, Am.J.Physiol. 225:930 (1973).

71. G. Pettersson, Effects of dinitrophenol on phosphorylase a activity, adenine nucleotide levels and tension in rabbit colon smooth muscle, Acta Pharmacol. Toxicol. 56:302 (1985).

72. T.M. Lincoln and R.M. Johnson, Possible role of cyclic GMP- dependent protein kinase in vascular smooth muscle function, Adv.Cyclic Nucleotide Res. 17:285 (1984).

73. P.J. Silver and J.T. Stull, Effect of the calmodulin antagonist, fluphenazine, on phosphorylation of myosin and phosphorylase in intact smooth muscle, Mol.Pharmacol. 23:655 (1983).

74. B.I. Kurganov, N.P. Sugrobova and L.S. Mil'man, Supramolecular organization of glycolytic enzymes, J.Theor.Biol. 116:509 (1985).

75. S.J.W. Busby and G.K. Radda, Regulation of the glycogen phosphorylase system. From physical measurements to biological speculations, Cur.Top.Cell.Regul. 10:89 (1976).

76. R.M. Lynch and R.J. Paul, Compartmentation of glycolytic and glycogenolytic metabolism in vascular smooth muscle, Science 222:1344 (1983).

77. R.J. Paul and R.M. Lynch, Integration of metabolism and contractility in vascular smooth muscle:Role of phosphorylation-dephosphorylation mechanisms in a functionally compartmented system, in:"Advances in Protein Phosphatases," W.Merlevede and J. DiSalvo, eds., Leuven University Press, Leuven (1985).

NOTE

When the preparation of this paper has been completed, Adreeva et al (Eur.J.Biochem., 158:99 (1986)), reexamining their previous results[23], presented evidence that phosphorylase kinase from chicken skeletal muscle can be phosphorylated by the catalytic subunit of cAMP-dependent protein kinase and by itself. Comparing these results with ours, concerning chicken gizzard phosphorylase kinase[26], we can suggest that the inability of gizzard kinase to be regulated by phosphorylation-dephosphorylation may be a characteristic property of smooth muscle of this primitive vertebrate.

DYNAMICS OF THE MEMBRANES OF CHROMAFFIN GRANULES

H.Winkler and R.Fischer-Colbrie

Dept. of Pharmacology, University of Innsbruck
A-6020 Innsbruck
Austria

INTRODUCTION

The chromaffin granules of the adrenal medulla store and secrete a complex mixture of compounds including catecholamines, nucleotides, proteins (chromogranins) and peptides (for review see Winkler and Carmichael, 1982). The membranes of these organelles have two main functions: (i) Transport of the secretory material from the Golgi region to the plasma membrane where secretion by exocytosis takes place (ii) Accumulation and modification of secretory products.
We will first discuss this latter point.

MOLECULAR FUNCTION OF THE CHROMAFFIN GRANULE MEMBRANE

Since this topic has recently been reviewed in great detail (Winkler et al., 1986) we will only discuss the main points. Fig. 1 demonstrates the main transport properties of the granule membranes for catecholamines, nucleotides and electrons. In addition the membrane is able to synthesize or modulate the secretory products. The conversion of dopamine to noradrenaline is catalysed within chromaffin granules by the enzyme dopamine ß-hydroxylase which is at least partly membrane-bound (Kirshner, 1975). Furthermore the membrane contains a carboxypeptidase H (enkephalin convertase: Fricker and Snyder, 1983; Hook, 1985), an enzyme involved in the formation of enkephalins, an important secretory product of adrenal medulla (Udenfriend and Kilpatrick , 1983). Recent studies (to be published) indicate that the so-called glycoproteins J and K (Wood et al., 1985) represent this enzyme.

Let us now return to the transport properties. Catecholamines are accumulated within chromaffin granules via a specific carrier which is driven by a chemiosmotic gradient built up by a H^+-pumping ATPase (see Njus et al., 1981). The ATPase has recently been purified by two groups (Percy et al., 1985; Nelson and Cidon, 1986). Both groups agree that subunits of this ATPase have apparent molecular weights of 70,000, 51,000 and 39,000, however additional subunits may be present.

The amine carrier has been also characterized by two groups (Gabizon and Schuldiner, 1985; Isambert and Henry, 1985). The original disagreement about the molecular weight of the carrier molecules (45,000 versus 70,000) seems at least partly resolved. Recent data of Henry's group indicate that the carrier may consist of at least two subunits (Henry et al., 1986).

Nucleotides are taken up into chromaffin granules by a saturable, carrier mediated mechanism as shown by studies with intact granules (Kostron et al., 1977a; Aberer et al., 1978; Weber and Winkler, 1981; Weber et al., 1983; Carmichael et al., 1980).

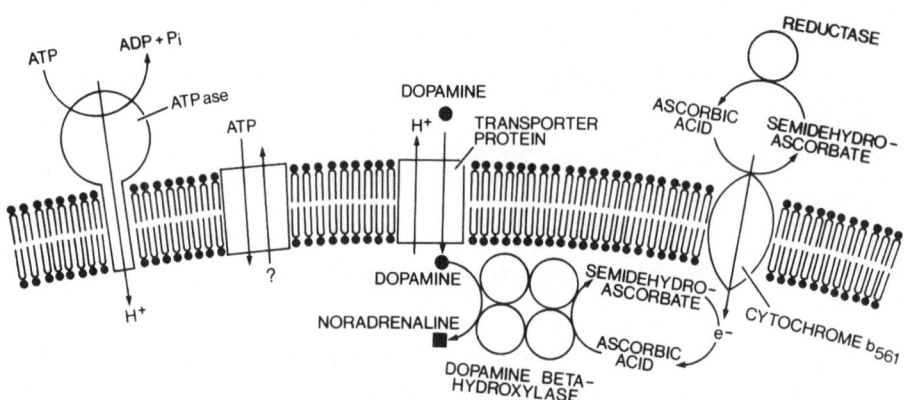

Fig. 1: Transport properties of the membranes of chromaffin granules. The membrane possesses a proton pumping ATPase which leads to an electrochemical gradient across the granule membrane. This gradient drives catecholamine and nucleotide uptake. The cytochrome b-561 acts as an electron carrier (see text for further discussion).

The carrier is relatively unspecific transporting not only various nucleotides but also phosphoenolpyruvate, phosphate and even sulphate (Weber et al., 1983). In intact granules the transport seems to depend on the membrane potential built up by the electrogenic proton transport (Weber and Winkler, 1981). Unfortunately the nucleotide transport does not work very well in ghost preparations since no gradient indicating active transport can be obtained (Grüninger et al., 1983). Apparently some co-factor or counter-ion is missing under these conditions (see Winkler, 1986, for a further discussion).

Calcium is taken up by chromaffin granules by a carrier mediated transport (Kostron et al., 1977b), which depends on a Na^+/Ca^{2+} exchange (Krieger-Bauer and Gratzl 1983; Phillips, 1981). However another group claims that calcium transport is ATP dependent (Burger et al., 1984).

Finally the membranes of chromaffin granules are likely to transport electrons via the cytochrome b-561 which is a major constituent of the granule membrane (Njus et al., 1983; Srivastava et al., 1984). The electrons are used to reduce semidehydroascorbate to ascorbate which is required by the enzymes dopamine ß-hydroxylase and probably also by the peptidyl ⍺-amidating monooxygenase (Diliberto et al., 1986).
It is interesting to note that the mechanisms just described are not confined to chromaffin granules but have apparently relevance for other storage organelles as well. For example cholinergic vesicles accumulate acetylcholine driven by a chemiosmotic mechanism (Anderson et al., 1983) and also nucleotides by an analogous transport (Luqmani, 1981). The cytochrome b-561 may be responsible for electron flow in vesicles which also require an ascorbate cycle for their ⍺ -amidating enzyme. This may be the case in the anterior hypophysis (Duong et al., 1984). Thus the membranes of transmitter and hormone storing organelles may use quite similar mechanisms to accumulate their secretory products. In analogy recent studies indicate that these storage vesicles also contain in common several neuropeptides (Udenfriend and Kilpatrick, 1983) and the so called chromogranins which originally were considered to be protein constituents typical only of chromaffin granules (see Winkler et al., 1986).

REGULATION OF SYNTHESIS OF MEMBRANES OF CHROMAFFIN GRANULES

Nervous stimulation of rat adrenal medulla by insulin or reserpine induces an increased synthesis of the enzyme dopamine ß-hydroxylase (Patrick and Kirshner, 1971; Ciaranello et al., 1975; Lima and Sourkes, 1986) which is a major constituent of chromaffin granules (Winkler, 1976). On the other hand, hypophysectomy causes a decline of dopamine ß-hydroxylase levels in the adrenal (Weinshilboum and Axelrod, 1970) probably by an increase in degradation of the enzyme (Ciaranello et al., 1975). It has also been established that secretory components of chromaffin granules i.e. enkephalins are regulated by nervous activities (Eiden et al., 1984; Wilson et al., 1984). In recent studies we have tried to answer the following two questions: (i) Are only the neuropeptides and dopamine ß-hydroxylase regulated in their synthesis or does this also apply to other granule constituents including further membrane components? (ii) What is the consequence of any such changes as far as the composition of a single granule is concerned? For these experiments rats were treated with insulin or reserpine or the effects of hypophysectomy were studied (to be published). The results are shown in Table 1.

These results demonstrate that major changes occur for the secretory products. Apparently the synthesis of the neuropeptides and chromogranin B can be significantly increased by nervous activitiy, whereas hypophysectomy induces a significant decline in chromogranin A levels. As far as the membrane components are concerned dopamine ß-hydroxylase goes up following nervous activity whereas the expected decline is seen after hypophysectomy. The second major membrane protein, i.e. cytochrome b-561 (see Winkler and Westhead, 1980) is apparently not changed under these conditions. It is interesting to note that cytochrome b-561, as outlined above, is functionally coupled to dopamine ß-hydroxylase by providing an electron flow for the ascorbate cycle. Apparently the electron flow is not rate limiting for the enzyme otherwise one would have expected both components to rise together. On the other hand recent studies (D.Sherman, M.Sietzen and H.Winkler, to be published) indicate that the levels of the amine carrier are increased after insulin treatment.

If the membrane proteins can be changed depending on various procedures this can have two consequences for the granule pool: (i) More granules are synthesized therefore e.g. more dopamine ß-hydroxylase is present.

Table 1: Levels of constituents of chromaffin granules in rat adrenals after various treatments. The results are expressed as a percentage of control levels. Insulin treatment: Rats were injected with 5 units/kg of insulin. After 2 hrs they were given glucose to stop hypoglycemia. They were killed 144 hrs later. Reserpine treatment: Rats were given three times 2.5 mg/kg reserpine on day 1, 3 and 5. 120 hrs after the last injection they were killed. Hypophysectomy: The rats were killed 14 days after hypophysectomy or after a control sham operation. Chromogranins were determined with specific antisera using the immunodot procedure. The figures are mean values of at least three animals. n.d. = not determined.

| | after | | |
	insulin	reserpine	hypophysectomy
Catecholamines	79	88	60
Chromogranin A	84	67	21
Chromogranin B	104	137	97
Dopamine ß–hydroxylase	145	236	44
Cytochrome b–561	n.d.	70	94
Total enkephalins	674	n.d.	83

(ii) The number of granules and the amount of granule membranes remain the same, however the enzyme concentration in the membranes is raised. The second possibility seems to be supported by two experimental facts. Since the levels of cytochrome b–561, a major membrane protein, do not change it seems likely that the total amount of granule membranes has remained the same and only dopamine ß–hydroxylase has been increased. A morphometric study is in agreement with this concept (Gagnon et al.,1976). After reserpine there was a slight increase in number of granules. Since they became smaller the total pool of granule membranes did not change (see Table 2) whereas the enzyme activity of isolated granule membranes was increased.

We can conclude, that under certain conditions the concentration of protein components in the granule membranes can be changed. This leads to chromaffin granules which have an altered composition. Thus the membranes of chromaffin granules show a certain degree of dynamic behaviour. On the other hand, as we will show in the next section, the chromaffin cells utilizes the same membrane for several secretory cycles.

MEMBRANES OF CHROMAFFIN GRANULES: RE-USABLE CONTAINERS

The total soluble content of chromaffin granules is secreted by exocytosis which is brought about by a process of fusion of the membranes of the granules with the plasma membranes. In order to prevent enlargement

Table 2: Morphometric analysis of chromaffin cells after reserpine. Rats were injected with reserpine (4 mg/kg i.p.) 7 times at 48 hrs intervals and killed 36 hrs after the last injection (Gagnon et al., 1976). Some of the morphometric results are shown.

	Controls	Reserpine
Chromaffin granules		
Volume density relative to cytoplasm (%)	32.1	27.6
Mean cross-sectional area (μm^2)	0.062	0.047
Surface density of membranes relative to cytoplasm ($\mu m^2/\mu m^3$)	5.1	5.5
Number of granules per unit volume cytoplasm (μm^{-3})	16.8	20.3

of the cell a mechanism must exist to retrieve membranes (Palade, 1975). Evidence for such a retrieval mechanism via coated vesicles was supplied by early studies (Diner, 1967; Nagasawa and Douglas, 1972). In 1972 we postulated (Winkler et al., 1972; see also Winkler, 1977) that these retrieved membranes are re-used for another secretory cycle. This concept was based on the observation, that the synthesis rate of the membrane proteins of chromaffin granules was significantly lower than that of the secretory proteins of the content. Decisive support for such a mechanism came from studies with exogenous tracers like cationized ferritin which during and after secretion were found to be endocytosed by cells of various endocrine organes and finally appeared in newly formed secretory vesicles (Herzog and Farquhar, 1977; Herzog and Reggio, 1980; Suchard et al., 1981). These studies demonstrated recycling of membranes but did not yet prove that the membranes of secretory vesicles were specifically retrieved and re-used. In order to demonstrate this specific markers had to be employed which were provided by antibodies against antigens present on the inner side of the granule membrane. Studies on isolated bovine chromaffin cells showed that during stimulation these antigens became exposed on the cell surface (Wildmann et al., 1981; Dowd et al., 1983; Phillips et al., 1983; Lingg et al., 1983; Patzak et al., 1984).
A further incubation of the cell after removal of the secretagogue led to a disappearance of the granule antigens from the cell surface (Phillips et al., 1983; Lingg et al., 1983; Patzak et al., 1984) with a half-life of about 10 to 20 minutes (Lingg et al., 1983; Patzak et al., 1984). Throughout this process the antigens were present on the cell surface as discrete patches (Patzak et al., 1984). Apparently the granule antigens did not spread translationally in the plasma membrane which represents a prerequisite for a specific and efficient retrieval of these membrane proteins. In order to study the further fate of these retrieved membranes, the cells were treated with gold-labelled antibodies against the specific granule protein and after further incubation were observed at the

ultrastructural level (Patzak and Winkler, 1986). It was shown, that the specifically immunolabelled membranes were retrieved via coated vesicles. After shedding their coat labelled vesicles reached the cytoplasm and the Golgi region, however label was never found within the Golgi stacks. Finally label appeared in newly formed chromaffin granules.
We can conclude:

During exocytosis membranes of chromaffin granules become part of the plasma membrane. Subsequently these membranes are specifically retrieved and finally recycled to new chromaffin vesicles. Thus these highly specialised membranes are used in an economical way as containers providing a shuttle service from the Golgi region to the plasma membrane.

CONCLUSIONS

The membranes of chromaffin granules are highly specialized to perform transport functions typical for transmitter storing organelles. These membranes are used in a very economical way as re-usable containers providing a shuttle service from the Golgi region to the plasma membrane. However these membranes are also subject to a regulated modification. Under certain conditions the composition of these membranes can be changed significantly, e.g. by increasing or decreasing the concentration of a major constituent, i.e. the enzyme dopamine ß-hydroxylase

ACKNOWLEDGEMENTS

Studies performed by H.W. quoted in this paper were supported by the Fonds zur Förderung der wissenschaftlichen Forschung.

REFERENCES

Aberer, W., Kostron, H., Huber, E., and Winkler, H., 1978, A characterization of the nucleotide uptake by chromaffin granules of bovine adrenal medulla, Biochem. J., 172:353.

Anderson, D. C., King, S. C., and Parsons, S. M., 1983, Pharmacological characterization of the acetylcholine transport system in purified torpedo electric organ synaptic vesicles, Mol. Pharmacol., 24:48.

Burger, A., Niedermaier, W., Langer, R., and Bode, U., 1984, Further characteristics of the ATP-stimulated uptake of calcium into chromaffin granules, J. Neurochem., 43:806.

Carmichael, S. W., Weber, A., and Winkler, H., 1980, Uptake of nucleotides and catecholamines by chromaffin granules from pig and horse adrenal medulla, J. Neurochem., 35:270.

Ciaranello, R. D., Wooten, G. F., and Axelrod, J., 1975, Regulation of dopamine ß-hydroxylase in rat adrenal glands, J. biol. Chem., 250:3204.

Diliberto, E. J. Jr., Menniti, F. S., Knoth, J., Viveros, O. H., and Kizer, S., 1986, Regeneration of Ascorbate for dopamine ß-hydroxylation and peptidyl-glycine ∝ -amidation, 3. Internat., Symp. Cell Biol., Coolfont, West Virgina, abstr. S13.

Diner, O., 1967, L'expulsion des granules de la médullosurrénale chez le hamster, C. r. hebd. Séane, Cada. Sci. Paris, 265:616.

Duong, L. T., Fleming, P. J., and Russell, J. T., 1984, An identical cytochrome b_{561} is present in bovine adrenal chromaffin vesicles and posterior pituitary neurosecretory vesicles, J. biol. Chem., 259:4885.

Dowd, D. J., Edwards, C., Englert, D., Mazurkiewicz, J. E., and Ye, H. Z., 1983, Immunofluorescent evidence for exocytosis and internalization of secretory granule membrane in isolated chromaffin cells, Neuroscience, 10:1025.

Eiden, L. E., Giraud, P., Affolter, H.-U., Herbert, E., and Hotchkiss, A. J., 1984, Alternative modes of enkephalin biosynthesis regulation by reserpine and cyclic AMP in cultured chromaffin cells, Proc. natn. Acad. Sci. USA, 81:3949.

Fricker, L. D., and Snyder, S. H., 1983, Purification and characterization of enkephalin convertase, an enkephalin-synthesizing carboxypeptidase, J. biol. Chem., 258:10950.

Gabizon, R., and Schuldiner S., 1985, The amine transporter from bovine chromaffin granules. Partial purification, J. biol. Chem., 260:3001.

Gagnon, C., Pfaller, W., Fischer, W. M., Schwab, M., Winkler, H., and Thoenen, H., 1976, Increased specific activity of membrane-bound dopamine ß-hydroxylase in chromaffin granules after reserpine treatment. J. Neurochem., 28:853.

Grüninger, H. A., Apps, D. K., and Phillips, J. H., 1983, Adenine nucleotide and phosphoenolpyruvate transport by bovine chromaffin granule "ghosts", Neuroscience, 9:917.

Henry, J.-P., Isambert, M.-F., Gasnier, B., Roisin, M.-P., and Scherman, D., 1986, Molecular pharmacology of the monoamine transporter, New York Acad. Sciences, Conf., June 1986, abstr. 14.

Herzog, V., and Farquhar, M. G., 1977, Luminal membrane retrieved after exocytosis reaches most Golgi cisternae in secretory cells, Proc. natn. Acad. Sci., 74:5073.

Herzog, V., and Reggio, H., 1980, Pathways of endocytosis from luminal plasmamembrane in rat exocrine pancreas, Eur. J. Cell. Biol., 21:141.

Hook, V. Y. H., 1985, Differential distribution of carboxypeptidase-processing enzyme activity and immunoreactivity in membrane and soluble components of chromaffin granules, J. Neurochem., 45:987.

Isambert, M.-F., and Henry, J.-P., 1985, Photoaffinity labelling of the tetrabenazine binding sites of bovine chromaffin granule membranes, Biochemistry, 24:3660.

Kirshner, N., 1975, Biosynthesis of the catecholamines, in: The Adrenal Gland, Hdb. of Endocrinology, eds Blaschko H., Sayers G. and Smith A.D., VI:341-355 Americ. Physiol. Soc, Washington.

Kostron, H., Winkler, H., Peer, L. J., and König, P., 1977a, Uptake of adenosine triphosphate by isolated chromaffin granules: a carrier-mediated process, Neuroscience, 2:159.

Kostron, H., Winkler, H., Geissler, D., and König, P., 1977b, Uptake calcium by chromaffin granules in vitro, J. Neurochem., 28:487.

Krieger-Bauer, H. I., and Gratzl, M., 1983, Effects of monovalent and divalent cations on Ca^{2+} fluxes across chromaffin secretory membrane vesicles, J. Neurochem., 41:1269.

Lima, L., and Sourkes, T. L., 1986, Reserpine and the monoaminergic regulation of adrenal dopamine ß-hydroxylase activity, Neuroscience, 17:235.

Lingg, G., Fischer-Colbrie, R., Schmidt, W., and Winkler, H., 1983, Exposure of an antigen of chromaffin granules on cell surface during exocytosis, Nature, 301:610.

Luqmani, Y. A., 1981, Nucleotide uptake by isolated cholinergic synaptic vesicles: evidence for a carrier of adenosine 5'-triphosphate, Neuroscience, 6:1011.

Nagasawa, J., and Douglas, W. W., 1972, Thorium dioxide uptake into adrenal medullary cells and the problem of recapture of granule membrane following exocytosis, Brain Res., 37:141.

Nelson, N., and Cidon, S., 1986, Chromaffin granule proton pump, Meth. Enzymol., in press.

Njus, D., Knoth, J., and Zallakian, M., 1981, Proton-linked transport in chromaffin granules, Curr. Top. Bioenerg., 11:107.

Njus, D., Knoth, J., Cook, C., and Kelley, P. M., 1983, Electron transfer across the chromaffin granule membrane, J. biol. Chem., 258:27.

Palade, G. E., 1975, Intracellular aspects of the process of protein synthesis, Science, N.Y., 189:347.

Patrick, R. L., and Kirshner, N., 1971, Effect of stimulation on the levels of tyrosine hydroxylase, dopamine ß-hydroxylase and catecholamines in intact and denervated rat adrenal glands, Mol. Pharmacol., 7:787.

Patzak, A., Böck, G., Fischer-Colbrie, R., Schauenstein, K., Schmidt, W., Lingg, G., and Winkler, H., 1984, Exocytotic exposure and retrieval of membrane antigens of chromaffin granules: quantitative evaluation of immunofluorescence on the surface of chromaffin cells, J. Cell Biol., 98:1817.

Patzak, A., and Winkler, H., 1986, Exocytotic exposure and recycling of membrane antigens of chromaffin granules: ultrastructural evaluation after immunolabeling, J. Cell. Biol., 102:510.

Percy, J. M., Pryde, J. G., and Apps, D. K., 1985, Isolation of ATPase I, the proton translocator of chromaffin granule membranes, Biochem. J., 231:557.

Phillips, J. H., 1981, Transport of Ca^{2+} and Na^+ across the chromaffin-granule membrane, Biochem. J., 200:99.

Phillips, J. H., Burridge, K., Wilson, S. P., and Kirshner, N., 1983, Visualization of the exocytosis/endocytosis secretory cycle in cultured adrenal chromaffin cells, J. Cell Biol., 97:1906.

Srivastava, M., Duong, L. T., and Fleming, P. J., 1984, Cytochrome b-561 catalyzes transmembrane electron transfer, J. biol. Chem., 259:8072.

Suchard, S. J., Corcoran, J. J., Pressmann, B. C., and Rubin, R. W., 1981, Evidence for secretory granule membrane recycling in cultured adrenal chromaffin cells, Cell Biol. Int. Rep., 5:953.

Udenfriend, S., and Kilpatrick, D. L., 1983, Biochemistry of the enkephalin-containing peptides, Archs Biochem. Biophys., 221:309-323.

Weber, A., Westhead, E. W., and Winkler, H., 1983, Specificity and properties of the nucleotide carrier in chromaffin granules from bovine adrenal medulla, Biochem. J., 210:789.

Weber, A., and Winkler, H., 1981, Specificity and mechanism of nucleotide transport in chromaffin granules, Neuroscience, 6:2269.

Weinshilboum, R., and Axelrod, J., 1970, Dopamine ß-hydroxylase activity in the rat after hypophysectomy, Endocrinology, 87:894.

Wildmann, J., Dewair, M., and Matthaei, D., 1981, Immunochemical evidence of exocytosis in isolated chromaffin cells after stimulation with depolarizing agents, J. Neuroimmunol., 1:353.

Wilson, S. P., Unsworth, C. D., and Viveros, O. H., 1984, Regulation of opoid peptide and processing in adrenal chromaffin cells by catecholamines and cyclic adenosine 3':5'-monophosphate, J. Neurosci., 4:2993.

Winkler, H., 1986, Composition and transport function of membranes of chromaffin granules: Established facts and unresolved topics, New York Acad. Sci., in press.

Winkler, H., 1976, The composition of adrenal chromaffin granules: an assessment of controversial results, Neuroscience, 1:65.

Winkler, H., 1977, The biogenesis of adrenal chromaffin granules, Neuroscience, 2:657.

Winkler, H., Apps, D. K., and Fischer-Colbrie, R., 1986, The Molecular function of adrenal chromaffin granules: established facts and unresolved topics, Neuroscience, 18:261.

Winkler, H., and Carmichael, S. W., 1982, The chromaffin granule, in: "The secretory Granule, eds Poisner A.M. and Trifaro J.M., pp. 3-79, Elsevier Biomedical Press, Amsterdam.

Winkler, H., Schöpf, J. A. L., Hörtnagl, H., and Hörtnagl, H., 1972, Bovine adrenal medulla: subcellular distribution of newly synthesized catecholamines, nucleotides and chromogranins, Naunyn-Schmiedebergs Arch. exp. Path. Pharmak., 273:43.

Winkler, H., and Westhead, E., 1980, The molecular organization of adrenal chromaffin granules, Neuroscience, 5:1803.

Wood, S. L., Apps, K. D., and Phillips, J., 1985, Purification of chromaffin granule membrane gylcoproteins by affinity chromatography on lectin columns, Biochem. Soc. Trans., 13:710.

THE ROLE OF ELECTROSTATIC FORCES IN REGULATING MEMBRANE CONFORMATIONAL

CHANGES INDUCED BY PROTEIN PHOSPHORYLATION

J. Barber

Department of Pure and Applied Biology
Imperial College of Science and Technology
London, SW7 2BB, U.K.

INTRODUCTION

One of the most striking demonstrations of membrane dynamics was made

sixteen years ago when it was shown that surface antigens of mouse and

human cultured cells were able to intermix when fusion was induced to form

heterokaryons (1). Since then it has been shown by a variety of techniques

that lateral diffusion of proteins occur in a wide range of membranes

including rhodopsin in photoreceptor membranes (2); lectin receptors in

myoblasts (3), fibroplasts (4), myotubes (5), glia (6) and neurons (7);

surface antigens in mast cells (8) and mouse eggs (9); acetylcholine

receptors in myoblasts and myotubes (10); hormone receptors in fibroblasts

(11) and integral proteins in erythrocytes (12), mitochondrial (13) and a

number of other membranes. For further information see reviews (14-18).

Of the many experimental approaches used to detect lateral protein

diffusion, the most powerful have been freeze-fracture electron microscopy

and Fluorescence Recovery After Photobleaching (FRAP). The latter

technique involves attaching a fluorescence marker to the membrane protein

(particularly convenient for lectin, hormone and acetylcholine receptors)

and subjecting a small area of the membrane (usually a spot of about 3 µm)

to an intense flash from a focused laser. The effect is to bleach the fluorochrome in the selected area and then monitor the fluorescence rise in this area as the fluorescence-labelled protein diffuses in (see refs. 18,19). In this way diffusion coefficients for intrinsic protein complexes can be calculated and usually fall into the range 10^{-9} to 10^{-12} $cm^{-2}s^{-1}$ for physiological temperatures (20,21).

The functional significance of lateral protein diffusion is becoming clear for a number of systems, particularly those associated with hormone-receptor triggering and cell-surface immunochemistry. Such movements may also control membrane permeability, interactions between transport components in mitochondrial and chloroplast membranes, and adenylate cyclase activity in growth control. Often associated with these various functional responses is clustering of membrane proteins to form domains. Moreover, protein aggregation within membranes can be triggered in a number of ways, such as pH changes, changing electrolyte levels, protein phosphorylation, mild action of proteases, treatments with lectins and by temperature changes. Reorganisation of membrane proteins in these ways will bring about changes in the properties of local regions on the membrane surface. For example, in the case of membrane fusion, Poste and Allison (22) have proposed that the associated redistribution and aggregation of intramembraneous particles might provide a mechanism whereby areas of membrane become devoid of internal proteins, with fusion taking place between the protein-free areas on adjacent surfaces. This proposal is, however, not applicable to those cases where close interactions occur between adjacent membranes without fusion, for example, tight and gap junctions usually have characteristic arrays of aggregated membrane protein complexes (23). Presumably these proteins impose surface characteristics which encourage membrane-membrane interaction but prevent fusion.

The brief survey above emphasises that biological membranes have considerable heterogeneity in their structures and can be highly fluid systems. The factors which control the conformation of a particular membrane may involve specific interactions or may be understood in more general terms of thermodynamics and long range non-specific forces. The entropy of mixing should favour a random distribution of proteins through-out the membrane but in practice such randomisation will be determined by other forces of interaction between the various components. In this article I want to focus attention on the structure and dynamics of the thylakoid membrane of higher plant chloroplasts which has many basic features which are likely to be highly applicable to other membrane systems.

THE THYLAKOID MEMBRANE

As shown in Fig. 1, the thylakoid membrane of higher plants is a complex folded system consisting of appressed and non-appressed regions. This membrane contains five different macroscopic protein complexes which interact together to bring about the light induced transfer of electrons and protons from water to NADP$^+$ and at the same time power the net synthesis of ATP.

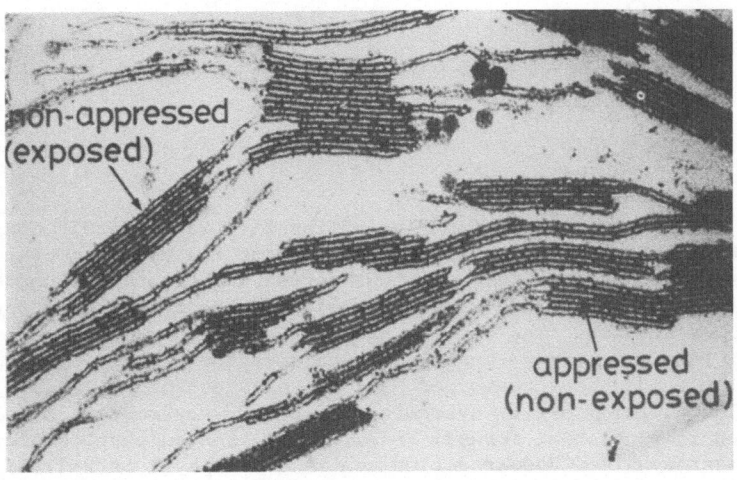

Fig.1.   Electron micrograph of thylakoids of pea chloroplasts showing appressed and non-appressed membrane regions. The non-appressed regions have their outer surface exposed to the aqueous medium while the appressed are non-exposed.

These products, $NADPH_2$ and ATP are then used to drive CO_2 fixation catalysed by the water soluble enzymes which are present in the stroma of the chloroplast. The by-product of the reaction is molecular oxygen (24). The five complexes are known as photosystem one (PS1), photosystem two (PS2), cyto—chrome b_6-f (cyt b_6-f), ATP synthase (CF_o-CF_1) and light harvesting chlorophyll a/b complex (LHC-2). As can be seen in Fig. 2 PS2 is the site of water oxidation while PS1 brings about the reduction of $NADP^+$ via ferredoxin (Fd) and a flavoprotein (Fp) having NADP-ferredoxin oxidoreductase activity. The two photosystems therefore operate cooperatively to drive electrons from H_2O to NADP and they do so with the help of the cyt b_6-f complex.

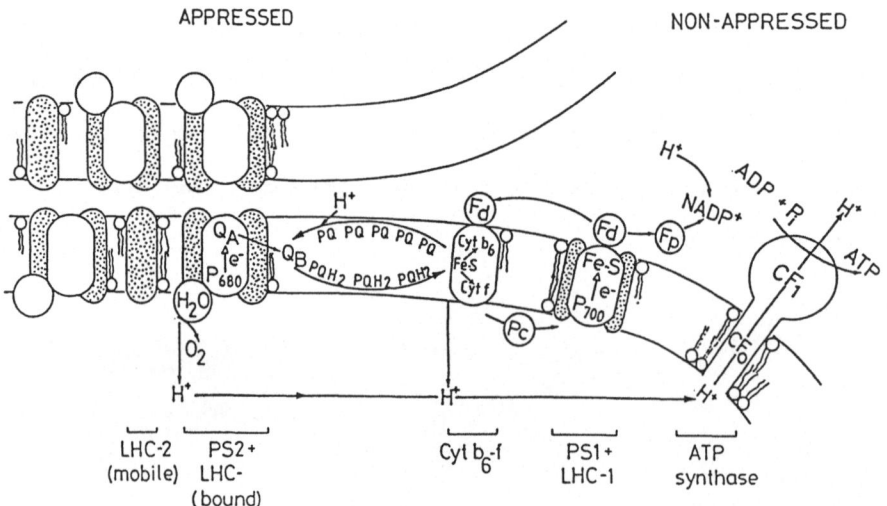

Fig.2. A diagrammatic representation of the five main protein complexes
showing how they interact to bring about the electron transfer from
water to $NADP^+$ and pump protons across the membrane in order to
create the electrochemical potential gradient necessary for driving
ATP synthesis. The symbols are: P680, reaction centre chlorophyll
of PS2; Q_A, primary stable quinone acceptor of PS2; Q_B, secondary
stable quinone acceptor of PS2; PQ, plastoquinone; PQH_2,
plastoquinol; cyt f, cytochrome f; cyt b_6, cytochrome b-653 (high
and low potential forms); Fe-S, Rieske iron sulphur centre; Pc,
plastocyanin; P700, reaction centre chlorophyll of PS1; Fe-S, iron-
sulphur centres of PS1; Fd, Ferredoxin; Fp, flavoprotein which
functions as a ferredoxin-NADP oxidoreductase; NADP, nicotinamide
adenine nucleotide phosphate; CF_o and CF_1, intrinsic and
extrinsic portions of ATP synthase respectively; LHC, light
harvesting chlorophyll a/b complex.

This complex oxidises plastoquinol (PQH_2) which is the product of the reduction of plastoquinone (PQ) by PS2, and reduces plastocyanin (Pc). The PQ/PQH_2 are localized in the lipid matrix of the membrane while Pc is an extrinsic protein associated with the inner surface of the thylakoid lumen. The role of Pc is to transfer electrons to PS1. The flow of electrons through these three complexes also brings about the vectorial movement of protons across the membrane as shown in Fig. 2. It is the resulting electrochemical potential gradient acting on the protons which is utilised to drive ATP synthesis at the CF_o-CF_1 complex. Under some circumstances reduced Fd can pass electrons to the cyt b_6-f complex and in this way promote a cyclic flow of electrons around PS1 solely for the purpose of pumping protons and synthesising additional ATP. This is known as cyclic phosphorylation and contrasts with noncyclic phosphorylation in that it does not involve the reduction of $NADP^+$ and the oxidation of H_2O. The fifth complex, LHC-2, shows no photochemical or enzymic activity but simply acts as a light harvesting system transferring energy mainly to PS2. Nevertheless, it is a major protein of the membrane and has important implications in the structural and regulatory control of photosynthetic efficiency.

Several techniques, especially the application of two phase separation procedures after mechanical disruption of isolated thylakoids, have revealed the interesting fact that the five complexes are not randomly distributed along the plane of the membrane (25). In fact as Fig 3 shows PS2 and LHC-2 are restricted mainly to the appressed membrane regions while PS1 and CF_o-CF_1 are located in the non-appressed lamellae. The position of the cyt b_6-f is a matter of debate, with most evidence suggesting that it is evenly distributed between both membrane regions. Clearly with such an organisation, lower molecular weight components, such as PQ, Pc and Fd can act as long range mobile electron carriers.

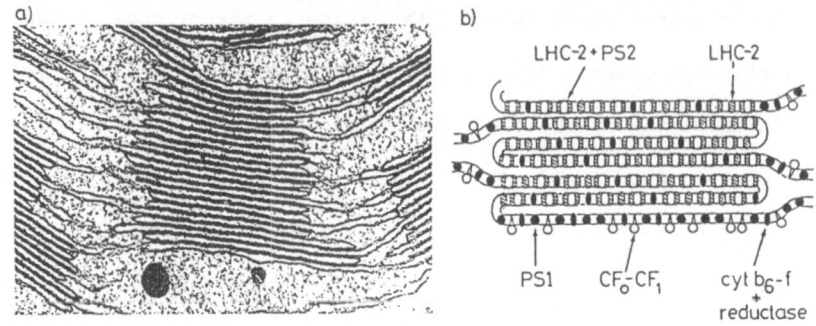

Fig.3. Distribution of the five main protein complexes in the thylakoids emphasising their lateral separation between appressed and non-appressed regions.

PHYSICAL BASIS OF THYLAKOID MEMBRANE ORGANISATION

Like all other biological membranes the surface of the thylakoid membrane carries net negative charge with the average charge density in the region of 0.025 C m^{-2} for the outer surface (26). These charges seem to be due entirely to glutamic and aspartic residues of exposed portions of integral proteins. The existence of net negative charges on the membrane give rise to the formation of a diffuse layer of counter-ions immediately adjacent to the surface. The thickness of the diffuse layer can be estimated from the Gouy-Chapman theory which states that the electrical potential (ψ_x) governing the distribution of the counter-ion falls off exponentially as a function of distance x from its maximum value at the surface (ψ_o) according to the approximate expression (27).

$$\psi_x = \psi_o \exp(-\kappa x) \qquad \cdots\cdots\cdots 1$$

The surface potential ψ_o is related to the surface charge density (σ) and concentration of salt $\Sigma_i C_{ib}$ where C_{ib} is the concentration of ion in the bulk medium. The relationship is expressed by combining the Poisson,

Boltzmann and Gauss equations to yield:

$$\sigma^2 = 2RT\varepsilon_r\varepsilon_o\Sigma_i C_{ib}(\exp\{-Z_i F\psi_o/RT\}-1) \qquad \ldots\ldots\ldots 2$$

where ε_r is the relative permittivity of the solution, ε_o is the permittivity of a vacuum, Z_i is charge carried by ion, F is the Faraday constant and T is the absolute temperature. The thickness of the diffuse layer is often defined as the distance when ψ_x drops to 1/e of ψ_o and is equal to $1/\kappa$ where:

$$\kappa^2 = \frac{2\ Z^2 F^2 \Sigma_i C_{ib}}{RT\varepsilon_o\varepsilon_r} \qquad \ldots\ldots\ldots 3$$

It is important to note that the value of ψ_o will vary with changes in the electrolyte concentration and will be highly dependent on valency. On the other hand, it is not sensitive to changes in salt levels unless specific ion binding or charge neutralization occurs. Another important parameter is the space charge density (ρ_x), since it is this quantity which relates directly to electrostatic screening at any point x in the diffuse layer and therefore to the control of coulombic repulsion between adjacent surfaces. Although changes in ρ_x and ψ_x often correlate there are situations with mixed valency electrolytes when they do not.

Some years ago it was shown that when the level of electrolytes was lowered in a suspension of chloroplast thylakoids the membranes unstacked (28) and there was a randomisation of components within the membrane as judged from freeze-fracture electron microscopy (29,30). Addition of cations to the medium induced restacking and lateral movements of intramembraneous particles to form domains which were either characteristic of appressed or non-appressed regions. Associated with these conformational modifications were changes in the yield of room temperature chlorophyll fluorescence indicative of changes in energy distribution between PS2 and PS1. I have previously given the general properties of these cation-induced

changes and argued how they can be understood in terms of electrostatic
screening of surface charge due to changes in ρ_x (26,27,31,32). The
concept presented was that as ρ_x increases the electrostatic repulsive
forces between the adjacent membrane and protein surface is decreased so that
relatively long-range van der Waals attractive forces are able to dominate.
However, although the van der Waals electrodynamic forces between macroscopic
systems can be relatively large (see 33-35) their ability to draw two
surfaces very close together will be restricted by the electrical double
layer repulsive force at short distances (36). Therefore to account for the
close inter-membrane distance of 4 nm or less within the appressed regions it
is necessary to have very low net surface charge density in this region. To
achieve this, it seems that charge displacement occurs as the two membranes
approach each other with increasing levels of cations (see Fig. 4).

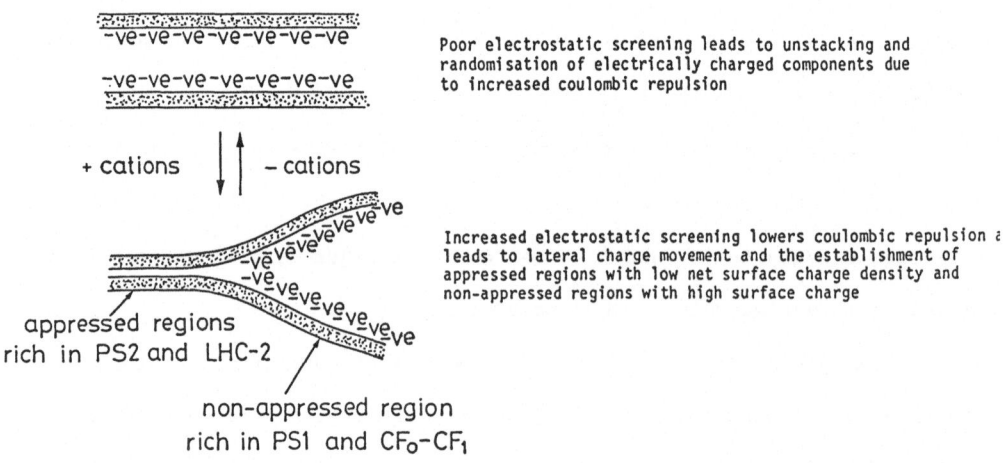

Poor electrostatic screening leads to unstacking and randomisation of electrically charged components due to increased coulombic repulsion

Increased electrostatic screening lowers coulombic repulsion and leads to lateral charge movement and the establishment of appressed regions with low net surface charge density and non-appressed regions with high surface charge

Fig.4. A charge migration model to account for the formation of appressed
and non-appressed membrane regions in response to an increase in the
electrostatic screening level of cations in the aqueous medium.
This phenomenon of "charge displacement" contrasts with charge
neutralization whereby the surface charge density is reduced by ion
binding or protonation (32). In the case of thylakoids, charge
neutralization, e.g. by lowering the pH to the membranes isoelectric
point of 4.3, leads to stacking without seggregation of complexes.

The above explanations for the cation induced changes in thylakoid
membrane organisation suggests, therefore, that the PS2-LHC-2 domains of the
appressed regions carry low net electrical charge while the non-appressed

regions containing PS1 and CF_o-CF_1 have high net electrical charge on the outer surface. This division of the thylakoid membrane into two separate domains based on differences in their surface charge character seems to have important implications for regulating photosynthetic electron flow via protein phosphorylation.

PROTEIN PHOSPHORYLATION AND LATERAL MIGRATION OF LHC-2

The formation of appressed and non-appressed membrane regions when cations are added to unstacked isolated thylakoids involves the lateral diffusion of pigment-protein complexes to two different domains. As a consequence of this energy transfer between them is reduced and the yield of chlorophyll fluorescence increased. Thus the kinetics of the fluorescence rise is indicative of the diffusion rates involved and can be compared with the fluorescence rise observed with the FRAP technique. With this concept in mind Rubin et al (37) and Briantais et al (38) were able to estimate diffusion coefficients for the lateral diffusion of the chlorophyll containing complexes to be in the region of 10^{-10} to 10^{-11} cm^2s^{-1} at room temperature. It is unlikely however that _in vivo_ such large changes in membrane organisation occur, since ionic levels probably do not fluctuate so dramatically within the intact chloroplast.

The physiological relevance of the potential for lateral diffusion of protein complexes in the thylakoid membrane did not become obvious until the discovery that some of the LHC-2 complexes can undergo a reversible phosphorylation on their outer surface (39). The importance of this phosphorylation/dephosphorylation process in relation to the regulation of energy distribution betweeen PS2 and PS1, and to the State 1-State 2 phenomenon observed in intact tissue has been extensively reviewed (26,40-43), and more recently by Barber (44,45). The phosphorylation occurs at one or two threonine residues at the N-terminal portion of the LHC-2 polypeptide (see Fig. 5) and is catalysed by a membrane bound kinase activated under conditions when the plastoquinone pool is over-reduced (either by excess PS2 light or artificially by adding appropriate chemical reductants).

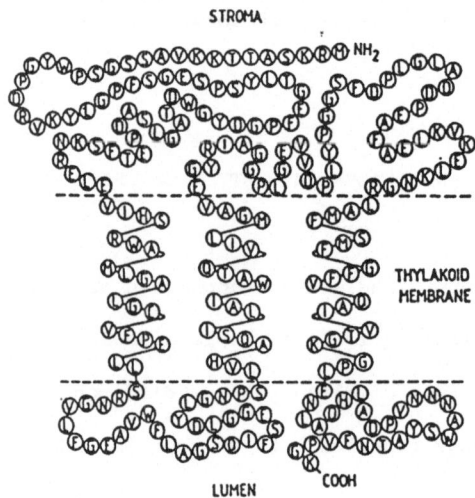

Fig.5. A folding model of pea LHC-2 polypeptide based on the amino acid sequence (46). The hydropathy determinations to obtain the model used the method of Eisenberg et al (47). At least one of the two threonine sites near the N-terminus can become phosphorylated.

When the kinase is not activated, for example, in the dark under oxidising conditions, or in excess PS1 light, a membrane bound phosphatase brings about the dephosphorylation of LHC-2.   Bennett et al (48) and Horton and Black (49) were the first to show that the phosphorylation processes were associated with changes in energy distribution between PS2 and PS1 (also see 50). From the work of Bonaventura and Myers (51) it was known that changes in energy distribution between PS2 and PS1 occurred within intact tissue so as to maximise photosynthetic efficiency under light limiting conditions (see 52). Such changes are known as State 1-State 2 transitions and were linked to the LHC-2 phosphorylation process by comparing chlorophyll fluorescence yield changes with the incorporation of radiophosphate into LHC-2 (53). From these and other experiments, it has become well accepted that the redox control of phosphorylation and dephosphorylation of LHC-2 underlies the molecular mechanism of the State transitions along the lines of that shown in Fig.6.

302

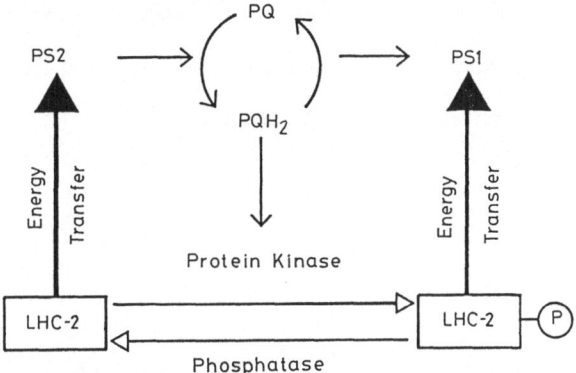

Fig.6. A scheme showing how the phosphorylation and dephosphorylation of a pool of LHC-2 dictates its ability to transfer energy to PS2 or PS1. The kinase responsible for the phosphorylation is activated when the plastoquinone pool (PQ) becomes reduced (PQH_2) due to an over-excitation of PS2 relative to PS1. The phosphatase is redox independent and thus brings about a net dephosphorylation of LHC-2 when the kinase is not active, that is under conditions when the PQ pool is oxidised due, for example, to excess PS1 relative to PS2 illumination.

How can phosphorylation of the surface of LHC-2 regulate energy distribution between PS2 and PS1? Bearing in mind the concepts presented in the previous section concerning the electrostatic control of thylakoid membrane organisation, it is logical to postulate that the phosphorylation/dephosphorylation would alter the surface electrical charge properties of LHC-2 and thus dictate its preference to partition into the appressed or non-appressed regions of the membrane (26,40). In its non-phosphorylated condition the mobile LHC-2 would be closely associated with PS2 complexes in the appressed regions while introduction of negative charges onto its surface by phosphorylation would create coulombic repulsive forces leading to its lateral migration into the PS1 enriched non-appressed membranes. In this way the phosphorylated and mobile forms of LHC-2 act either as an antenna for PS2 or PS1 (see Fig. 7). Support for this mobile antenna model has been obtained from a variety of experiments ranging from freeze-fracture electron microscopy (43,54) and biochemical analyses (55,56) to energy transfer and quantum yield studies (57,58). However, as yet the quantitative details of

303

this regulatory process are not firmly established. We do not know

precisely how much of the LHC-2 in the membrane is involved or to what degree

the LHC-2 becomes functionally linked to PS1.

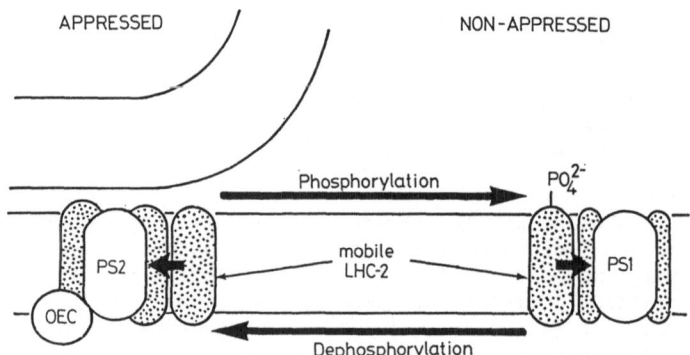

Fig.7. Diagrammatic model of how a pool of LHC-2 can act as a mobile antenna
able to transfer energy either to PS2 or PS1 depending on its state
of phosphorylation. The preferred partitioning of LHC-2 into either
appressed or non-appressed regions of the thylakoids is probably due
to the change in its surface charge properties resulting from the
phosphorylation/dephosphorylation process.

It is unclear whether only LHC moves or whether some PS2 complexes also

laterally diffuse into the non-appressed regions. If the latter occurs then

there will be changes in energy transfer betwen PS2 and PS1 (spillover) as

well as changes in the absorption cross-sections of the two photosystems.

Indeed with isolated thylakoids the degree of LHC-2 and PS2 lateral diffusion

in response to protein phosphorylation is sensitive to the level of cations

in the suspension medium. At sub-optimal levels of cations (e.g. below 5 mM)

the effect of LHC-2 phosphorylation is to cause a greater lateral mixing of

complexes as monitored by changes in the extent of chlorophyll fluorescence

(59) and by the degree of membrane unstacking (60). Such a dependency on the

background level of cations is in agreement with the concept that the

regulatory process relies on changes in electrostatic forces. Within the

intact chloroplast cation levels as well as phosphorylation changes may occur

to some extent in response to changes in light quality and intensity.

However, those experiments which have set out to investigate this possibility

have concluded that the State transitions are due almost entirely to absorption cross-section changes indicative of only LHC-2 migration (61,62).

The lateral shuffling of LHC-2 complexes in the thylakoid membrane in response to surface phosphorylation requires the lipid matrix to be fluid. In fact the acyl chains of the thylakoid polar lipids are very unsaturated (over 80% being linolenic acid) indicating that they are likely to give rise to a membrane having a hydrophobic interior with a high degree of motion (63). Indeed, experimental evidence for this contention has been obtained from time-resolved anisotropy measurements using the fluorescence probe 1,6-diphenyl-1,2,5-hexatriene (DPH) (64). Such measurements indicate that the microviscosity of this membrane was about 0.34 P at 25°C, a value which contrasts with 0.82 P at 35° for rat liver mitochondrial membranes (65). Evidence that fluidity properties of the thylakoid membrane are important for the lateral diffusion of pigment proteins has been shown both for the salt- (66) and for the phosphorylation- (67) induced movements. Thus the maintenance of a fluid lipid matrix of the thylakoids is not only important for those electron transfer processes involving long range movements (e.g. PQ and PQH_2 diffusion between PS2 and cyt b_6-f) but also for the regulation of energy distribution.

CONCLUSION

The structure of higher plant canopies and the formation of algal blooms in the aquatic environment means that the photosynthetic apparatus is often light limited and exposed to changes in the relative levels of different wavelengths of light. Such short-term changes in light quantity and quality requires flexibility in the light gathering properties of the organism so as to maintain a maximum efficiency of photosynthesis for a given irradiation condition. To achieve maximum efficiency under light limiting conditions it is necessary to excite each photosystem (PS1 and PS2) equally. In an attempt to achieve this end the chlorophyll b containing organisms seem to use a pool of LHC-2 as a mobile antenna which can either transfer energy

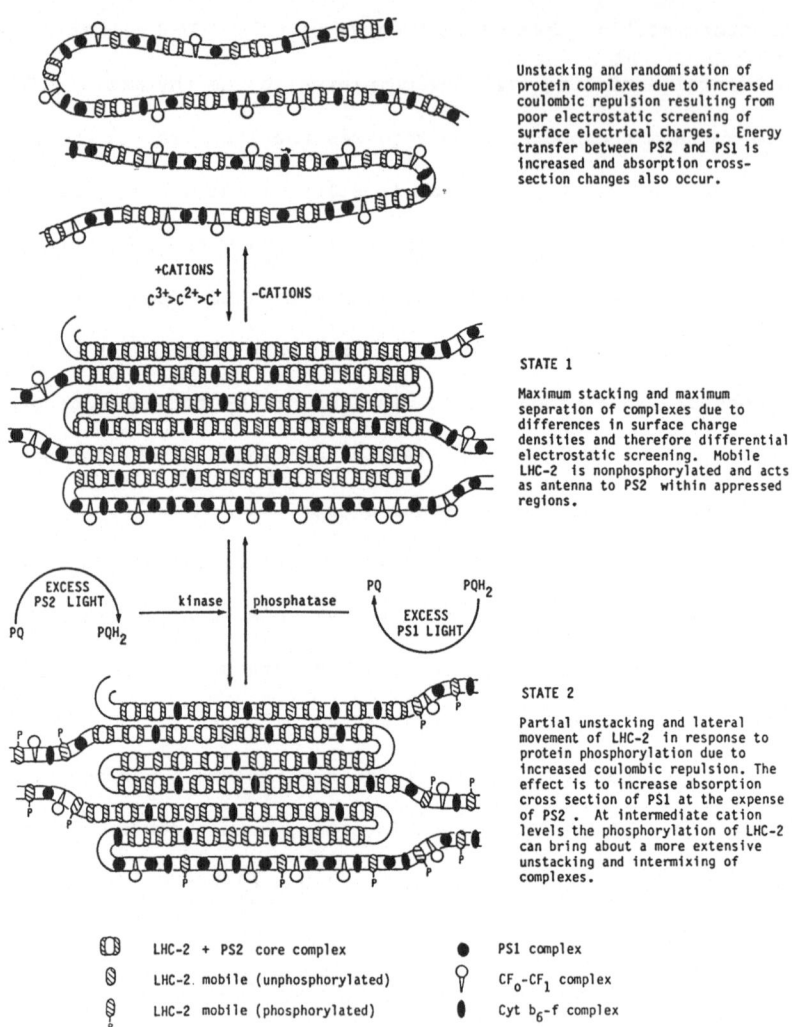

Unstacking and randomisation of protein complexes due to increased coulombic repulsion resulting from poor electrostatic screening of surface electrical charges. Energy transfer between PS2 and PS1 is increased and absorption cross-section changes also occur.

+CATIONS
$c^{3+} > c^{2+} > c^+$
-CATIONS

STATE 1

Maximum stacking and maximum separation of complexes due to differences in surface charge densities and therefore differential electrostatic screening. Mobile LHC-2 is nonphosphorylated and acts as antenna to PS2 within appressed regions.

EXCESS PS2 LIGHT   kinase   phosphatase   PQ   PQH₂
PQ   PQH₂   EXCESS PS1 LIGHT

STATE 2

Partial unstacking and lateral movement of LHC-2 in response to protein phosphorylation due to increased coulombic repulsion. The effect is to increase absorption cross section of PS1 at the expense of PS2. At intermediate cation levels the phosphorylation of LHC-2 can bring about a more extensive unstacking and intermixing of complexes.

LHC-2 + PS2 core complex   ● PS1 complex
LHC-2 mobile (unphosphorylated)   CF₀-CF₁ complex
LHC-2 mobile (phosphorylated)   Cyt b₆-f complex

Fig.8. Diagrammatic representation of changes in thylakoid membrane organisation in response either to LHC-2 phosphorylation/ dephosphorylation or to modification of cation levels.

to PS2 or PS1 depending on the redox condition of the PQ pool. The PQ pool is a sensitive redox detector of the relative excitation of PS1 and PS2 and is therefore an ideal trigger for controlling the activity of the membrane kinase responsible for phosphorylating LHC-2. The effect of changing the phosphorylation state of LHC-2 is to alter its surface electrical properties which controls partitioning between the PS2 enriched appressed or PS1 enriched non-appressed regions of the thylakoid membrane. The concept that lateral movement of LHC-2 in response to phosphorylation is due to changes in surface electrical charge comes from detailed studies of the properties of

306

salt induced changes in thylakoid membrane organisation. The basic concepts
of the two effects are summarised in Fig. 8. Thus in the thylakoid membrane
system we have an interesting example of the regulation of the functional
interaction of intrinsic protein complexes simply by changing the surface
charge properties. This feature may not be unique to this particular
membrane and may underlie the reason for phosphorylation of components in
other membrane systems.

ACKNOWLEDGEMENTS

I wish to thank the Agricultural and Food Research Council (AFRC) and
the Science and Engineering Research Council (SERC) for financial support.

REFERENCES

1.   L. D. Frye, and M. Edidin, J. Cell Sci., 7:319-335 (1970).

2.   M. M. Poo and R. A. Cone, Nature 274: 438-440 (1974).

3.   J. Schlessinger, D. E. Koppel, D. Axelrod, K. Jacobson, W. W. Webb, and
          E. L. Elson, Proc. Natl. Acad. Sci. U.S.A., 73: 2409-2413
          (1976).

4.   B. A. Smith, W. R. Clarke, and H. M. McConnell, Proc. Natl. Acad. Sci.
          U.S.A., 76:5641-5644 (1979).

5.   D. Axelrod, P. Raudin, D. E. Koppel, J. Schlessinger, W. W. Webb, E. L.
          Elson, and T. R. Podleski, Proc. Natl. Sci. U.S.A., 73:4594-4598
          (1976).

6.   Y. Zagyansky, P. Brenda, and J. C. Bisconte, FEBS Lett., 77:206-208
          (1977).

7.   S. W. DeLaat, P. T. van der Saag, E. L. Elson, and J. Schlessinger,
          Biochim. Biophys. Acta, 558:247-250 (1979).

8.   J. Schlessinger, W. W. Webb, E. L. Elson, and H. Metzger, Nature,
          264:550-552 (1976).

9.   M. Johnson, and M. Edidin, <u>Nature</u> 272:448-450 (1978).

10.   M. M. Poo, J. W. Lam, N. Orida, and A. W. Chao, <u>Biophys. J.</u>, 26:1-22 (1979).

11.   J. Schlessinger, Y. Schechter, P. Cuatreeasas, M. C. Willingham, and I. Pastan, <u>Proc. Natl. Acad. Sci. U.S.A.</u>, 75:5353-5357 (1978).

12.   V. Fowler, and D. Branton, <u>J. Cell Biol.</u>, 63:100-108 (1977).

13.   M. Hochli, and C. R. Hackenbrock, <u>Proc. Natl. Acad. Sci. U.S.A.</u>, 76:1236-1240 (1979).

14.   M. Edidin, <u>in</u>: "New Comprehensive Biochemistry" vol 1, pp. 37-82, Elsevier, Amsterdam, (1981).

15.   J. Schlessinger and E. L. Elson, <u>Meth. Exp. Phys.</u>, 20:197-227 (1982).

16.   R. Peters, <u>Naturwissenschaften</u>, 70:294-302 (1983).

17.   D. Axelrod, D. <u>J. Membrane Biol.</u>, 75:1-10 (1983)

18.   H.-G. Kapitza and K. A. Jacobson, <u>in</u>: "Techniques for the Analysis of Membrane Proteins" C. I. Ragen and R. J. Cherry, eds. pp. 345-375, Chapman-Hull, London, New York, (1986).

19.   D. Axelrod, D. E. Koppel, J. Schlessinger, E. L. Elson and W. W. Webb, <u>Biophys. J.</u>, 16:1055-1069 (1976).

20.   R. J. Cherry, <u>Biochim. Biophys. Acta</u>, 559:289-327 (1979).

21.   R. J. Cherry, <u>in</u>: "Membranes and Transport" A. N. Martonosi, ed., Vol. 1, pp. 145-152, Plenum Publishing Co., N.Y. (1982).

22.   G. Poste and A. C. Allison, <u>Biochim. Biophys. Acta</u>, 300:421-465 (1973).

23.   S. Weinbaum, <u>J. Theor. Biol.</u>, 83:63-92 (1980).

24.   J. Barber and N. R. Baker, <u>in</u>: "Photosynthesis and the Environment", Vol. 5, Topics in Photosynthesis, Elsevier, Amsterdam (1985).

25.   J. M. Anderson, <u>FEBS Lett.</u>, 124:1-10 (1981).

26.   J. Barber, <u>Ann. Rev. Plant Physiol.</u>, 33:261-295 (1982).

27.   J. Barber, <u>Biochim. Biophys. Acta</u>, 594:253-308 (1980).

28.   S. Izawa and N. E. Good, <u>Plant Physiol.</u>, 41:544-552 (1966).

29.   A. Y. I. Wang, and L. Packer, <u>Biochim. Biophys. Acta</u>, 305:488-492 (1973).

30. L. A. Staehelin, J. Cell Biol., 71:136-138 (1976).

31. J. Barber, J. Mills and A. Love, FEBS Lett., 74:174-181 (1977).

32. J. Barber, FEBS Lett., 188:1-10 (1980).

33. V. A. Parsegian, Ann. Rev. Biophys. Bioeng., 2:221-255 (1973).

34. J. N. Israelachvili, Q. Rev. Biophys., 4:341-387 (1974).

35. D. Langbein, in "Theory of van der Waals Attraction" Springer-verlag, Berlin (1974).

36 M. J. Sculley, J. T. Duniec, S. W. Thorne, W. S. Chow and N. K. Boardman, Arch. Biochem. Biophys., 201:339-346 (1980).

37. B. T. Rubin, J. Barber, G. Paillotin, W. S. Chow and Y. Yamamoto, Biochim. Biophys. Acta, 638:69-74 (1981).

38. J.-M. Briantais,C. Vernotte,J. Lavorel, J. Olive and F. A. Wollman in: "6th Int. Congr. Photosynth.,"Brussels, Abstract 1, pp 284 (1983)

39. J. Bennett, Nature 269:344-346 (1977).

40. J. Barber, Photobiochem. Photobiophys., 5:181-190 (1983).

41. J. Bennett, Biochem. J., 212:1-13 (1983)

42. P. Horton, FEBS Lett., 152:47-52 (1983).

43. L. A. Staehelin, and C. J. Arntzen, J. Cell Biol., 97:1327-1337 (1983).

44. J. Barber, in: "Photosynthesis III" Encyc. Plant Physiol., L. A. Staehelin and C. J. Arntzen eds. Vol. 19, pp. 653-664, Springer-Verlag, Berlin.

45. J. Barber, Photosyn. Res. (1986), in press.

46. A. R. Cashmore, Proc. Natl. Acad. Sci. U.S.A., 81:2960-2964 (1984).

47. D. Eisenberg, R. M. Weiss, T. C. Terwilliger and W. Wilcox, Faraday Symp. Chem. Soc., 17:109-120 (1982).

48. J. Bennett, K. E. Steinback and C. J. Arntzen, Proc. Natl. Acad. Sci. U.S.A., 77:5253-5257 (1980).

49. P. Horton and M. T. Black, FEBS Lett., 119:141-144 (1980).

50. J. F. Allen, J. Bennett, K. E. Steinback and C. J. Arntzen, Nature, 291:21-25 (1981).

51. C. Bonaventura and J. Myers, Biochim. Biophys. Acta, 189:366-383
        (1969).

52. J. Barber, in: "The Intact Chloroplast" Vol. 1, Topics in
        Photosynthesis, J. Barber ed., pp 89-134, Elsevier, Amsterdam
        (1975).

53. A. Telfer, J. F. Allen, J. Barber and J. Bennett, Biochim. Biophys.
        Acta, 722:176-181 (1983).

54. D. J. Kyle, L. A. Staehelin and C. J. Arntzen, Arch. Biochem. Biophys.,
        222:527-541 (1983).

55. B. Andersson, H.-E. Akerlund, B. Jergil and C. Larsson, FEBS Lett.
        149:181-185 (1982).

56. D.J. Kyle, T.-Y. Kuang, J. L. Watson and C. J. Arntzen, Biochim.
        Biophys. Acta, 765:89-96 (1984).

57. J. Farchaus, R. A. Dilley and W.A. Cramer, Biochim. Biophys. Acta,
        809:17-26 (1985).

58. A. Telfer, H. Bottin, J. Barber and P. Mathis, Biochim. Biophys. Acta,
        764:324-330 (1984).

59. A. Telfer, M. Hodges and J. Barber, Biochim. Biophys. Acta, 724:167-175
        (1983).

60. A. Telfer, M. Hodges, P. A. Millner and J. Barber Biochim. Biophys.
        Acta, 766:554-562, (1984).

61. M. Hodges and J. Barber, Plant Physiol., 72:1119-1122 (1983).

62. S. Malkin, A. Telfer and J. Barber, Biochim. Biophys. Acta, 848:42-48
        (1986).

63. K. Gounaris and J. Barber Trends in Biochem. Sci., 8:378-381 (1983).

64. R. C. Ford and J. Barber Biochim. Biophys. Acta, 722: 341-348 (1983).

65. K. Kinosita, R. Kataoka, Y. Kimura, O. Gotoh and A. Ikagami,
        Biochemistry, 20:4270-4277 (1981).

66. J. Barber, W. S. Chow, C. Scoufflaire and R. Lannoye, Biochim. Biophys.
        Acta, 591:92-103 (1980).

67. P. Haworth, Arch. Biochem. Biophys., 226:145-154 (1983).

OSCILLATIONS IN WHEAT CHLOROPLAST PHOTOCHEMICAL ACTIVITY:

EFFECT OF UNCOUPLERS

Shabbir A. Sayeed and Prasanna Mohanty

School of Life Sciences
Jawaharlal Nehru University
New Delhi - 110067, India

INTRODUCTION

Temporal oscillations in morphological, physiological and biochemical parameters of plants are ubiquitous. Rhythmic phenomena in photosynthesis have been, found to be endogenous in nature and have been studied in algae as well as in higher plants (Edmunds, 1984; Sweeney, 1979; Lonergan, 1984b; Sweeney and Prezelin, 1978). Among the O_2 evolving photosynthetic organisms, with the exception of cyanobacteria a large number of algal species representing almost all major taxonomic groups seem to exhibit rhythmic property in their photosynthetic capacity. However, the number of reports on photosynthesis rhythms in higher plants is limited to only a few species. Observations on the rhythmicity in photosynthesis have been made both under field and laboratory growth conditions. Time-dependent reversible variations in the chloroplast shape, movement and in the ultrastructure of thylakoid membranes in the case of algal cells have been investigated in detail (Sweeney, 1979). While many algal species did not show any ocillations in the chlorophyll (Chl) content (Harding et al., 1981; Prezelin and Matlick, 1980),a few species have been shown to exhibit diurnal variations in their cellular pigment contents (Laval-Martin et al., 1979; Owens et al., 1980).

The fact that none of the enzymes involved in the Calvin cycle reactions exhibited rhythmic changes in their activity profile has led to the suggestion that a mechanism for the origin of the observed rhythms may exist in the light reactions of photosynthesis. Photosynthetic capacity measured either as CO_2 fixation or whole cell O_2 evolution in algae (Edmunds, 1984) or in higher plants (Lonergan, 1981; Wagner, 1977) was rhythmic. Though whole chain electron transport oscillated rhythmically in many systems (Lonergan, 1981; 1984a; Lonergan and Sargent, 1979; Samuelsson et al., 1983), the partial electron transport reactions pertaining to photosystem II (PS II) and photosystem I (PS I) did not always exhibit similar rhythmic changes (Lonergan, 1981; Lonergan and Sargent, 1979). However, there are two reports on the oscillations in PS II activity (Okada and Horie, 1979; Samuelsson et al., 1983). While most of the reported photosynthesis rhythms are circadian in nature (Lonergan, 1984b), some non-circadian type of rhythms with shorter period lengths have also been reported (Wagner, 1977).

Many attempts have been made to understand the mechanism of the photosynthesis-rhythms and its possible localization to the electron transport chain. Studies on the rhythms in Gonyaulax (Prezelin and Sweeney, 1977), Euglena (Lonergan and Sargent, 1979; Lonergan, 1983) and Pisum (Lonergan, 1981) have ascribed a role for the thylakoid membranes in the generation of rhythms. Temporal changes in the membrane conformation is suggested to be responsible for the rhythmic variations in photochemical activity of the chloroplasts. In the present report, we describe our studies on the oscillations in the photochemical activity of chloroplasts isolated from the primary leaves of wheat seedlings grown under continuous light. Endogenous oscillations of about 12-16 h periodicity have been observed in PS II and PS I activity in the absence as well as in the presence of photophosphorylation uncouplers such as ammonium chloride and carbonyl cyanide 4-(trifluoromethoxy·) phenylhydrazone (FCCP). The results suggest that rhythmic changes in the physical state of the thylakoid membranes may be responsible for the oscillations in the electron transport activity.

MATERIALS AND METHODS

Healthy seeds of Triticum aestivum L. cv. Kalyansona obtained from Indian Agricultural Research Institute, New Delhi, were germinated on moist germination paper in petri-dishes under continuous illumination ($\simeq$ 20 W m^2) at $25° \pm$ 1° C. The seedlings were supplied with half-strength Hoagland nutrients twice in the 12-15 days growth period. About 30 fully expanded primary leaves from 12 to 15 day old seedlings were sampled at regular intervals (four hours) for isolation of chloroplasts as described elsewhere (Sayeed et al., 1983). The isolation medium contained 5 mM TES (pH 7.8), 0.4 M sucrose and 1 mM Na_2 EDTA. All operations were carried out at 4° C under dim green light. Chlorophyll was estimated according to Arnon (1949). Measurements of photochemical activity were started at sun rise time and is denoted as zero hour in the figures.

PS II and PS I activity measurements were made polarographically using a Clark-type O_2 electrode at 25° ± 0.1°C as described in Sayeed et al. (1985). PS II activity was estimated as H_2O to PD_{ox} + FeCN assay (O_2 evolution). Three ml of the reaction medium contained 50 mM TES (pH 7.5), 10 mM KCl, 5 mM $MgCl_2$, 1 mM $K_3Fe(CN)_6$ and 100 µM phenylene-diamine (PD_{ox}). When present the final concentrations of FCCP and ammonium chloride were 5 µM and 5 mM, respectively. Chloroplasts equivalent to 40 to 50 μg Chl were taken for each measurement. PS I activity was assayed as O_2 consumption due to photoreduction of methylviologen (MV) and its subsequent autooxidation using 2,6-dichloro-phenol indophenol (DCPIP) and sodium ascorbate as electron donors. The assay medium consisted of 50 mM TES (pH 7.5), 10 mM KCl, 5 mM Mg Cl_2, 100 μM DCPIP, 2 mM sodium ascorbate, 1 mM sodium azide, 0.5 mM MV and 10 μM DCMU. Concentration of the uncouplers and the amount of chloroplasts were same as in the PS II assay condition.

RESULTS

Oscillations in Photosystem II Activity

Measurements of electron flow from water to ferricya-nide (FeCN) do not constitute an exclusive PS II assay since, FeCN is known to accept electrons at both PS II and PS I reducing sides reflecting mostly the whole chain electron transport (Trebst, 1974). However, oxidized phenylenediamines, the lipophilic PS II electron acceptors (Saha et al., 1971), when used in combination with FeCN mediate the electron flow effectively to the latter, thus constituting a reliable PS II assay system. We have, there-fore, used H_2O to PD_{ox} + FeCN as an assay system for the estimation of PS II catalyzed O_2 evolution.

PS II catalyzed O_2 evolution in the chloroplasts isola-ted from mature primary leaves of wheat seedlings grown under continuous light exhibited rhythmic oscillations as a function of time of day (Fig. 1). The oscillations in PS II activity measured in the absence of any uncouplers of photophosphorylation showed a periodicity of about 12 h. The activity peaks appeared both in the presumptive day and night phases of the cycles with a minimum to maximum variation of 1.5 to 1.6 fold. Addition of the uncoupler FCCP to the reaction medium did not change the basic pattern of the oscillation. However, no uncoupler mediated enhance-ment in PS II activity was noticed at any time during the two day long observation period, instead inhibition of the activity was observed at most of the measurement points. The amplitude of the oscillation in the presence of FCCP ranged between 1.3 and 1.9 fold. Another uncoupler of photophosphorylation, NH_4Cl also elicited a similar res-ponse, though the extent of inhibition of PS II activity was much less as compared to FCCP. The lack of uncoupler mediated enhancement in PS II activity was probably due to the presence of PD_{ox} in the assay system which is known to enhance the rates of PS II catalyzed O_2 evolution to maximum levels, nearly equalizing the rates obtained with uncouplers (Saha et al., 1971).

Oscillations in Photosystem I Activity

PS I activity of chloroplasts isolated from the primary leaves of wheat seedlings grown under continuous light was rhythmic in the absence as well as in the presence of the uncouplers (Fig. 2) with a periodicity of about

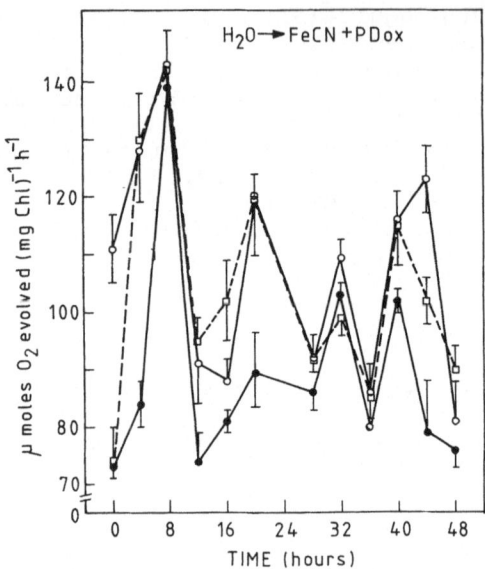

Fig. 1.Oscillations in photosystem II supported O_2 evolution activity of
chloroplasts isolated from the primary leaves of continuous light
grown wheat seedlings. O, without
any uncoupler; ●, with FCCP; □ ,
with NH_4Cl. Other details are given
in the text.

16 h. A distinct pattern of oscillation in the PS I activity is being reported for the first time, in wheat. The
amplitude of the oscillation observed in the absence of
uncouplers varied between 2.2 and 4.7 fold. Presence of
either FCCP or NH_4 Cl in the reaction medium did not
affect the periodicity of the rhythm, though the amplitude
was different and ranged between 1.4 and 1.9 fold. Unlike
in the case of PS II activity, presence of the uncouplers
enhanced the PS I activity considerably throughout the
observation period. The extent of uncoupler mediated enhancement in PS I activity was similar whether FCCP or
NH_4Cl was used. Maximum enhancement effect of the uncouplers
was observed only when the activity measured in the absence
of the uncouplers was at the minimal levels. During the
peak phases of PS I mediated electron flow the thylakoid
membranes seemed to have a reduced sensitivity to the
chemicals. That the time dependent sensitivity of the chlo-
ropolast membranes to the uncouplers is not due to differen-
tially coupled chloroplasts isolated at various times of
day is demonstrated by the synchronous oscillations in
the presence and absence of the uncouplers.

DISCUSSION

Partial electron transport reactions catalyzed by
PS II and PS I in chloroplasts isolated from the primary

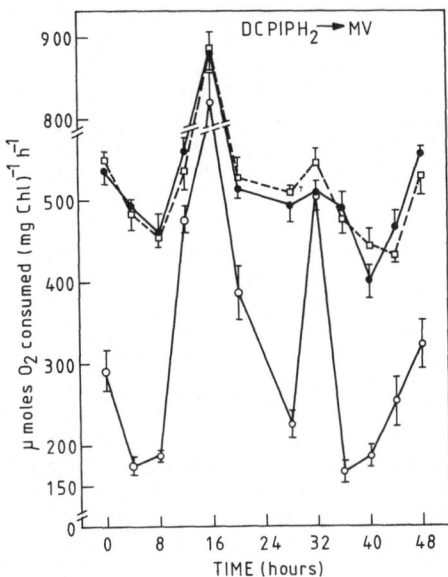

Fig. 2. Oscillations in photosystem I activity of chloroplasts isolated from the primary leaves of continuous light grown wheat seedlings. O, without any uncoupler; ●, with FCCP; □, with NH_4Cl. Other details are given in the text.

leaves of continuous light grown wheat seedlings exhibit rhythmic oscillations. While three species of legumes and maize have been shown to exhibit rhythms in their electron transport activity (Lonergan, 1981; 1984a), our results presented here constitute the first observation on oscillations in the photochemical reactions of wheat, a C_3 cereal crop plant. PS II and PS I activity monitored at regular intervals in wheat chloroplasts manifested a rhythmic pattern with a non-circadian periodicity of about 12 and 16 h, respectively. This observation is in contrast to most of the previously reported photosynthesis rhythms where the period was close to 24 h (Lonergan, 1984b), typical of circadian type rhythms. However, similar to our observation, non-circadian type of rhythms in the net photosynthesis (Cumming, 1972), and Chl accumulation pattern (Wagner and Cumming, 1970) with 15 h periodicity have been reported for Chenopodium rubrum.

The 12 to 16 h rhythmicity in the partial photochemical reactions was not influenced by the constant high light (~20 W m^2) provided from the germination stage onwards. The persistence of these oscillations under constant light and temperature (25°C) conditions indicates that the oscillatory phenomena in wheat chloroplast photochemical activity have an endogenous origin.

Both the PS II and PS I rhythms exhibited peak phases of activity in the subjective day and night phases of the cycle in the case of continuous light grown plants similar to the rhythmicity observed in light-dark cycle grown plants (data not shown). The oscillations in wheat chloroplast activity were clearly discernible when monitored in the presence or in the absence of photophosphorylation uncouplers. The oscillations observed in the presence of the uncoupler were in phase with those observed in the absence of the chemicals. Earlier, Lonergan(1981) reported the presence of oscillations in Pisum chloroplasts only when measured in the presence of uncouplers such as gramicidin and NH_4Cl. The synchronous oscillations in the presence and absence of the uncouplers suggest that the observed rhythmicity is not due to differential degree of coupling between electron transport and photophosphorylation in the chloroplasts isolated at various times of day, instead it is an endogenous property of the chloroplast membranes to modulate electron transport in a time dependent manner. The thylakoid membranes exhibited temporal sensitivity to the chemicals such that the highest degree of enhancement in PS I photochemical activity was observed only when the basal activity was at the minimal level. The uncoupler effect seemed to be restricted to a few hours of the day, an observation similar to that of Lonergan (1981), in Pisum chloroplasts. Moreover, both NH_4Cl and FCCP, though have different mechanisms of uncoupling (Good, 1977), produced similar effects on the electron transport rhythms. This observation further indicates that the rhythmic vartiations in the membrane functions are due to endogenous changes in the thylakoids rather than a response to the chemicals.

The oscillatory patterns observed in partial photochemical reactions suggest that they are the constituent parts of the whole chain electron transport rhythm (data not shown) observed in wheat chloroplasts. The rhythmicity in PS II and PS I activities measured under saturating light conditions, however, exhibited a difference in the peak timings. The peak phase of PS II activity seemed to alternate with that of PS I activity, thus suggesting an out-of-phase relationship between the activity profiles of the two photosystems during the day. This may be due to time-dependent changes in the conformation of the thylakoid membranes, thus regulating the relative amounts of electron flow through PS II and PS I units. Prezelin and Sweeney (1977) have also suggested that a reversible conformational change in the thylakoid membranes of Gonyaulax polyedra may be responsible for the rhythm in O_2 evolution in whole cells. It is possible that in wheat chloroplasts the relative populations of active PS II and PS I units undergo a reversible temporal change in order to regulate the electron flow, and also possibly the excitation energy distribution between the two photosystems.

In conclusion, the partial electron transport activity measured as PS II and PS I assays in wheat chloroplasts isolated from the continuous light grown seedlings showed an endogenous non-circadian type of oscillations with periods of about 12 to 16 h. The oscillations observed in the absence of uncouplers also persisted when monitored in their presence. The thylakoid membranes exhibited a time dependent sensitivity to the chemical uncouplers suggesting possible oscillations in the degree of coupling between electron transport and photophosphorylation in the chloroplasts in situ. The peak timings of PS II rhythm alternated with those of PS I rhythm. A reversible change in the number of active photosystem units may be responsible for such oscillations. The observed rhythmic phenomena in chloroplast functions are suggested to be due to temporal modulation of the physico-chemical state of the thylakoid membranes.

ACKNOWLEDGEMENTS

The work was supported by an ICAR-USDA (FG-IN-575) grant to PM. SAS thanks the Council of Scientific and Industrial Research for a senior fellowship.

REFERENCES

Arnon, D.I., 1949, Copper enzymes in isolated chloroplasts. Polyphenol oxidase in Beta vulgaris, Plant Physiol. 24: 1-15.

Cumming, B.G., 1972, The role of circadian rhythmicity in photoperiodic induction in plants, in: Proceedings of International Symposium on Circadian Rhythmicity (Wageningen), pp 33-85.

Edmunds, Jr., L.N., 1984, Physiology of circadian rhythms in microorganisms, in: Advances in Microbial Physiology, A.H. Rose and D.W. Tempest, eds., 25: 61-148, Academic Press, London.

Good, N.E., 1977, Uncoupling of electron transport from phosphorylation in chloroplasts, in: Encyclopeida of Plant Physiology, Photosynthesis-I, A.Trebst and M. Avron, eds., New Series, Vol. 5, pp. 429-436, Springer-Verlag, Berlin.

Harding, Jr., L.W., Meeson, B.W., Prezelin, B.B., and Sweeney, B.M., 1981, Diel periodicity of photosynthesis in marine phytoplankton, Mar. Biol., 61: 95-105.

Laval- Martin, D.L., Shuck, D.J., and Edmunds, Jr., L.N., 1979, Cell cycle related and endogenously controlled circadian photosynthetic rhythm in Euglena, Plant Physiol., 63: 495-502.

Lonergan, T.A., 1981, A circadian rhythm in the rate of light induced electron flow in three leguminous species, Plant Physiol., 68: 1041-1046.

Lonergan, T.A., 1983, Regulation of cell shape in Euglena gracilis. I. Involvement of the biological clock, respiration, photosynthesis and cytoskeleton, Plant Physiol., 71: 719-730.

Lonergan, T.A., 1984a, A diurnal rhythm in the rate of light induced electron flow in maize bundle sheath strips, Photochem. Photobiol., 39: 89-93.

Lonergan, T.A., 1984b, Regulation of the photosynthetic light reactions by the biological clock, Photochem. Photobiol., 40: 789-793.

Lonergan, T.A., and Sargent, M.L., 1979, Regulation of
    the photosynthetic rhythm in Euglena gracilis, II.
    Involvement of electron flow through both photosystems,
    Plant Physiol., 64: 99-103.
Okada, M., and Horie, H., 1979, Diurnal rhythm in the Hill
    reaction in cell-free extracts of the green alga
    Bryopsis maxima, Plant Cell Physiol., 20: 1403-1406.
Owens, T.G., Falkowski, P.G., and Whitledge, T.E. 1980,
    Diel periodicity in cellular chlorophyll content in
    marine diatoms, Mar. Biol., 59: 71-77
Prezelin, B.B., and Matlick, H.A., 1980, Time course of
    photoadaptation in the photosynthesis-irradiance rela-
    tionship of a dinoflagallate exhibiting photosynthetic
    periodicity, Mar. Biol., 58: 85-96.
Prezelin, B.B., and Sweeney, B.M., 1977, Characterization
    of photosynthetic rhythms in marine dinoflagellates.
    II. Photosynthesis irradiance curves and in vivo
    chlorophyll fluorescence, Plant Physiol., 60: 388-
    392.
Saha, S., Ouitrakul, R., Izawa, S., and Good, N.E., 1971,
    Electron transport and photophospshorylation in chloro-
    plasts as a function of the electron acceptor. J.
    Biol. Chem.,246: 3204-3209.
Samuelsson, G., Sweeney, B.M., Matlick, H.A., and Prezelin,
    B.B., 1983, Changes in photosystem II account for
    the circadian rhythm in photosynthesis in Gonyaulax
    polyedra, Plant Physiol., 73: 329-331.
Sayeed, S.A., Behera, B.K., and Mohanty, P., 1983, Electron
    transport in chloroplasts from detached senescing
    leaves of wheat (Triticum aestivum), Trop. Plant Sci.
    Res., 1:81-87.
Sayeed, S.A., Behera, B.K., and Mohanty, P., 1985, Inter-
    action of light quality and benzimidazole on photo-
    chemical activity of chloroplasts isolated from deta-
    ched senescing wheat (Triticum aestivum) leaves.
    Physiol. Plant.,64: 383-388.
Sweeney, B.M., 1979, Endogenous rhythms in the movement
    of plants, in: Encyclopedia of Plant Physiology, W.
    Haupt, and M.E. Felnleib, eds., New Series, Vol. 7,
    pp.71-93, Springer-Verlag, Berlin.
Sweeney, B.M., and Prezelin, B.B., 1978, Circadian rhythms,
    Photochem. Photobiol., 27: 841-847.
Trebst, A., 1974, Energy conservation in photosynthetic
    electron transport of chloroplasts, Ann. Rev. Plant
    Physiol., 25: 423-458.
Wagner, E., 1977, Molecular basis of physiological rhythms,
    in: Integration of Activity in the Higher Plant,
    D.H. Jennings, ed., pp. 33-72, Cambridge University
    Press, Cambridge.
Wagner, E., and Cumming, B.G., 1970, Betacyanian accumu -
    lation, chlorophyll content, and flower initiation
    in Chenopodium rubrum as related to endogenous rhythmi-
    city and phytochrome action, Can. J. Bot., 48: 1-18.

RELATIONSHIP OF CARDIOLIPIN TO CYTOCHROME $\underline{c}$ OXIDASE KINETICS AS PROBED BY

ADRIAMYCIN

A. Trivedi, M. Schwab, D. Fantin and E. Reno Tustanoff

Department of Biochemistry, University of Western Ontario
and Department of Clinical Pathology, Victoria Hospital
London, Ontario, Canada, N6A 4G5

INTRODUCTION

The interaction of cytochrome $\underline{c}$ oxidase with mitochondrial phospholipids recently has received a great deal of scrutiny (Dennis, 1986). Much of this work has been focused specifically on the functional relationship between cardiolipin and oxidase, however, this dependence is now being questioned (Watts et al., 1978; Powell and Abramovitch, 1985). Utilizing the observation that the anthracycline antibiotic, adriamycin, forms undissociable complexes with cardiolipin, investigations have been carried out using this drug to illustrate the involvement of this acidic phospholipid in various mitochondrial membrane-related processes (Goormaghtigh and Ruysschaert, 1984). Adriamycin first recognized for its antineoplastic activity (Gianni et al., 1983), has been shown to have cellular toxicological effects, causing cardiac toxicity (Demant and Wassermann, 1985) and impaired mitochondrial function (Praet et al., 1984). Ruysschaert and his co-workers (1982) demonstrated that this drug blocked mitochondrial electron flow by inactivating cytochrome $\underline{c}$ oxidase. Their data suggested that adriamycin sequestered closely associated cardiolipin molecules from the enzyme by forming an adriamycin-cardiolipin complex and thereby inhibiting oxidase activity. Results from our laboratory have shown that altering the configuration of yeast mitochondrial membrane phospholipid acyl chains dramatically modify the kinetic activity of the oxidase (Trivedi et al.,1986). We wish to extend our kinetic investigation of this enzyme's activity by using the ability of adriamycin to complex cardiolipin $\underline{in}$ $\underline{situ}$ in order to assess the involvement of this negatively charged phospholipid with the oxidase.

METHODS

The parental strain of Saccharomyces cerevisiae (ade5a) and a mutant strain VAL2C (MATa chol leu2 ade6) designated chol, were obtained from Dr. S. A. Henry, Albert Einstein College of Medicine, New York, U.S.A. These strains were maintained and cultured as previously reported (Atkinson et al., 1980). For experiments, cells were grown in synthetic medium supplemented with either 1 mM choline or 1 mM ethanolamine and grown to their late logarithmic phase at 30°C. After harvesting the cells, mitochondria were isolated mechanically and analyzed for their phospholipid and fatty acid content as previously described (Trivedi et al., 1986).

Cytochrome $\underline{c}$-depleted mitochondria and reduced horse heart cytochrome $\underline{c}$ were prepared as previously described (Trivedi et al., 1986).The oxidase concentrations (heme aa$_3$) were determined according to Van Gelder (1966).

Cytochrome $\underline{c}$ oxidase activity was measured polarographically at 25°C. Where indicated, adriamycin (30 µM) was pre-incubated with mitochondria in temperature controlled polarographic chambers for a preset time (20 min) prior to measuring the rate of oxidation of cytochrome $\underline{c}$ using a Clark oxygen electrode (Brautigan et al., 1978). Since it has been reported by Dethmers et al. (1979) that cytochrome $\underline{c}$ may bind to glass surfaces, all polarographic assays were carried out in repeatedly acid-washed chambers to ensure optimal assay conditions. The oxidase assay buffer contained 20 mM Tris-acetate (pH 7.9), 250 mM sucrose, 7 mM ascorbate (from a stock solution of 0.5 M sodium ascorbate containing 1 mM EDTA), 0.7 mM TMPD, cytochrome $\underline{c}$ (0.05 to 50 µM) and yeast mitochondria (4 nM cytochrome aa$_3$). The oxygen uptake data were represented in the form of Eadie-Hofstee plots (v/S versus v) where v is the velocity expressed as nmole O_2. min^{-1} and S is the substrate concentration in µM. These biphasic curves were obtained as lines of best fit using a computer programme to plot the calculated data. The apparent Km values were calculated from the negative reciprocal slopes of the high and low affinity phases of the plots. The extension of the abscissa intercept values of each phase of the curves when divided by the cytochrome $\underline{c}$ oxidase concentration gave the maximum turnover number (TN_{max}).

RESULTS AND DISCUSSION

Since it has been suggested that adriamycin selectively binds negatively charged phospholipids, we have embarked on a series of experiments to investigate this drug's involvement with other membrane phospholipids such as phosphatidylcholine (PC) and phosphatidylethanolamine (PE) in order to resolve the question of cytochrome $\underline{c}$ oxidase's specific reliance on cardiolipin for its optimal activity. Using a choline/ethanolamine auxotroph of S. cerevisiae, chol, we were able to manipulate the PC and PE levels in mitochondrial membranes in these cells (Table I). This mutant is unable to synthesize phosphatidylserine (PS) in

Table I. Mitochondrial Phospholipid Composition

The yeast strains were grown to their late exponential phase in medium supplemented with 1 mM choline or 1 mM ethanolamine as indicated. Isolated mitochondria were analyzed for their phospholipid content as described in Methods Section. PI, phosphatidylinositol; PS, phosphatidylserine; PC, phosphatidylcholine; PE, phosphatidylethanolamine; CL, cardiolipin; UC uncharacterized phospholipid; ND, undetected.

Strain	Medium Supplemented	Polar Head Group Ratio of Total Phospholipid (%)					
		PI	PS	PC	PE	CL	UC
ade5a	none	16.7	7.4	43.3	15.2	11.5	5.9
chol[a]	choline	23.1	ND	58.3	4.6	8.6	5.4
chol[a]	ethanolamine	20.8	ND	36.0	28.5	8.9	5.8

a) This strain has been shown to lack any detectable phosphatidylserine (Atkinson et al., 1980)

vivo (Atkinson et al., 1980). Hence supplementing the synthetic minimal medium with either choline or ethanolamine, it was possible to rescue mutant growth by utilizing a salvage pathway (Kennedy and Weiss, 1956) and thereby enriching the mitochondrial membranes with increased levels of PC or PE. The mutant when deprived of either of the two phospholipid bases was unable to grow and therefore it was impossible to compare the mitochondrial levels of phospholipids in supplemented and unsupplemented cells. To overcome this impasse, the mutant data presented in this paper was compared with its parental strain, ade5a.

As can be seen in Table I, the mitochondrial phospholipid composition of the mutant reflected the growth supplement added. Mitochondria isolated from mutant cells grown in ethanolamine-supplemented medium had a level of PE which was approximately 2-fold greater than those seen in the wild-type strain. In a similar manner, a 37% increase in PC level was achieved in choline-supplemented cells. These phospholipid constituents increased at the expense of falls in the levels of PC and PE, respectively. No significant changes in cardiolipin content were noted. It may be mentioned that addition of either organic base to the growth medium did not alter the mitochondrial phospholipid profiles of the wild-type strain (data not shown). Fatty acid analyses of these mitochondrial membranes showed relatively little change in composition (Table II) with the ratio of unsaturated to saturated fatty acids varying marginally between 0.9 and 1.1.

Recently, several studies have been focused on the interaction of adriamycin with phospholipid vesicles in order to determine the thermodynamic and structural parameters of the drug's interactions with specific lipid components (Burke and Tritton, 1985; Samuni et al., 1986). It appears that the antibiotic binds strongly with cardiolipin followed by PS, phosphatidic acid, PE and lastly PC (Demant, 1984). However, some inconsistencies in characterizing the binding of the drug to charged and neutral lipids have arisen in these different studies. Using our in vivo system, in which the phospholipid composition of the mitochondrial membranes can be manipulated, a unique opportunity is afforded to study these drug-lipid interactions under more dynamic conditions.

To assess the effect of adriamycin on mitochondrial function, mitochondria isolated from the wild-type strain, PC- and PE-enriched mutant cells were incubated in the presence of increasing concentrations of the

Table II: Mitochondrial Fatty Acid Composition

The yeast strains were grown to their late exponential phase in medium supplemented with 1 mM choline or 1 mM ethanolamine as indicated. Cells were harvested and their mitochondria were isolated as described in Methods Section. The fatty acids were quantitated by gas chromatography.

Strain	Medium Supplemented	Total Fatty Acid (%)						U/S[*]
		12:0	14:0	16:0	16:1	18:0	18:1	
ade5a	none	4.5	2.3	33.0	29.7	8.5	22.0	1.1
chol	choline	5.9	2.3	32.4	25.1	14.1	20.2	0.9
chol	ethanolamine	3.3	2.6	30.8	26.4	14.1	22.8	1.0

[*] U/S, ratio of the percent of unsaturated and saturated fatty acids.

drug (0-50 µM) and then monitored for their cytochrome c oxidase activities (Fig. 1A). The oxidase activity of the parental strain was affected the most, followed in decreasing order by PE- and PC-enriched mitochondria. On increasing the concentration of the drug, it was not possible to completely inhibit oxidase activity. This may be explained on the basis that the cardiolipin binding sites were not completely accessible to adriamycin or that the drug was bound by other phospholipid components. When the oxidase activity was measured at a predetermined concentration of drug (30 µM) as a function of time, similar inhibition patterns were observed (Fig. 1B) indicating that the membrane lipid composition indeed control the degree of drug interaction.

Goormaghtigh et al. (1982) using adriamycin found a good correlation between mitochondrial impairment and cytochrome c oxidase activity. They proposed that cardiolipin was extracted from the lipid surrounding environment of the oxidase by the drug and then segregated in a separate phase inaccessible to the enzyme, thus accounting for the inhibition. Using the wild-type and phospholipid altered mitochondria, experiments were carried out to monitor the kinetics of the oxidase after exposure to adriamycin in order to substantiate the proposed mechanism of action (Fig. 2). When studied under optimal polarographic conditions, steady-state

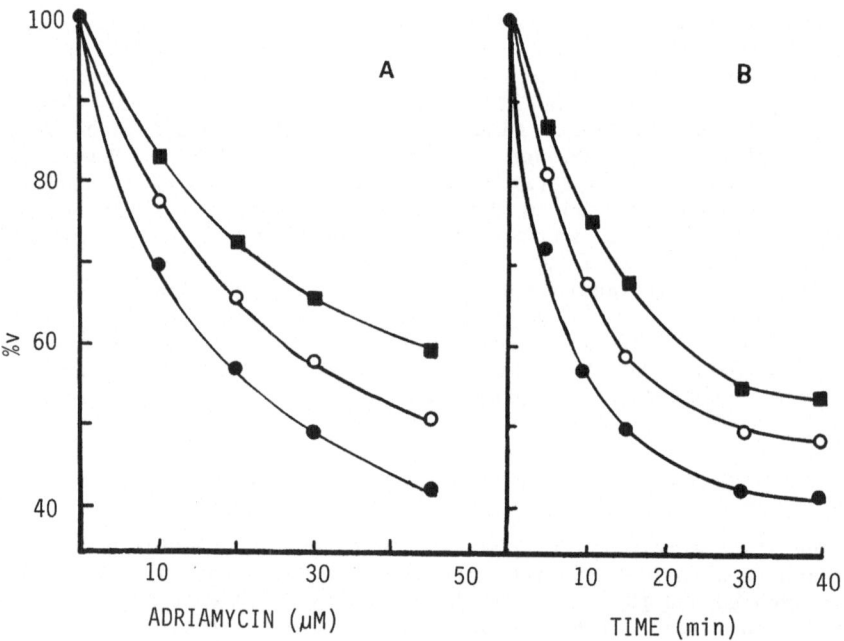

Fig. 1A: Effect of Increasing Concentrations of Adriamycin on Cytochrome c Oxidase Activity. Mitochondria were incubated with adriamycin (0-50 µM) for 15 min prior to monitoring their oxidase activity polarographically.

Fig. 1B: Effect of Incubation Time on Cytochrome c Oxidase Activity. Mitochondria were incubated with 30 µM adriamycin for indicated periods prior to assaying their oxidase activity. ade5a (●), choline-(■), ethanolamine-supplemented (O), chol mitochondria.

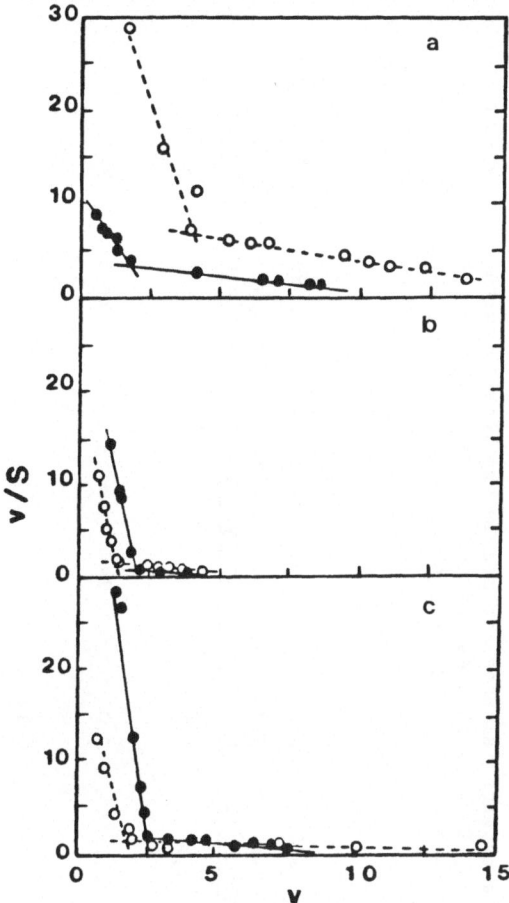

Fig. 2: Eadie-Hofstee Plots of Cytochrome c Oxidase Activity.
These series of graphs represent the kinetic properties of membrane cytochrome c oxidase preparations
which were obtained from A) wild-type, B) choline-
supplemented chol and C) ethanolamine supplemented
chol cells. The oxidase assay was carried out as
described in the Methods Section using 4 nM enzyme
based on heme aa_3 content. Adriamycin-treated mitochondria (O), untreated mitochondria (●).

biphasic kinetics for this reaction have been shown to be first order and
can be explained by two simultaneous reactions proceeding between
cytochrome c and its oxidase, i.e., high and low affinity reaction sites
(Trivedi et al., 1986). The data from these adriamycin-treated membrane
preparations is documented in Table III after being calculated from the
biphasic Eadie-Hofstee plots shown in Fig. 2. It can be seen that
phospholipid manipulated membranes had no significant effect on the
"apparent Km" values for the enzyme's low or high affinity binding sites.
However, the maximum turnover rate (TN_{max}), defined polarographically as
transfer rate of electrons from the oxidase to molecular oxygen, was
affected deferentially at the enzyme's low affinity binding site. This
impaired electron flow may be due to the altered membrane environment.

After incubating these mitochondrial preparations for 20 min in the

Table III

Kinetic Parameters of Cytochrome c Oxidase

The yeast strains were grown to their late exponential phase in medium supplemented with 1 mM choline or ethanolamine as indicated. Isolated mitochondria were assayed for cytochrome c oxidase activity as described in Methods Section. The data calculated from Eadie-Hofstee plots (Fig. 2) using 4 nM enzyme (based on heme $\underline{aa_3}$ content).

Strain	Media Supplemented	Apparent K_m, µM				TN_{max}, electron · s^{-1}			
		Untreated		ADR-Treated*		Untreated		ADR-Treated*	
		Low Affinity	High Affinity	Low Affinity	High Affinity	Low Affinity	High Affinity	Low Affinity	High Affinity
ade5a	none	4.6	0.25	2.6	0.14	184	40	273	81
chol	choline	4.8	0.18	2.1	0.09	79	33	78	24
chol	ethanolamine	4.3	0.12	2.0	0.12	136	44	125	32

* Adriamycin-Treated

presence of 30 μM adriamycin, significant changes were observed in the oxidase kinetics. When compared to the untreated mitochondria, marked decreases were noted in both the high and low affinity "apparent Km" values which approached 50% (Table III). With the exception of the high affinity phase in PE-enriched mitochondria, which remained unchanged, the other kinetic parameters of the drug-induced membranes were affected to the same degree. Since the Eadie-Hofstee plots of the adriamycin-treated oxidase activities were still biphasic, it may be implied that cardiolipin is not directly involved in the dimerization of the oxidase since segregation of this phospholipid should have generated monophasic kinetics. This assumption is based on the report that the oxidase's biphasic kinetics are due exclusively to the dimerization of the enzyme's monomeric subunits since each monomer has either a specific high or low affinity binding site for cytochrome c (Bolli et al., 1985). Our results suggest that cardiolipin either is not totally participating in the binding of cytochrome c to the enzyme or else it has not been pulled far enough away from the enzyme complex by adriamycin to inactivate its function. Recently, Powell and Abramovitch (1985) indicated a similar possibility. After isolating a cardiolipin-free oxidase preparation which still manifested biphasic kinetics, they proposed that the phospholipid was not involved in the binding of cytochrome c to the enzyme.

It remains to be seen whether PC or PE membrane enrichment has affected the binding between adriamycin and cardiolipin. This appears to be unlikely since similar kinetics were observed in both the wild-type and phospholipid manipulated mitochondria (Table III). The only difference in the drug-treated mitochondria was seen in their TN_{max} values. While the rate of electron transfer for both the high and low affinity phases had increased in the treated wild-type mitochondria, no changes were seen in the phospholipid manipulated organelles. These striking differences in kinetic behaviour could indicate a role for phospholipids in drug-membrane interactions.

It has been shown that there are two types of binding which can take place between adriamycin and cardiolipin and each determines the location of the drug in the membrane (Henry et al., 1982). The first type involves the fixation of the α-amino group of the sugar residue of adriamycin to the ionized phosphate residue of cardiolipin with the dihydroanthraquinone moiety lying outside the bilayer. The second type also involves inter-action with the phosphate group but in addition, the dihydroanthraquinone is now embedded in the lipid bilayer. Our results can be interpreted on the basis that the binding of adriamycin to mitochondrial membranes perturbs the enzyme's attraction for electrons at both its high and low affinity as a result of the penetration of the drug into the membrane bilayer. It could be possible that the segregation or binding of cardiolipin by the drug, in close proximity to the enzyme, is a major factor in modulating the catalytic centre of the oxidase. In a recent report dealing with $F_0 \cdot F_1$ -ATPase, Laird et al. (1986) classified the primary role of phospholipids in intact mitochondria as stabilizing the conformation of the enzyme. Taking this into account, the modulation of cytochrome c oxidase activity in adriamycin-treated mitochondria may also be explained as a result of destabilization of the enzyme complex. Another possible explanation for the drug-induced enzyme kinetics which appear more cogent is based on an earlier ^{31}P-NMR study by Goormaghtigh et al. (1982) which showed that cytochrome c binds cardiolipin and induces non-bilayer structures in order to reach a region of the enzyme complex buried in the membrane bilayer, however, adriamycin inhibits the formation of these non- bilayer structures. They also showed that adriamycin can effectively displace both reduced and oxidized cytochrome c from cardiolipin containing liposomes when the drug concentration is sufficient

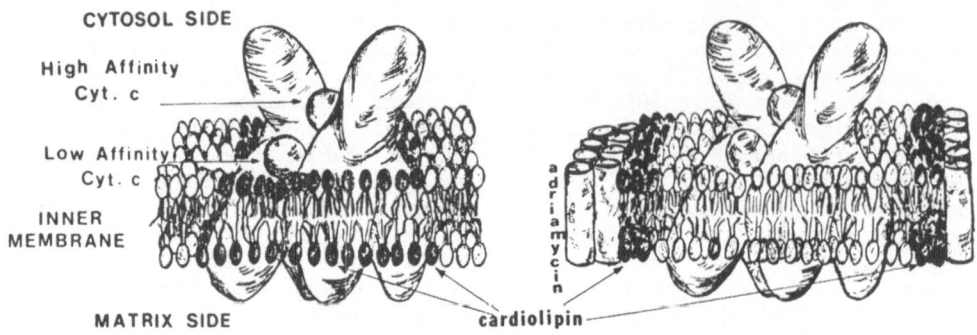

CYTOSOL SIDE

High Affinity Cyt. c

Low Affinity Cyt. c

INNER MEMBRANE

MATRIX SIDE

adriamycin

cardiolipin

Fig. 3: Model of Cytochrome c Oxidase Dimer Interaction with Adriamycin-Cardiolipin Complex. Left side of model shows the natural protein relationship with cardiolipin membrane phospholipids. Right side shows the segregation of cardiolipin by the added adriamycin.

to complex all the phospholipid molecules. These factors can indeed have a bearing on the enzyme kinetics observed in adriamycin-treated mitochondria. It is clear from the data presented here that PC or PE enrichment of mitochondrial membranes has controlled electron flow (TN_{max}) significantly in both untreated and treated organelles. It has also been reported that the binding of adriamycin to PC and to other phospholipids in their gel state are weak or non-existent, whereas its binding in the liquid-crystal state is stronger (Karczmar and Tritton, 1979). It is indeed possible that we have affected the phase state of the membrane near the enzyme by our enrichment procedures and this could account for our drug results.

We have recently proposed that the high affinity site on the oxidase lies farther away from the membrane bilayer while the low affinity site is either juxtaposed or is embedded in the membrane bilayer (Fig. 3). The observed effects of adriamycin on both phases of the oxidase's kinetics can be interpreted successfully using this model. On addition of the drug, the segregated cardiolipin-adriamycin complex may pull away from the enzyme but still remain in the vicinity of the oxidase (Fig. 3). The formation of the complex presumably nullify the charge contribution by cardiolipin for cytochrome c binding to its enzyme. It is also possible that this complex remains near or surrounds the enzyme and thus influences the stability of the membrane architecture. Since PC- and PE-enriched mitochondria have shown similar kinetic patterns to those seen in the wild-type organelles, it may be inferred that these phospholipids are not involved in the formation of the drug-lipid complex. Therefore, cytochrome c interaction with cardiolipin may have an important role in the transfer of electrons to cytochrome c oxidase.

ACKNOWLEDGEMENT
    This work was supported by a grant-in-aid to E.R.T. by the Medical Research Council of Canada (MT-1460).

REFERENCES

Atkinson, K.D., Jensen, B., Kolat, A.I., Storm, E.M., Henry, S.A., and

Fogel, S., 1980, Yeast mutants auxotrophic for choline or ethanolamine, J. Bacteriol., 141: 558-564.

Bolli, R., Nalecz, K.A., and Azzi, A., 1985, The aggregation state of bovine heart cytochrome c oxidase and its kinetics in monomeric and dimeric forms, Arch. Biochem. Biophys., 240, 102-116.

Brautigan, D .L., Ferguson-Miller, S., and Margoliash, E., 1978, Mitochondrial cytochrome c: preparation and activity of native and chemically modified cytochromes c, Methods Enzymol., 53: 128-164.

Burke, T.G. and Tritton, T.R., 1985, Structural basis of anthracycline selectivity for unilamellar phosphatidylcholine vesicles: an equilibrium binding study, Biochemistry, 24, 1768-1776.

Demant, E. J. F., 1984, Binding of adriamycin-Fe^{+3} complex to membrane phospholipids, Eur. J. Biochem., 142, 571-575.

Demant, E.J.F., and Wassermann, K., 1985, Doxorubicin induced alterations in lipid metabolism of cultured myocardial cells, Biochem. Pharmacol., 34: 1741-1746.

Dennis, M., 1986, Structure and function of cytochrome c oxidase, Biochimie 68: 459-470.

Dethmers, J.K., Miller-Ferguson, S., and Margoliash, E., 1979, Comparison of yeast and beef cytochrome c oxidase, kinetics and binding of horse, fungal and euglena cytochrome c, J. Biol. Chem., 254: 11973 - 11981.

Gianni, L., Corden, B.J., and Myers, C.E., 1983, The biochemical basis of anthracycline toxicity and antitumor activity, in: "Reviews in Biochemical Toxicology," E. Hodgson, J. Bend, and R.M. Philpot, eds., Elsevier Biomedical, New York.

Goormaghtigh, E., Brasseur, R., and Ruysschaert, J.M., 1982, Adriamycin inactivates cytochrome c oxidase by exclusion of the enzyme from its cardiolipin essential environment, Biochem Biophys Res Commun, 104: 314-320.

Goormaghtigh, E., Vandenbranden, M., Ruysschaert, J.M., and de Kruijff, B., 1983, Adriamycin inhibits the formation of non-bilayer lipid structures in cardiolipin-containing model membranes, Biochim. Biophys. Acta, 685: 137-143.

Goormaghtigh, E., and Ruysschaert, J.M., 1984, Anthracycline glycoside-membrane interactions, Biochim. Biophys. Acta, 779: 271-288.

Henry, N., Fantine, E.D., Bolard, J., and Garnier-Suillerot, A., 1985, Interaction of adriamycin with negatively charged model membranes: evidence of two types of binding sites, Biochemistry, 24: 7085-7092.

Karczmar, G.S., and Tritton, T.R., 1979, The interaction of adriamycin with small unilamellar vesicle liposomes, Biochim. Biophys. Acta, 557, 306-319.

Kennedy, E.P., and Weiss, S.B., 1956, The function of cytidine coenzymes in the biosynthesis of phospholipids, J. Biol. Chem., 222: 193-214.

Laird, D.M., Parce, J.W., Montgomery, R.I., and Cunningham, C.C., 1986, Effect of phospholipids on the catalytic subunits of the mitochondrial $F_o \cdot F_1$-ATPase, J. Biol. Chem., 261, 14851-14856.

Powell, G.L., and Abramovitch, D.A., 1985, Function of cardiolipin in cytochrome c oxidase, Fed. Proc., 478.

Praet, M., Pollakis, G., Goormaghtigh, E., and Ruysschaert, J.M., 1984, Damages of the mitochondrial membrane in adriamycin treated mice, Cancer Letts. 25: 89-96.

Samuni, A., Chong, P.L-G., Barenholz, Y., and Thompson, T.E., 1986, Physical and chemical modifications of adriamycin:iron complex by phospholipid bilayers, Cancer Res., 46: 594-599.

Trivedi, A., Fantin, D.J., and Tustanoff, E.R., 1986, Role of phospholipid fatty acids on the kinetics of high and low affinity sites of cytochrome c oxidase, Biochem. Cell Biol, 64: 1195-1209.

Van Gelder , B.F., 1966, On cytochrome $\underline{c}$ oxidase, I. The extinction coefficient of cytochrome a and cytochrome a_3, <u>Biochim. Biophys. Acta</u>, 118: 36–46.

Watts, A., Marsh, D., and Knowles, P.F., 1978, Lipid-substituted cytochrome oxidase: no absolute requirement of cardiolipin for activity, <u>Biochem. Biophys. Res. Commun.</u> 81: 403–409.

MOLECULAR DYNAMICS AND SELECTIVITY IN BIOMEMBRANES

Anthony Watts

Biochemistry Department
University of Oxford
South Parks Road
Oxford, OX1 3QU, U.K.

INTRODUCTION

A wide range of dynamic interactions take place within biological membranes. For example, photon incidence on the chromophore of rhodopsin induces electron transfer in picoseconds whereas lipid flip-flop across membranes, which is responsible for membrane asymmetry, takes place in hours, if not days. The task before the membrane spectroscopist is therefore to identify the correct approach suitable for the dynamic range of interest and then to relate the information gained to the function of the membrane. It is becoming clear that such function is determined by molecular interactions which involve recognition of many species for their neighbours (for example, proteins for lipids) and their substrate or substrates (ions, metabolites, etc.). It has been stated by Williams (1985) that the relationship:-

Chemical composition ----> Structure ----> Mobility ----> Function

is essential to all biological processes. To define and describe these relationships therefore requires detailed molecular information, partly available through spectroscopic means.

In this brief review of membrane dynamics written for the biologist, a range of different techniques will be mentioned, with a certain amount of emphasis on magnetic resonance methods because of their versatility. In particular, the role of intermolecular interactions in determining the function of biomembranes will be demonstrated for the cases investigated to date, with special reference to the most recent results.

THE MEMBRANE BILAYER

It is now generally accepted that the unit structure of virtually all biomembranes is the bilayer. This satisfies thermodynamic and stability criteria used to explain the macromolecular arrangement of amphiphilic lipids, both phospholipid and non-phosphate containing lipids. Such complexes can be described by two fundamental physical properties, their order and dynamics. Thus, the membrane "fluidity", which itself cannot be described in any physical way, can now be quantified in terms of this molecular order and rate of motion.

Bilayer Order

Biological membranes, as well as being highly dynamic, are ordered structures. Both X-ray and neutron diffraction studies of a very wide range of biomembranes give a well defined electron density profile which can be assigned to a bilayer arrangement of the lipid molecules, the highest electron density in the regular repeating unit being the phosphate groups of the phospholipid molecules (Shipley, 1973; Blaurock, 1986). The proteins are not normally ordered in membranes (except for the case of bacteriorhodopsin) and therefore may alter the scattering profile in an averaged, but asymmetric way, without altering the general features, namely two peaks of higher electron density about 5 nm apart with a trough in the middle of the unit cell and a plateau in between these features (Figure 1).

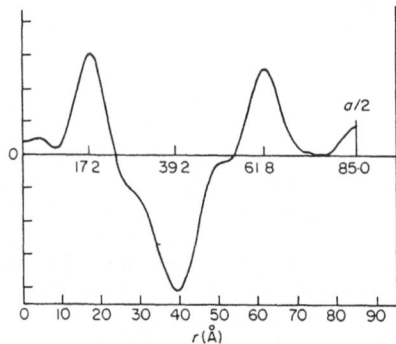

Fig. 1. The electron density profile from frog sciatic nerve myelin (adapted from Kirschner & Casper, 1971).

Neutron diffraction studies of bilayer membranes show similar features in the scattering density profile but have the added advantage that individual deuterons, incorporated into specific positions in a lipid (or protein) molecule, can be identified unambiguously within the bilayer profile. Deuterons have a scattering amplitude of opposite sign and of 6 times the magnitude of protons which makes them clearly visible in neutron diffraction profiles. Furthermore, by contrast matching the membrane scattering density by carrying out the experiment in various proportions of heavy water and water, the penetration of water into the membrane can be directly visualized in the difference scattering profiles between different solvent mixtures (Figure 2).

The results from both X-ray and neutron diffraction experiments prove that pure lipid dispersions, and biomembranes, are bilayers of phospholipid molecules as their major structural element. The scattering profiles are essentially snap-shot pictures of the membrane, the spectroscopic time-scale being determined approximately by the frequency of the radiation being used, that is, about 10^{-18}s for a wavelength of 0.1 - 0.5nm.

Other spectroscopic methods can be used to reveal the order of bilayer membranes. Deuterium NMR is one method which has proved particularly successful in that a deuterium placed instead of a proton at a specific place in a phospholipid, can be detected directly without

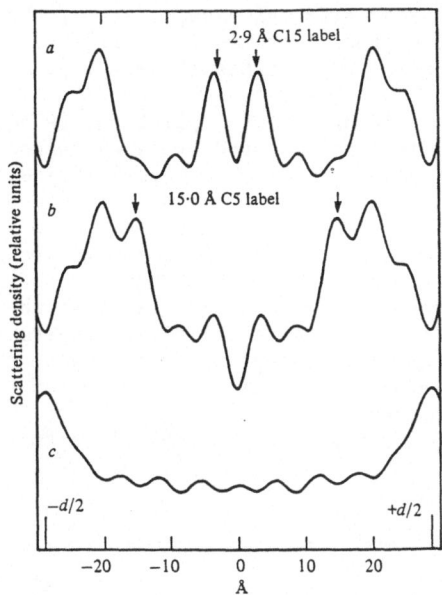

Fig. 2. Neutron diffraction profiles of a synthetic
lipid (dipalmitoyl phosphatidyl choline)
deuterated at the C15 (a) and C5 (b) positions
in the chains. The water distribution across
the bilayer is shown in (c) and is the difference
pattern for bilayers in H_2O and D_2O (adapted
from Seelig & Seelig, 1980).

assignment problems. The deuterium NMR method, and its application to
membranes, will not be reviewed in detail here (see Seelig, 1977; Seelig &
Seelig, 1980, Davis, 1983; Bloom & Smith, 1985; Watts et al., 1985; Watts,
1986; for reviews). In essence, a deuteron (nuclear spin = 1), placed
specifically within a biomolecule as a CD-group, gives rise to an NMR
spectrum which contains information about the time-averaged amplitude and
rate of CD-group motion. In "fluid" or liquid crystalline bilayers, the
CD-bond within a lipid molecule undergoes very fast motion (with a
correlation time, τ_c of typically nanoseconds or faster), but of limited
amplitude within a cone of angle α. This motion gives rise to a rather
broad (compared to proton NMR spectra) NMR spectrum from which the observed
anisotropy can be determined from the quadrupole splitting, $\Delta\nu_Q$, which is
the distance between the major peaks in the spherically averaged powder
spectrum; these peaks are the spectral lines from the $90°$-orientation
spectral lines of the CD-bond.

The order parameter, S(CD), is then defined as the ratio of the
(observed anisotropy)/(maximum anisotropy). The amplitude in degrees, α, of
CD-bond motion about the axis of averaging is given by $S = 1/2(3\overline{\cos^2\alpha} - 1)$.
Values of the modulus of S therefore have limits of 0 and 1. Additionally
S-values cannot be determined for motion which is slow on the time-scale
sensitivity of the method being used to measure it because the bar above
the $\cos^2$ term denotes the time-averaged nature of the motion.

If the axis of molecular averaging is at an angle β and does not
coincide with the axis of CD-bond motion averaging, as may be the case for

331

CD–groups at the polar head–group of phospholipids, then the measured value of $\Delta\nu_Q$ is related to the product of two (or more) order parameters, S(amplitude) X S(conformation) or alternatively S(α) X S(β).

Values for the static quadrupole splitting, (e^2qQ/h), for methylene CD–groups are close to 167kHz; measured averaged values are typically up to 100kHz for rapidly moving CD–groups in the acyl chains of membrane bilayers and between 1 and 65kHz for CD–groups in the polar head–groups at the surface of a membrane. The order of a membrane bilayer is shown in the order parameter profile (Figure 3) in which the values of S (rather than α) for acyl chain labelled phospholipids are shown as a function of the position of the deuteron in the chain. Similarly to the X–ray and neutron diffraction profiles, a plateau with relatively constant order in the upper half (between the 2nd and 10th methylene segments) of the bilayer is seen. This feature is clearly due to the tighter packing of the lipid acyl chains in this part of the membrane contributing greatly to the bilayer permeability barrier properties essential to all cytoplasmic membranes.

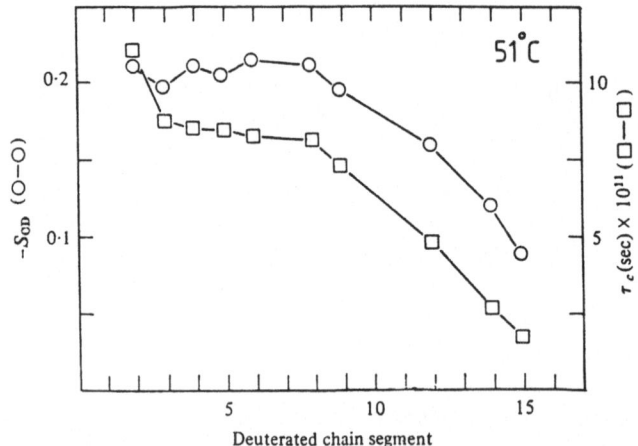

Fig. 3. Comparison of the rotational correlation times, τ_c, determined from T_1 data and the deuterium order parameters, S, as a function of segment position across a lipid bilayer (from Seelig & Seelig, 1980).

Only one membrane protein, the photoreceptor protein from a photosynthetic bacterium R. viridis, has been crystallized to date and the resultant X–ray information interpreted in terms of a high resolution (0.28nm) structure (Deishofer et al., 1985). There is every reason to anticipate that, in principal, more proteins will be crystallized since membrane proteins have no less internal order than globular ones, of which many are known in atomic detail. The problem clearly is in the methodology, the use of the correct detergents, the correct purification methods, etc. So while more structures are awaited, it is to be assumed that integral and peripheral proteins are intrinsically highly ordered; an essential feature for their function. Helices are the major structural element of these proteins, assigned by analogy to bacteriorhodopsin which itself in known to about 0.5nm detail from electron diffraction images. The peptide backbone in such ordered structures is expected to relatively immobile with some motion in the side groups (see below). Nonetheless, the order and structure of a protein is bound to be less than for a cooperative array of lipid acyl chains; the influence of proteins on lipid order will be discussed below.

Rates of intramolecular motion in membranes

As well as being highly ordered, lipids and proteins in membranes have a high degree of internal molecular motion. The acyl chains of lipids and the amino acid residues of proteins move with relatively fast motions which can be measured by non-perturbing NMR techniques, especially deuterium NMR. Protons, although ubiquitous in nature, only give informative proton NMR spectra for rapidly tumbling, small biomolecules without the use of specialized techniques for solids which are still being developed for membranes (Deese & Dratz, 1986).

The rate of motion is determined from relaxation time measurements of an NMR spectrum. Here, the rate at which a particular nucleus gives up its excitation energy to either the surroundings (T_1) or to another similar nucleus (T_2), can be determined to give a correlation times, τ_c, for the CD-bond being examined. The standard ways of determining these relaxation times are used. Both T_1 and T_2 times are anisotropic and relatively fast (typically T_1 is $100 - 20$ms and T_2 is $5 - 0.1$ ms) (see Davis, (1983) for a review of the details). For small order parameters, the anisotropy of the spin-lattice relaxation time can be accounted for and a value of τ_c for the CD-group is given by (Brown et al., 1979):-

$$\frac{1}{T_1} = \frac{3\pi^2}{2} \left(\frac{e^2qQ}{h} \right)^2 (1 + \frac{1}{2} S_{CD} - \frac{3}{2} S_{CD}^2) \tau_c$$

The gradient of order along the acyl chain of a phospholipid in a bilayer is similarly reflected by the rate of acyl chain motion (Figure 3). Spin-lattice relaxation times are sensitive to motions which are of the same motional range as the methylene trans-gauche rotational isomerisms which give rise to membrane "fluidity". Values for the spin-spin relaxation times are much more difficult to determine for broad NMR spectra and special methods, such as echo techniques, are needed (Davis, 1983). This relaxation time is sensitive to slower motions (τ_c around 10^{-7} s) than those that affect T_1. The best way to treat such motional values is in a comparative way. The reason for this is that the absolute values vary with the technique used to measure then. Even by NMR methods, the value of τ_c (and S) depend upon the frequency of the spectrometer being used. The time-window for observation can therefore, in principal, be selected from the method employed.

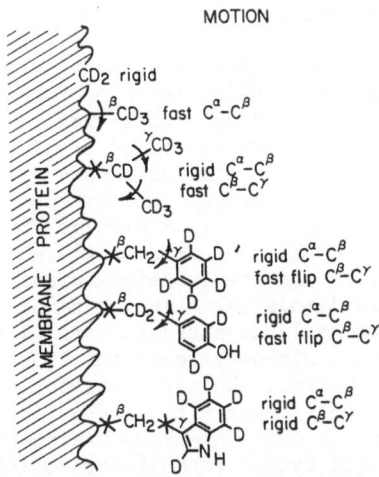

Fig. 4. Qualitative rates and types of motions of a variety of aliphatic and aromatic amino acid side groups of bacteriorhodopsin in the purple membrane (from Oldfield et al., 1982).

Intramolecular protein motion is difficult to quantitate even by proton NMR for small globular proteins. Again, deuterium NMR can be used to investigate bacteriorhodopsin in H. halobacterium grown on deuterated amino acids (Kinsey, et al., 1981; Bowers et al., 1986). Fortunately, little scrambling of the side-chains during growth meant that specific residues could be studied. The qualitative results from this work are shown in Figure 4, in which it can be seen that the more remote a residue is from the protein backbone, the more mobile the residue is. One exception to this is the rather rigid tryptophan and phenylalanine residues which are presumably constrained within the protein such that they can only undergo flips about two positions.

EFFECTS OF PROTEINS ON BILAYERS

There are a number of cases where "non-fluid" membranes do not support functional activity of the proteins in the membrane (Abney & Owicki, 1985; Sanderman, 1986). It is possible that allosteric synergism between the bilayer lipids and the protein residues can maintain the protein in a dynamic state essential for biological function. Similarly, the interfacial lipids between the integral proteins and the hydrophobic core of the bilayer, are clearly those responsible for sealing a protein into the membrane. Their properties in particular are of some significance for maintaining membrane integrity.

Proteins, both integral and peripheral, exert different effects on bilayer order and dynamics. Proteins themselves possess a high degree of internal motion and order (see above) and this must in some way interface with the lipid bilayer to form the stable, functional entity, that biomembranes clearly are. Reconstitution experiments have proved a very valuable initial approach to studying the intermolecular associations between proteins and lipids in bilayers. Any experimentally determined changes are then attributed to the known components in the system. Of all the proteins studied to date (see Marsh & Watts, 1982; Seelig et al., 1982; Seigneuret & Devaux, 1985 for reviews) one example only will be selected, that of cytochrome oxidase. This protein has been studied by many techniques, is familiar to most biochemists and has been used to demonstrate specific points about how information from physical methods relate to the biomembrane function.

Cytochrome oxidase (CO) has been studied by electron diffraction methods and shows the protein to be a distorted Y-shape (Fuller et al., 1979). It was the first to be studied with a view to demonstrate direct lipid-protein interactions in membranes (Jost et al., 1973a). Here, lipids labelled in their acyl chains with nitroxide spin-labels were detected by electron spin resonance (ESR) to probe inner mitochondrial membrane fragments enriched in CO. Subsequently, a similar ESR method was applied to purified active enzyme, reconstituted into single component phosphatidylcholine bilayers and the protein-lipid interactions studied by varying the protein to lipid mole ratio in the complex. In addition, a wide range of chemically different lipid nitroxide spin-labels, which were analogues of the lipid types commonly found in mitochondrial membranes, such as cardiolipin, phosphatidyl choline, phosphatidyl ethanolamine and phosphatidyl serine and others, were also studied (Knowles et al., 1979, 1981). Using this approach it could be demonstrated that a stronger degree of association of some lipid types, notably cardiolipin, is revealed in the recorded ESR spectrum from lipids at the protein-lipid interface. Here, the protein induces a broad spectral component, not seen in protein-free bilayers, which is proportional to the amount of protein in the bilayer (Marsh & Watts, 1982). Computer spectral subtraction methods are required to quantify how many spin–labels (and by extension, bilayer lipids) are motionally restricted at the protein interface. In quantifying such information, it must be realized that one spin–label is competing against a background of 99 other phospholipids (phosphatidyl choline in this case) because of the nature of the spin–label probe experiment; labelling levels are typically 1–2 mole % of the unlabelled lipids. It was shown that the relative binding energy for a labelled and unlabelled lipid is 1, and therefore relative binding energies for the range of different lipid types can be estimated; these are given in Table I for CO (see Marsh, (1985) for a comprehensive list for other proteins). The higher binding energies consequently imply that lipids can reside at the interface for longer than other lipids; about 5 times and 2 times longer for cardiolipin and phosphatidic acid respectively.

To analyse lipid selectivity for the interface of an integral protein in this way in terms of binding energies, it must be assumed that the number of sites (n) on the protein for lipid binding is the same for each lipid type. Indeed the opposite may be true (as suggested for the Ca/Mg-ATPase which may have highly specific sites for particular lipids (East et al., 1985)) and in Table I the number of extra sites for each lipid type is given on the assumption that they all have the same value of ΔG for association with the protein interface.

Table 1. Head group selectivity for the motionally restricted lipids in cytochrome oxidase–phosphatidyl choline bilayers. Relative association constants, K_r, or first shell occupancies, n, are given.

L^a	K_r^L/K_r^{PC}	$\Delta G_L^\circ - \Delta G_{PC}^\circ$ (cal/mol)	n_1^L	n_1^{PC}
CL*	5.4	1020	77	43
PA*	1.9	390	59	44
PE*, PS*, PG*	1.0	0	$n_1^L = n_1^{PC}$	

From Knowles et al. (1981).
^{a}Asterisk indicates that the phospholipid is a spin label.

The chemical composition of a lipid, cardiolipin in this case, appears to determine a structural association. The motional properties of the lipids and protein have been examined from an analysis of the ESR spectra from the spin–labelled lipids at the protein interface. The motional restriction of lipid acyl chains at the protein interface has been studied to show that the exchange of lipids onto and off the interface is about ten times slower than between two lipids (Knowles et al., 1981). Complete immobilization of lipids does not therefore take place, rather simply a reduction of lipid motion by the protein. On the spin–label ESR time–scale, therefore, it is possible to detect two distinct spectral components which are exchanging slowly with each other, within the time window of anisotropy averaging of the ESR spectrum, that is, slower than $10^7 s^{-1}$.

It had been suggested some time ago (Aswathi et al., 1971) that cardiolipin is essential for the oxidising capacity of cytochrome oxidase. More recently another report has confirmed this idea (Robinson et al., 1980) but in both studies, residual cardiolipin after partial lipid extraction was assumed to be maintaining the oxidase activity. The phopholipid composition and organization of CO preparations has been determined using phosphorus–31 NMR (Seelig & Seelig, 1985). In lipid–depleted complexes, rich in CO after detergent solubilization, not only cardiolipin, but also phosphatidyl choline and phosphatidyl ethanolamine were found, as shown in Table 2. The amount of cardiolipin was, however, proportionatly higher than in the lipid-rich preparations, confirming partly the earlier suggestions than the four chain lipid has some degree of preference for CO than other lipids. These results show that only complete exchange of lipids from a protein can be used as a good criteria for invoking lipid-protein selectivity.

Table 2. Phosphorus-31 NMR analysis of phospholipid-rich and phospholipid-depleted cytochrome oxidase from bovine heart (from Seelig & Seelig, 1985).

Protein preparation	PC (mol%)	PE (mol%)	CL (mol%)
Lipid-rich preparation [a]	50 ± 1 [b]	41 ± 1	9 ± 1
Lipid-depleted preparations			
Sample I [a]	47	41	12
Sample II	28	58	14
Sample III	31	52	17

We used this approach to compared the activity of cytochrome oxidase in the complete absence and presence (by adding back) of cardiolipin in phosphatidyl choline bilayers (Watts et al., 1979). Thus, only by exchanging out all the residual lipid by a large (1000 fold) excess of a single lipid, is it possible to produce bilayers with no endogenous mitochondrial lipid as shown by gas chromatography. The activity of such complexes in oxidizing reduce cytochrome c, with lipid to protein ratios of about 50:1, was the same within experimental error regardless of whether cardiolipin was present or not. The most important aspect of this approach however, is to measure electron pumping and use uncouplers to identify whether cardiolipin is responsible for efficient sealing of the protein in the membrane bilayer. This aspect of lipid-protein interactions clearly needs considerable further efforts, with this and other proteins.

Since much work had been carried out using spin-labels to study cytochrome oxidase-lipid interactions, it was obvious that deuterium NMR methods should be applied to the same system. Initial studies demonstrated that the terminal methyl group, which was deuterated, gave spectra which were broader, but with a smaller quadrupole splitting than for protein-free bilayer (Oldfield, 1978 . These results suggest that fast exchange, within the deuterium quadrupolar anisotropy time-scale, of the lipids occurs between the protein interface and the rest of the bilayer. The exchange rate must therefore be faster than $10^4 s^{-1}$, which does not contradict the ESR spin label results (Rice et al., 1979). It has been concluded that the orientational order of lipid chains next to the protein is lower than in the rest of the bilayer, from both spin-label (Jost et al., 1973b) and the NMR results. More recent work has determined the average orientation disorder of all the acyl chains in a bilayer by CO, using a moment analysis of the order parameter distribution revealed by the deuterium NMR spectrum from per-deuterated lipids (Paddy et al., 1981). This work suggests that no overall average order change occurs due to CO. These authors also anlaysed the relaxation rates for the deuterons placed at the ends of the acyl chains of a unsaturated phospholipid. It was concluded that the exchange of lipids between the protein interface and the rest of the bilayer has a minimum values of $10^6 - 10^7 s^{-1}$, again in keeping with the ESR spin-label work.

Similar conclusions have been reached from experiments designed to observe whether the polar head groups of bilayer lipids interact with CO.

Here, deuterium, nitrogen–14 and phosphorus–31 NMR were used to study complexes with phosphatidyl choline as the bilayer lipid (Tamm & Seelig, 1983). These complexes were active in reducing cytochrome c. Essentially no restriction of the fast head group motions were observed from T_1 measurements, but some line–broadening, probably due to slower motion affecting T_2, were caused by the protein when compared with the protein–free bilayers. Also, no change in the values measured for the quadrupole splittings for deuterons in the polar head groups of the lipid was observed at any protein–lipid ratio. This is in contrast to similar studies with two other integral proteins, rhodopsin and band 3 from erythrocytes, both of which are monomeric or dimeric in the membrane. Both of these proteins caused a successive reduction in the $\Delta\nu_Q$ values for CD–groups in the polar head groups of phosphatidyl choline bilayers with increasing protein content in the bilayers. These results were analysed using a simple fast exchange model for lipids diffusing between the bulk phase and the protein interface. "Fast" in this context implies a rate faster than 0.1ms, the rate of motion needed to average the measured $\Delta\nu_Q$ values of around 1kHz (Dempsey et al., 1986; Ryba et al., 1986); a similar method was used for the binding of a peripheral protein, the basic protein from myelin, to charged bilayers (Sixl et al., 1984). For both the integral proteins, line–broadening, caused by slow lipid motions, was also observed.

Expressions for T_2 can be written in terms of the molecular reorientational time, τ_c, and for CD–bonds, these formulations also include terms relating to the anisotropy of motion, through the order parameter. For CO reconstituted into phosphatidyl choline bilayers with unsaturated chains, T_1 is greater than T_2, or more precisely, the T_2 measured in an echo of the deuterium FID signal. (Values of 120–96ms for T_1 and 1.8–0.8ms for T_{2e} were measured at two protein–lipid ratios). For this long correlation time limit, where the rate of molecular motion is much slower than the frequency of the exciting radio frequency, the rate of lipid motion is calculated to have a lower limit of about $10^7 s^{-1}$ and may reflect either the exchange of lipids between the bulk and the protein interface, or motions of the lipid on the protein itself. Interestingly, it was also calculated that the order parameter of lipids at the protein surface is reduced to zero; this has been suggested experimentally in three other cases more recently by us from comprehensive lipid–protein titration studies and examination of the extensive data directly (Sixl et al., 1984; Dempsey et al., 1986; Ryba et al., 1986).

Some caution must however be used when discussing integral membrane proteins in bilayers and due regard must be paid to the biochemistry and protein activity. Firstly, cytochrome oxidase has a propensity to self–associate in reconstituted membranes. It has now been shown (Dempsey et al., 1986; Ryba et al., 1986) that aggregation of proteins in bilayers reduces the protein–lipid interfacial interactions such that in an averaged view of the membrane order and dynamics (as shown by D–NMR experiments (Davis, 1983)), it appears that no such interactions take place. ESR on the other hand, detects very short–range interactions (Marsh & Watts, 1982) usually over just one lipid "shell" remote from the protein. This difference is a direct result of the different time–scales for motional averaging by the ESR and NMR methods needed to affect the spectra; motional averaging of the spectral anisotropy required for nitroxides and deuteriums requires lipid exchange rates of 10^9 and 10^4 Hz respectively. It is clear however, that in the D–NMR spectra, some spectral broadening (related to T_2 and thus motional rates of about 10^7 Hz) do occur. It is therefore suggested that a closer inspection of T_2 (and T_2 related parameters) is required as attempted by Paddy et al., (1981), since this is the appropriate spectral parameter with sensitivity to the motional time–scale of 10^8–10^6 Hz suggested to be the important one in lipid–protein interactions (Marsh & Watts, 1982).

CONCLUSIONS

The studies described above for one system on molecular dynamics of the membrane surface and hydrophobic core, and of molecular selectivity, in bilayer membranes have led to some tentative conclusions, namely:-

1). the rates of fast intramolecular motions of lipids in bilayer membranes is decreased by proteins;

2). the intrinsic lipid order in membranes is disrupted by proteins which presumably have highly invaginated and rough surfaces;

3). the molecular selectivity between particular lipid types and a protein is probably determined, not simply be electrostatic interactions, but also through highly specific steric associations involving the protein residues and the chemistry of the lipid itself.

4). a high degree of intra- and intermolecular motion encompassing wide motional time-scales of the lipids, proteins, substrates etc., is needed for full functional activity in membranes.

It is still too early to describe membranes in detail from the structural and dynamic aspect. In addition, it may well be that there are no generalized rules; each membrane may be different. Due regard has not always been paid to the differences in the conclusions reached from the results of various techniques on even the same system. For example, electron microscopic evidence for membrane structure cannot always be related directly to results from calorimetric studies or magnetic resonance experiments. Even within what we may regard as a static structure, the cytoskeleton-membrane association or bacteriorhodopsin trimeric lattice in H.halobium as examples, there are dynamics and motions of such wide range that any specific method which can define just one range of motions is too narrow in its definition and scope such that the quantitative information is essentially independent of other motions. Often within one technqiue, the information content may not be unambiguous; as one example of this, the correlation times measured from magnetic resonance studies are different depending upon the excitation frequency and also the parameter being measured.

How can the biologist comprehend such a diversity of physical parameters? The answer is clearly in the biology itself. The wide range of motions which occur in biology are essential to function. What we are doing as experimentalists is to focus into any one type of motion with one particular method. Biology uses all such motions synergistically and only with a complete description of them all will we actually be capable of comprehending the biology and then manipulate it to our own needs or design; this will be a real test of our comprehension. Clearly terms such as "fluidity" are far too crude; more precise terminology is required with due regard to a functional aspect which itself must be defined, either ATP hydrolysis or ion transport, for example. For a complete description of membrane function, a wide range of techniques will be needed, some of which may not as yet have obvious application, followed by computer simulations which themselves are not yet capable of coping with the diversity of dynamic ranges in biology.

References

Abney, J.R. & Owicki, J.C. (1985) in:- "Prog. in Protein-lipid Interactions" Vol. 1, (Watts, A. & de Pont, J.J.H.H.M. eds) Elsevier, Amsterdam.

Aswathi, Y.C., Chuang, T.F., Keenan, T.W. & Crane, F.L. (1971) Biochim. Biophys. Acta 226, p42-52.

Blaurock, A. (1986) in:- "Prog. in Protein-Lipid Interactions", 2, (Watts, A. and de Pont, J.J.H.H.M. eds). Elsevier, Amsterdam.

Bloom, M. and Smith, I.C.P. (1986) in:- "Prog. in Protein-Lipid Interactions" Vol. 1, (Watts, A. and de Pont, J.J.H.H.M. eds). Elsevier, Amsterdam.

Bowers, J.L., Smith, R.L., Coretsopoulos, C., Kunwar, A.C., Keniry, M., Shan, X., Gutowsky, H.S., & Oldfield, E. (1986) in:- "Prog. in Protein-lipid Interactions" Vol. 2, (Watts, A. & de Pont, J.J.H.H.M. eds) Elsevier, Amsterdam.

Brown, M.F., Seelig, J. & Haberlen, U. (1979) J. Chem. Phys. 70, p5045-5053.

Caspar, D.L.D. & Kirschner, D.A. (1971) Nature New Biol., 231, p46-52.

Davis, J. (1983) Biochim. Biophys. Acta 737, p117-171.

Deese, A.J. & Dratz, E.A. (1986) in:- "Prog. in Protein-Lipid Interactions" Vol. 2, (Watts, A. and de Pont, J.J.H.H.M. eds). Elsevier, Amsterdam.

Deishofer, J., Epp, O., Miki, K., Huber, R. & Michel, H. (1985) Nature, 318, p618.

Dempsey, C.E., Ryba, N.J.P. & Watts, A. (1986) Biochemistry, 25, p2180-2187.

Devaux, P.F. & Seigneuret, M. (1985) Biochim. Biophys. Acta 822, p63-126.

East, J.M., Melville, D., & Lee, A.J. (1985) Biochemistry, 24, p2615-2623

Fuller, S.D., Capaldi, R.A. & Henderson, R. (1979) J. Mol. Biol. 134, p305-327.

Jost, P.C., Griffith, O.H., Capaldi, R.A. & Vanderkooi, G. (1973a) Proc. Natl. Acad. Sci. U.S.A., 70, p480-484.

Jost, P.C., Griffith, O.H., Vanderkooi, G. & Capaldi, R.A. (1973b) J. Supermol. Structure, 1, p269-280.

Kinsey, R.A., Kintanar, A., Smith, R.L., Gutowsky, H.S. & Oldfeild, E. (1981) J. Biol. Chem. 256, p4146-4149.

Knowles, P.F., Watts, A. & Marsh, D. (1979) Biochemistry, 18, p4480-4487.

Knowles, P.F., Watts, A. & Marsh, D. (1981) Biochemistry, 20, p5888-5894.

Marsh, D. (1985) in:- "Prog. in Protein-lipid Interactions" Vol. 1, (Watts, A. & de Pont, J.J.H.H.M. eds) Elsevier, Amsterdam.

Marsh, D. and Watts, A. (1982) in:- "Lipid-protein Interactions" (Jost, P.C. and Griffiths, O.H. eds). Wiley Interscience.

Oldfield, E., Gilmore, R., Glaser, M., Gutowsky, H.S., Hshung, J.C., Kang, S.Y. King, T.E., Meadows, M. and Rice, D. (1978) Proc. Natl. Acad. Sci. U.S.A., 75, p4657-4660.

Oldfield, E., Kinsey, R.A. & Kintanar, A. (1982) Meth. in Enzymology 88, p310-326.

Paddy, M.R., Dahlquist, F.W., Davis, J.H. and Bloom, M. (1981). Biochemistry, 20, 3152-3162.

Rice, D.M., Meadows, M.D., Scheiman, A.O., Goni, F.M., Gomez-Fernandez, J.C., Moscarello, M.A., Chapman, D. & Oldfield, E. (1979) Biochemistry, 18, p5893-5903.

Robinson, N.C., Strey, F., Talbert, F. (1980) Biochemistry, 19, p3656-3661.

Seelig, J. (1977) Qu. Rev. Biophys. 10, p353-418.

Seelig, J. & Seelig, A. (1980) Qu. Rev. Biophys. 13, p9-61.

Shipley, G.G. (1973) in-: "Biological Membranes" 2, p1-90.

Watts, A. (1985) Studia Biophy. 110, p149-154.

Watts, A., Knowles, P.F. & Marsh, D. (1979) Biochem. Biophys. Res. Comm. 81, p403-409.

Watts, A., Sixl, F., Ryba, N.J.P., Dempsey, C.E. & Brophy, P.J. (1985) in: "Mag Res. in Biology and Medicine" (Govil, Khetrapal & Saran, eds) Tata-McGraw Hill, New Delhi.

Williams, R.J.P. (1985) in:- "Rec. Adv. in Biological Membrane Studies" (Packer L. ed) Plenum.

STRUCTURE AND FUNCTION OF PHOSPHOINOSITIDES

IN MEMBRANES AND CELLS

Karel W.A. Wirtz, Peter A. van Paridon,
Anton J.W.G. Visser and Ben de Kruijff

Laboratory of Biochemistry and
Institute of Molecular Biology
State University of Utrecht
3508 TB Utrecht, The Netherlands

INTRODUCTION

Phosphatidylinositol (PI) and its phosphorylated derivatives phosphatidylinositol 4-phosphate (PIP) and phosphatidylinositol 4,5-bisphosphate (PIP_2) have caught the eye and
fired the imagination of the scientific community. In a
recent review (1) Hawthorne underscored the air of excitement
by stating :"In 1960 inositol phospholipids were only for the
connoiseur or the addict, but thanks to the association with
receptor activation and the resulting interest of the pharma-
cologists they now represent something of a growth industry".
Indeed, particularly during the past ten years we have seen
an ever-increasing flow of studies dealing with the key role
of phosphoinositides in the stimulus-induced signal transduction across the plasma membrane (2-4). One is referred to the
paper by Williamson and Hansen for a current view on the subject of receptor-linked breakdown of PIP_2 yielding inositol
1,4,5-trisphosphate (IP_3) and diglyceride as the intracellular second messengers. We will touch upon various aspects of
this process only as far as they are related to the properties and structural functions of phosphoinositides in membranes which form the main topics of discussion in this chapter.

As for phosphoinositides in membranes, many questions
come to mind that wait for an answer. Does PI behave diffe-
rently from other phospholipids in the membrane; does it show
any preference for particular membrane proteins? Is PIP_2
subject to lateral mobility in the plane of the membrane or
does it persist in different pools as part of a particular
receptor-phosphodiesterase complex? What other structural
functions may phosphoinositides fulfil? Here we will discuss
and present some recent studies that have a bearing on these
questions.

The molecular order and rotational mobility of phospholipids in artificial and biological membranes has been extensively studied by way of spectroscopic techniques (nuclear magnetic resonance (NMR), electron spin resonance and fluorescence spectroscopy). So far, PI in membranes has not been subject of spectroscopic investigation most likely for reasons that probed analogues were not available. Recently a semisynthetic method was developed by which the 2-fatty acyl chain in yeast PI was substituted for its natural fluorescent analogue, cis-parinaric acid (i.e. cis, trans, trans, cis-9,11,13,15-octadecatetraenoic acid) yielding the probe molecule 2-parinaroyl-PI (5). Parinaric acid was selected since it structurally resembles most closely the polyunsaturated fatty acids found in mammalian PI. In addition, this conjugated polyene fatty acid has very attractive spectroscopic properties which make it a useful probe of membrane structure (6-12) and lipid-protein interactions (13-15). To be more specific, its absorption spectrum, fluorescence quantum yield and lifetime are very sensitive to environment (16). Its lifetime is in the ns-range which puts it in the order of the molecular motion of phospholipids in membranes. Fluorescence anisotropy decay as measured by polarized, time-resolved fluorescence spectroscopy has provided information about the effects of membrane lipid composition and integral membrane proteins on the order and dynamics of free parinaric acid incorporated in membranes (17,18). We have performed similar measurements on 2-parinaroyl-PI and 2-parinaroyl-PC incorporated in phospholipid vesicles and in the acetylcholine receptor-rich plasma membranes from the electric organ of Torpedo Marmorata. The latter membrane was selected in an attempt to obtain additional proof for the evidence that PI specifically interacts with the acetylcholine receptor (19). The probe lipids were incorporated into the Torpedo membranes by incubation with PI- and PC transfer proteins that carried these lipids (20). The time-resolved fluorescence measurements were performed with the equipment described before (21). Samples were excited at 306 nm by 4 psec laser pulses. Fluorescence and anisotropy decays were analyzed with a nonlinear least-squares iterative reconvolution approach as described (21).

The fluorescence decay curve for 2-parinaroyl-PI in Torpedo membranes is shown in Fig.1. Similar curves were obtained for 2-parinaroyl-PC and for both probe lipids (1 mole%) in vesicles consisting of PC-PA (95:5, mole%). The measurements were performed at $4^{o}C$ to reduce the mobility of the probes , thereby accentuating differences in interactions experienced by these probes. The experimental data points could be well fitted to a decay model based on two fluorescence lifetimes. As one sees from Table 1 the fluorescence decay is described by a short and a long lifetime component. In both vesicles and Torpedo membranes 2-parinaroyl-PI and 2-parinaroyl-PC displayed a rather similar fluorescence decay behaviour. Both probes had an identical short lifetime component in the vesicles (4.8 ns) and in the Torpedo membranes (6.6 ns). As for the long lifetime component ranging between 16.6 and 22.5 ns, the fluorescence decay tended to be slightly faster for 2-parinaroyl-PI than for 2-parinaroyl-PC in both membrane systems. This is most noticable in the Torpedo

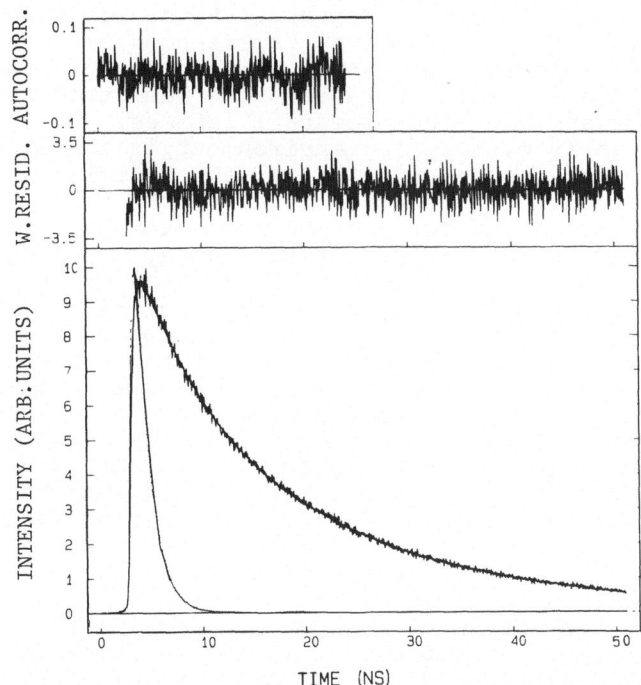

Fig.1. Fluorescence decay analysis of 2-parinaroyl-PI incorporated in the acetylcholine receptor-rich membranes from Torpedo marmorata. Both the experimental and calculated fluorescence decay patterns are indicated. The quality of the analysis is given by the weighted residuals and the autocorrelation of the residuals (see top of Fig.). The fluorescence decay parameters and standard deviations are given in Table 1.

Table 1. Fluorescence lifetimes of 2-parinaroyl-PI (PnA-PI) and 2-parinaroyl-PC (PnA-PC) in membranes

Sample	Probe	α_1	τ_1 (ns)	α_2	τ_2 (ns)	$<\tau>$ (ns)
Vesicles	PnA-PI	0.25	4.8	0.75	16.6	13.6
(PC-PA, 95:5)	PnA-PC	0.20	4.9	0.80	17.3	14.8
Torpedo	PnA-PI	0.38	6.6	0.62	19.2	14.4
membranes	PnA-PC	0.31	6.6	0.69	22.5	17.6

α is the preexponential factor (s.d.$\pm$0.01); τ is the lifetime component (s.d.$\pm$0.2 ns); $<\tau>$ is the average fluorescence lifetime at 4°C ($\alpha_1\tau_1 + \alpha_2\tau_2$)

membranes suggesting that 2-parinaroyl-PI senses an environment slightly different from that experienced by 2-parinaroyl-PC. Both lifetime components were shorter in the vesicles than in the Torpedo membranes. The same trends were observed for the average fluorescence lifetimes. This could signify a slightly more apolar environment in the bilayer of the Torpedo membranes than in that of the vesicles.

The fluorescence anisotropy decay curves of 2-parinaroyl-PI in vesicles and Torpedo membranes are shown in Fig.2. Both curves demonstrate a fast initial decay leveling off to a constant anisotropy value. The fast initial decay is a measure for the rotational correlation time (ϕ) of the parinaroyl-chain; the initial and residual anisotropy value (r_∞) are directly related to the order parameter S. Similar curves were obtained for 2-parinaroyl-PC in these membrane systems. The parameters ϕ, r_∞ and S derived from these decay curves are presented in Table 2. The correlation times for both probes are very similar in vesicles (3.5 ns) and in the Torpedo membranes (4.0 ns). As for the limiting anisotropy and the order parameter, the values for 2-parinaroyl-PI and 2-parinaroyl-PC were much higher in the Torpedo membranes (see Fig.2B) than in the vesicles (see fig.2A). This indicates that other membrane components (e.g. proteins and cholesterol) restrict the motion of the probe lipids in the Torpedo membranes. It is to be noted that in the vesicles the PI probe displayed a higher degree of order than the PC probe (S of 0.42 and 0.38, respectively). We assume that this difference in S is an intrinsic property of the two probe lipids.

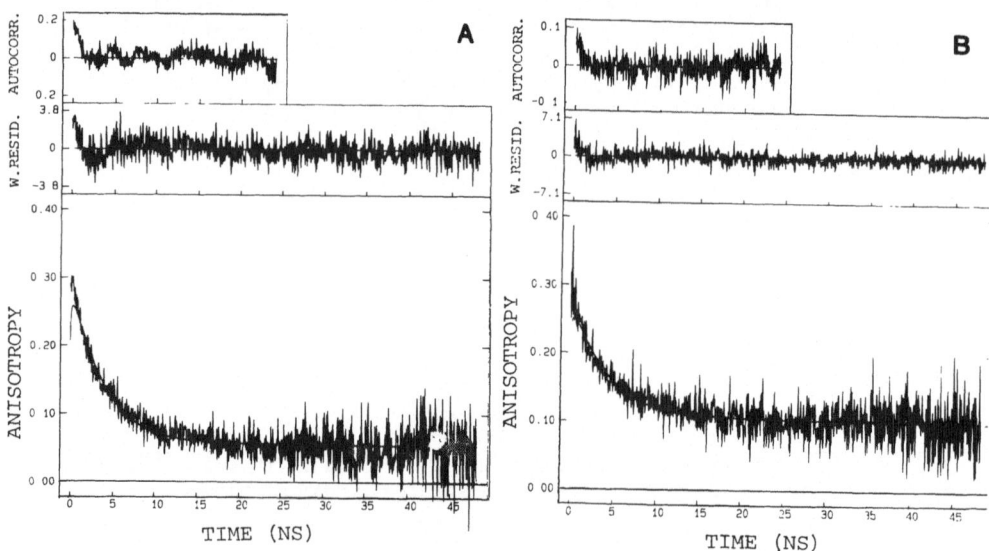

Fig.2. Fluorescence anisotropy decay analyses of 2-parinaroyl PI incorporated in phospholipid vesicles (A) and Torpedo membranes (B). The vesicles consisted of PC-PA (95:5,mole%). The experimental anisotropy is indicated by the noisy curve; the smooth line represents the fitted function. The weighted residuals and the autocoorrelation of the residuals are indicated (see top of Fig.). The parameters from the data analysis are given in Table 2.

Table 2. Mobility and order parameters of 2-parinaroyl-PI
(PnA-PI) and 2-parinaroyl-PC (PnA-PC) in membranes.

Sample	Probe	Φ (ns)	r_∞	S
Vesicles	PnA-PI	3.6	0.056	0.42
(PC-PA, 95:5)	PnA-PC	3.5	0.045	0.38
Torpedo	PnA-PI	3.9	0.106	0.58
membranes	PnA-PC	4.2	0.122	0.62

Φ is the rotational correlation time (+0.1 ns); r_∞ is the residual anisotropy (+0.001); S is the order parameter (+0.007)

In view of this property, the lower degree of order for 2-parinaroyl-PI than for 2-parinaroyl-PC in Torpedo membranes (S of 0.58 and 0.62, respectively) is highly significant. One interpretation would be that part of the PI probe preferentially interacts with the acetylcholine receptor, thereby sensing a somewhat different, less ordered environment than the PC probe. On the other hand, it is possible that the different lipid composition of the Torpedo membrane (e.g. the cholesterol content) has a differential effect on the order of the PI- and PC probes. However, this remains to be confirmed by fluorescence measurements on reconstituted membranes of well-defined composition.

PI AS AN ANCHOR FOR MEMBRANE PROTEINS

It has been known for many years that treatment of mammalian cells with PI-phospholipase C selectively released certain proteins from the plasma membrane without causing cell lysis (23). These proteins included alkaline phosphatase (24), acetylcholinesterase (25-27) and 5 -nucleotidase (28-29). More recently, it was observed that distinct membrane surface proteins such as the Thy-1 glycoprotein from murine thymocytes (30), the variant surface glycoprotein (VSG) from Trypanozoma brucei (31) and the neural cell adhesion molecule from brain membranes and astroglial cells (32), were released by this enzyme treatment. These observations strongly suggested that these proteins were attached to the membranes by covalent linkage to PI. Recent structural studies on VSG have shown that the membrane-anchoring domain consists of a glycolipid composed of dimyristoyl-PI, non-acetylated glucosamine glycosidically bonded to the inositol moiety, several sugars and ethanolamine linked to the C-terminal-carboxyl group of VSG by an amide bond (for a brief review, see Ref.33). Similar moieties have been identified in the membrane-anchoring domain of thymus Thy-1 (34) and human erythrocyte acetylcholine esterase (35). The PI-phospholipase C used in these studies, was purified from Staphylococcus aureus culture filtrates. Recently, it was reported that Trypanosoma brucei contains a phospholipase C which cleaves dimyristoyl glycerol from the glycolipid attached to VSG (36). Interestingly, this enzyme failed to degrade 1-stearoyl-2-arachidonoyl-PI which is the most abundant species in mammalian cells (37). It remains to be established whether a similar phospholipase C occurs in mammals. A point of great interest is the extent to which dimyristoyl glycerol will activate protein kinase C. To date, it is not known whether 1-stearoyl-2-arachidonoyl-glycerol generated by the agonist-induced breakdown of PIP_2, is responsible for the intracellular stimulation of protein kinase C.

CHARGE STATE OF PIP AND PIP_2 AS A FUNCTION OF pH

To date few studies have appeared on the physical properties of PIP_2 and its immediate precursor PIP (38-40). It is generally assumed that the very acidic nature of PIP and PIP_2 is an important factor in dictating the behaviour and function of these phospholipids in the plasma membrane. From measuring calcium binding and electrophoretic properties it was inferred that at pH 5.55 PIP_2 in egg yolk PC vesicles contains five negative charges (41). Since the charge state of PIP and PIP_2 at physiological pH may have important implications for their physiological function, we have investigated the ionization as a function of pH by use of ^{31}P-NMR spectroscopy.

Spectra of PIP and PIP_2 isolated from bovine brain in pure micellar form are presented in Fig.3. The spectrum for PIP_2 consists of three well separated resonance peaks. From a comparison with other phospholipid spectra we conclude that the high-field peak represents the phosphodiester moiety, and

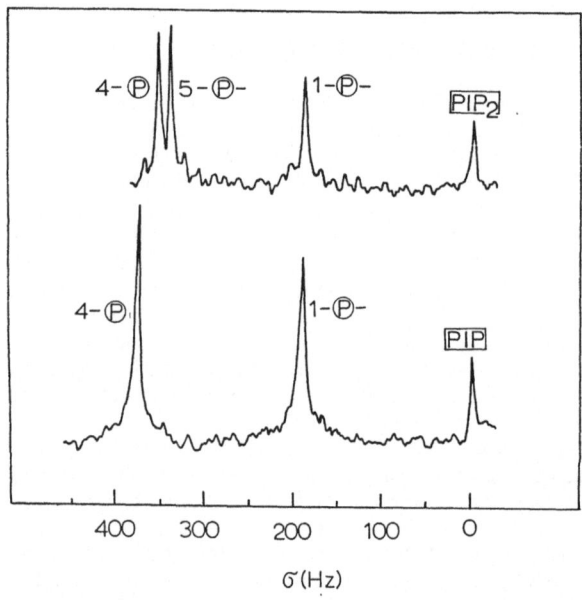

Fig.3. ^{31}P-NMR spectra of micellar PIP_2 and PIP in H_2O/D_2O (1:1) containing 20 mM Tris-maleate, 5mM EDTA and 100 mM NaCl, pH 8.0. Spectra were recordedon a Bruker WH 90 spectrometer operating at 36.4 MHz on ^{31}P-NMR as described before (42). Triphenylphosphine was used as external standard. The phospholipids were converted into the sodium form. For all experimental details, see Ref. 43.

the two low-field peaks the 4- and 5-phosphate residues. In
agreement with the indicated assignment, the spectra of IP_3
yielded three phosphorus resonances with chemical shifts of
5.41 (4-phosphate), 5.22 (5-phosphate) and 4.84 ppm (1-phos-
phate) at pH 9.5 (44). The spectrum for pure PIP gives two
resonances with the low-field peak representing the 4-phos-
phate moiety. The spectrum for PIP_2 (2.5 mole%) in mixed PC-
vesicles (Fig.4) again shows two well separated phosphomonoester resonance peaks while the phosphodiester resonance is
obscured by the PC phosphodiester signal. Under comparable
conditions PIP has one low-field resonance peak representing
the 4-phosphate residue. The spectra of Fig.3 were recorded
at pH 8.0 and of Fig.4 at pH 5.9. At identical pH these positions were found to be similar for vesicles and micelles.

It was previously observed that the ionization of the
phosphate groups of lipids affects the chemical shift (45).
We have used this property to determine the pK-values of the
4- and 5-phosphate moiety in PIP and PIP_2. A typical titration curve for PIP_2 in micellar form is shown in Fig.5. The
resonance position of the phosphodiester moiety is not shifted over the pH-range between 4 and 13. This is in agreement
with the pK values of 2.5 and 3.1 measured for the phosphodiester residues of PI and PG, respectively (46,47). As for
the phosphomonoester resonance signals a clear pH dependence
of the chemical shift was observed yielding a single inflection point at pH 6.7 for the 4-phosphate and at pH 7.6 for
the 5-phosphate. This shows that in the physiological pH-

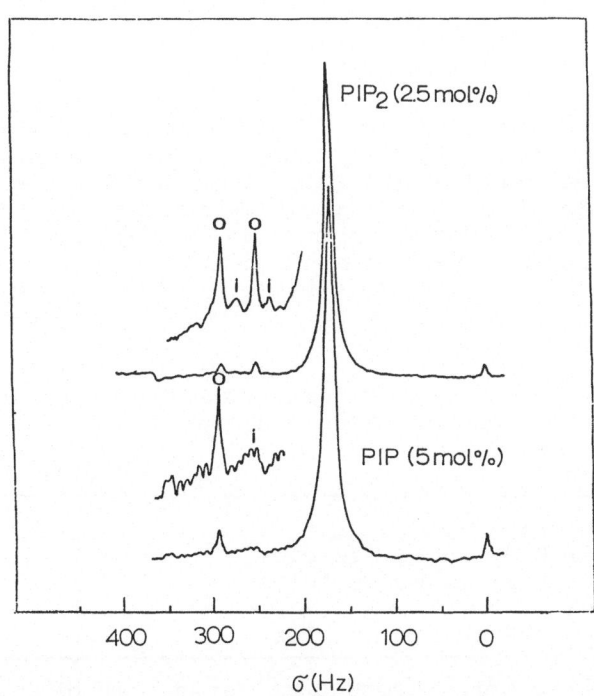

Fig.4. ^{31}P-NMR spectra of PIP and PIP_2 in small unilamellar
egg-yolk PC vesicles at pH 5.9. For further details
see Fig.1.

range, the 5-phosphate is more easily protonated than the 4-phosphate. For comparison, the first protonation of the phosphate moiety in PA has a pK value of 8.0 and the second protonation a pK of 3.0 (46). Similar titration curves were determined for PIP$_2$ in mixed vesicle form yielding the pK values as presented in Table 3. It is evident that the pK values for 4- and 5-phosphate are not affected by whether the measurements were performed on pure PIP$_2$ or on PIP$_2$ present in mixed vesicles. Titration of PIP in pure or mixed-vesicle

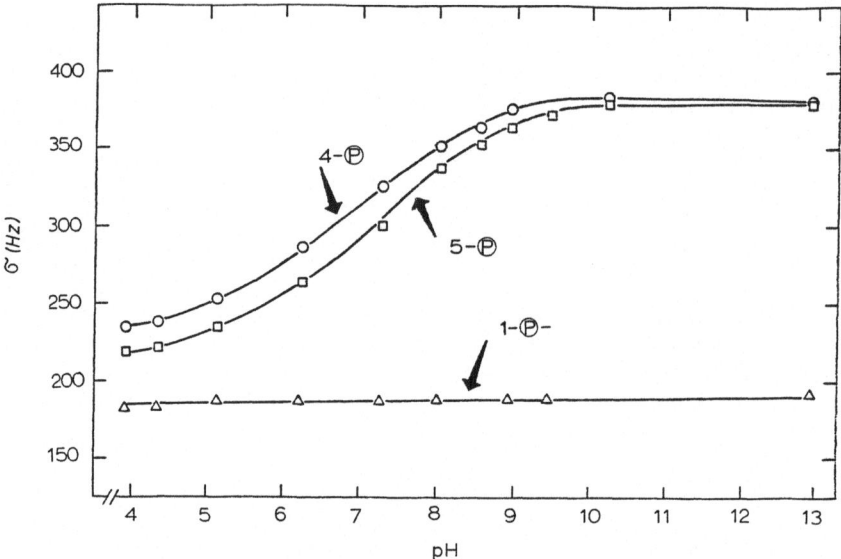

Fig.5. Chemical shift as a function of pH for pure PIP$_2$ micelles. The chemical shift position of the phosphate resonances relative to the internal standard were determined as a function of pH. The pK values for the protonation of the phosphate residues were derived from the inflection point in the titration curve. For further details see Ref. 43.

Tabel 3. The pK values forthe first protonation of the phosphomonoester residues in PIP and PIP$_2$

Phospholipid	pK(4-P)	pK(5-P)
100% PIP$_2$	6.7	7.6
2.5% PIP$_2$ in PC vesicles	6.5	7.7
1.0% PIP$_2$ in PC vesicles	6.7	7.7
100% PIP	6.3	
5.0% PIP in PC vesicles	6.1	

^{31}P-NMR spectra were recorded at various pH of PIP and PIP$_2$, either pure or in mixed vesicles with egg PC. From the titration curves as shown in Fig.5 the pK values were calculated. The pK values in this table were corrected for the presence of 50% D$_2$O (+ 0.2 pH unit).

form yielded pK values of 6.3 and 6.1, respectively. It is apparent from Table 3 that the 4-phosphate on PIP is slightly more acidic than the 4-phosphate on PIP_2. From the pK-values one may calculate that PIP and PIP_2 undergo a charge shift of -2 to -3 and of -3 to -5, respectively, in the pH-range of 6.0 to 8.5. At the pH of the cytosol of resting cells (i.e. pH 7.1-7.3) PIP_2 has four and PIP three negative charges. At present it is hard to say whether fluctuations in the intracellular pH may be sufficient to affect the charge state of these phosphoinositides. An intracellular pH-rise has been observed upon activation of the Na+/H+-exchanger (48). Since this proton pump is present in the plasma membrane one cannot preclude the possibility that at the immediate environment of thes pumps the local pH change is sufficient for changes in the charge state of PIP and PIP_2 to occur, thereby affecting crucial cellular processes.

THE STRUCTURAL ROLE OF PIP_2 IN ERYTHROCYTES

It is well established that mammalian erythrocytes have the ability to convert PI into PIP and PIP_2 (49). Moreover, phosphomonoesterases are present to dephosphorylate these phosphoinositides under formation of PI (50). Erythrocytes have a Ca^{2+}-dependent phospholipase C that degrades PIP and PIP_2 to diglyceride (51). However, this enzyme is inactive under the physiological conditions of intracellular Ca^{2+} (<1 uM). The relative levels of PI, PIP and PIP_2 depend on the energy state of the erythrocyte (52). ATP-depletion lead to reduced levels of PIP_2 which is compensated by increased levels of PI; PIP-levels are not affected.
It is generally accepted that PIP and PIP_2 and the bulk of PI are located in the inner leaflet of the erythrocyte membrane where they comprise 3.4% of the total phospholipid (52). Shifts in the PIP_2/PI ratio by ATP-depletion are accompanied by shape changes of the erythrocyte from smooth discocytes to crenated echinocytes (52). These shape changes may be due to the fact that the conversion of PIP_2 to PI will lead to a shrinkage of the membrane inner monolayer, thereby generating a bilayer imbalance. On the other hand, red cell shape may be controlled by the interaction between the membrane skeleton and the membrane proper. The membrane skeletal protein, protein 4.1, is known to be involved in this interaction by its ability to bind to glycophorin (53). Very interestingly, PIP_2 forms a strong and specific complex with glycophorin (54,55). It is only to this complex that protein 4.1 binds (55). These observations suggest that a reduction of PIP_2-levels may lead to a disruption of the interaction between protein 4.1 and glycophorin and as a direct consequence, to a change of erythrocyte shape. It remains to be established whether agonist-stimulated breakdown of PIP_2 has any effect on membrane organization and, for this matter, on the shape of other cells.

REFERENCES

1. J.N. Hawthorne, Bioscience Rep. 3, 887-904 (1983).
2. Y. Nishizuka, Nature 308, 693-698 (1984).
3. M.J. Berridge, Biochem. J. 220, 345-360 (1984).
4. L.E. Hokin, Ann. Rev. Biochem. 54, 205-235 (1985).

5. P. Somerharju and K.W.A. Wirtz, <u>Chem. Phys. Lip.</u> 30, 81-91 (1982).
6. L.A. Sklar, B.S. Hudson and R.D. Simoni, <u>Proc. Natl. Acad. Sci. U.S.A.</u> 72, 1649-1653 (1975).
7. L.A. Sklar, B.S. Hudson and R.D. Simoni, <u>Biochemistry</u> 16, 819-828 (1977)
8. L.A. Sklar, G.P. Miljanich and E.A. Dratz, <u>Biochemistry</u> 18, 1707-1716 (1979)
9. F. Schroeder, J.F. Holland and P.R. Vagelos, <u>J. Biol. Chem.</u> 251, 6739-6746 (1976).
10. F.Schroeder, <u>Eur. J. Biochem.</u> 132, 509-516 (1983).
11. A.J. Waring, P. Glatz and J.M. Vanderkooi, <u>Biochim. Biophys. Acta</u> 557, 391-398 (1979).
12. R. Welti and D.F. Silbert, <u>Biochemistry</u> 21, 5685-5689 (1982).
13. L.A. Sklar, B.S. Hudson and R.D. Simoni, <u>Biochemistry</u> 16, 5100-5108 (1977).
14. C.B. Berde, B.S. Hudson, R.D. Simoni and L.A. Sklar, <u>J. Biol. Chem.</u> 254, 391-400 (1979)
15. D. Kimelman, E.S. Tecoma, P.K. Wolber, B.S. Hudson, W.T. Wickner and R.D. Simoni, <u>Biochemistry</u> 18, 5874-5880 (1979).
16. L.A. Sklar, B.S. Hudson, M. Petersen and J. Diamond, <u>Biochemistry</u> 16, 813-819 (1977).
17. P.K. Wolber and B.S. Hudson, <u>Biochemistry</u> 20, 2800-2810 (1981).
18. P.K. Wolber and B.S. Hudson, <u>Biophys. J.</u> 37, 253-262 (1982).
19. H.W. Chang and E. Bock, <u>Biochemistry</u> 18, 172-179 (1979).
20. A. Rousselet, P.F. Devaux and K.W.A. Wirtz, <u>Biochem. Biophys. Res. Commun.</u> 90, 871-877 (1979).
21. A. van Hoek, and A.J.W.G. Visser, <u>Anal. Instr.</u> 14, 359-378 (1985)
22. A.J.W.G. Visser, T. Ykema, A. van Hoek, D.J. O Kane and J. Lee, <u>Biochemistry</u> 24, 1489-1496 (1985)
23. H. Ikezawa, M. Yamanegi, R. Taguchi, T. Miyashita and T. Ohyabu, <u>Biochim. Biophys. Acta</u> 450, 154-164 (1976).
24. M.G. Low and J.B. Finean, <u>Biochem. J.</u> 167, 281-284 (1977).
25. M.G. Low and J.B. Finean, <u>FEBS Lett.</u> 82, 143-146 (1977).
26. A.H. Futerman, M.G. Low and I. Silman, <u>Neurosci. Lett.</u> 40, 85-89 (1983).
27. A.H. Futerman, M.G. Low, D.M. Michaelson and I. Silman, <u>J. Neurochem.</u> 45, 1487-1494 (1985).
28. M.G, Low and J.B, Finean, <u>Biochim. Biophys. Acta</u> 508, 565-570 (1978).
29. S.D. Shukla, R. Coleman, J.B. Finean and R.H. Michell, <u>Biochem. J.</u> 187, 277-280 (1980).
30. M.G. Low and P.W. Kincade, <u>Nature</u> 318, 62-64 (1985).
31. M.A.J. Ferguson, M.G. Low and G.A.M. Cross, <u>J. Biol. Chem.</u> 260, 14547-14555 (1985).
32. H.T. He, J. Barbet, J.C. Chaix and C. Goridis, <u>EMBO J.</u> 5, 2489-2494 (1986).
33. M.G. Low, M.A.J. Ferguson, A.H. Futerman and I. Silman, <u>Trends Biochem. Sci.</u> 11, 212-215 (1986).
34. A.G.D. Tse, A.N. Barclay, A. Watts and A.F. Williams, <u>Science</u> 230, 1003-1008 (1985).
35. W.L. Roberts and T.L> Rosenberry, <u>Biochem. Biophys. Res. Commun.</u> 133, 621-627 (1985).
36. D. Hereld, J.L. Krakow, J.D. Bangs, G.W. Hart and P.T. Englund, <u>J. Biol. Chem.</u> 261, 13813-13819 (1986).

37. B.J. Holub _in_: "Inositol and Phosphoinositides",
    J.E. Bleasdale, J. Eichberg and G. Hauser, eds.,
    pp. 31-47, Humana Press, Clifton, NJ (1985).
38. Y. Sugiura, Biochim. Biophys. Acta 641, 148-159 (1981).
39. T. Shibata, J. Uzawa, Y. Sugiura, K. Hayashi and
    T. Takizawa, Chem. Phys. Lipids 34, 107-113 (1984).
40. K. Hayashi, M. Muehleisen, W. Probst and H. Rahman, Chem.
    Phys. Lipids 34, 317-322 (1984).
41. H. Hauser and R.M.C. Dawson, Eur. J. Biochem. 1, 61-69
    (1967)
42. B. de Kruyff, P.R. Cullis and G.K. Radda, Biochim.
    Biophys. Acta 436, 729-740 (1976).
43. P.A. van Paridon, B. de Kruyff, R. Ouwerkerk and K.W.A.
    Wirtz, Biochim. Biophys. Acta 877, 216-219 (1986).
44. S. Cerdan, C.A. Hansen, R. Johanson, T. Inubushi and
    J.R. Williamson, J. Biol. Chem. 261, in press (1986)
45. M. Koter, B. de Kruyff and L.L.M. van Deenen, Biochim.
    Biophys. Acta 514, 255-263 (1978).
46. M.B. Abramson, G. Colacicco, R. Cursi and M.M. Rapport,
    Biochemistry 7, 1692-1698 (1968).
47. P.W.M. van Dijck, B. de Kruyff, A.J. Verkley, L.L.M. van
    Deenen and J. de Gier, Biochim. Biophys. Acta 512,
    84-96 (1978).
48. W.H. Moolenaar, L.G.J. Tertoolen and S.W. de Laat, Nature
    312, 371-374 (1984).
49. J.T. Buckley and J.N. Hawthorne, J. Biol. Chem. 247,
    7218-7223 (1972)
50. C.P. Downes, P.T. Hawkins and R.H. Michell, Biochem. Soc.
    Trans. 10, 250-251 (1982)
51. D. Allan and R.H. Michell, Biochim. Biophys. Acta 508,
    277-286 (1978)
52. J.E. Ferrell and W.H. Huestis, J. Cell Biol. 98,
    1992-1998 (1984)
53. R.A. Anderson and R.E. Lovrien, Nature 307, 655-658
    (1984)
54. I.M. Armitage, D.L. Shapiro, H. Furthmayr and
    V.T. Marchesi, Biochemistry 16, 1317-1320 (1977)
55. R.A. Anderson and V.T. Marchesi, Nature 318, 295-298
    (1985)

TIME-RESOLVED FLUORESCENCE DEPOLARIZATION STUDIES
OF PARINAROYL PHOSPHATIDYLCHOLINE IN TRITON X-100 MICELLES
AND RAT SKELETAL MUSCLE MEMBRANES

Antonie J.W.G. Visser, Arie van Hoek and
Peter A. van Paridon

Department of Biochemistry
Agricultural University
Wageningen, The Netherlands

INTRODUCTION

Fluorescence anisotropy under constant illumination is a technique
that provides information on the local order of cylindrical probes like
1,6-diphenyl-1,3,5-hexatriene (DPH) or cis/trans-parinaric acid in membrane
vesicles, which as such can be considered as a macroscopically isotropic
system. It was shown that the steady-state fluorescence anisotropy, at a
time scale much longer than the average fluorescence lifetime, is connected
to the second-rank order parameter for rod-shaped probe molecules in
membranes [1,2].

Time-dependent fluorescence anisotropy measurements of cylindrical
probes (mainly DPH) have been widely used in order to obtain dynamic and
orientational order information on artificial and biological lipid bilayers
[3-15] and on the interaction between proteins and membranes, e.g.
cytochrome oxidase [16], M13 virus coat protein [17,18], apolipoprotein C-I
[19] and blood clotting factor Va [20].

Considerable progress has been made in a theoretical description of
fluorescence depolarization of cylindrical and other probes in membrane
bilayers. The concepts of order parameter (second and fourth rank) and
correlation function are needed to describe the orientational distribution
and dynamics of the probe in the membrane. Several models have been
advanced such as the cone model, in which the probe symmetry axis wobbles
within a cone of certain semiangle [21-23] and the Gaussian model, which
has been applied either to strong collision or diffusion problems [24,25].
A general expression of the fluorescence anisotropy decay on the basis of
the Smoluchowski equation for hindered rotational diffusion has been
derived showing general applicability to any particular distribution model
[26]. Experimental verification was achieved for DPH in dimyristoylphosphatidylcholine (DMPC) or dipalmitoylphosphatidylcholine (DPPC)
vesicles and for DPH in DMPC vesicles with α-lactalbumin attached at acidic
pH [27]. A comprehensive and unified formalism encompassing several physical situations has been developed to describe time-dependent fluorescence
depolarization in macromolecules and membranes [28].

Probes like DPH (and possibly parinaric acid) have the disadvantage
that they have a nonneglizable orientational distribution in a plane perpendicular to the membrane normal [27]. Fluorescent parinaroyl phospholi-

pids are very appropriate probes for artificial and biological membrane structures, because the parinaric acyl chain of the lipid is uniquely oriented along the fatty acids of the membranes. The fluorescent lipid probe can be considered as a minimal perturbing agent. A DPH-phosphatidyl choline probe has also been designed [29], The spectral properties of parinaric acid and its incorporation into artificial bilayers have been reported [30,31].

In this paper we describe the fluorescence anisotropy results obtained with parinaroyl phosphatidylcholine incorporated into micelles and rat skeletal muscle membranes (sarcolemma and sarcoplasmic reticulum). The lipid used was sn-2-cis-parinaroyl-phosphatidylcholine (PnA-PC) This probe was also used in vesicles and microsomal membranes by other workers [32-34].

The fluorescent lipid can be incorporated into the outer monolayer of the biological membranes without any disturbance by incubating the membrane preparations with phospholipid transfer proteins loaded with the particular lipid [35].

The interpretation of steady-state fluorescence anisotropy of probes in natural membranes has been reviewed by Zannoni et al. [25]. Schroeder [36] has reviewed fluorescence applications of a variety of probes incorporated into different membranes. The latter author also described the incorporation of parinaric acid into plasma membranes from rat liver and results from time-resolved fluorescence and steady-state fluorescence anisotropy [37]. The application of time-resolved fluorescence of lipid probes in biological membranes has been reported previously [4,12].

METHODOLOGY

The fluorescent lipid was synthesized as described previously [38]. Time-resolved fluorescence measurements were carried out using the laser excitation source and time-correlated single photon counting as described in detail elsewhere [39-41]. Data analysis has been described in detail by Vos et al. [42]. In the latter paper advantage of the utilization of a reference compound, in order to remove the color effect of the photomultiplier response, has been outlined. For parinaroyl fluorescence POPOP (p-bis[2-(5-phenyloxazolyl)]benzene) is a suitable reference compound (τ=1.3 ns). The excitation wavelength was 305 nm, the emission was viewed through a K45 band-pass filter (Balzers). All experiments were carried out at 20°C.

We will briefly summarize the main points of the data analysis of anisotropy decay. The two intensity components, $i_{//}(t)$ and $i_{\perp}(t)$, were analyzed according to the total fluorescence $s(t)$ to yield the fluorescence lifetime components τ_i with amplitudes α_i:

$$s(t) = i_{//}(t) + 2i_{\perp}(t) = \sum_i \alpha_i \exp(-t/\tau_i).$$

For anisotropy decay $i_{//}(t)$ and $i_{\perp}(t)$ were fitted simultaneously to yield the parameters describing anisotropy decay $r(t)$:

$$i_{//}(t) = s(t)/3 \ [1 + 2r(t)]$$

$$i_{\perp}(t) = s(t)/3 \ [1 - r(t)]$$

Details of this procedure and its associated advantage (preservation of Poissonian statistics and usual fitting criteria) have been outlined [27,42,43].

Anisotropy decay data were fitted according to various models:
For Triton X-100 micelles:

$$r(t) = \beta_{int}\exp(-t/\phi_{int}) + \beta_{mic}\exp(-t/\phi_{mic})$$

where the subscript 'int' refers to internal motion and the subscript 'mic' to overall micellar rotation. The amplitude of the internal motion, expressed in the angle θ_0, is given by:

$$\beta_{mic}/(\beta_{int} + \beta_{mic}) = [1/2\cos\theta_0(1 + \cos\theta_0)]^2$$

For the natural membranes we restrict ourselves to the cone model:

$$r(t) = (r_0 - r_\infty)\exp(-t/\phi) + r_\infty$$

where r_∞ is the residual anisotropy at the end of the fluorescence experiment (about 50 ns), ϕ is the relaxation time in the cone model.
Other models consisting of 1,2 or 3 exponentials and a constant anisotropy have been tested by Ameloot et al. [27].
The second-rank order parameter $<P2>$ can be obtained from:

$$r_\infty/r_0 = <P_2>^2$$

The cone angle θ_c, describing the range of acyl motion in the cone model is determined from:

$$<P_2> = 1/2\cos\theta_c(1 + \cos\theta_c)$$

The wobbling diffusion constant D_w, conceived as the rate of probe motion in the cone model, is calculated from:

$$D_w\phi(r_0 - r_\infty)/r_0 = f(\cos\theta_c) = f(x)$$

$$f(x) = -x^2(1+x^2)\left[\ln\{(1+x)/2\}+(1-x)/2\right]/\{2(1-x)\}+(1-x)(6+8x-x^2-12x^3-7x^4)/24$$

Also the diffusion constant $D_\perp$ for rotation around an axis perpendicular to the probe symmetry axis can be evaluated from

$$D_\perp = (r_0 - r_\infty)/(6r_0\phi)$$

$D_\perp$ is a model-independent parameter and is equal to the derivative of $r(t)$ with respect to time at t=o.
It should be noted that $D_{//}$ cannot be obtained, since such rotation would not depolarize the fluorescence.

In the biological membranes the anisotropy shows an initial rapid decay followed by a subsequent gradual increase. The anisotropy decay can then be approximated to arise from two populations A and B characterized by different average lifetimes, correlation times and order:

$$r(t) = [r_A(t) + r_B(t)]/[s_A(t) + s_B(t)]$$

For example population A:

$$s_A(t) = \exp(-t/\tau_A)$$

$$r_A(t) = s_A(t)[(r_0 - r_\infty)\exp(-t/\phi) + r_\infty]_A$$

where the subscript A under the square brackets denotes the parameters r_0, r_∞ and ϕ belonging to population A.

RESULTS AND DISCUSSION

We will present fluorescence anisotropy decay examples of PnA-PC in three different environments. The probe, dispersed in micelles, is located in a medium that is comparable to an apolar solvent as regard to polarity and fluidity. In biological membranes the parinaroyl lipid is embedded in lipid regions with or without cholesterol or in the vicinity of membrane proteins. In lipid extracts of biological membranes the probe is only in a lipid environment with the possible presence of cholesterol. The results are presented in three tables: Table 1 for fluorescence lifetimes, Table 2 for correlation times and order parameters and Table 3 for a limited population analysis of intact membranes. In the paper by Wirtz et al. some time-resolved fluorescence results are presented of PnA-PC and PnA-PI in egg PC vesicles and membranes from the electric organ of fish (Torpedo marmorata).

Triton X-100 Micelles

We have carried out this experiment in order to assess the average fluorescence lifetime of a parinaroyl lipid in an apolar environment. The fluorescence lifetimes are collected in Table 1. The lifetimes are shorter than those found in membranes or liposomes (see below). This phenomenon might be related to the lack of rigidity of the micellar environment. The correlation between rigidity and lifetimes was also noted by others [3,11,31]. On the other hand, the presence of water molecules in the vicinity of the probe, thus a polarity effect, cannot be ruled out.

The initial anisotropy amounted to 0.31, which is the same value as found for the parinaroyl lipids bound to transfer proteins [44]. Because of the small micellar size the anisotropy decayed to zero within the time scale of the experiment. In Figure 1 the initial fluorescence anisotropy decay of PnA-PC and the calculated one are shown. The best fit consisted of a biexponential function. The shorter correlation time of 2.1 ns reflects internal motion and the longer one of 7.8 ns micellar rotation. The latter correlation time is similar to the correlation times found by others using different probes and techniques [45,46]. The internal motion as observed here was not found earlier. From the preexponential factors, the amplitude of the motion can be estimated (see preceding section). The angle of internal motion amounted to 32°.

Sarcolemma and Sarcoplasmic Reticulum Membranes

The average fluorescence lifetime of the lipid probe is not drastically different in the membrane preparations (Table 1).

In both lipid extracts the anisotropy decay could be well represented by an exponential and a residual anisotropy. A typical example is shown in Figure 2.

In the intact membranes, both from sarcolemma and sarcoplasmic reticulum, the striking observation is that the time-dependent fluorescence anisotropy decays initially, but increases after having reached a minimum (Figure 3). The latter behavior has been noted also by Wolber and Hudson [18] and assigned to partitioning of probe molecules between membrane regions of different fluidity and ordening. While Wolber and Hudson simulated several examples, we have fitted the anisotropy data by an analysis in two populations. The shorter fluorescence lifetime component of the biexponential decay was linked to one class with a particular residual anisotropy and correlation time and the longer fluorescence lifetime to another class with different parameters.

Table 1. Fluorescence lifetimes of parinaroyl phosphatidyl-choline in different media.

Medium	α_1	τ_1 (ns)	α_2	τ_2 (ns)	$\bar{\tau}$ (ns)
Triton X-100	0.51	3.0	0.49	6.4	4.7
Sarcolemma	0.73	7.1	0.27	16.0	9.5
Sarcolemma lipid extract	0.57	5.0	0.43	11.4	7.7
Sarcoplasmic reticulum	0.78	4.8	0.22	12.2	6.4
Sarcoplasmic r. lipid extract	0.52	4.4	0.48	9.3	6.7

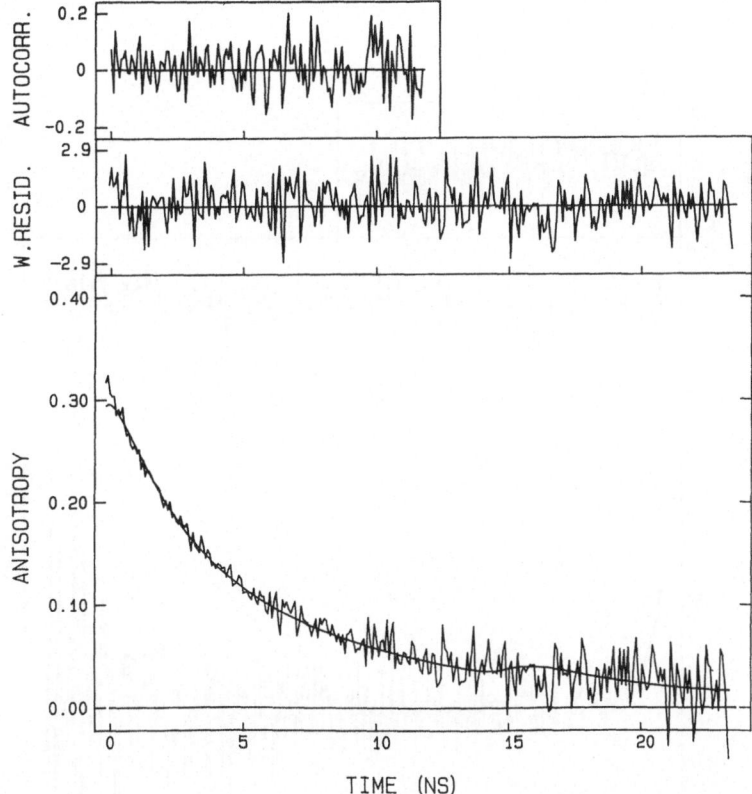

Figure 1. Fluorescence anisotropy decay analysis of PnA-PC (13μM) in Triton X-100 micelles. Analysis was carried out over the whole decay curves (1024 channels with 79 ps time spacing) Shown is a replot over a limited time range of the experimental anisotropy (noisy curve). The initial anisotropy is 0.31 and the fitted curve (smooth line) is a biexponential:0.12exp(-t/2.1) + 0.19exp(-t/7.8) with t in ns. The quality of the fit is indicated by the weighted residuals and autocorrelation function on top of the curves. Fitting criteria: χ^2=1.08 and Durbin-Watson parameter DW=2.1

Table 2. Fluorescence anisotropy decay parameters of parinaroyl phosphatidylcholine in lipid extracts of sarcolemma and sarcoplasmic reticulum membranes.

Membrane	ϕ (ns)	r_∞	$\langle P_2 \rangle$	θ_C (o)	D_W (GHz)	$D_\perp$ (GHz)
Sarcolemma	1.4	0.091	0.56	48	0.114	0.084
Sarcoplasmic reticulum	1.9	0.068	0.51	51	0.085	0.064

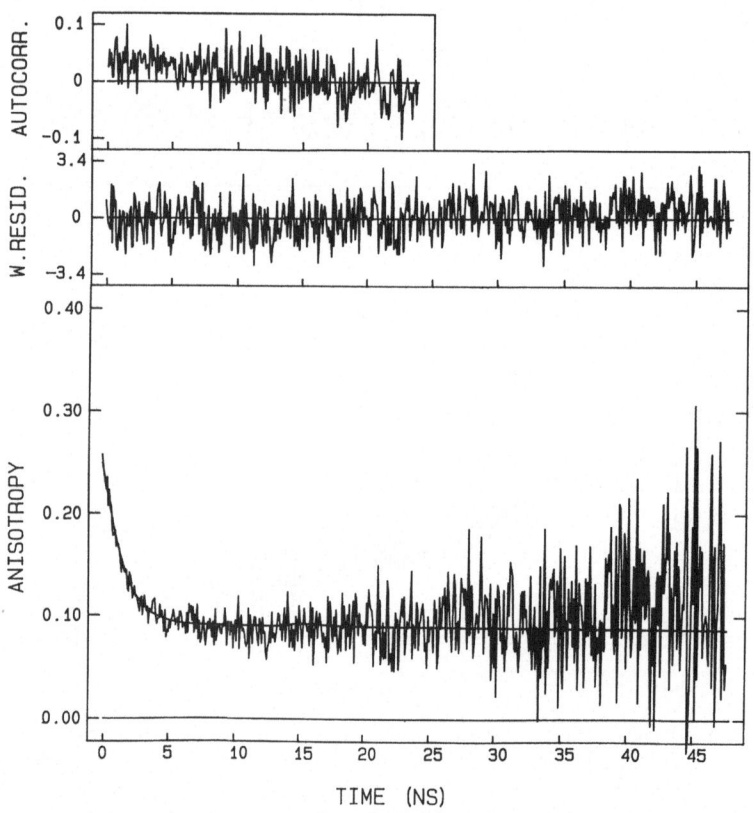

Figure 2. Fluorescence anisotropy decay analysis of PnA-PC (1.3µM) in liposomes extracted from sarcolemmal rat skeletal muscle membranes. Details as in Figure 1. The noisy curve is the experimental anisotropy and the smooth curve a fit according to the function 0.20exp(-t/1.4) + 0.091. Fitting criteria: χ^2=1.35 and DW=2.1

Table 3. Population analysis of the fluorescence anisotropy decay of parinaroyl phosphatidylcholine in intact sarcolemmal and sarcoplasmic reticulum membranes.

Membrane	Shorter lifetime[a]			Longer lifetime[a]		
	ϕ (ns)	r_∞	$<P_2>$	ϕ (ns)	r_∞	$<P_2>$
Sarcolemma	1.2	0.035	0.48	3.1	0.040	0.54
Sarcoplasmic reticulum	1.0	0.028	0.55	2.2	0.021	0.33

a) Listed in Table 1

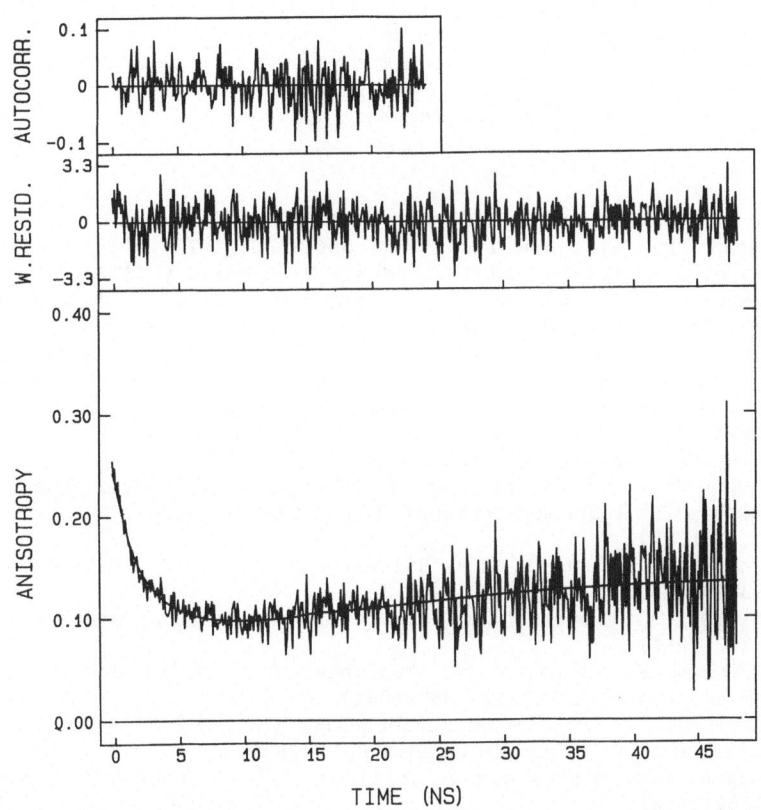

Figure 3. Fluorescence anisotropy decay analysis of PnA–PC (1.3μM) in intact sarcolemmal membranes from rat skeletal muscle. The noisy curve indicates experimental data, while the fitted decay function is represented by the smooth line. The fluorescence anisotropy clearly does not decay to a constant value, but rises after passage through a minimum. The figure shows the results obtained with a model in which two fluorescence lifetimes were coupled to two separate restricted rotational motions. The parameters are given in Table 3. Fitting criteria: $\chi^2 = 1.38$ and DW=1.8.

This approach is a simplification since the probe fluorescence decay is already nonexponential in more homogeneous media like apolar solvents [47] and micelles (Table 1). The average fluorescence lifetime is probably composed of more components. Results of these analyses for the two membranes are given in Table 3. Based on the anisotropy results both coexisting phases must be considered as fluid, the one belonging to the shorter lifetime is more fluid than the other. Schroeder [37] used the fluorescence lifetimes of parinaric acid to assign the coexisting phases in rat liver membranes as fluid and solid. Our anisotropy data are in contrast with the distinction between fluid and solid phases, since neither probe exists in a gel-type of environment at $20^{\circ}C$. Ellena et al. [48] have confirmed the latter conclusion by ^{31}P NMR spectroscopy of rod outer segments and sarcoplasmic reticulum membranes showing that at least 90% of the phospholipids are in the liquid-crystalline phase at $22^{\circ}C$. At most 10% of the phospholipids could be immobilized to the Ca^{2+}-ATPase in sarcoplasmic reticulum membranes.

Although slightly increasing anisotropy of DPH-probes in sarcoplasmic reticulum can be noticed in their figure, Stubbs et al. [12] analyzed the data in terms of a single wobbling in cone. A single wobbling-in-cone model gave poor fits for our experimental data. Because of the large number of parameters involved in the population analysis of a single anisotropy decay curve, the correlation between the various parameters is very high and firm conclusions about possible differences are not possible yet.

REFERENCES

1. M.P. Heyn, FEBS Lett. 108,359-364(1979)
2. F. Jahnig, Proc.Natl.Acad.Sci.USA 76,6361-6365(1979)
3. S. Kawato, K. Kinosita and A. Ikegami, Biochemistry 16,2319-2324(1979)
4. K. Kinosita, R. Kataoka, Y. Kimura, O. Gotoh and A. Ikegami, Biochemistry 20,4270-4277(1981)
5. L.A. Chen, R.E. Dale, S. Roth and L. Brand, J.Biol.Chem. 252,2163-2169 (1977)
6. R.E. Dale, L.A. Chen and L. Brand, J.Biol.Chem. 252,7500-7510(1977)
7. R.E. Dale in Time-Resolved Fluorescence Spectroscopy (R.B. Cundall and R.E. Dale, Eds.) pp.555-605, Plenum, New York, 1983.
8. M.G. Badea, R.P. DeToma and L. Brand, Biophys.J. 24,197-212(1978)
9. J.R. Lakowicz, F.G. Prendergast and D. Hogan, Biochemistry 18,508-519(1979)
10. J.R. Lakowicz, H. Cherek, B.P. Maliwal and E. Gratton, Biochemistry 24, 376-383(1985)
11. C.D. Stubbs, T. Kouyama, K. Kinosita and A. Ikegami, Biochemistry 20, 4257-4262(1981)
12. C.D. Stubbs, K. Kinosita, F. Munkonge, P.J. Quinn and A. Ikegami, Biochim.Biophys.Acta 775,374-380(1984)
13. L.W. Engel and F.G. Prendergast, Biochemistry 20,7338-7345(1981)
14. P.K. Wolber and B.S. Hudson, Biochemistry 20,2800-2810(1981)
15. S.R. Meech, C.D. Stubbs and D. Phillips, IEEE J.Quantum Electr. QE 20, 1343-1352(1984)
16. K. Kinosita, S. Kawato, A. Ikegami, S. Yoshida and Y. Orii, Biochim. Biophys.Acta 647,7-17(1981)
17. D. Kimelman, E.S. Tecoma, P.K. Wolber, B.S. Hudson, W.T. Wickner and R.D. Simoni, Biochemistry 18:5874-5880(1979)
18. P.K. Wolber and B.S. Hudson, Biophys.J. 37,253-261(1982)
19. A. Jonas, J.P. Privat and Ph. Wahl, Eur.J.Biochem. 133,173-177(1983)
20. P. van de Waart, A.J.W.G. Visser, H.C. Hemker and T. Lindhout, Eur.J. Biochem. (in press)
21. K. Kinosita, S. Kawato and A. Ikegami, Biophys.J. 20,289-305(1977)
22. K. Kinosita, A. Ikegami and S. Kawato, Biophys.J. 37,461-464(1982)

23. G. Lipari and A. Szabo, Biophys.J. 30,489-506(1980)
24. C. Zannoni, Mol.Phys. 42,1303-1320(1981)
25. C. Zannoni, A. Arcioni and P. Cavatorta, Chem.Phys.Lipids 32,179-250 (1983)
26. W. van der Meer, H. Pottel, W. Herreman, M. Ameloot, H. Hendrickx and H. Schroder, Biophys.J. 46,515-523(1984)
27. M. Ameloot, H. Hendrickx, W. Herreman, H. Pottel, F. van Cauwelaert and W. van der Meer, Biophys.J. 46,525-539(1984)
28. A. Szabo, J.Chem.Phys. 81,150-167(1984)
29. C.G. Morgan, E.W. Thomas, T.S. Moras and Y.P. Yianni, Biochim.Biophys. Acta 692,196-201(1982)
30. L.A. Sklar, B.S. Hudson, M. Peterson and J. Diamond, Biochemistry 16, 813-819(1977)
31. L.A. Sklar, B.S. Hudson and R.D. Simoni, Biochemistry 16,819-828(1977)
32. E.L. Pugh, M. Kates and A.G. Szabo, Chem.Phys.Lipids 30,50-69(1982)
33. R. Welti and D.E. Silbert, Biochemistry 21,5685-5689(1982)
34. R. Welti, Biochemistry 21,5690-5693(1982)
35. T.A. Berkhout, A.J.W.G. Visser and K.W.A. Wirtz, Biochemistry 23, 1505-1513(1984)
36. F. Schroeder, Subcellul.Biochem. 11,51-101(1985)
37. F. Schroeder, Eur.J.Biochem. 132,509-516(1983)
38. P. Somerharju, H. Brockerhoff and K.W.A. Wirtz, Biochim.Biophys.Acta 649,521-528(1981)
39. A. van Hoek, J. Vervoort and A.J.W.G. Visser, J.Biochem.Biophys.Methods 7,243-254(1983)
40. A.J.W.G. Visser, T. Ykema, A. van Hoek, D.J. O'Kane and J. Lee, Biochemistry 24,1489-1496(1985)
41. A. van Hoek and A.J.W.G. Visser, Anal.Instrum. 14,359-378(1985)
42. K. Vos, A. van Hoek and A.J.W.G. Visser, submitted
43. A.J. Cross and G.R. Fleming, Biophys.J. 46,45-56(1984)
44. P.A. van Paridon, A.J.W.G. Visser and K.W.A. Wirtz, Biochim.Biophys. Acta, in press
45. E. Blatt, K.P. Ghiggino and W.H. Sawyer, Chem.Phys.Lett. 114,47-52(1985)
46. M.D. Ediger, R.P. Domingue and M.D. Fayer, J.Chem.Phys. 80,1246-1253 (1984)
47. T. Parasassi, F. Conti and E. Gratton, Biochemistry 23,5660-5664(1984)
48. J.F. Ellena, R.D. Pates and M.F. Brown, Biochemistry 25,3742-3748(1986)

GENETIC BASIS OF MEMBRANE PHOSPHOLIPID DIVERSITY:

A SUMMARY

Christian R.H. Raetz

Department of Biochemistry
420 Henry Mall
University of Wisconsin-Madison
Madison, Wisconsin 53706

A. INTRODUCTION

The long-range goal of this research is the elucidation of the organization and properties of the genes coding for the enzymes of membrane phospholipid synthesis in animal cells. When this work was initiated in 1978, animal cell mutants defective in membrane phospholipid synthesis had not been described.[1] Much of the initial effort in my laboratory was directed towards the development of procedures for the isolation of such mutants.[1-3] The discovery of the ability of Chinese hamster ovary cells to form macroscopic colonies on filter paper[1] or polyester cloth[3] has made it possible to perform high resolution replica plating and rapid biochemical sorting of animal cell colonies. This methodology has yielded several biochemically defined mutants, including strains defective in the biosynthesis of phosphatidylcholine,[4-6] phosphatidylethanolamine,[7] phosphatidylserine[8,9] and plasmalogens.[10]

In the coming years, some specific goals of my laboratory will be:

(1) Development of schemes for isolation of animal cell mutants deficient in phosphatidylinositol metabolism. Recent interest in the role of phosphatidylinositol turnover in receptor mediated events[11-16] calls for the construction and phenotypic characterization of mutants altered in phosphatidylinositol synthesis, phosphorylation and turnover.

(2) Biochemical and genetic characterization of animal cell mutants deficient in the biosynthesis of plasmalogens. Recently ten CHO cell mutants deficient in the peroxisomal dihydroxyacetone phosphate acyltransferase, the first step in formation of ether lipids,[17] were isolated. These strains will be analyzed biochemically and phenotypically. They will be fused with each other and with fibroblasts from Zellweger's patients (peroxisome-plasmalogen deficient)[18-20] to determine complementation groups. Additional CHO mutants defective in plasmalogens will be isolated.

(3) Characterization of the interaction of Escherichia coli lipid A precursors with the membranes of animal cells. Studies in my laboratory with E. coli lipid mutants have led to the discovery of a new class of biologically active lipids, derived from glucosamine rather than glycerol.[21-25] These substances are precursors of lipid A, a mitogenic

and toxic material that constitutes the outer layer of gram-negative bacteria.[23] Lipid A molecules of high specific radioactivity will be prepared enzymatically in order to study their uptake and metabolism by responsive animal cells in culture. The possibility that lipid A is mimicking a minor animal cell lipid that functions as a mediator or second messenger will also be explored.

B. SIGNIFICANCE AND BACKGROUND

Structural Diversity of Membrane Lipids and Related Substances

All biological membranes contain a very large number of lipids and lipid-derived substances.[2,11,26] Procaryotic membranes, like those of E. coli, consist primarily of glycerophospholipids.[26] However, the outer surface of the outer membrane is composed of lipid A, a glucosamine-derived lipid that serves as the hydrophobic anchor of lipopoly-saccharide.[23] Eucaryotic membranes, such as those of Chinese hamster ovary (CHO) cells, are much more complex in their lipid composition.[27] They also contain sterols, sphingolipids, and plasmalogens, as well as inositol- and choline-linked glycerolipids.[11,27] If only the phospho-lipids are considered, then the membranes of E. coli are estimated to contain about ten major molecular species (chemically distinct combina-tions of polar headgroups and fatty acid), and eucaryotic membranes possess approximately ten times as many.[2] If the many (poorly charac-terized) minor phospholipids and metabolic intermediates are also counted (as in ref. 24), then procaryotic membranes are estimated to contain about 100 distinct phospholipid structures, while eucaryotic membranes must have about 1,000.[2]

All cells are equipped with enzymatic systems responsible for the generation of diverse lipid molecules. In bacteria, these reactions occur on the cytoplasmic membrane.[2] In animal cells (Table I), many of the reactions of phospholipid synthesis (Figs. 1-2) take place on the cyto-plasmic face of the endoplasmic reticulum.[28] However, mitochondria are capable of generating cardiolipin,[11] peroxisomal membranes are responsible for the generation of ether lipids[29,30], and plasma membranes may be involved in sphingolipid synthesis.[31] The enzymes of phospholipid metabolism are minor integral membrane proteins.[2,11] Many of them are essential for membrane biogenesis and cell growth, since mutants defective in these enzymes are often conditionally lethal.[2] In animal cell systems very few of the lipid synthetic enzymes have been purified to homo-geneity,[11,32] and consequently antibodies are still not available.

The biological significance of lipid diversity is poorly understood. Almost nothing is known about the regulation of lipid metabolism or its coordination with membrane protein synthesis.[2,26] Some lipids clearly have specific functions, such as the prostaglandins[33] or the fat-soluble vitamins,[34,35] but the biological role of most membrane lipids remains elusive. Nevertheless, as illustrated by the recent discovery of the regulatory role of 1,2-diglycerides in relation to protein kinase C (Fig. 3)[12-16] and by the discovery of platelet activating factor[36-38] (an ether-linked phospholipid), important biological functions of lipid molecules have been overlooked. Undoubtedly there are many more. A substantial collection of lipid mutants should facilitate the elucida-tion of lipid function.

Genetic Basis of Membrane Lipid Synthesis

From the considerable structural heterogeneity of membrane lipids, it follows that there must exist a significant amount of genetic

Table 1. Phospholipid Composition of Animal Cells

Phospholipid	Percent of total phospholipid			
	CHO cells	Bovine liver[a]	Bovine brain[a]	Bovine heart[a]
Phosphatidylcholine	52.4	54.2	29.2	24.2
Choline plasmalogen	0.6	1.5	trace	17.5
Phosphatidylethanolamine	11.7	9.4	12.1	16.5
Ethanolamine plasmalogen	7.6	3.6	21.1	11.0
Sphingomyelin	9.4	5.8	12.5	11.5
Phosphatidylinositol	8.4	7.9	6.6[b]	4.1
Phosphatidylserine	5.0	4.2	16.6	2.4
Phosphatidic acid	1.2	2.2	0.5	2.2
Phosphatidylglycerol	0.7	trace	trace	trace
Cardiolipin	2.5	4.1	0.7	8.9
Other[c]	0.5	6.6	0.7	1.2

[a] Data collected from (11).
[b] Includes mono-, di-, and triphosphoinositides.
[c] Includes lysophospholipids and unidentified compounds.

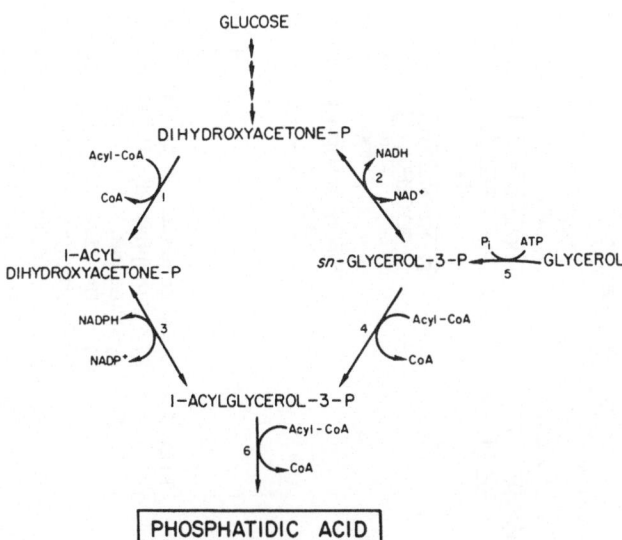

Fig. 1. Biosynthesis of phosphatidic acid: 1. Dihydroxyacetone phosphate acyltransferase. 2. sn-Glycerol-3-phosphate dehydrogenase. 3. Acyl(alkyl)dihydroxyacetone phosphate oxidoreductase. 4. sn-Glycerol-3-phosphate acyltransferase. 5. Glycerol kinase. 6. 1-Acyl-sn-glycerol-3-phosphate acyltransferase.

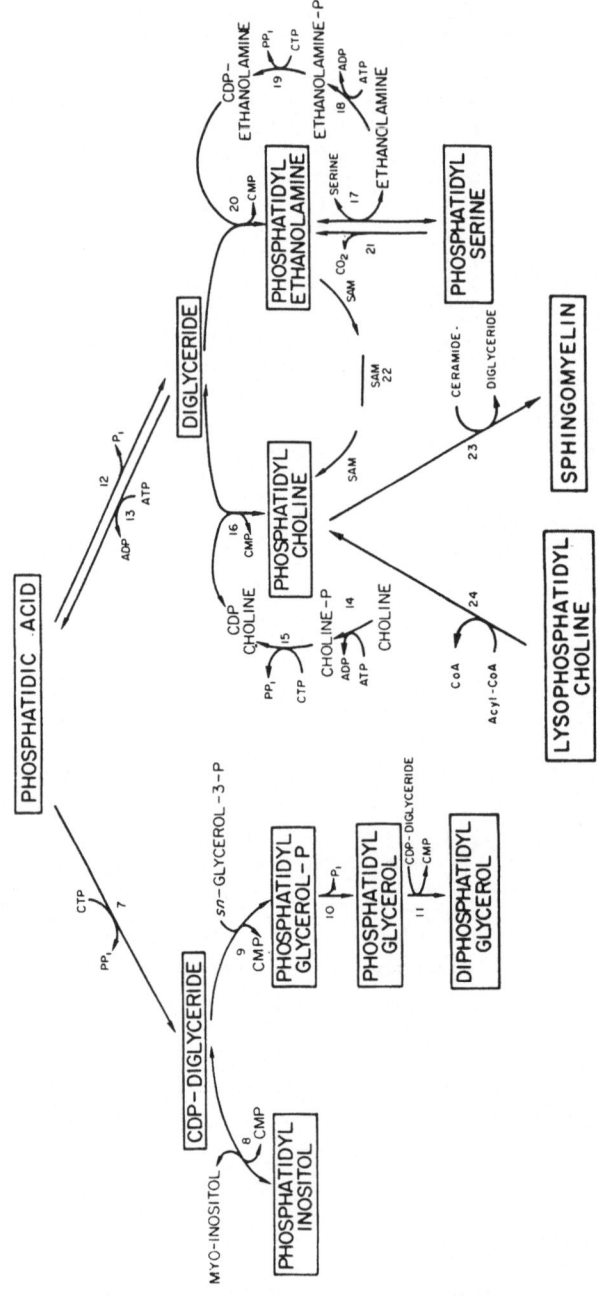

Fig. 2. Biosynthesis of the major phospholipids: 7. CDP-diglyceride synthetase.
8. Phosphatidylinositol synthetase. 9. Phosphatidylglycerol phosphate
synthetase. 10. Phosphatidylglycerol phosphate phosphohydrolase.
11. Cardiolipin synthetase. 12. Phosphatidic acid phosphohydrolase.
13. Diglyceride kinase. 14. Choline kinase. 15. CDP-choline synthetase.
16. Diacylglycerol cholinephosphotransferase. 17. Phosphatidylethanolamine
serine transferase. 18. Ethanolamine kinase. 19. CDP-ethanolamine
synthetase. 20. Diacylglycerol ethanolaminephosphotransferase. 21. Phosphatidylserine decarboxylase. 22. Phosphatidylethanolamine N-methyltrans-
ferase(s). 23. Sphingomyelin synthase. 24. 1-acyl-sn-glycerol-3-phos-
phorylcholine acyltransferase.

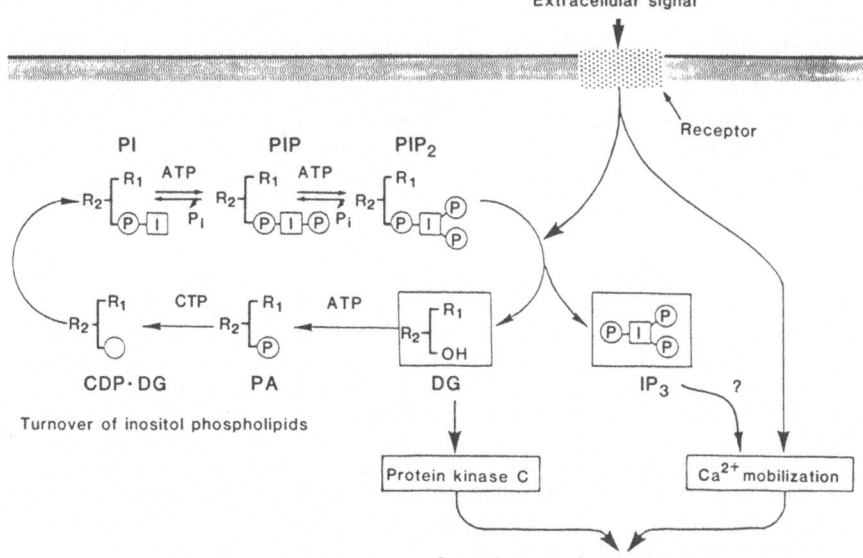

Fig. 3. Turnover of inositol phospholipids and signal transduction.
Inositol phospholipids appear to be in equilibrium in
membranes, although polyphosphoinositides are normally very
minor components. Recent evidence obtained for several
tissues, such as iris smooth muscle, hepatocytes, parotid
gland, brain, adrenal gland, platelets, and blowfly salivary
gland, indicates that phosphatidylinositol 4,5-bisphosphate is
the immediate target of the signal-dependent hydrolysis.
The enzymological basis and tissue variation of this reaction
remain to be determined. PI, phosphatidylinositol; PIP,
phosphatidylinositol 4-phosphate; PIP_2, phosphatidylinositol
4,5-bisphosphate; DG, 1,2-diacylglycerol; IP_3, demyo-
inositol 1,4,5-trisphosphate; I, inositol; P, phosphoryl
group; and R_1 and R_2, fatty acyl groups.

information to direct the synthesis of these compounds. In E. coli, as
in animal cells, it is estimated that 3-6% of the genome must be involved
in membrane lipid assembly.[2] The application of genetics and molecular
biology to the problem of lipid regulation and function is still
relatively unexploited considering the enormous scope of the problem.
Substantial progress has been made with the isolation of E. coli lipid
mutants,[2,26] but even in this simple system no more than half of all the
lipid genes that must exist based on the known biochemistry have been
identified. Studies in the laboratories of Yamashita[24,40] and Henry[41-43]
have advanced yeast lipid genetics considerably in the last few years, but
relatively little work has been done at the level of enzymology, metabolic
regulation or phenotypic analysis. Only a few of the yeast lipid genes
have been cloned.[40,43]

In the case of animal cells, a few mutants (mainly derived from the
CHO line) have become accessible because of the development of rapid
colony screening techniques applicable to the late steps of glycerolipid
synthesis.[1-3] These are: a) inositol auxotrophs;[1,44] b) temperature-
sensitive CDP-choline synthetase mutants;[4-6] c) ethanolamine phosphotransferase mutants;[7] d) dihydroxyacetone phosphate acyltransferase
mutants;[10] e) arachidonoyl CoA synthase mutants;[45] and f) mutants in
certain enzymes catalyzing polar headgroup exchange,[46] including strains
deficient in phosphatidylserine formation.[8,9,47] None of these mutants
has been examined (as yet) by classical technique of somatic cell genetics,

such as complementation by fusion or chromosome mapping. None of the genes coding for these enzymes in higher eucaryotic cells has been cloned. Progress in this direction has been limited by the fact that the lipid enzymes have not been purified to homogeneity. Consequently, antibodies and protein sequences (both helpful in cloning) are lacking. In the case of E. coli[2,11,48-51] and yeast,[40,43,52] the lipid enzymes were not generally purified to homogeneity prior to the availability of over-producing clones, all of which were found by correction of specific mutations.

The available lipid mutants of Chinese hamster ovary cells are nevertheless beginning to provide evidence that the biosynthetic and catabolic pathways deduced from earlier enzymatic studies are physiologically relevant.[11] They also provide an alternative approach to gene cloning by opening the door to transfection experiments.[53-57] The available somatic cell mutants have provided an opportunity to assess the effects of membrane lipid composition on cell growth.[2-11] It is already clear that a proper balance of phosphatidylcholine and phosphatidyl-ethanolamine is essential for cell growth,[4-6,58,59] as is the presence of phosphatidylserine in CHO cells.[9,10] In contrast, the level of ether-linked lipids can be reduced by a factor of 20 without any apparent effect on growth in tissue culture.[10,19,20] These generalizations will certainly be extended as additional mutants become available and must be taken into account in any model of lipid function based on physical or chemical studies.[60-63]

Mutants of animal cells defective in sterol biogenesis have also been described.[64-67] Much is now known about the structure and regulation of HMG CoA reductase.[68-70] In the case of sterol biosynthesis, specific selection procedures based on resistance to various drugs have facilitated the molecular biology.[64-71] Mutants in sterol biosynthesis are under intensive investigation in several laboratories,[64-71] and have not been studied by the PI.

Structure and Biosynthesis of the Lipid A Component of Lipopolysaccharide and Its Role in the Outer Membrane of E. coli

In the last three years, a vast amount has been learned about the lipid A component of lipopolysaccharide of gram-negative bacteria.[23,72-78] A previously unknown class of glucosamine-derived phospholipids has been discovered,[21,22,79,80] together with a set of enzymes for their biosynthesis.[24,25,81] As discussed below, these new findings in bacterial lipid metabolism have important implications for lipid function in animal cell membranes.

Briefly, lipopolysaccharide consists of a family of related molecules that are found on the outer surface of the outer membrane of gram-negative bacteria, such as E. coli.[23,72] Lipopolysaccharide can be divided into three domains (Fig. 4). These are:[23,72] a) the O-antigen region, consisting of repeating oligosaccharide units that represent the outermost sugars; b) the core sugars, consisting of nonrepeating and unusual structures, such as heptoses, heptose phosphates and 2-keto-3-deoxy-octanoate (KDO); and, c) lipid A, also known as gram-negative endotoxin, that serves as a hydrophobic anchor holding lipopolysaccharide in place. The lipid A molecule[23] minimally consists of a disaccharide of glucosamine, linked β,1$\rightarrow$6, substituted with monophosphate moieties at positions 1 and 4' (Fig. 5), and bearing four to seven fatty acyl moieties depending on the bacterium.[23,74-76] The four fatty acids invariably attached to the glucosamine backbone are β-hydroxymyristoyl moieties.[23,74-76] These are not present in the glycerophospholipid fraction.[26] Additional saturated fatty acids, usually laurate and myristate when present, are esterified in

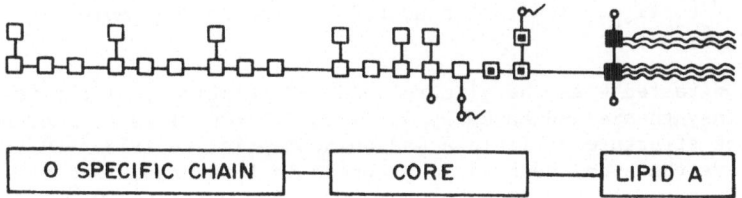

Fig. 4. The three domains of lipopolysaccharide. Squares indicate various sugar residues, as discussed in ref. 88.

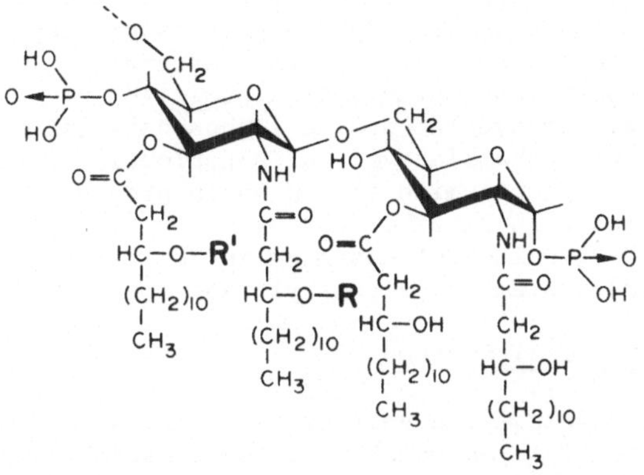

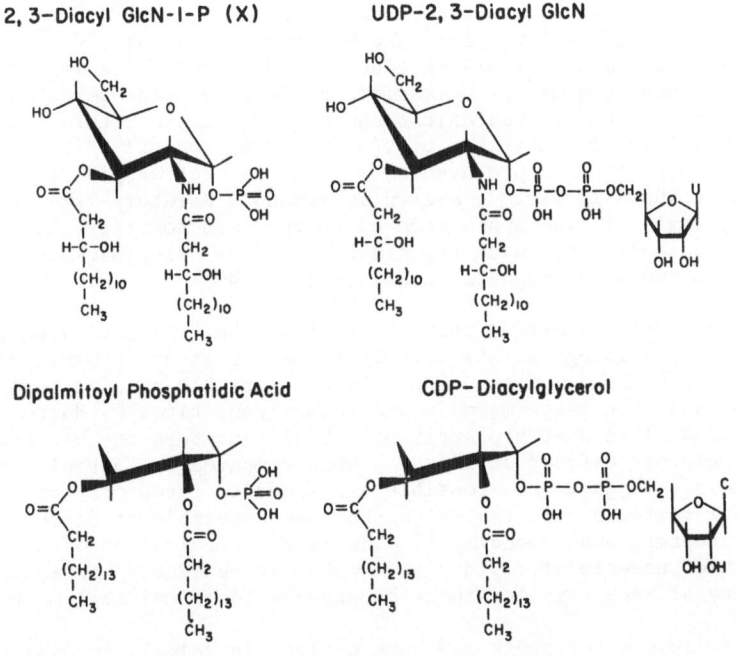

Fig. 5. Covalent structure of E. coli lipid A.[21-23]

2, 3–Diacyl GlcN–I–P (X)   UDP–2, 3–Diacyl GlcN

Dipalmitoyl Phosphatidic Acid   CDP–Diacylglycerol

Fig. 6. Formal resemblance of lipid A precursors to glycerophospholipids.

a "piggyback" configuration to the nonreducing end of the molecule
(Fig. 5).[23,74-76]

When contrasted with the glycerolipids,[26] relatively little is known
about the biosynthesis and assembly of lipid A.[21,22] This is because the
true covalent structure of lipid A was unknown prior to 1983.[72,82] In the
last two years there has been rapid progress in this area, made possible
by the application of fast atom bombardment mass spectrometry[74] and NMR
spectroscopy to the problem.[75-80] The lipid A structure shown in Fig.5
not only has been verified by careful studies of the natural product,[72-78]
but it also has been chemically synthesized by Shiba and coworkers.[83-85]

An entirely independent approach to the lipid A question has focused
on its biosynthesis.[24,25] This was made possible by the discovery in my
laboratory of novel monosaccharide lipid A precursors, especially 2,3-
diacyl-glucosamine 1-phosphate (Fig. 6), that accumulate in certain
phosphatidylglycerol-deficient mutants of E. coli.[79,80,86] The elucidation
of the structure of 2,3-diacylglucosamine 1-phosphate (also referred to as
lipid X),[79] together with the discovery of a nucleotide form of this
material, UDP-2,3-diacylglucosamine,[24] led us to propose the biosynthetic
pathway for lipid A shown in Fig. 7.[24,25,81,87] Recently, it has been
demonstrated that UDP-2,3-diacylglucosamine is derived from UDP-N-acetyl-
glucosamine by a novel system of fatty acyltransferases.[81]

Effects of Lipid A on Animal Systems

Lipid A is responsible for a variety of important pathophysiological
responses when injected into mammals.[88] These include induction of
endotoxic shock, fever, activation of B lymphocytes, stimulation of
macrophages to release prostaglandins, and regression (in some settings)
of experimental tumors. The pharmacological basis of lipid A action on
mammalian systems must be re-evaluated, given that the covalent structure
of lipid A has finally been solved.

The use of chemically synthesized lipid A analogs,[83-85] together with
natural precursors available from the biosynthetic studies[25,79,80,89-92]
has facilitated the structure-function analysis of lipid A mediated
responses. This data may be summarized as follows. The monosaccharide
precursors are virtually nontoxic to animals,[93] but do retain some
immunostimulatory activity when provided in high doses.[94,95]
The disaccharide lipid A precursors, such as the tetraacyldisaccharide
bisphosphate shown in Fig. 7 , are fully immunostimulatory[84,92] and
moderately toxic.[84] The appearance of extreme endotoxicity, however,
appears to be correlated with the presence of the "piggybacked" fatty acyl
moieties and two monophosphate functions.[74,83,84]

An interesting question that arises from the extensive studies of
lipid A pathophysiology is how and why animal cells are capable of
responding to such molecules. There is considerable species variability,
but humans and most large mammals are rather sensitive.[88] Mouse mutants
have been described that are strikingly resistant both to the toxic and
immunostimulatory effects of lipid A, when compared to "normal" mice.[96,97]
These findings suggest the possible existence of receptors for lipid A on
animal cell surfaces. So far it has not been possible to demonstrate
lipid A receptors unequivocally.[98] The recent elucidation of the structure and biosynthesis of lipid A[21-23] makes it necessary to reexamine the
possibility of receptors for these substances in animal cell membranes.

While lipid A receptors may have evolved in animals to deal with the
presence of gram-negative bacteria in the environment, it is also possible
that animal cells respond to lipid A, because lipid A mimicks a minor lipid

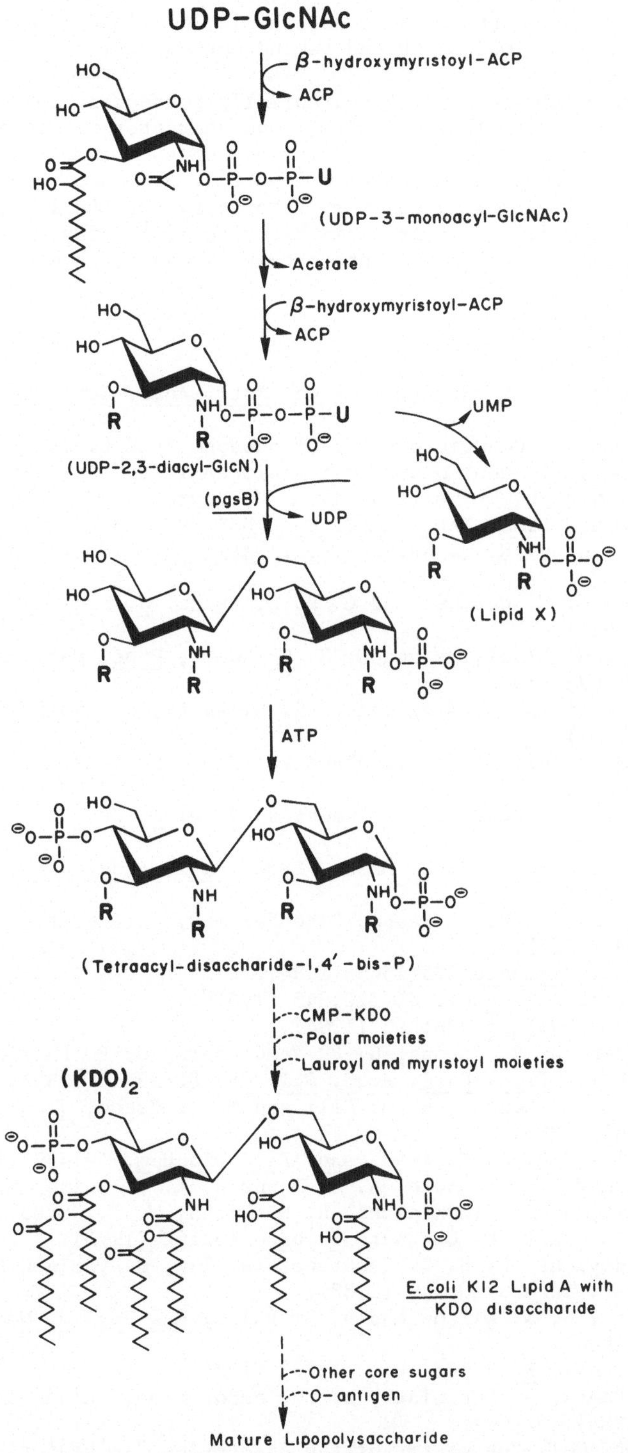

Fig. 7. Biosynthesis of lipid A in E. coli. UDP-2,3-diacylglucosamine is generated from UDP-GlcNAc by a novel system of acyltransferases.[81]

mediator normally present in these cells. Certainly, the discoveries of prostaglandins,[33] platelet activating factors,[36-38] and biologically active diglycerides[12-16] document the inherent ability of lipid molecules to function as biological mediators and second messengers.

If lipid A is mimicking a ligand normally present in animal cell membranes, it may be possible to isolate and characterize lipid A-like molecules from animal systems. The existence of such compounds (presumably minor ones on a mass basis) might have important pharmacological implications. The techniques developed in my laboratory for the isolation of animal cell mutants deficient in lipid biosynthesis[1-3] may be applicable to these compounds, or to the receptors that recognize them.

C. REFERENCES

1. J. D. Esko and C. R. H. Raetz, Proc. Natl. Acad. Sci. USA 75:1190 (1978).
2. C. R. H. Raetz, Genetic control of phospholipid bilayer assembly, in "Elsevier's New Comprehensive Biochemistry" 4:435-577 (1982).
3. C. R. H. Raetz, M. M. Wermuth, T. M. McIntyre, J. D. Esko and D. C. Wing, Proc. Natl. Acad. Sci. USA 79:3223 (1982).
4. J. D. Esko and C. R. H. Raetz, Proc. Natl. Acad. Sci. USA 77:5192 (1980).
5. J. D. Esko, M. M. Wermuth and C. R. H. Raetz, J. Biol. Chem. 256: 7388 (1981).
6. J. D. Esko, M. Nishijima and C. R. H. Raetz, Proc. Natl. Acad. Sci. USA 79:1698 (1982).
7. M. A. Polokoff, D. C. Wing and C. R. H. Raetz, J. Biol. Chem. 256: 7687 (1981).
8. O. Kuge, M. Nishijima and Y. Akamatsu, Proc. Natl. Acad. Sci. USA 82:1926 (1985).
9. M. Nishijima, O. Kuge and Y. Akamatsu, J. Biol. Chem., in press (1986).
10. R. A. Zoeller and C. R. H. Raetz, Proc. Natl. Acad. Sci. USA, submitted (1986).
11. J. D. Esko and C. R. H. Raetz, "The Enzymes," 3rd edition, Vol. XVI, 208-253 (1983).
12. Y. Nishizuka, Science 225:1365 (1984).
13. M. J. Berridge, Biochem. J. 220:345 (1984).
14. Y. Nishizuka, Nature 308:693 (1984).
15. A. M. Jetten, B. R. Ganong, G. R. Vandenbark, J. E. Shirley and R. M. Bell, Proc. Natl. Acad. Sci. USA 82:1941 (1985).
16. R. Davis, B. R. Ganong, R. M. Bell and M. P. Czech, J. Biol. Chem. 260:1562 (1985).
17. R. M. Bell and R. A. Coleman, Ann. Rev. Biochem. 49:459 (1980).
18. S. Goldfisher, C. L. Moore, A. B. Johnson, A. J. Spiro, M. P. Valsamis, H. K. Wiesniewski, R. H. Ritch, W. T. Norton, I. Rapin and L. M. Gartner, Science 182:62 (1973).
19. H. S. A. Heymans, R. B. H. Schutgens, R. Tan, H. van den Bosch and P. Borst, Nature 306:69 (1983).
20. N. S. Datta, G. N. Wilson and A. K. Hajra, New Eng. J. Med. 311: 1080 (1984).
21. C. R. H. Raetz, Rev. Inf. Dis. 6:463 (1984).
22. C. R. H. Raetz, in "Chemistry of the endotoxins," Elsevier, 248- 268 (1984).
23. E. T. Rietschel, "Chemistry of the endotoxins," Elsevier, Amsterdam (1984).
24. C. E. Bulawa and C. R. H. Raetz, J. Biol. Chem. 259:4846 (1984).
25. B. L. Ray, G. Painter and C. R. H. Raetz, J. Biol. Chem. 259:4852 (1984).

26. C. R. H. Raetz, Microbiol. Revs. 42:614 (1978).
27. F. Snyder, "Lipid metabolism in mammals," Plenum Press, New York (1977).
28. R. M. Bell, L. M. Ballas and R. E. Coleman, J. Lipid Res. 22:391 (1981).
29. C. L. Jones and A. K. Hajra, Biochem. Biophys. Res. Commun. 76: 1138 (1977).
30. A. K. Hajra and J. E. Bishop, Ann. N.Y. Acad. Sci. 386:170 (1982).
31. D. R. Voelker and E. P. Kennedy, Biochem. 21:2753 (1982).
32. H. Kanoh, H. Kondoh and T. Ono, J. Biol. Chem. 258:1767 (1983).
33. B. Samuelsson, Angen. Chem. 22:805 (1983).
34. H. F. DeLuca and H. K. Schnoes, Ann. Rev. Biochem. 52:411 (1983).
35. J. Stenflo and J. W. Suttie, Ann. Rev. Biochem. 46:157 (1977).
36. B. B. Vargafig and J. Benveniste, Trends Pharmacol. Sci. 3:341 (1983).
37. C. A. Demopoulos, R. N. Pinckard and D. J. Itanahan, J. Biol. Chem. 254:9355 (1979).
38. J. Benveniste and B. Arnoux, eds., "Platelet activating factor and structurally related ether-lipids," INSERM Symposium 23, Elsevier, Amsterdam (1983).
39. J. Nikawa and S. Yamashita, Eur. J. Biochem. 125:445 (1982).
40. J. Nikawa and S. Yamashita, Eur. J. Biochem. 143:251 (1984).
41. S. A. Henry, "The molecular biology of the yeast, Saccharomyces," Cold Spring Harbor Press, New York, 101-158 (1982).
42. K. Atkinson, B. Jensen, A. I. Kolat, E. M. Storm and S. A. Henry, J. Bacteriol. 141:558 (1980).
43. V. A. Letts, L. S. Klig, M. Bae-Lee, G. M. Carman and S. A. Henry, Proc. Natl. Acad. Sci. USA 80:7279 (1983).
44. J. D. Esko and C. R. H. Raetz, J. Biol. Chem. 255:4474 (1980).
45. E. J. Neufeld, T. E. Brass and P. W. Majerus, J. Biol. Chem. 259: 1986 (1984).
46. M. Nishijima, O. Juge, O. Maeder, M. Nakano and Y. Akamatsu, J. Biol. Chem. 259:7101 (1984).
47. D. R. Voelker and J. L. Frazier, J. Biol. Chem. 261, in press (1986).
48. C. R. H. Raetz, T. J. Larson and W. Dowhan, Proc. Natl. Acad. Sci. USA 74:1412 (1977).
49. V. A. Lightner, T. J. Larson, P. Tailleur, G.D. Kantor, C. R. H. Raetz, R. M. Bell and P. Modrich, J. Biol. Chem. 255:9413 (1980).
50. R. J. Tyhach, E. Hawrot, M. Satre and E. P. Kennedy, J. Biol. Chem. 254:627 (1979).
51. C. P. Sparrow and C. R. H. Raetz, J. Biol. Chem. 260, in press (1985).
52. M. S. Bae-Lee and G. M. Carman, J. Biol. Chem. 259:10857 (1984).
53. F. L. Graham and A. J. Van der Eb, Virology 52:456 (1973).
54. M. Wigler, S. Silverstein, L. S. Lee, A. Pellicer, Y. C. Cheng and R. Axel, Cell 11:223 (1977).
55. R. C. Mulligan and P. Berg, Proc. Natl. Acad. Sci. USA 78:2072 (1981).
56. A. Westerveld, J. H. J. Hoeijmakers, M. van Duin, J. deWit, H. Odijk, A. Pastink, R. D. Wood and D. Bootsma, Nature 310:425 (1984).
57. D. R. Littman, Y. Thomas, P. J. Madden, L. Chess and R. Axel, Cell 40:237 (1985).
58. M. Glaser, K. Ferguson and P. R. Vagelos, Proc. Natl. Acad. Sci. USA 71:4072 (1974).
59. J. D. Esko, J. F. Gilmore and M. Glaser, Biochem. 16:1881 (1977).
60. L. Rilfors, G. Lindblom, A. Wieslander and A. Christiansson, in "Membrane fluidity," M. Kates, ed., Plenum Press, New York, 205-245 (1984).

61. A. Wieslander, A. Christiansson, L. Rilfors, A. Kan, L. B. Johansson and G. Lindblom, FEBS Lett. 124:273 (1981).

62. P. R. Cullis et al., in "Membrane fluidity in biology," Vol. 2, R. C. Aloe, ed., Academic Press, New York (1982).

63. J. Gallay and B. deKruijff, FEBS Lett. 143:133 (1982).

64. Y. Saito, S. M. Chou and D. F. Silbert, Proc. Natl. Acad. Sci. USA 74:3730 (1977).

65. M. Sinenski, G. Duwe and F. Pinkerton, J. Biol. Chem. 254:4482 (1979).

66. T. Y. Chang and C. C. Y. Chang, Biochem. 21:5316 (1982).

67. M. Krieger, J. Martin, M. Segal and D. Kingsley, Proc. Natl. Acad. Sci. USA 80:5607 (1983).

68. D. J. Chin, K. L. Luskey, R. G. W. Anderson, J. R. Faust, J. L. Goldstein and M. S. Brown, Proc. Natl. Acad. Sci. USA 79:1185 (1982).

69. D. J. Chin et al., Nature 308:613 (1984).

70. G. Gil, J. R. Faust, D. J. Chin, J. L. Goldstein and M. S. Brown, Cell 41:249 (1985).

71. E. C. Hardeman, H. S. Jenke and R. D. Simoni, Proc. Natl. Acad. Sci. USA 80:1516 (1983).

72. L. Anderson and F. M. Unger, eds., "Bacterial lipopolysaccharides," ACS Symposium Series 231 (1983).

73. M. Imoto et al., Tetrahedron Letters 24:4017 (1983).

74. N. Qureshi, K. Takayama and E. Ribi, J. Biol. Chem. 257:11808 (1982).

75. K. Takayama, N. Qureshi and P. Mascagni, J. Biol. Chem. 258:12801 (1983).

76. N. Qureshi, K. Takayama, D. Heller and C. Fenselau, J. Biol. Chem. 258:12947 (1983).

77. S. M. Strain, S. W. Fesik and I. M. Armitage, J. Biol. Chem. 258: 2906 (1983).

78. S. M. Strain, S. W. Fesik and I. M. Armitage, J. Biol. Chem. 258: 13466 (1983).

79. K. Takayama, N. Qureshi, P. Mascagni, M. A. Nashed, L. Anderson and C. R. H. Raetz, J. Biol. Chem. 258:7379 (1983).

80. K. Takayama, N. Qureshi, P. Mascagni, L. Anderson and C. R. H. Raetz, J. Biol. Chem. 258:14245 (1983).

81. M. S. Anderson, C. E. Bulawa and C. R. H. Raetz, Fed. Proc. 44:487 (1985).

82. E. Th. Rietschel et al., in "Bacterial lipopolysaccharides," L. Anderson and F. Unger, eds., ACS Symposium Series 231, 195-218 (1983).

83. C. Galanos et al., Eur. J. Biochem. 148:1 (1985).

84. M. Imoto, H. Yoshimura, M. Yamamoto, T. Shimamoto, S. Kusomoto and T. Shiba, Tetrahedron Letters 25:2667 (1984).

85. S. Kusomoto, M. Yamamoto and T. Shiba, Tetrahedron Letters 25: 3727 (1984).

86. M. Nishijima and Raetz, C.R.H., J. Biol. Chem. 254:7837 (1979).

87. B. L. Ray and C. R. H. Raetz, Fed. Proc. 44:487 (1985).

88. C. Galanos, O. Lüderitz, E. Th. Rietschel and O. Westphal, Int. Rev. Biochem. 14:239 (1977).

89. P. D. Rick, L. W. - M. Rung, C. Ho and M. J. Osborn, J. Biol. Chem. 252:4904 (1977).

90. V. Lehmann and E. Rupprecht, Eur. J. Biochem. 81:443 (1977).

91. C. R. H. Raetz, K. Takayama, L. Anderson, I. M. Armitage and S. M. Strain, Fed. Proc. 43:1567 (1984).

92. C. R. H. Raetz, S. Purcell, M. V. Meyer, N. Qureshi and K. Takayama, J. Biol. Chem. 260, in press (1985).

93. R. A. Proctor, J. A. Will, K. E. Burhop and C. R. H. Raetz, Infection and Immunity, submitted (1986).

94. C. R. H. Raetz, S. Purcell and K. Takayama, Proc. Natl. Acad. Sci. USA 80:4624 (1983).

95. M. Nishijima, F. Amano, Y. Akamatsu, K. Akayawa, T. Tokunaga and
    C. R. H. Raetz, <u>Proc</u>. <u>Natl</u>. <u>Acad</u>. <u>Sci</u>. <u>USA</u> 82:282 (1985).
96. B. M. Sultzer, <u>Nature</u> 219:1253 (1968).
97. D. C. Morrison and J. L. Ryan, <u>Adv</u>. <u>Immunol</u>. 28:293 (1978).
98. L. Forni and A. Coutinho, <u>Nature</u> 273:304 (1978).

INOSITOL LIPID METABOLISM : A TOPOLOGICAL POINT OF VIEW

G. Mauco, Ph. Dajeans, H. Chap and L. Douste-Blazy

INSERM Unité 101, Biochimie des Lipides

Hôpital Purpan, 31059 Toulouse Cédex, France

INTRODUCTION

Since its discovery, the so-called "phospholipid effect" has deserved a great deal of investigations (for reviews see J. Williamson's and K.W.A. Wirtz's chapters in this book and ref. [1-4]), leading to the widely accepted scheme of inositol-1,4,5-trisphosphate (IP_3) and diacylglycerol (DG) production as second messengers of hormone action. As discussed elsewhere in this book, this dual pathway leads to increase the cytosolic free calcium concentration on one hand and to the activation of protein kinase C on the other. However, most of the studies described the overall effect of hormone-receptor interaction at the plasma membrane on the metabolic changes of inositol-phospholipids of the whole cell. It was then considered that inositol lipids were a homogeneous entity, ignoring possible subcellular compartmentation of metabolically different pools. The existence of micro-domains on the plasma membrane is not documented either even if some data suggest the existence of agonist-linked pools. The aim of this chapter is to review the presently available data concerning the topology of inositol-lipid metabolism.

I - SUBCELLULAR LOCALIZATION OF INOSITOL-LIPIDS

Owing to the fact that ligand-receptor interactions generate a trans-membrane signal, it is assumed that phospholipase C activation occurs at the plasma membrane. Therefore, phosphatidylinositol 4,5 bisphosphate (PIP_2), phosphatidylinositol 4 phosphate (PIP) and phosphatidylinositol (PI) should be localized in the plasma membrane. Indeed available data support this hypothesis.

Due to the low level of inositol lipids (about 5 % of total phospholipids) and of the lower content of the phosphorylated derivatives (PI =
16 nmoles, PIP = 3.5 nmoles and PIP_2 = 1.6 nmoles/10^9 platelets for
instance)[5], these studies have been based on labelling whole or broken
cells. We will consider here the data obtained by radiolabelling prior to
cell disruption. Griffin and Hawthorne[6] found ^{32}P labelled PIP and PIP_2 in
the synaptosomal plasma membrane, while they were absent from synaptic
vesicles. An exclusive localization of PIP and PIP_2 in the plasma membranes
of Friend erythroleukaemic cells was also demonstrated by Rawyler et al.[7],
who even suggested to use (poly)phosphoinositide phosphorylation as a
specific marker for plasma membrane. A similar finding was obtained by
Seyfred and Wells[8], who labelled hepatocytes with ^{32}Pi prior a rapid subcellular fractionation. Their kinetic studies show that the label is incorporated in plasma membranes 5-10 times faster into PIP and 25-50 times
faster into PIP_2 compared to other cell compartments. However, significant
amounts of PIP were recovered in lysosomes. Since they reached a radioactive equilibrium between $\gamma-^{32}P-ATP$ and the 4'and 5' phosphates of PIP and
PIP_2, radioactivity reflects the actual masses and it is possible to calculate that PIP and PIP_2 are present in almost equal amounts in the plasma
membrane of hepatocytes. Using this model, the same authors[9] demonstrated
that vasopressin induces a loss of phosphoinositide from the plasma membrane.

However, PI localization is difficult to predict with such experimental
model since the turnover of the phosphodiester is much lower and this phosphorus does not reach isotopic equilibrium in reasonable times. This is
illustrated by our most recent experiences on subcellular localization of
platelet inositol lipids[10]. We designed a rapid subcellular fractionation
procedure on Percoll gradients at alkaline pH to investigate on the
distribution of labelled lipids from ^{32}P or ^3H-inositol preincubated
platelets. $^{32}P-PIP_2$, PIP and PI were found exclusively in the plasma
membrane, isotopic equilibrium being reached only for PIP and PIP_2. On the
other hand, we observed a bimodal distribution of ^3H-PI. Since in that
case isotopic equilibrium was apparently reached between the three inositol
lipids, we assumed that radioactivity paralleled the actual masses. Thirty-
five percent of the total ^3H-PI were recovered in the plasma membrane,
confirming previously reported data obtained by Fauvel et al.[11], who
measured the major phospholipid contents of subcellular fractions from
blood platelets. They showed a PI content higher in endoplasmic reticulum
than in plasma membranes. On the opposite, all the ^3H-labelled PIP and PIP_2
were recovered in the plasma membrane[10].

Starting from these observations, it is tempting to speculate that these two lipids are confined to the inner leaflet of the plasma membrane, since they are in rapid equilibrium with PI. It is well known indeed that PI are almost exclusively localized on the cytoplasmic side of platelet plasma membrane[12,13] as well as erythrocyte plasma membrane[14]. However this hypothesis is documented only for human erythrocyte membranes. Garrett and Redman[15] showed that only inside out vesicles incorporated ^{32}P from γ-^{32}P-ATP, while right side out vesicles were unable to form ^{32}P-PIP and ^{32}P-PIP$_2$. This is not a definitive argument however, since one can reply that only the kinases are on the cytoplasmic side but the substrate can be on both sides.

In contrast to poly-PI, PI distribution is more documented. It appears to be localized in all the membranes found in the cell, with some enrichment in the intracellular membranes, specifically the endoplasmic reticulum [11,16,17] and/or Golgi apparatus[17].

II - LOCALIZATION OF THE ENZYMES OF INOSITOL-LIPID METABOLISM

In the precedent chapter, we focused on data obtained in prelabelled whole cells. Another approach of the problem is to study the subcellular localization of the enzymes involved in inositol-lipid metabolism. However one must consider that a metabolic step can proceed in one cell compartment while the final destination of the product is in another compartment. Therefore this approach can lead to erroneous localizations and should be compared to data obtained in whole cells.

1. Kinases

In most cases, kinases are detected by incubating subcellular fractions with γ-^{32}P-ATP, extracted lipid radioactivity reflecting the balance between kinases and phosphatases. For example, Smith and Wells[18] showed the occurrence of PI kinase in nuclear envelopes, Lundberg et al.[19] in Golgi membranes and Campbell et al.[20] in endocytic vesicles. We found also a phosphatidylinositol kinase in the lysosomes/endoplasmic reticulum fractions of blood platelets (unpublished data). PIP-kinase, as well as PI-kinase, appears to be localized in the plasma membrane in most cells[7,8,15,18,21] including human platelets[22] (and Mauco et al., unpublished data). The diacylglycerol kinase, which is also involved in inositol lipid metabolism, was found both in cytosol and plasma membranes by Cail and Rubert[23] and ourselves (unpublished data).

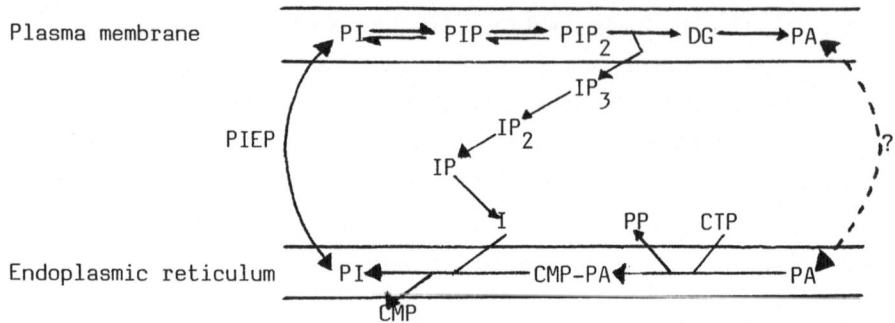

Fig. 1. Subcellular localization of the PI cycle (PIEP, PI exchange protein; IP_2, inositol 1,4 bisphosphate; IP, inositol 1 phosphate; I, inositol; ?, PA transfer protein, putative; PP, inorganic pyrophosphate).

2. PI resynthesis

After phosphodiesteratic cleavage, the recycling of the diacylglycerol moiety via the PI cycle is achieved through a two-step process of nucleotide activation by CTP. The first step involves a CTP:phosphatidate cytidylyl transferase producing CMP-PA (also called CDP-DG) from PA and CTP. CMP is then exchanged with inositol by a CMP:phosphatidate:myo-inositol 3 phosphatidyl transferase. Both enzymes are localized in the microsomal fraction of liver[24-26]. The second enzyme has been purified to near homogeneity by Takenawa and Egawa[27]. Even if this is a well admitted concept, some discrepancies can be found, at least in platelet fractions. For instance, recent studies performed in our laboratory revealed an incorporation of ^{3}H-inositol into PI, PIP and PIP_2 by pure plasma membranes and into PI and PIP by endoplasmic reticulum (unpublished data). This sheds a new light on the classical subcellular compartmentation of the PI cycle shown in Fig. 1.

In order to close the cycle, this scheme calls for two transfer proteins which should be able to shuttle PA and PI between the two intracellular compartments. The PI exchange protein has been shown in many cells[28], including platelets[29,30]; however no data are available on the existence of such an activity for PA, even if actively searched for by Laffont et al.[29]. Our preliminary data suggest now that the entire cycle could take place in the plasma membrane, since an inositol incorporating system exists in pure platelet plasma membranes. This discrepancy with the admitted data could come from a term ambiguity. Very often, microsomes are considered as enriched fractions of endoplasmic reticulum, however other membranes such as plasma membrane, but also Golgi apparatus and nuclear envelopes can significantly contribute to this subcellular fraction. We thus propose a modification of the classical pathway shown in Fig. 2 and suggest a dual mechanism for PI synthesis : the plasma membrane pathway could be involved in the PI-cycle, whereas the endoplasmic reticulum would achieve only de novo synthesis. However, this has to be documented further even if it is known that de novo synthesis of CDP-DG takes place in the intracellular membranes (for a review see [24]).

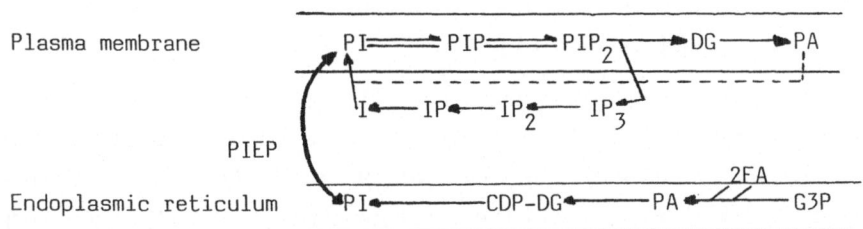

Fig. 2. The hypothesis of a dual mechanism of PI resynthesis
(2FA, fatty acids in acyl CoA form; G3P, glycerol 3 phosphate).

3. Inositol-lipid hydrolysis

This aspect is developped in other chapters of this book, however we shall review some important data on this metabolic step. Early studies on phospholipase C showed an almost exclusive localization in the cytosol[31-35]. However plasma membranes, under certain circumstances show a phosphodiesteratic activity maybe regulated by lipidic environment[36] and/or physical state[37]. Another very interesting possibility, discussed elsewhere in this book, is given by the existence of GTP-binding proteins which regulate the phosphodiesteratic activity of liver[38] or platelet[39] plasma membranes. Cytosolic phospholipase C, which displays little activity if any against membrane organized substrates was shown to be activated by GTP^{Y}S in the absence[40] or in the presence of micromolar calcium concentrations[41]. This could give an interesting insight on the way an external signal can trigger extremely rapidly phospholipid hydrolysis by a soluble enzyme...

III - DO INOSITOL-LIPID MICRODOMAINS EXIST ?

The next step in the comprehension of agonist-receptor induced hydrolysis of phospholipids is the investigation on micro-domains, i.e. on areas or functional areas of the cell which are directly linked to a kind of receptor. The obvious and easiest way is to look for subcellular compartmentation and we have seen that at least two domains can be found : an intracellular pool of PI and a plasma membrane pool of PI, PIP and PIP_2. It is predictable that some differences would be observed between the metabolic responses of these two pools upon agonist stimulation.

Indeed, some studies have confirmed this hypothesis. For instance, ^{3}H -inositol prelabelled pancreatic islets display a secretagogue labile

PI pool which is membrane bound and an hormone insensitive intracellular pool. Furthermore only islets preincubated with ^{3}H -inositol in the presence of glucose, a secretagogue, showed a loss of radioactivity from PI upon subsequent stimulation with glucose or other secretagogues[43]. It could well be that inositol labelling at the plasma membrane is independent from the labelling in endoplasmic reticulum, supporting therefore the hypothesis illustrated in Fig. 2.

The independence of these two subcellular sites of synthesis (or turn-over) is supported also by the observations of Schoepp[44] who labelled rat cerebral cortex slices with ^{3}H - inositol. By adding Mn^{2+}, he obtained an important (5 times) increase in the total amount of radioactivity incorporated in the slice phospholipids. Moreover, upon carbachol or norepinephrine stimulation, the release of soluble ^{3}H -inositol phosphates was identical in both control and Mn^{2+} treated cells. One can conclude that without Mn^{2+} the hormone sensitive pool is already replenished in inositol and that the addition of Mn^{2+} only increases the labelling of the hormone insensitive pool.

A similar kind of compartmentation was demonstrated in WRK1 rat mammary tumor cells[45]. Furthermore, Koreh and Monaco[46] showed that these cells have two distinct pools of hormone sensitive and insensitive PIP_2. Unfortunately no data are available on the subcellular localization of these two pools, but if in WRK1 cells as in other cells (see above) PIP_2 are almost exclusively localized on the plasma membrane, one must admit a mosaicism of the plasma membrane with some areas which are not accessible to phosphodiesteratic cleavage.

Another explanation could be a chemical difference between PIP_2 from the two pools. In platelets also some data are in favour of a microheterogeneity of PI, PIP or PIP_2, since with all the major agonists, thrombin[5], ADP^{47}, PAF-acether[48] for instance, the hydrolysis of PIP_2 culminated at 35-40 % of the total PIP_2 and that of PI at 50-60 % of the total PI. However no heterogeneity was observed in the fatty acid composition of PIP_2, PIP nor PI and the related compounds PA and DG[5,49]. More than 90 % of these molecules display a stearoyl-arachidonoyl glycerol backbone... No data are available on the ability of inducing more than 40 % hydrolysis of PIP_2 by adding simultaneously more than one agonist...

CONCLUSION

Even if the data briefly reviewed here are only partial, it is clear that subcellular compartmentation of inositol lipids is an important phenomenon. Most if not all of the agonist sensitive inositol lipids are on the plasma membrane and it may well be that plasma membrane is in some way and between some limits autonomous for the phospholipid response. This has to be confirmed by careful studies which are under progress in our laboratory. On the other hand, it is reasonable to admit the existence of microdomains whose existence has been suggested by functional evidence. Nevertheless, the role of inositol lipids is still growing with the comprehension of more intimal mechanisms of phosphodiesterase activation. Few other fields could also be of interest like the role of these lipids in cytoskeleton anchorage[50] for instance.

REFERENCES

1. R.H. Michell, Inositolphospholipids and cell surface receptor function, Biochim. Biophys. Acta 415:81 (1975).
2. M.J. Berridge, Inositol trisphosphate and diacylglycerol as second messenger, Biochem. J. 220:345 (1984).
3. J.R. Williamson, Role of inositol lipid breakdown in the generation of intracellular signals. State of the art lecture, Hypertension 8:II 140 (1986).
4. L.E. Hokin, Receptors and phosphoinositide-generated second messengers, Annu. Rev. Biochem. 54:205 (1985).
5. G. Mauco, C.A. Dangelmaier and J.B. Smith, Insotiol lipids, phosphatidate and diacylglycerol share stearoyl-arachidonoyl-glycerol as a common backbone in thrombin-stimulated human platelets, Biochem. J. 224:933 (1984).
6. H.D. Griffin and J.N. Hawthorne, Calcium activated hydrolysis of phosphatidyl-myo-inositol 4-phosphate and phosphatidyl-myo-inositol 4,5-bisphosphate in guinea pig synaptosomes, Biochem. J. 176:541 (1978).
7. A.J. Rawyler, B. Roelofsen, K.W.A. Wirtz and J.A.F. Op den Kamp, (poly) Phosphoinositide phosphorylation is a marker for plasma membrane in Friend erythroleukaemic cells, FEBS Lett. 148:140 (1982).
8. M.A. Seyfred and W.W. Wells, Subcellular incorporation of ^{32}P into phosphoinositides and other phospholipids in isolated hepatocytes, J. Biol. Chem. 259:7659 (1984).
9. M.A. Seyfred and W.W. Wells, Subcellular site of vasopressin-stimulated hydrolysis of phosphoinositides in rat hepatocytes, J. Biol. Chem. 259:7666 (1984).
10. G. Mauco, Ph. Dajeans, H. Chap and L. Douste-Blazy, Subcellular localization of inositol lipids in blood platelets as deduced from the use of labelled precursors. Submitted for publication.
11. J. Fauvel, H. Chap, V. Roques, S. Lévy-Tolédano and L. Douste-Blazy, Biochemical characterization of plasma membranes and intracellular membranes isolated from human platelets using Percoll gradients, Biochim. Biophys. Acta 236, 176 (1986).
12. H. Chap, R.F.A. Zwaal and L.L.M. van Deenen, Action of highly purified phospholipases on blood platelets. Evidence for asymmetric distribution of phospholipids in the surface membrane, Biochim. Biophys. Acta 467:146 (1977).
13. B. Perret, H. Chap and L. Douste-Blazy, Asymmetric distribution of arachidonic acid in the plasma membrane of human platelets. A determination using purified phospholipases and a rapid method for membrane isolation, Biochim. Biophys. Acta 556:434 (1979).
14. R.F.A. Zwaal, R. Roelofsen, P. Comfurius and L.L.M. van Deenen, Organization of phospholipid in human red cell membranes as detected by the action of various purified phospholipases, Biochim. Biophys. Acta 406:83 (1975).

15. R.J.B. Garret and C.M. Redman, Localization of enzymes involved in polyphosphoinositide metabolism on the cytoplasmic surface of the human erythrocyte membrane, Biochim. Biophys. Acta 382:58 (1975).

16. M. Lagarde, M. Guichardant, S. Menashi and N. Crawford, The phospholipid and fatty acid composition of human platelet surface and intracellular membranes isolated by high voltage free flow electrophoresis, J. Biol. Chem. 257:3100 (1982).

17. F. Zambrano, S. Fleisher and B. Fleisher, Lipid composition of the Golgi apparatus of rat kidney and liver in comparison with other subcellular organelles, Biochim. Biophys. Acta 380:357 (1975).

18. C.D. Smith and W.W. Wells, Characterization of a phosphatidylinositol 4-phosphate-specific phosphomonoesterase in rat liver nuclear envelopes, Arch. Biochem. Biophys. 235:529 (1984).

19. B. Jergil and R. Sundler, Phosphorylation of phosphatidylinositol in rat liver Golgi, J. Biol. Chem. 258, 7969 (1983).

20. C.R. Campbell, J.B. Fishman and R.E. Fine, Coated vesicles contain a phosphatidylinositol kinase, J. Biol. Chem. 260:10948 (1985).

21. G.A. Lundberg, B. Jergil and R. Sundler, Subcellular localization and enzymatic properties of rat liver phosphatidylinositol-4-phosphate kinase, Biochim. Biophys. Acta 846:379 (1985).

22. H.D. Kaulen and R. Gross, Metabolic properties of human platelet membranes. II. Thrombin-induced phosphorylation of membrane-lipids and demonstration of phosphorylating enzymes in the platelet membrane, Thromb. Haemost. 35:364 (1976).

23. F.L. II Call and M. Rubert, Diglyceride kinase in human platelets, J. Lipid Res. 14:466 (1973).

24. J.D. Esko and C.R.H. Raetz, Synthesis of phospholipids in animal cells, Enzymes 16:206 (1983).

25. W. Thompson and G. MacDonald, Synthesis of molecular classes of cytidine diphosphate diglyceride by rat liver in vivo and in vitro, Can. J. Biochem. 55:1153 (1977).

26. H.Paulus and E.P. Kennedy, The enzymatic synthesis of inositol monophosphatide, J. Biol. Chem. 235:1303 (1960).

27. T. Takenawa and K. Egawa, CDP-diglyceride: inositol transferase from rat liver, J. Biol. Chem. 252:5419 (1977).

28. G. M. Helmkamp Jr., Phosphatidylinositol transfer protein : structure, catalytic activity and physiological function, Chem. Phys. Lipids 38:3 (1985).

29. F. Laffont, H. Chap, G. Soula and L. Douste-Blazy, Phospholipid exchange proteins from platelet cytosol possibly involved in phospholipid effect. Biochem. Biophys. Res. Commun. 102:1366 (1981).

30. P.Y. George and G.M. Helmkamp Jr., Purification and characterization of a phosphatidylinositol transfer protein from human platelets, Biochim. Biophys. Acta 836:176 (1985).

31. D. Allan and R.H. Michell, Phosphatidylinositol cleavage catalysed by the soluble fraction from lymphocytes. Activity of pH 5.5 and pH 7.0, Biochem. J. 142:599 (1974).

32. G. Mauco, H. Chap and L. Douste-Blazy, Characterization and properties of a phosphatidylinositol phosphodiesterase (phospholipase C) from platelet cytosol, FEBS Lett. 100:367 (1979).

33. S.E. Rittenhouse-Simmons, Production of diglyceride from phosphatidylinositol in activated human platelets, J. Clin. Invest. 63:580 (1979).

34. S.E. Rittenhouse, Human platelets contain phospholipase C that hydrolyses polyphosphoinositides, Proc. Natl. Acad. Sci. US 80:5417 (1983).

35. W. Siess and E.G. Lapetina, Properties and distribution of phosphatidylinositol - specific phospholipase C in human and horse platelets, Biochim. Biophys. Acta 752:329 (1983).

36. M. Plantavid, L. Rossignol, H. Chap and L. Douste-Blazy, Studies on endogenous polyphosphoinositide hydrolysis in human platelet membranes. Evidence that polyphosphoinositide remain unaccessible to phosphodiesterase in the native membrane, Biochim. Biophys. Acta 875:147(1986).

37. H. M'Zali and F. Giraud, Phosphoinositide reorganization in human erythrocyte membrane upon cholesterol depletion, Biochem. J., 234:13 (1986).
38. R.J. Uhing, H. Jiang, V. Prpic and J.H. Exton, Regulation of a liver plasma membrane phosphoinositide phosphodiesterase by guanine nucleotides and calcium, FEBS Lett. 118:317 (1985).
39. J.J. Baldassare and G.J. Fisher, GTP and cytosol stimulate phosphoinositide hydrolysis in isolated membranes, Biochem. Biophys. Res. Commun. 137: 801 (1986).
40. J.J. Baldassare and G.J. Fisher, Regulation of membrane-associated and cytosolic phospholipase C activities in human platelets by guanosine triphosphate, J. Biol. Chem. 261:11942 (1986).
41. Y. Banno, S. Nakashima, T. Tohmatsu, Y. Nozawa and E.G. Lapetina, GTP and GDP will stimulate platelet cytosolic phospholipase C independently of Ca2+, Biochem. Biophys. Res. Commun. 140, 128 (1986).
42. R.S. Rana, R.J. Mertz, A. Kowluru, J.F. Dixon, L.E. Hokin and M.J. Mac Donald, Evidence for glucose-responsive and un-responsive pools of phospholipid in pancreatic islets, J. Biol. Chem. 260:7861 (1985).
43. R.S. Rana, A. Kowluru and M.J. MacDonald, Secretagogue-responsive and unresponsive pools of phosphatidylinositol in pancreatic islets, Arch. Biochem. Biophys. 245:411 (1986).
44. D.D. Schoepp, Manganese stimulates the incorporation of (^{3}H) inositol into a pool of phosphatidylinositol in brain that is not coupled to agonist-induced hydrolysis, J. Neurochem. 45:1481 (1985).
45. M.E. Monaco, The phosphatidylinositol cycle in WRK-1 cells, evidence for a separate, hormone-sensitive phosphatidylinositol pool. J. Biol. Chem. 257:2137 (1982).
46. K. Koreh and M.E. Monaco, The relationship of hormone-sensitive and hormone-insensitive phosphatidylinositol to phosphatidylinositol 4,5-bisphosphate in the WRK-1 cell, J. Biol. Chem. 261:88 (1986).
47. J.D. Vickers, M. Kinlough-Rathbone and J.F. Mustard, Changes in phosphatidylinositol-4,5-bisphosphate 10 seconds after stimulation of washed rabbit platelets with ADP, Biochem. Biophys. Res. Commun. 133: 98 (1982).
48. G. Mauco, H. Chap and L. Douste-Blazy, Platelet activating factor (PAF-acether) promotes an early degradation of phosphatidylinositol-4,5-bisphosphate in rabbit platelets, FEBS Lett. 153:361 (1983).
49. J.M. Broekman, J.W. Ward and A.J. Marcus, Fatty acid composition of phosphatidylinositol and phosphatidic acid in stimulated platelets. Persistence of arachidonoyl-stearyl structure, J. Biol. Chem. 256: 8271 (1981).
50. I. Lassing and U. Lindberg, Specific interaction between phosphatidylinositol 4,5-bisphosphate and profilactin, Nature 314:472 (1985).

METABOLISM OF LYSO(BIS)PHOSPHATIDIC ACID IN RABBIT ALVEOLAR MACROPHAGES

Todd Thornburg, Felicia R. Cochran, Vicki L. Roddick, and
Moseley Waite

Department of Biochemistry
Bowman Gray School of Medicine
Winston-Salem, N.C.

INTRODUCTION

Lyso(bis)phosphatidic acid [L(bis)PA] is a phosphatidyl glycerol
isomer with one fatty acid esterified to each glycerol moiety. This
novel phospholipid was first identified by Body and Gray (1), and has
since been found in a variety of mammalian tissues (2-5), normally
comprising less than 1% of the total cellular lipid phosphorus. It
appears that the stereochemistry of the lipid is unique, having the
phosphoester bonds attached to the sn-1 carbons of each glycerol (6,7).
The possible functional role of this lipid is only beginning to be
understood.

Early interest in L(bis)PA was spurred by the discovery that
substantial increases in the lipid were found in the tissues of patients
with lipid storage disorders such as the infantile form of Niemann-Pick
disease (8). Wherrett and Huterer (9), observing an increase in cellular
lysosomes in related disorders, examined the subcellular distribution of
the lipid in rat liver. They found an enrichment of L(bis)PA in the
lysosomal fraction. Similar results were found in cultured baby hamster
kidney (BHK) cells (10).

Interest in this lipid increased when Mason et al. (11) showed that
L(bis)PA constituted 15% of total phospholipid in rabbit alveolar
macrophages and 25% of the lipid mass found in phagocytic vesicles.
Also of importance is the fact that arachidonic acid constitutes a major
portion of the fatty acid esterified to L(bis)PA (11-13), and its
incorporation is rapid during cell labeling (12). Even more intriguing
is the demonstration that phosphatidylinositol (PI) is capable of
stimulating synthesis of the lipid by acting as an acyl donor in an
acyl-CoA independent transacylation reaction (14,15). A simple transacylation reaction, however, does not explain the peculiar sn-1 to sn-1
configuration of this lipid.

Our laboratory is currently working to determine the physiologic
role of L(bis)PA in both resident and Bacillus Calmette Guerin-activated
rabbit alveolar macrophages. It is of interest to determine how it is
made, its importance in the production of eicosanoids and its metabolic
relationship with PI. We are also interested in how this role may
change with challenges by differing stimuli.

PRODUCTION OF LYSO(BIS)PHOSPHATIDIC ACID

The first step in elucidating the function of L(bis)PA was to determine how the lipid is made within the cell. It has been demonstrated that in rat liver lysosomes this lipid could be formed from lysophosphatidylglycerol (LPG) via a transacylation reaction involving PI (15). Using cultured hamster fibroblasts, exogenous phosphatidyl-glycerol (PG) was shown by Somerharju and Renkonen (16) to be converted specifically to L(bis)PA. To determine if a similar pathway may be used in rabbit alveolar macrophages, cells were incubated with [^{32}P]PG for up to 28 h. [^{32}P]Phosphatidic acid (PA) and [^{32}P]phosphatidylethanolamine (PE) were used as controls for the experiment. After a 28 h incubation much of the [^{32}P]PG label was found in L(bis)PA; other phospholipids showed no such incorporation. At the end of the 28 h period the labeled L(bis)PA comprised one-third of the lipid subclass in the cell and about one-half of the original substrate remained (17).

These results showed that the macrophages use exogenous PG as a substrate in the production of L(bis)PA. The use of PG is not surprising considering the enrichment of this lipid in lung surfactant which surrounds the alveolar macrophage. This hypothesis is strengthened by the observation that macrophages from other sources demonstrate a markedly depressed metabolism of L(bis)PA (12).

The relationship between L(bis)PA production and PI degradation was studied using resident macrophages labeled with [^{3}H]arachidonic acid. Prelabeled cells were then incubated with tetradecanoyl-phorbol-acetate (TPA). We found that the TPA-induced deacylation of L(bis)PA was consistently accompanied by a loss of [^{3}H]arachidonic acid from PI (13). Similar results have been reported by Matsuzawa et al. (15) using rat liver lysosomes. This indicates the metabolism of the lipid may occur as a deacylation-reacylation reaction. Further support of this hypothesis came from studies showing that in contrast to the marked incorporation of [^{32}P] into cellular PI, phosphatidylcholine (PC), and PE, the de novo synthesis of L(bis)PA is very low.

LYSO(BIS)PHOSPHATIDIC ACID AS A SOURCE OF ARACHIDONIC ACID

Macrophages are abundant in tissues responding in an inflammatory reaction. Bioactive mediators such as eicosanoids are secreted by these cells and must be synthesized from the precursor molecule arachidonic acid. The production of prostaglandins and leukotrienes is dependent on the release of this fatty acid from its storage site in cellular phospholipid pools.

Analysis of cellular lipid phosphorous composition was carried out on resident alveolar macrophages and L(bis)PA was shown to comprise over 18% of the total phospholipid phosphorus. Of the total fatty acid associated with this lipid, nearly 20% was shown to be arachidonic acid (13,18).

When these cells were stimulated with the metabolic modulators, TPA and calcium ionophore A23187, there was a different response in the release of arachidonic acid. Calcium ionophore A23187 was shown to release arachidonic acid from PC and PE, while the stimulation of cells with TPA caused the release of this fatty acid from PE, phosphatidylserine (PS)/PI, PC, and L(bis)PA. Lipoxygenase products were shown to be the major arachidonic acid metabolites synthesized in TPA-stimulated cells. The secretion of leukotriene B$_4$ (LTB$_4$), leukotriene C$_4$ (LTC$_4$) and 5-hydroxyeicosatetraenoic acid (5-HETE) by these cells was confirmed by high pressure liquid chromatography (HPLC) analysis of eicosanoids isolated from cell media after stimulation. The secretion of these lipoxygenase products was found to be low in cells treated with calcium ionophore A23187. Indeed the prostaglandin 6-keto $F_{1\alpha}$ was shown to be the major metabolite. This metabolite is not secreted by TPA-stimulated cells (18).

As previously stated, it is our hypothesis that the mechanism of L(bis)PA synthesis in these macrophages may involve PI as a donor phospholipid for arachidonic acid in a transacylation reaction using LPG as the fatty acid acceptor. We further postulate that the turnover of these two lipids may be related to the synthesis of leukotriene C_4 (LTC_4) and the lipoxygenase pathway. If this relationship exists, the production of this metabolite is dependent on the presence of exogenous PG. This could account for differences noted by other investigators in arachidonic acid metabolites from alveolar- and peritoneal-derived macrophages.

LOCALIZATION OF ARACHIDONIC ACID IN LYSO(BIS)PHOSPHATIDIC ACID

Alveolar macrophages prelabeled with [³H]arachidonic acid were extracted and the cellular lipids incubated for 4 h with porcine pancreatic phospholipase A_2. Although label was lost from cardiolipin, PA, PS/PI, and PC, the phospholipase was unable to release label from L(bis)PA. This finding would indicate either the absence of labeled fatty acid at the sn-2 position, or that the lipid is not of the sn-3 type (7,13,19). Treatment of the [³H]arachidonic acid L(bis)PA with Rhizopus delmar phospholipase (lipase) A_1 released less than one-third of the [³H]arachidonic acid. The Rhizopus delmar lipase is capable of hydrolyzing acyl esters at positions sn-1 and sn-3 of triglyceride as well as the sn-1 acyl ester of phosphoglycerides. Control lipids cardiolipin, PG, and L(bis)PA showed complete hydrolysis. A [³H]-labeled compound that comigrated with LPG was formed during the incubation. This is an indication that the Rhizopus delmar enzyme removed a fatty acid from the sn-1 position of one glycerol moiety. A consistently large percentage of the [³H]L(bis)PA was left unhydrolyzed with as much as 80% remaining intact. These results show that most fatty acids in L(bis)PA are unable to be released by the classical phospholipases A_1 and A_2 (13). Why the Rhizopus delmar enzyme releases only a limited portion of labeled fatty acid is not clear. We postulate that the action of a novel phospholipase A_1 is responsible for the removal of arachidonic acid from L(bis)PA yet verification remains a challenge.

It is our conclusion that L(bis)PA is produced in rabbit alveolar macrophages via the use of extracellular PG. We propose that this PG may come from the PG-rich lung surfactant surrounding the cell. The metabolism of this novel lipid seems coupled to the metabolism of another important cellular phospholipid PI. The deacylation-reacylation reaction proposed between PI and LPG may be essential in the production of LTC_4 and seems to be unique to certain cellular stimuli — most notably TPA. The exact way in which arachidonic acid is released from L(bis)PA remains an intriguing question. It is clear that L(bis)PA has an important role in the immune inflammatory response as an important source of arachidonic acid for the production of bioactive mediators.

REFERENCES

1. D. R. Body, and G. M. Gray, The isolation and characterization of phosphatidylglycerol and a structural isomer from pig lung, Chem. Phys. Lipids 1:254 (1967).

2. C. F. Baxter, G. Rouser, and G. Simon, Variations among vertebrates of lung phospholipid class composition, Lipids 4:243 (1969).

3. A. N. Siakotos, G. Rouser, and S. Fleisher, Isolation of highly purified human and bovine brain endothelial cells and nuclei and their phospholipid composition, Lipids 4:234 (1969).

4. G. Rouser, G. Simon, and G. Kritchevsky, Species variations in phospholipid class distribution of organs. I. Kidney, liver and spleen, Lipids 4:599 (1969).

5. G. Simon, and G. Rouser, Species variations in phospholipid class distribution of organs. II. Heart and skeletal muscle, Lipids 4:607 (1969).

6. J. Brotherus, O. Renkonen, J. Herrman, and W. Fischer, Novel stereoconfiguration in lyso-bis-phosphatidic acid of cultured BHK-cells, Chem. Phys. Lipids 13:178 (1974).

7. J. R. Wherrett, and S. Huterer, Bis-(monoacylglyceryl)-phosphate of rat and human liver: fatty acid composition and NMR spectroscopy, Lipids 8:531 (1973).

8. G. Rouser, G. Kritchevsky, A. Yamamoto, and G. Simon, Accumulation of glycerophospholipid in classical Niemann-Pick disease, Lipids 3:287 (1968).

9. J. R. Wherrett, and S. Huterer, Enrichment of bis-(monoacylglyceryl)-phosphate in lysosomes from rat liver, J. Biol. Chem. 247:4114 (1972).

10. J. Brotherus, and O. Renkonen, Subcellular distributions of lipids in cultured BHK cells: evidence of the enrichment of lysobis-phosphatidic acid and neutral lipids in lysosomes, J. lipid Res. 18:191 (1977).

11. R. J. Mason, T. P. Stossel, and M. Vaughan, Lipids of alveolar macrophages, polymorphonuclear leukocytes, and their phagocytic vesicles, J. Clin. Invest. 51:2399 (1972).

12. S. Huterer, and J. Wherrett, Metabolism of bis(monoacylglycero)-phosphate in macrophages, J. Lipid Res. 20:966 (1979).

13. F. R. Cochran, V. Roddick, J. Connor, J. T. Thornburg, and M. Waite, Regulation of arachidonic acid metabolism in resident and BCG-activated alveolar macrophages: role of lyso(bis)-phosphatidic acid, J. Immunol., in press (1986).

14. B. J. H. M. Poorthuis, and K. Y. Hostetler, Studies on the localization and properties of bis(monoacylglyceryl)phosphate biosynthesis in rat liver, J. Biol. Chem. 251:4596 (1976).

15. Y. Matsuzawa, B. J. H. M. Poorthuis, and K. Hostetler, Mechanism of phosphatidylinositol stimulation of lysosomal bis(monoacyl-glyceryl)phosphate synthesis, J. Biol. Chem. 253:6650 (1978).

16. P. Somerharju, and O. Renkonen, Conversion of phosphatidylglycerol lipids to bis(monoacylglycero)phosphate in vivo, Biochim. Biophys. Acta 618:407 (1980).

17. M. Waite, V. Roddick, L. King, and F. Cochran, Conversion of phosphatidylglycerol to lyso(bis)phosphatidic acid by rabbit alveolar macrophages, J. Biol. Chem., submitted (1986).

18. F. Cochran, J. Connor, V. Roddick, and M. Waite, Lyso(bis)-phosphatidic acid: a novel source of arachidonic acid for oxidative metabolism by rabbit alveolar macrophages, Biochem. Biophys. Res. Commun. 130:800 (1985).

19. G. H. deHaas, N. M. Postema, W. Nieuwenhuizen, and L. L. M. van Deenen, Purification and properties of phospholipase A from porcine pancreas, Biochim. Biophys. Acta 159:103 (1968).

INDEX

Tetanotoxin
  binding to gangliosides,
      133-135
    circular dichroism, 140
    complex formation, 138
  interaction with cells,
      35-38
  properties, 32
Thy-1 glycoprotein, 331
Thylakoid membrane
  ATP synthase, 204
  EDTA wash, 206-208
  microviscosity, 292
  protein complexes
    in electron transfer,
        283-284, 292
    lateral diffusion,
        288-292
  protein phosphorylation,
      288-292
  proton flow, 207-210
    unit conductance, 210
  role in photosynthesis
      rythms, 302-303
  surface charge density,
      287
  surface potential, 285-286
Time-resolved neutron
      diffraction, 15-19
Triton X-100 micelles, 341
Tyrosine kinase, 56

Valinomycin, 195
Vanadate, 252
Variant surface glycoprotein,
    331
Vasopressin, 29, 34, 249
    361
Visual transduction cascade,
    6
  amplification steps, 8-9
  arrestin, 5, 20
  ATP-dependent kinase, 5
  cGMP phosphodiesterase, 5
    inhibition, 20
  near infrared light
      scattering, 10
  rhodopsin, 4
    phosphorylation, 20
  transducin, 5
  triggering mechanism, 7